Molybdenum and Tungsten Enzymes

Biochemistry

RSC Metallobiology Series

Titles in the Series:
1: Mechanisms and Metal Involvement in Neurodegenerative Diseases
2: Binding, Transport and Storage of Metal Ions in Biological Cells
3: 2-Oxoglutarate-Dependent Oxygenases
4: Heme Peroxidases
5: Molybdenum and Tungsten Enzymes: Biochemistry

How to obtain future titles on publication:
A standing order plan is available for this series. A standing order will bring delivery of each new volume immediately on publication.

For further information please contact:
Book Sales Department, Royal Society of Chemistry, Thomas Graham House, Science Park, Milton Road, Cambridge, CB4 0WF, UK
Telephone: +44 (0)1223 420066, Fax: +44 (0)1223 420247
Email: booksales@rsc.org
Visit our website at www.rsc.org/books

Molybdenum and Tungsten Enzymes
Biochemistry

Edited by

Russ Hille
University of California, Riverside, CA, USA
Email: russ.hille@ucr.edu

Carola Schulzke
University of Greifswald, Germany
Email: carola.schulzke@uni-greifswald.de

and

Martin L. Kirk
University of New Mexico, Albuquerque, NM, USA
Email: mkirk@unm.edu

ROYAL SOCIETY OF CHEMISTRY

THE QUEEN'S AWARDS FOR ENTERPRISE: INTERNATIONAL TRADE 2013

RSC Metallobiology Series No. 5

Print ISBN: 978-1-78262-089-1
PDF eISBN: 978-1-78262-391-5
EPUB eISBN: 978-1-78262-881-1
Three-volume set print ISBN: 978-1-78262-879-8
ISSN: 2045-547X

A catalogue record for this book is available from the British Library

Published by The Royal Society of Chemistry,
Thomas Graham House, Science Park, Milton Road,
Cambridge CB4 0WF, UK

Registered Charity Number 207890

For further information see our website at www.rsc.org

Printed in the United Kingdom by CPI Group (UK) Ltd, Croydon, CR0 4YY, UK

Preface

In the late 1950s and early 1960s, evidence was accumulating that molybdenum was not simply present in the enzyme xanthine oxidase from cow's milk but that it was required for its activity and changed its oxidation state in the course of the reaction with substrate. In a *tour-de-force* isotopic substitution study reported in *Nature* in 1966, R. C. Bray and L. S. Meriwether demonstrated unequivocally that the EPR signals elicited by the enzyme upon treatment with xanthine arose from a molybdenum-containing active site. It is a happy coincidence but altogether fitting that this volume marks the 50th anniversary of this seminal work.

For many years, only five enzymes were recognized as possessing molybdenum in their active sites: nitrogenase from bacteria such as *Klebsiella pneumoniae* and *Azotobacter vinelandii*; xanthine oxidase from bovine milk (and other vertebrate sources); aldehyde oxidase from vertebrate as well as bacterial sources; the vertebrate sulfite oxidase; and the assimilatory nitrate reductase from plants (and algae and fungi). That began to change in the 1980s with the demonstration by K. V. Rajagopalan that an organic cofactor accompanied the molybdenum in the active sites of these enzymes (with the exception of nitrogenase), and with the contemporaneous discovery that tungsten was also found in the active sites of enzymes in certain bacteria.

There are now several dozen molybdenum- and tungsten-containing enzymes that have been crystallographically characterized, along with most of the enzymes responsible for the biosynthesis of the organic cofactor variously known as molybdopterin, tungstopterin and pyranopterin. The active site metal centres of these enzymes have proven to be fascinating and challenging targets for synthetic inorganic chemists, and both enzymes and synthetic models have proven fertile ground for the application of a range of physicochemical and spectroscopic methods probing their physical and electronic structures as well as their intrinsic reactivity. At present, well

RSC Metallobiology Series No. 5
Molybdenum and Tungsten Enzymes: Biochemistry
Edited by Russ Hille, Carola Schulzke, and Martin L. Kirk

Published by the Royal Society of Chemistry, www.rsc.org

over 50 molybdenum- and tungsten-containing enzymes have been isolated and characterized, and these have been found to catalyze a broad range of oxidation–reduction reactions, and even reactions that (at least formally) do not involve oxidation–reduction of substrate. These enzymes are found in a wide range of metabolic pathways and play particularly prominent roles in the global cycling of nitrogen, sulfur and carbon. Many have vital roles in bacterial bioenergetics, catalyzing crucial energy-conserving reactions under a variety of growth conditions. Indeed, they seem to have been among the earliest enzyme systems to have arisen, as reflected in their near-universal distribution in the biosphere. Finally, genomics analyses have led to the identification of hundreds of genes encoding putative new proteins that are likely to possess one or another metal. These systems represent an enormous frontier of new enzymes that remains to be explored.

This title provides an up-to-date account of the state of our understanding of molybdenum and tungsten enzymes and is divided into three volumes, dealing with: (1) the enzymes themselves, along with pyranopterin cofactor biosynthesis and incorporation of the mature cofactor into apoprotein (*Molybdenum and Tungsten Enzymes: Biochemistry*), (2) inorganic complexes that model the structures and/or reactivity of the active sites of each major group of molybdenum and tungsten enzymes (*Molybdenum and Tungsten Enzymes: Inorganic Chemistry*) and (3) spectroscopic and related methods of physical chemistry (including computational work) that have been applied to both enzymes and model compounds (*Molybdenum and Tungsten Enzymes: Physical Methods*). Each volume is introduced by an overview chapter written by a leading expert in the field, followed by the individual chapters that detail specific topics associated with each volume. The intent of these overview chapters is to provide an overarching and unifying theme that places each of the three major subject areas in proper context.

We are deeply indebted to each of the contributors for their efforts, which lay out the current state of our understanding in each of the many subject areas considered. The coverage of these volumes is inevitably incomplete due to space constraints, however, and for this we apologize. However, the topics that are covered are presented to the reader in considerable detail; written in a style and spirit that will be fully accessible by current researchers in the field as well as those who wish to learn more about these fascinating metalloproteins. We sincerely hope that these volumes will underscore how rapid the progress has been over the past decade or so, and also how rapidly the field is expanding. The ultimate goal is to stimulate further research on molybdenum and tungsten enzymes, and especially to encourage new investigators to take up one or another aspect of these systems. It seems inevitable that many exciting new discoveries lie in wait.

Russ Hille
Carola Schulzke
Martin L. Kirk

Dedication

It is all too fitting that these volumes dealing with the bioinorganic chemistry of molybdenum and tungsten be dedicated to three outstanding chemists whose contributions to the field over many years continues to inform, illuminate and inspire: Richard H. Holm, C. David Garner and John H. Enemark.

Prof. Holm has over 500 research publications (cited over 35 000 times) covering a wide range of nickel, iron and molybdenum chemistry (among other transition metals). He is perhaps most widely recognized for studies, beginning in the 1970s, that describe the synthesis and characterization of iron–sulfur clusters. This work came to include modelling the M and P clusters of nitrogenase, which perhaps provided the motivation to investigate models of mononuclear molybdenum-containing enzymes. His molybdenum work achieved great success with the synthesis of MoO_2 models for enzymes of the sulfite oxidase, and later the DMSO reductase family, and the characterization of their properties as oxygen atom transfer catalysts. A key contribution was his use of bulky ligands to the metal that prevented μ-oxo

RSC Metallobiology Series No. 5
Molybdenum and Tungsten Enzymes: Biochemistry
Edited by Russ Hille, Carola Schulzke, and Martin L. Kirk

Published by the Royal Society of Chemistry, www.rsc.org

dimerization, which had long stymied work in the field. He is Higgins Professor of Chemistry at Harvard University, a member of the National Academy of Sciences and the recipient of many other awards.

Prof. Garner already had a strong track record in the synthesis of copper and molybdenum complexes when, beginning in the late 1970s, he became one of the first researchers to apply the then-new analytical method of X-ray absorption spectroscopy not only to models of molybdenum enzymes but also to the enzymes themselves. The discovery of thiolate-like sulfur, Mo=O and Mo=S ligands to the metal in the active sites of enzymes such as sulfite oxidase, xanthine oxidase and DMSO reductase was critical in establishing the molybdenum coordination environment in these enzymes and greatly focused efforts to synthesize accurate structural and functional mimics of the enzymes. With over 300 publications (having over 8000 citations), he is presently Professor Emeritus at the University of Nottingham and a Fellow of the Royal Society. He is also past President of the Royal Society of Chemistry.

Prof. Enemark was already well recognized for his work on metal nitrosyls and related systems when he began to exploit the tris-pyrazolylborate ligand as a scaffold on which to construct and study MoO_2 and MoO complexes. This work led to the synthesis and characterization of the first model that fully mimicked the catalytic cycle of oxotransferase enzymes such as sulfite oxidase. Enemark also played an instrumental role in the work that led to the first crystal structure of sulfite oxidase. Since that time, Enemark has pioneered the application of pulsed EPR methods to molybdenum enzymes and synthetic models of their active sites; work that has led to a deep understanding of not simply the physical but also the electronic structures of these systems. With over 250 publications and 10 000 citations, he is Regents Professor of Chemistry at the University of Arizona, a former Fulbright Scholar and recipient of the Humboldt Research Prize, among other national and international recognitions.

Contents

RSC Metallobiology Series No. 5
Molybdenum and Tungsten Enzymes: Biochemistry
Edited by Russ Hille, Carola Schulzke, and Martin L. Kirk

Published by the Royal Society of Chemistry, www.rsc.org

CHAPTER 1

Molybdenum and Tungsten-Containing Enzymes: An Overview

LUISA B. MAIA[a], ISABEL MOURA[a], AND JOSÉ J. G. MOURA*[a]

[a]UCIBIO, REQUIMTE, Departamento de Química, Faculdade de Ciências e Tecnologia, Universidade Nova de Lisboa, 2829-516 Caparica, Portugal
*E-mail: jose.moura@fct.unl.pt

1.1 Introduction

Molybdenum and tungsten are heavy metallic elements, belonging to the sixth group of the "d-block" of the periodic table, with electronic configurations [Kr] $4d^5$ $5s^1$ and [Xe] $4f^{14}$ $5d^4$ $6s^2$, respectively (atomic numbers 42 and 74). They are trace elements, either in the Universe or in Earth crustal rocks and oceans (Table 1.1). In spite of that scarcity, molybdenum is essential to most organisms,[1,2] from archaea and bacteria to higher plants and mammals, being found in the active site of enzymes that catalyze oxidation–reduction reactions involving carbon, nitrogen and sulfur atoms of key metabolites.[3–10] Some of the molybdenum-dependent reactions constitute key steps in the global biogeochemical cycles of carbon, nitrogen, sulfur and oxygen, with particular emphasis on the atmospheric dinitrogen fixation (reduction) into organic ammonium (nitrogen cycle/nitrogenase enzyme).

RSC Metallobiology Series No. 5
Molybdenum and Tungsten Enzymes: Biochemistry
Edited by Russ Hille, Carola Schulzke, and Martin L. Kirk

Published by the Royal Society of Chemistry, www.rsc.org

Table 1.1 Abundances of molybdenum, tungsten and some other elements with biological relevance in different environments[675]

Location	Abundance (ppb by atoms)						
	Mo	W	Fe	H	C	N	O
Universe	0.1	0.003	20×10^3	930×10^6	500×10^3	90×10^3	800×10^3
Crustal rocks	230	120	23×10^6	31×10^6	3.1×10^3	29×10^3	600×10^6
Oceans	0.64	0.004	0.33	662×10^6	14.4×10^3	220	331×10^6
Human body	7	—	6.7×10^3	620×10^6	120×10^6	12×10^6	240×10^6

Presently, more than 50 molybdenum-containing enzymes are known. The great majority are prokaryotic, with eukaryotes holding only a restricted number of molybdoenzymes.[4–10] Noteworthy, while all higher eukaryotic organisms use this element, many unicellular eukaryotes, including *Saccharomyces* and most other yeasts, have lost the ability to use molybdenum.[1,2] Tungsten, probably because of its limited bioavailability (Table 1.1),[11] is far less used, being present predominantly in thermophilic anaerobes,[3,12,13] although it is also found in some strictly aerobic bacteria (*e.g.* certain methylotrophs[14–19]).

This chapter provides an overview on the molybdo- and tungstoenzymes. Their physiological context and significance will be described in Section 1.2, where the recent hypothesis that the lack of molybdenum could have been the limiting factor for the life evolution and expansion on early Earth will receive special attention (Section 1.2.1). A brief introduction to the chemical properties that shape the catalytically competent molybdenum/tungsten centres will be made in Section 1.3. In Section 1.4, the enzymes will be grouped in five main families (Sections 1.4.1 to 1.4.5), according to their metal/cofactor structure, and a general view on the structural (section (a)) and mechanistic (section (b)) versatility of each family will be presented. A brief account of novel heteronuclear centres containing molybdenum, whose physiological function is not yet fully understood, will be made in Section 1.4.6. A final outlook on our present knowledge about these enzymes will conclude this chapter.

1.2 Living with Molybdenum and Tungsten

The *human history* of molybdenum began in the 18th century, when Carl Wilhelm Scheele isolated molybdic acid ($MoO_3{\cdot}H_2O$) and Peter Jacob Hjelm subsequently found a dark metallic powder that he named "molybdenum".[20] Nevertheless the successful and widespread use of molybdenum only took place in the 20th century and nowadays it is used in bridges and buildings (I.M. Pei's steel pyramid entrance for the Musée du Louvre is an elegant example), pipes and power plants, cars and computers, paints, plastics, catalysts and medical procedures.[21–23] By contrast, the *biological history* of molybdenum is almost as old as life on Earth.

When we think about the elements that are essential for life on Earth, we hardly ever consider molybdenum. The biological role of molybdenum

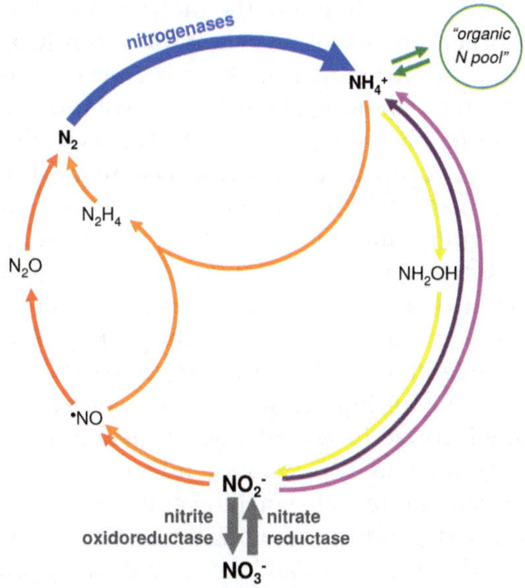

Figure 1.1 Biochemical cycle of nitrogen. Dinitrogen fixation, blue arrow; "organic nitrogen pool", green arrows; assimilatory ammonification, pink arrow; dissimilatory nitrate reduction to ammonium, violet arrow; nitrification, yellow arrows; denitrification, red arrows; anaerobic ammonium oxidation (AnAmmOx), orange arrows. The steps catalyzed by molybdenum-containing enzymes are highlighted with thick arrows, nitrogenase (blue), nitrate reductase and nitrite oxidoreductase (grey). Adapted with permission from ref. 62. Copyright 2014 American Chemical Society.

can only be appreciated when put in perspective. Nitrogen is the fourth most abundant element in living organisms (only behind hydrogen, oxygen and carbon) and life on Earth depends on the nitrogen biogeochemical cycle to keep this element in forms that can be used by the organisms.[24–33] Noteworthy, the "closing" of the nitrogen cycle, with the atmospheric dinitrogen fixation into ammonium[30,34–36] (Figure 1.1, blue arrow), depends virtually entirely on the trace element molybdenum: nitrogenase, a prokaryotic enzyme responsible for dinitrogen reduction to ammonium, requires one molybdenum atom in its active site[†] (Figure 1.3b; see Section 1.4.5 and ref. 55). The few organisms possessing this enzyme are capable of producing their own reduced ("fixed") nitrogen forms, using directly the atmospheric dinitrogen, the largest nitrogen source (biological nitrogen fixation is the main route by which nitrogen enters the biosphere).[56–58] All other organisms, the vast majority of life on Earth, depend on the availability of ammonium and nitrate (from soils, oceans and other organisms).[30,36,59–62]

[†]Note that, besides the molybdenum/iron-dependent enzyme, there are also other nitrogenases that depend on vanadium/iron and only on iron, but they exhibit different catalytic efficiencies and products stoichiometry.[37–54].

With this wide perspective in mind, the molybdenum biological role certainly assumes another dimension. In fact, it was recently proposed that the lack of molybdenum, while hampering the existence of an efficient nitrogenase, could have been the limiting factor for life evolution and expansion on early Earth, as described below (Section 1.2.1). However, the involvement of molybdenum in the nitrogen cycle is not restricted to the dinitrogen fixation, as the element is also essential for the reduction of nitrate to nitrite and for the oxidation of nitrite to nitrate (Figure 1.1, grey arrows), both processes being exclusively dependent (as far as is presently known) on the molybdenum-containing enzymes nitrate reductases (from both prokaryotic and eukaryotic sources) and nitrite oxidoreductases (from prokaryotes only).[62]

Noteworthy, molybdenum has also been suggested to be essential for nitrite reduction to nitric oxide for biological signalling purposes. Nitric oxide is a signalling molecule involved in several physiological processes, in both prokaryotes and eukaryotes, and nitrite is presently recognized as a nitric oxide source particularly relevant to cell signalling and survival under challenging conditions.[62,63] Nitrite-dependent signalling pathways have been described in mammals, plants and also bacteria, and are carried out by proteins present in cells to carry out other functions, including several molybdoenzymes (which thus form a new class of "non-dedicated" nitric oxide-forming nitrite reductases): mammalian xanthine oxidoreductase, aldehyde oxidase,[64,65] sulfite oxidase[66] and mitochondrial amidoxime reducing component,[67] plant nitrate reductase[62] and bacterial aldehyde oxidoreductase[64] and nitrate reductases.[62]

Molybdenum is also involved in the carbon cycle. The first example that comes to mind is provided by the formate dehydrogenases that are used by acetogens to fix carbon dioxide (reduce it) into formate and eventually form acetate; but molybdenum is also present in carbon monoxide dehydrogenases (catalyzing the oxidation of carbon monoxide to carbon dioxide), aldehyde oxidoreductases (catalyzing the interconversion between aldehydes and carboxylic acids) and in other formate dehydrogenases (that are involved in physiological pathways where formate is oxidized to carbon dioxide). The primitive carbon cycle would have also been dependent on molybdenum, as the metal (together with tungsten) would have been essential for the earliest, strictly anaerobic, organisms to handle aldehydes and carboxylic acids, catalyzing their interconversion.[68]

Molybdenum also plays several other "carbon-related" roles in modern higher organisms. The aldehyde oxidase of higher plants is responsible for the oxidation of abscisic aldehyde to abscisic acid (a plant hormone involved in development processes and in a variety of abiotic and biotic stress responses)[69,70] and has been implicated in the biosynthesis of indole-3-acetic acid (an auxin phytohormone).[71] The mammalian aldehyde oxidases have been suggested to participate in the formation of retinoic acid (a metabolite of retinol (vitamin A) that is involved in growth and development) and in the metabolism of xenobiotic compounds, where they would catalyze the hydroxylation of carbon centres of heterocyclic aromatic compounds and the oxidation of aldehydic groups.[72–76]

The dependence of higher plants and animals on molybdenum is also observed in purine catabolism, where xanthine oxidoreductase is involved in the hydroxylation of hypoxanthine and xanthine into urate.[77–79] Noteworthy, involvement of molybdenum in purine metabolism is common to virtually all forms of life and only a small number of organisms use other mechanisms to oxidize xanthine (*e.g.* some yeasts[80]), thus confirming the essential role of molybdenum for life on Earth.

Another important aspect of molybdenum in biology can be seen in sulfite-oxidizing enzymes, which are used by almost all forms of life in the catabolism of sulfur-containing amino acids and other sulfur-containing compounds, oxidizing sulfite to sulfate. Certainly, sulfite oxidase is one of the most striking examples of the human dependence on molybdenum.[81–86] Sulfite (derived not only from the catabolism of sulfur-containing amino acids, but also from sulfur-containing xenobiotic compounds) is toxic and its controlled oxidation to sulfate is critical for survival. Underscoring this vital role, human sulfite oxidase deficiency results in severe neonatal neurological problems, including attenuated growth of the brain, mental retardation, seizures and early death.[‡81–86] Molybdenum-dependent sulfite-oxidizing enzymes are also important for some prokaryotes that are able to generate energy from the respiratory oxidation of inorganic sulfur compounds[87–90] – hence, extending the role of molybdenum to the sulfur cycle.

Tungsten was likely an essential element for the earliest life forms (see Section 1.2.1 below for some details about Earth's history). Under euxinic conditions (sulfidic and anoxic conditions), tungsten forms relatively soluble salts (WS_4^{2-}) and it was therefore probably more available in the euxinic ocean than molybdenum (which would have been present as the water-insoluble MoS_2). The same reasoning explains the higher tungsten availability in today's marine hydrothermal vent waters, precisely where most of the hyperthermophilic organisms were discovered that were found to possess tungstoenzymes.[91] As with molybdenum, it is believed that tungsten would have carried out much the same chemistry as it does today in the enzymes of contemporary organisms. The reversible handling of aldehydes and carboxylic acids by primitive strict anaerobes is plausibly matched by the aldehyde : ferredoxin oxidoreductase of today's *Pyrococcus furiosus* (one of the benchmark tungstoenzymes). Still, today only relatively few organisms utilize tungsten, which might seem puzzling if one considers the chemical similarities between tungsten and molybdenum and the fact that both metals are coordinated by the same organic cofactor (Figure 1.3a; described below). Indeed, it seems that for each tungstoenzyme there is a homologous molybdoenzyme, either in the same or in different organisms, and there are several examples of molybdo- and tungstoenzymes that catalyze the same reaction (*e.g.* aldehydes, oxidoreductases and formate dehydrogenases that can contain molybdenum or tungsten). Could the modern scarcity of tungstoenzymes reflect the early Earth scenario? Were the tungstoenzymes widespread in early Earth and subsequently lost

---

‡Sulfite oxidase deficiency can be caused by protein point mutations, but also by the inability to synthesize the cofactor that holds the molybdenum atom in the active site (Figure.1.3a; described below); the last case results in deficiency in all four human molybdoenzymes. However, only the sulfite oxidase deficiency is a serious life threat.

with the "pollution" of the atmosphere by dioxygen, forcing organisms to use molybdenum (available as the highly water-soluble MoO_4^{2-}) instead? That is, did molybdenum become dominant only later in the Earth's history, due to its availability and properties? This is a plausible scenario if one takes into account the higher availability of tungsten under euxinic conditions and the chemical singularities of tungsten (instead of the similarities between the two metals): tungsten compounds exhibit lower reduction potentials, higher bond strengths and enhanced thermal stability compared to iso-structural molybdenum counterparts, but are more sensitive to dioxygen.[3,12,92–96] These differences support the idea that tungsten would have been a better choice for anaerobic low reduction potential reactions carried out under euxinic conditions. As the environmental conditions were changing and the Earth became increasingly oxygenated, tungsten could have been replaced by molybdenum: the chemical similarities between the two metals could have been exploited by the surviving organisms to evolve enzymes that enabled them to continue catalyzing the same old reactions and new reactions dictated by the needs imposed by the new environment.[§]

Regardless, both molybdo- and tungstoenzymes probably existed in the last universal common ancestor (LUCA).[106,107] Therefore, the two cofactors that hold the metals in the enzymes active site would also have to have been present. This is particularly remarkable when we realize how elaborated the two cofactors are (particularly the nitrogenase one; Figure 1.3) and how "limited" their utilization compared to, for instance, porphyrin-related structures. Why do living organisms expend so much effort to use these metals in a (comparatively) small number of reactions? This effort (including synthesizing the protein machinery to scavenge the metals from the environment, producing and inserting the specialized cofactors and regulating the whole process) underscores how important both metals would have been, and still are to extant organisms, particularly in the case of molybdenum.

1.2.1 The Nitrogen-to-Molybdenum Bio-to-Inorganic Bridge Hypothesis

The atmosphere of early Earth held no dioxygen (if present, it would be less than 10^{-5} times the current atmospheric level). Only in the second half of the Earth's 4560 million years (Myr) history, between 2450 and 2220 Myr ago, did dioxygen levels rise in the oceans and atmosphere as a consequence of the

---

[§]A note of caution in this simplistic scenario, where molybdenum "simply" took the place of tungsten: the differences between molybdenum and tungsten are sufficient to interfere with the properties of the vast majority of today's enzymes. In fact, tungsten is regarded as an antagonist and inhibitor of molybdoenzymes and the substitution of molybdenum by tungsten results in metal-free and tungsten-substituted enzymes, both with no enzymatic activity.[97–105] This outcome arises from differences in the metals' uptake and/or incorporation into the enzymes, but also from differences in the properties of the enzymes themselves. However, it should be noted that there are some prokaryotic enzymes that are active with either molybdenum or tungsten, as will be discussed in Section 1.4.4.

"invention" of oxygenic photosynthesis by cyanobacteria[108–111] – the so-called "Great Oxidation Event". Recent geochemical data[112,113] are changing that time frame, however, suggesting that small amounts of dioxygen were present in the environment more than 50 Myr before the start of the Great Oxidation Event, supporting the hypothesis that primitive organisms had learned to produce dioxygen much earlier than previously thought[112,114] – future work will determine if the dioxygen biogenesis is even more ancient than presently thought. Presently, several geoscientists defend the idea that the Earth oxygenation proceeded in two broad steps, near *ca.* 2500 and *ca.* 540 Myr ago.[109,115–121] (Readers not familiar with geochemical studies are referred to note ¶ for a brief explanation.)

During the first oxygenation phase, probably only the ocean surface was affected by photosynthesizing bacteria. Although the dioxygen would have started to increase, only *ca.* 2150 Myr ago, more than 300 Myr after the initial

---

¶The history of Earth oxygenation is written in the geological record of redox-sensitive transition metal elements preserved in ancient authigenic sediments. The principle is that the amount of those elements present in sedimentary rocks is determined by the dioxygen availability during formation of the sediments.

On early, anoxic (strongly reducing) Earth, molybdenum would have been largely retained in crustal sulfide minerals (it would have not been weathered, solubilized) and its presence would have been small in the oceans and sediments. Under low dioxygen pressures, the rate of dissolution of submarine and sub-aerial sulfide minerals (such as molybdenite, MoS_2) would have been enhanced. Hence, after the rise of dioxygen, oxidative weathering of molybdenum-containing sulfide minerals in crustal rocks would have led to the molybdenum accumulation in oceans (molybdenum dissolution, in the form of the soluble molybdate ion, MoO_4^{2-}).

In fact, today, molybdenum is the most abundant transition metal element in the oceans (present at a concentration of ≈110 nM). Under oxygenated conditions, the oceanic organic-rich sediments would, consequently, show a high authigenic molybdenum enrichment (today, typically values are >100 ppm in sediments *versus* <1 ppm in average crust). Under euxinic, *i.e.* sulfidic and anoxic, conditions (conditions created after the rise of dioxygen; see below), the hydrogen sulfide would have reduced the dissolved molybdenum to Mo^{4+} and the highly insoluble molybdenum sulfide, MoS_2, would have been formed; thus molybdenum would have been removed from the sea water solution. Noteworthy, under euxinic conditions, tungsten and vanadium form relatively soluble salts and they were probably more available in the euxinic ocean. Following this reasoning, the amount of molybdenum in oceanic sedimentary rocks rich in organic matter (black shales) should be a good indicator of the amount of dioxygen in sea water and atmosphere.

A similar analysis would apply to sulfur. On early, anoxic Earth, sulfur would have been largely retained in minerals, and sulfate (and, consequently, hydrogen sulfide) would have been sparse in pre-oxic oceans (before the appearance of dioxygen). The rise of dioxygen would have weathered the minerals (solubilized and oxidized them to release sulfate) and sulfate would have been accumulated in the oceans (where, today, it is the second most abundant anion). In a post-oxic era (after the appearance of dioxygen), under anoxic conditions, the mobilized sulfate would have been reduced to hydrogen sulfide by sulfate-reducing bacteria, creating, in this way, euxinic conditions.

On the contrary, iron would have behaved in an opposite way. Iron would have been easily mobilized during anoxic weathering, thus enriching the sulfur-poor oceans (as aqueous ferrous) and the sediments, leading to banded iron formations. Oxygenation of surface environments would have oxidized the aqueous ferrous iron to insoluble ferric oxides, thus reducing the availability of this element. Euxinic conditions would have also reduced the availability of iron, but in the form of insoluble iron sulfides. Hence, both dioxygen and hydrogen sulfide may have pulled the dissolved iron from solution and be responsible for the disappearance of banded iron formations.

rise in atmospheric dioxygen, the dioxygen pressure would have been sufficiently high to cause the persistent and vigorous oxidative weathering of the molybdenum-containing sulfide minerals in crustal rocks (note that sulfide minerals weather rapidly, and a very low pressure of dioxygen would be enough to account for this effect). Molybdenum, released in this way, would have accumulated (dissolved) in oceans, eventually resulting in the enrichment of the authigenic ocean sediments.[109,115,121] However, the greater oxidative weathering would have also increased the delivery of sulfur to the ocean. The consequent increase in the oceanic hydrogen sulfide would have, in its turn, removed the molybdenum from solution (as the insoluble MoS_2).[116,121–123] Accordingly, expansion of the euxinic (sulfidic and anoxic) conditions, after *ca.* 1800 Myr ago, would have kept molybdenum availability below 10–20% of the modern value. Those same conditions could have promoted the removal of dissolved iron from the sea water (as insoluble iron sulfides) and kept the iron availability low.[116] (In the classic model, the disappearance of banded iron formations is explained invoking oxygenation of the deep ocean in this early time frame, as a consequence of the formation of insoluble ferric oxides;[124,125] according to this new hypothesis, the disappearance of banded iron formations was due to the precipitation of iron under euxinic conditions (anoxic) and not to deep ocean oxygenation, which is suggested to have taken place only more than 1000 Myr later).

The second oxygenation phase is suggested to mark the time when the entire ocean became oxygenated. By *ca.* 660–550 Myr ago, the sediments content suggests an extreme molybdenum presence in the oceans, pointing to the oxygenation of the deep ocean and to the corresponding decrease in sulfidic conditions.[119–121] The process responsible for this sudden change in the molybdenum record and dioxygen presence, however, is not yet fully understood.[126,127] Thus, according to this hypothesis, atmospheric dioxygen levels did not increase monotonically to their modern value but rather proceeded in two phases, separated by ≈2000 Myr. Interestingly, the same time frame is suggested to separate the emergence and subsequent expansion of eukaryotes.[114,123,128] This coincidence has raised the hypothesis that the molybdenum oceanic bioavailability could have played a major role in the ≈2000 Myr delay in the development of early life.[121,123,129]

Plausibly, early forms of life would have employed ferrous iron, abundant on the anoxic oceans, to handle abiotic ammonium, nitrite and nitric oxide and also to fix atmospheric dinitrogen through a primitive iron-containing nitrogenase.[130,131] The rise of dioxygen, with subsequent oxidation and precipitation of iron, would have turned the oceanic dissolved iron into a scarce element. Concurrently, the dioxygen-triggered mobilization of molybdenum would have allowed the evolution of a molybdenum-containing nitrogenase, a new enzyme able to efficiently fix (reduce) dinitrogen into ammonium. However, the subsequent onset of deep ocean euxinia would have acted to remove the dissolved molybdenum (and iron), maintaining the oceanic molybdenum (and iron) concentration low. This molybdenum scarcity would have hampered the expansion of molybdenum-containing nitrogenase,[131] limiting the availability of "bio-nitrogen" for the early organisms (limiting the rate of dinitrogen fixation and the supply of fixed nitrogen inter-organisms). Ultimately, the molybdenum scarcity could have restricted the evolutionary path of eukaryotes[121,123,129,132] – *the nitrogen-to-molybdenum bio-to-inorganic bridge hypothesis*.

Nevertheless, the molybdenum shortage would have also contributed to limit, spatially and temporally, the extent of sulfidic conditions, because molybdenum would be also essential (directly or indirectly) for the bacteria that carry out the reduction of sulfate to sulfide, and because organic matter is required for the exhaustion of dioxygen.[121] This negative feedback mechanism could explain the apparent decline in euxinic deposition after 1400 Myr ago. By *ca.* 660–550 Myr ago, the deep ocean oxygenation and the increased molybdenum availability would have favoured the diversification of dinitrogen fixing organisms; this would have boosted the photosynthetic dioxygen production in the oceans and, in this way, the ocean-atmosphere thorough oxygenation (like a vicious cycle).[I] With a more efficient respiratory substrate and "bio-nitrogen" source, the stage was set for the subsequent evolution and proliferation of structurally complex forms of life, leading to the Cambrian explosion of metazoan life – remarkably, only in the last tenth of the Earth's history.

In this scenario, Earth and life on it would have evolved together, with different key events being closely interrelated:[134–136] the accumulation of (biologically produced) dioxygen triggered the geological molybdenum release, but the subsequent environmental removal of molybdenum retarded its biological utilization and, eventually, delayed the evolution of complex life for *ca.* 2000 Myr. This scenario exemplifies how biology, geology and the atmosphere might have interacted and the knowledge gathered may be helpful to understand the environmental and climate issues we are facing today (*e.g.* the greenhouse effect gas carbon dioxide scavenging by an iron-starved ocean). More important in the context of this book, the hypothesis described above emphasizes the critical biological role of molybdenum for the life on Earth. At the same time, this hypothesis also raises the question as to why there is (as far as is presently known) no tungsten-dependent nitrogenase. If the organisms were able to develop iron and vanadium-/iron-dependent nitrogenases, why did they not also use tungsten?

1.3 Chemistry Relevant to Molybdenum and Tungsten Biochemistry

Organisms use molybdenum and tungsten for the most part to catalyze oxidation–reduction reactions, most of which involve oxygen atom transfer to/from a carbon, nitrogen and sulfur atom of key metabolites.[3–10] Certainly, both metals exhibit the chemical properties appropriate for redox biochemistry:[26]

[I]But the "revolution" in dinitrogen fixation initiated by the dioxygen rise had not been finished yet: the dioxygen that triggered (*indirectly*) the evolution of molybdenum-dependent nitrogenase also inhibits (*directly*) the *new* enzyme. This forced the nitrogen-fixing organism to evolve mechanism to protect the enzyme from dioxygen.[133] The structural protection in aerobic organisms continues to the present.

they are redox-active under physiological conditions and their oxidation state can range from 6+, 5+ and 4+, and even 3+, in the molybdenum of nitrogenase. This versatility allows molybdenum- and tungsten-containing enzymes to catalyze either two-electron ($M^{6+} \leftrightarrow M^{4+}$) or one-electron ($M^{6+} \leftrightarrow M^{5+}$, $M^{5+} \leftrightarrow M^{4+}$) oxidation-reduction reactions, and thus couple obligatory two-electron and one-electron systems, *e.g.* the reduction of a two-electron respiratory substrate with a one-electron transfer protein. In addition, their chemistry is dominated by the formation of oxides and sulfides and a very versatile first coordination sphere. The strong tendency of molybdenum to bind oxo groups is balanced by its ability to easily lose a single oxygen atom, a property that makes molybdenum complexes excellent "oxygen atom exchangers",[137–156] as long as the thermodynamics of the reaction of "oxygen exchange" is favourable (eqn (1.1) (M stands for metal))[139,143,152] – the "oxo transfer hypothesis" coined by Holm and others in the 1980s.

$$\text{M–O} + \text{X} \rightleftharpoons \text{M} + \text{X–O} \qquad (1.1)$$

Organisms explore this rich chemistry to carry out oxo transfer reactions: typically, the molybdo- and tungstoenzymes catalyze the transfer of an oxygen atom from water to product – oxygen atom insertion (eqn (1.2); Figure 1.2, blue arrows) – or from substrate to water –oxygen atom abstraction (eqn (1.3); Figure 1.2, green arrows) – in reactions that entail a net exchange of two electrons, in which the molybdenum/tungsten atom cycle between Mo^{6+}/W^{6+} and Mo^{4+}/W^{4+}, and, most importantly, where the metal is the direct oxygen atom acceptor or donor.[3–10,154–159] (The detailed mechanisms through which the enzymes catalyze these reactions will be discussed in Sections 1.4.1–1.4.5.) It is based on these catalytic features that these enzymes are commonly, although inaccurately, referred to as oxotransferases, since there are several noteworthy exceptions to the oxo transfer activity. The versatile chemistry of molybdenum and tungsten has allowed the evolution of enzymes that catalyze reactions, for example, of (i) proton abstraction (formate dehydrogenase-catalyzed formate oxidation to carbon dioxide; eqn (1.25) in Section 1.4.3), (ii) sulfur atom transfer (polysulfide reductase-catalyzed inorganic sulfur reduction to sulfide (eqn (1.26) in Section 1.4.3) or MOSC-catalyzed sulfurations) or (iii) even non-redox hydration reaction (acetylene hydratase-catalyzed hydration of acetylene to acetaldehyde; eqn (1.28) in Section 1.4.3).

$$M^{6+} + R + H_2O \rightarrow M^{4+} + \text{R–O} + 2H^+ \qquad (1.2)$$
$$M^{4+} + \text{Q–O} + 2H^+ \rightarrow M^{6+} + Q + H_2O \qquad (1.3)$$

Another interesting exception is the group of enzymes able to catalyze both oxygen atom insertion and abstraction during the same catalytic cycle. This is the case of the prokaryotic molybdenum-containing pyrogallol : phloroglucinol transhydroxylase (eqn (1.27) in Section 1.4.3)). This enzyme catalyzes the "simple" hydroxyl transfer between two hydroxylated benzene compounds:[160] the reduced molybdenum core accepts one hydroxyl group

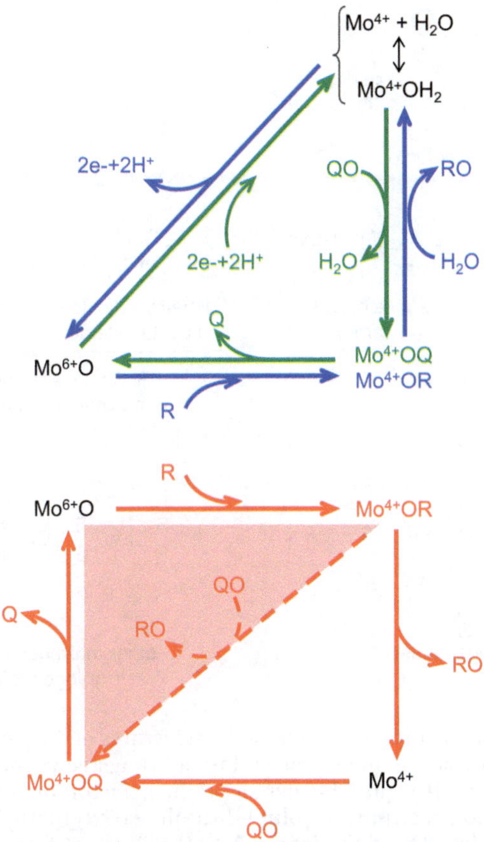

Figure 1.2 Mono oxo transfer (blue and green) and double oxo transfer (red) hypothesis. This schematic representation highlights that Mo^{6+} cores can be thought of as competent oxo donors, while the Mo^{4+} cores would act as oxo acceptors. The mono oxo transfer path for oxygen atom insertion reactions is represented in blue (*e.g.* for the sulfite oxidase reaction (eqn (1.16)), R is sulfite and RO is sulfate). The mono oxo transfer path for oxygen atom abstraction reactions is represented in green (*e.g.* for the nitrate reductase reaction (eqn (1.17)), QO is nitrate and Q is nitrite). The double oxo transfer path is represented in red (*e.g.* for the simultaneous oxygen atom insertion and abstraction reaction of the pyrogallol : phloroglucinol transhydroxylase (eqn (1.27)), R represents pyrogallol that is hydroxylated to tetrahydroxybenzene, represented by RO, and tetrahydroxybenzene, QO, is reduced to phloroglucinol, represented by Q. The shaded triangle illustrates the possibility that substrate QO displaces the product RO, without the formation of a reduced "free" molybdenum centre.

from one of the substrates, to become itself oxidized and hydroxylated; subsequently the molybdenum core transfers the hydroxyl group to the second substrate in an oxidative hydroxylation (thus becoming reduced and "dehydroxylated"). Therefore, this enzyme uses its molybdenum centre to directly transfer the hydroxyl group from one substrate to the second one,

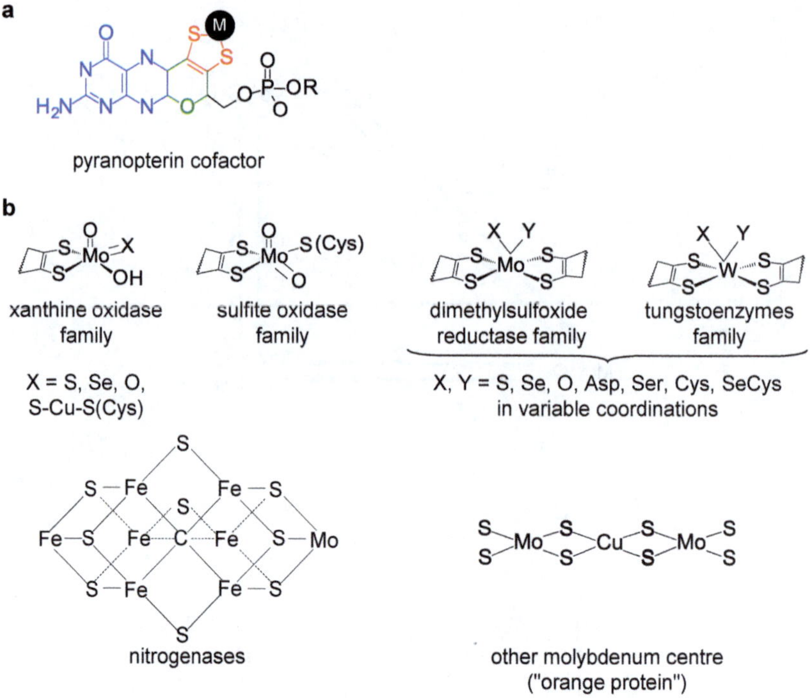

Figure 1.3 Schematic representation of the structures of the pyranopterin cofactor (a), the active site of molybdo- and tungstoenzymes and the "orange protein" (b). (a) The pyranopterin cofactor molecule is formed by pyrano(green)-pterin(blue)-dithiolene(red)-methylphosphate(black) moieties. The dithiolene (–S–C=C–S–) group forms a five-membered ene-1,2-dithiolate chelate ring with the molybdenum/tungsten atom. In eukaryotes, the cofactor is found in the simplest monophosphate form (R is a hydrogen atom), while in prokaryotes it is most often found esterified with several nucleotides (R can be one cytidine monophosphate, guanosine monophosphate or adenosine monophosphate). M, stands for metal, molybdenum or tungsten. (b) Structures of the molybdenum and tungsten centres of the five families of molybdo- and tungstoenzymes in the oxidized form and of the "orange protein". For simplicity, only the *cis*-dithiolene group of the pyranopterin cofactor is represented in the xanthine oxidase, sulfite oxidase, dimethylsulfoxide reductase and tungstoenzymes families. In carbon monoxide dehydrogenase, where X represents S–Cu–S(Cys), the terminal hydroxyl (Mo–OH) is replaced by an oxo group (Mo=O). The images were produced with Accelrys Draw 4.0 (Accelrys Software Inc.).

but without using water as the ultimate oxygen acceptor/donor** (eqn (1.4); Figure 1.2, red arrows).

$$R + Q\text{–}O \rightarrow R\text{–}O + Q \tag{1.4}$$

**There are other mechanistic proposals, more complex, for this enzyme,[161] but all must explain the hydroxyl transfer activity, with the hydroxyl group not being derived from solvent.

Another example of the ability to catalyze oxygen atom insertion and abstraction in the same catalytic cycle is provided by the nitric oxide-forming nitrite reductase activity (eqn (1.5)) of some molybdoenzymes (see Section 1.2 for the physiological context of this activity).[62,63] Mammalian xanthine oxidoreductase, aldehyde oxidase[64,65] and sulfite oxidase[66] and bacterial aldehyde oxidoreductase[64] catalyze the oxygen atom insertion into their "classic" substrates (xanthine (eqn (1.6) in Section 1.4.1)), aldehyde (eqn (1.7) in Section 1.4.1)), sulfite (eqn (1.16) in Section 1.4.2)) and aldehyde (eqn (1.26) in Section 1.4.3)), respectively); the resulting reduced molybdenum centres are then able to catalyze the abstraction of the oxygen atom of nitrite to yield nitric oxide (eqn (1.4); Figure 1.2, red arrows). Hence, nitrite, the second substrate, acts as oxidizing substrate and oxo group donor, while the "classic" substrate functions as a reducing substrate and oxo group acceptor.[63] This description does not intend to mean that it is necessarily the oxygen atom of the second substrate that is inserted into the "classic" substrate (which was not yet confirmed experimentally, because the molybdenum labile oxo/hydroxyl group can be easily exchanged with solvent water), although, as represented in Figure 1.2, red arrows, it is possible that this is the case.

$$NO_2^- + e^- + 2H^+ \rightarrow {}^{\bullet}NO + H_2O \tag{1.5}$$

These molybdenum-containing enzymes[††] demonstrate that, in the presence of an oxo donor and oxo acceptor, the molybdenum centre can catalyze the oxygen transfer between the two as long as the thermodynamics of the reaction is favourable.[139,143,152] In other words, the nitrite reductase activity can be common to other molybdoenzymes and, more interestingly, the "double oxo transfer" reaction should be possible for substrates other than nitrite.

The versatile chemistry displayed by the molybdenum- and tungsten-containing enzymes is also a consequence of the "environment" (first and second coordination spheres) of the metals inside the protein, which would shape the final catalytically competent metal centre. As will be discussed in Section 1.4, with very few exceptions, these enzymes hold the molybdenum or tungsten atom coordinated by a unique cofactor: a pyranopterin moiety modified with a *cis*-dithiolene group (–S–C=C–S–) and a methylphosphate (which can be further esterified) (Figure 1.3a), herein designated as the pyranopterin cofactor.[‡‡] The pyranopterin cofactor is not thought to be an "innocent scaffold",[162–166] but has been suggested to facilitate intramolecular

[††]Of note, the nitric oxide-forming nitrite reductase activity was also described in other molybdoenzymes, namely in mammalian mitochondrial amidoxime reducing component (eqn (1.21))[67] and plant and bacterial nitrate reductases[62] (eqn (1.16)). However, the "classic" activity of those enzymes is already an oxygen atom abstraction reaction and, hence, no concomitant oxygen insertion was demonstrated.

[‡‡]The pyranopterin cofactor molecule is still commonly referred to as "molybdopterin", because, historically, the cofactor was first identified in molybdenum-containing enzymes. However, because the same cofactor molecule is used to coordinate tungsten in tungstoenzymes (described under Section 1.4.1), this denomination is not clear.

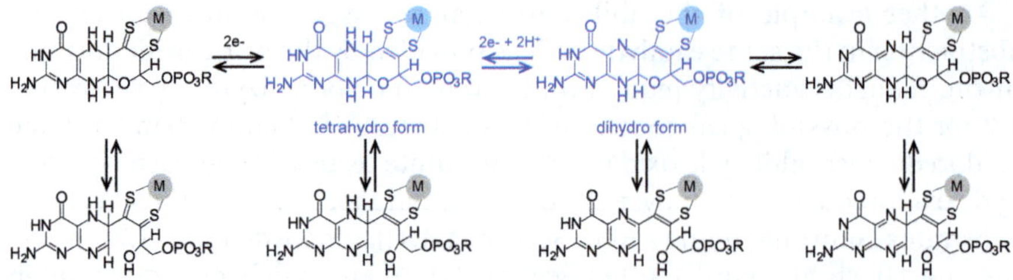

Figure 1.4 Schematic representation of some isomers and redox forms of the pyranopterin cofactor. The dihydro and tetrahydro forms are highlighted in blue. The structures represented below correspond to the scission reaction of the pyran ring (the "open pyran ring forms"). M, stands for metal.

electron transfer, acting like a "wire" to conduct the electrons to other redox-active centres that are most often found in these enzymes. Moreover, because the pyranopterin cofactor has several potential structural isoforms and oxidation states (Figure 1.4),[153,167–172] each adopting a different geometry, it has recently been suggested that each enzyme holds a binding pocket that selectively controls the pyranopterin conformation, mainly in the dihydro and tetrahydro reduced forms, in order to tune the metal centre oxidation state and facilitate electron transfer.[173] Of note, two crystal structures have been reported in which the pyran ring of the cofactor is in the open form, that of the respiratory *E. coli* NaR and that of *Aromatoleum aromaticum* ethylbenzene reductase.[174–176] Although it can be suggested that the opening and closing of the pyran ring might provide protons for the reaction,[170,177,178] its physiological relevance remains to be established. In fact, it has recently been suggested that the as-isolated periplasmatic nitrate reductase from *Rhodobacter sphaeroides* holds its proximal cofactor molecule in a fully oxidized and ring-opened form; this enzyme form has been suggested to require an activation mechanism that involves both the cyclization of the pyran ring and the reduction of the pterin to yield a catalytically competent tricyclic tetrahydropyranopterin cofactor and recovery of the physiologic intramolecular electron transfer between the molybdenum centre and its [4Fe–4S] centre.[179]

Besides the pyranopterin cofactor, the protein and the other atoms of the metals' coordination sphere can influence reactivity. In fact, even small alterations of the metal centre geometry can alter the energy level of the metal d orbitals, in particular of the d_{xy} orbital in the ground state, thus modulating the reduction potential of the molybdenum centre.[180–185] The influence of the protein can be envisaged, for example, in enzymes where the metal is coordinated by two pyranopterin cofactors: the trigonal prismatic coordination geometry observed in the enzymes contrasts with the octahedral geometry found in model compounds, suggesting that the geometry in the enzymes is imposed by the polypeptide chain,[186–188] apparently with the objective of creating an entatic state that would labilize the oxo group and facilitate its dissociation.[189] The role of the first coordination sphere in modulating enzyme reactivity can be appreciated when the effects of sulfo or thiolate groups are compared: in xanthine oxidoreductase, the presence of a highly covalent terminal sulfur atom (Mo=S) with an available S π-bond, instead of thiolate (Mo–S(Cys)), allows the enzyme to accept a proton plus two electrons (hydride), thus facilitating the cleavage of a C–H bond. On the other hand, in sulfite oxidase, which holds a thiolate (Mo–S(Cys)), it is an apical terminal oxygen, with a formal triple bond (Mo≡O), that would act to labilize the oxo group to facilitate its dissociation and the metal re-oxidation.[190–196] In the following Section (1.4), the description of the structural and mechanistic properties of an array of molybdenum- and tungsten-containing enzymes will illustrate how the organisms exploit the versatile chemistry of these two metals.

1.4 Molybdenum- and Tungsten-Containing Enzymes

The biological relevance of molybdenum was demonstrated in the early 1950s–1960s, by Bray, Beinert, Lowe, Massey, Palmer and others, with ground-breaking studies performed by electron paramagnetic resonance spectroscopy that demonstrated the reduction of molybdenum to Mo^{5+} under different conditions, using mainly the molybdoenzymes xanthine oxidase and sulfite oxidase.[197–212] Noteworthy, xanthine oxidase had been already purified in 1924.[213] Tungsten only attracted the attention of the "bio-community" much later. Although it had been known since the 1970s that tungsten stimulated the growth of some acetogens and methanogens,[14,15,214–217] only in the 1980s was the first tungsten enzyme purified, one formate dehydrogenase,[218] and only in the 1990s was interest in these enzymes really boosted,[219–221] in particular after the first crystal structure determination in 1995, of the aldehyde : ferredoxin oxidoreductase from *Pyrococcus furiosus*.[222] The simultaneous determination of the first crystal structure of a molybdoenzyme, the aldehyde oxidoreductase from *Desulfovibrio gigas*,[223] and, subsequently, of dimethylsulfoxide reductase from *Rhodobacter sphaeroides*,[224] allowed the direct comparison between molybdo- and tungstoenzymes and highlighted how these enzymes are strikingly related.

Presently, more than 50 molybdoenzymes are known, many of which have already been biochemically and structurally characterized, and the number is increasing every year, with several more being foreseen to be identified in the near future based on genomic analyses.[2,225,226] With the exception of the unique heteronuclear [$MoFe_7S_9C$] cofactor of nitrogenase, and a few other heteronuclear centres whose physiological function is not yet fully understood (Figure 1.3b; described under Sections 1.4.5 and 1.4.6), molybdenum is found in the enzymes active sites in a mononuclear form, [§§] hereafter designated as molybdenum centre.[4–10] In these centres, one molybdenum atom

[§§]The carbon monoxide dehydrogenase from *Oligotropha carboxidovorans* or *Hydrogenophaga pseudoflava*, with its unique binuclear Mo/Cu cofactor, is another exception. Nevertheless, this enzyme is grouped under the mononuclear molybdoenzymes classification, as a member of the xanthine oxidase family (discussed below).

is coordinated by the *cis*-dithiolene (–S–C=C–S–) group of one or two pyranopterin cofactor molecules (Figure 1.3a) and by oxygen and/or sulfur and/or selenium atoms in a diversity of arrangements that established the innovator classification of the mononuclear molybdoenzymes into three large families, denominated after one benchmark enzyme (Figure 1.3b):[4] the xanthine oxidase family, the sulfite oxidase family and the dimethylsulfoxide reductase family (described under Sections 1.4.1–1.4.3).

Tungsten is also found in the active site of enzymes in a mononuclear form (herein denominated tungsten centre), where it is coordinated by the *cis*-dithiolene group of two molecules of the same pyranopterin cofactor found in molybdoenzymes (Figure 1.3a).[3,5,227] The tungsten coordination sphere is completed with oxygen and/or sulfur and/or selenium atoms in a diversity of arrangements, most often in the same geometry as found in members of the dimethylsulfoxide reductase family of molybdenum enzymes (Figure 1.3b). The classification of tungstoenzymes has been a matter of some controversy. Some authors pool all the tungsten-containing enzymes into a single family, characterized by the presence of a tungsten-bis-pyranopterin system. Others prefer to classify the tungsten-containing aldehyde : ferredoxin oxidoreductases separately, in a distinct family, with the other tungstoenzymes in the molybdenum-containing dimethylsulfoxide reductase family. This latter classification is based on the fact that the great majority of the prokaryotic molybdo- and tungstoenzymes harbour an active site with the metal coordinated by two pyranopterin molecules esterified to guanosine monophosphate, forming a pyranopterin guanosine dinucleotide (PGD) (which is not the case of the aldehyde : ferredoxin oxidoreductases, whose pyranopterin cofactor molecules are in the mononucleotide form) (Figure 1.3a). As such, some authors place the molybdenum-containing dimethylsulfoxide reductase family enzymes and those tungstoenzymes containing PGD together in a unique super-family denominated "molybdenum/tungsten-bis pyranopterin guanosine dinucleotide-containing enzymes", with the acronym Mo/W-bis PGD[228] (where the family name is not misleading, the name stressing the presence of both molybdo- and tungstoenzymes). This classification is particularly useful when one aims to highlight the structural/functional similarities between, for example, the homologous molybdenum-containing and tungsten-containing formate dehydrogenases.

We suggest that a systematic organization, based on the metal/cofactor structure, should be followed and all tungsten–pyranopterin-containing enzymes should be grouped together. This tungstoenzymes family can then be sub-divided to account for differences between members, as some authors do with other families (the dimethylsulfoxide reductase family is sometimes organized in subfamilies based on the amino acid residue that coordinates the molybdenum atom; also the sulfite oxidase family has become extremely heterogeneous, with the identification of new members). Following the above criteria based on the metal/cofactor structure, the molybdo- and tungstoenzymes will be here organized into five families (Figure 1.3b), described in Sections 1.4.1 to 1.4.5: xanthine oxidase family, sulfite oxidase family, dimethylsulfoxide reductase family, tungstoenzymes family and nitrogenases.

1.4.1 The Xanthine Oxidase Family

The active site of the xanthine oxidase (XO) family enzymes (in its oxidized form) has a molybdenum ion coordinated in a square-pyramidal geometry by an apical oxo group (Mo=O) and, in the equatorial plane, by the two sulfur atoms of the *cis*-dithiolene group of one pyranopterin cofactor molecule,[¶¶] one catalytically labile –OH group (in most cases) and one terminal sulfo (Mo=S; in most cases) or seleno (Mo=Se) or oxo (Mo=O) group (Figure 1.3b). This family comprises enzymes such as mammalian XO (eqn (1.6)) and aldehyde oxidase (AO; eqn (1.7)), prokaryotic *Desulfovibrio* aldehyde oxidoreductases (AOR, eqn (1.7)), *Eubacterium barkerii* nicotinate dehydrogenase (eqn (1.8)), *Pseudomonas putida* quinoline 2-oxidoreductase (eqn (1.9)) and *Thauera aromatica* 4-hydroxybenzoyl-CoA reductase (eqn (1.10)) (Table 1.2). The prokaryotic *Oligotropha carboxidovorans* carbon monoxide dehydrogenase (eqn (1.11)), with its unique binuclear Mo/Cu cofactor coordinated to the polypeptide chain by one cysteine residue through the Mo–S–Cu–S(Cys) motif, is also included in the XO family.[229,230] Although the active site of this enzyme is not mononuclear and its sulfur atom is not terminal (as is the case in other XO family enzymes), its classification under the XO family is supported by the considerable sequence and structural homology with the bovine XO and common reactivity, and also its sensitivity to inhibition by cyanide.[||||,229–233]

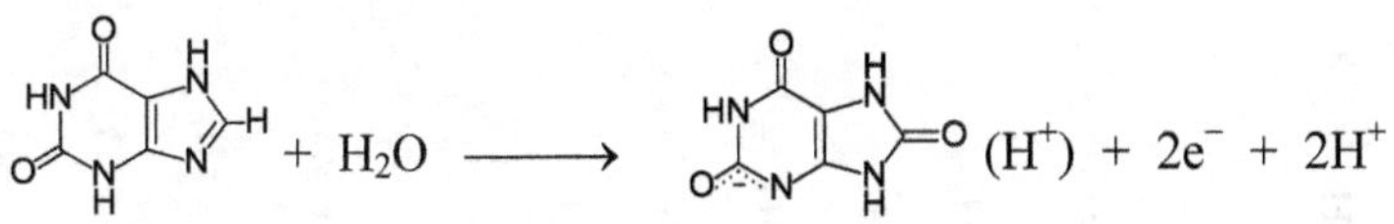

(1.6)

xanthine urate

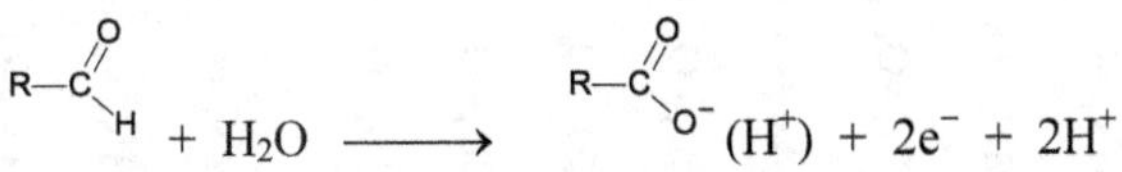

(1.7)

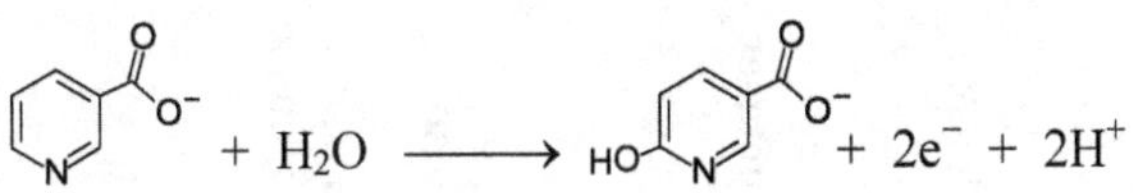

(1.8)

nicotinate 6-hydroxynicotinate

[¶¶]The pyranopterin cofactor molecule in prokaryotic enzymes is often found esterified, typically with a cytidine monophosphate (Figure 1.3a).

[||||]Cyanide reacts with the equatorial Mo=S group of the molybdenum centre, abstracting its sulfur atom in the form of thiocyanide and leaving a Mo=O group in its place. The desulfo forms thus formed are non-functional.

Table 1.2 Structural characteristics of some molybdoenzymes and tungstoenzymes.

Enzyme	Molybdenum/tungsten coordination[a]	Subunit composition[b]	Notes (ref.)
Xanthine oxidase family			
Mammalian xanthine oxidase	PMN, =O, =S, –OH	α_2, α: Mo, 2 [2Fe–2S], FAD	Eqn (1.6), (1.12) and (1.13), Figures 1.5, 1.6, see text for details
R. capsulatus xanthine oxidoreductase	PMN, =O, =S, –OH	$(\alpha\beta)_2$, a: Mo, b: 2 [2Fe–2S], FAD	Eqn (1.6) ([257])
Mammalian aldehyde oxidase	PMN, =O, =S, –OH	α_2, α: Mo, 2 [2Fe–2S], FAD	Eqn (1.7), (1.14), Figure 1.6, see text for details
Desulfovibrio aldehyde oxidoreductase	PCD, =O, =O, –OH	α_2, α: Mo, 2 [2Fe–2S]	Eqn (1.7), (1.15), Figures 1.5, 1.6, see text for details
E. coli aldehyde oxidoreductases	PCD, =O, =S, –OH	αβγ, α: Mo, β: FAD, [4Fe–4S], γ: 2 [2Fe–2S]	Eqn (1.7), Figure 1.6, ([310])
E. barkerii, nicotinate dehydrogenase	PCD, =O, =Se, –OH	$(\alpha\beta\gamma\delta)_2$, α, β: Mo, γ: 2 [2Fe–2S], δ: FAD	Eqn (1.8), see text for details
P. putida quinoline 2-oxidoreductase	PCD, =O, =S, –OH	$(\alpha\beta\gamma)_2$, α: Mo, β: 2 [2Fe–2S], γ: FAD	Eqn (1.9), see text for details
T. aromatica, 4-hydroxybenzoyl-CoA reductase	PCD, =O, =O, –OH	$(\alpha\beta\gamma)_2$, α: Mo, β: 2 [2Fe–2S], γ: FAD, [4Fe–4S]	Eqn (1.10), Figure 1.5, see text for details
Oligotropha carboxidovorans, carbon monoxide dehydrogenase	PCD, =O, –S–Cu–S(Cys), =O	$(\alpha\beta\gamma)_2$, α: Mo, β: 2 [2Fe–2S], γ: FAD	Eqn (1.11), Figure 1.5, see text for details
Sulfite oxidase family			
Chicken sulfite oxidase	PMN, =O, –S(Cys), =O	α_2, α: Mo, haem	Eqn (1.16), (1.18), Figures 1.7, 1.8, see text for details
Plant sulfite oxidase	PMN, =O, –S(Cys), =O	α_2, α: Mo	Eqn (1.16), (1.19), Figures 1.7, 1.8, see text for details
S. novella sulfite dehydrogenase	PMN, =O, –S(Cys), =O	αβ, α: Mo, β: Haem	Eqn (1.16) Figures 1.7, 1.8, see text for details
Sinorhizobium meliloti sulfite dehydrogenase	PMN, =O, –S(Cys), =O	α_2, α: Mo	Eqn (1.16), see text for details

Enzyme	Molybdenum/tungsten coordination[a]	Subunit composition[b]	Notes (ref.)
Eukaryotic assimilatory nitrate reductases	PMN, =O, –S(Cys), =O	α_2, α: Mo, haem, FAD	Eqn (1.17) and (1.20), Figures 1.7, 1.8, see text for details
Mammalian mitochondrial amidoxime reducing component	PMN, =O, –S(Cys), =O	Three protein chain complex with cyt. b_5 plus cit. *b* reductase	Eqn (1.21), Figure 1.7, see text for details
E. coli YedY	PMN, =O, –S(Cys), =O	αβ, α: Mo, β: Haem	([396])
Dimethylsulfoxide reductase family			
R. sphaeroides dimethylsulfoxide reductase	PGD, –O(Ser), =O	α, α: Mo	Eqn (1.22), Figures 1.9, 1.10, see text for details
E. coli dimethylsulfoxide reductase	PGD, –O(Ser), =O?	(αβγ)?, α: Mo, [4Fe–4S]?, β: 4 [4Fe–4S]?, γ: none?	Eqn (1.22), Figures 1.9, 1.10, see text for details
D. desulfuricans periplasmatic nitrate reductase	PGD, –S(Cys), =S	α, α: Mo, [4Fe–4S]	Eqn (1.17), Figures 1.9, 1.10, see text for details
C. necator periplasmatic nitrate reductase	PGD, –S(Cys), =S	αβ, α: Mo, [4Fe–4S], β: 2 haem	Eqn (1.17), see text for details
E. coli periplasmatic nitrate reductase	PGD, –S(Cys), –OH	α, α: Mo, [4Fe–4S]	Eqn (1.17), see text for details
E. coli, respiratory nitrate reductase (NaRGHI)	PGD, –O(Asp), =O or PGD, –O(Asp)O–	$(\alpha\beta\gamma)_2$, α: Mo, [4Fe–4S], β: 3 [4Fe–4S], [3Fe–4S], γ: 2 haem	Eqn (1.17), Figure 1.9, see text for details
E. coli formate dehydrogenase H	PGD, –Se(Cys), =S	α, α: Mo, [4Fe–4S]	Eqn (1.25), Figures 1.9, 1.11, see text for details
E. coli formate dehydrogenase N	PGD, –Se(Cys), =S	$(\alpha\beta\gamma)_3$, α: Mo, [4Fe–4S], β: 4 [4Fe–4S], γ: 2 haem	Eqn (1.25), Figures 1.9, 1.11, see text for details
E. coli formate dehydrogenase O	PGD, –Se(Cys), =S	$(\alpha\beta\gamma)_3$, α: Mo, [4Fe–4S], β: 4 [4Fe–4S], γ: 2 *b* haems	Eqn (1.25), [455,469,470]
D. desulfuricans formate dehydrogenase	PGD, –Se(Cys), S?	αβγ, α: Mo, [4Fe–4S], β: [4Fe–4S], γ: 4 *c* haems	Eqn (1.25), [471 and 472]
R. eutropha formate dehydrogenase	PGD, –S(Cys), S?	αβγδ?, α: Mo, 2 [2Fe–2S], 3 [4Fe–4S], β: [4Fe–4S], FMN, γ: 2 [2Fe–2S]	Eqn (1.25), [473 and 475]
R. capsulatus formate dehydrogenase	PGD, –S(Cys), =S	$(\alpha\beta\gamma)_2$, α: Mo, [2Fe–2S], 4 [4Fe–4S], β: [4Fe–4S], FMN, γ: [2Fe–2S]	Eqn (1.25), ([476])

(*continued*)

Table 1.2 (*continued*)

Enzyme	Molybdenum/tungsten coordination[a]	Subunit composition[b]	Notes (ref.)
M. formicicum formate dehydrogenase	PGD, –S(Cys), S?	αβ, FAD, several Fe/S, Zn	Eqn (1.25), ([676–681])
T. thermophilus polysulfide reductase	PGD, –S(Cys), –OH	$(\alpha\beta\gamma\delta)_2$, α: Mo, [4Fe–4S], β: 4 [4Fe–4S], γ: none	Eqn (1.26), see text for details
A. faecalis arsenite oxidase	PGD, =O, –OH	αβ, α: Mo, [3Fe–4S], β: [2Fe–2S]	Eqn (1.23), [404–407]
P. acidigallici pyrogallol:phloroglucinol transhydroxylase	PGD, –O(Ser), –OH/–OH_2?	αβ, α: Mo, β: 3 [4Fe–4S]	Eqn (1.27), ([161])
A. aromaticum ethylbenzene dehydrogenase	PGD, –O(Asp), =O?	αβγ, α: Mo, [4Fe–4S], β: 3 [4Fe–4S], [3Fe–4S], γ: haem	([176,682–685])
Shewanella massilia trimethylamine-*N*-oxide reductase	PGD, –O(Ser), =O	α, α: Mo	([686,687])
tungstoenzymes family			
P. furiosus aldehyde:ferredoxin oxidoreductase	PMN, =O, =O/–OH?	α_2, α: W, [4Fe–4S]	Eqn (1.7), Figure 1.12, see text for details
P. furiosus formaldehyde:ferredoxin oxidoreductase	PMN, =O, =O/–OH?	α_4, α: W, [4Fe–4S]	Eqn (1.7), see text for details
D. gigas formate dehydrogenase	PGD, –Se(Cys), =S	αβ, α: W, [4Fe–4S], β: 3 [4Fe–4S]	Eqn (1.25), Figure 1.12, see text for details
Methylobacterium extorquens formate dehydrogenase	PGD, –S(Cys), S?	αβ, α: W, ≥1 Fe/S, β: [4Fe–4S], FMN	Eqn (1.25), ([18])
G. metallireducens benzoyl-CoA reductase	PMN, –S(Cys)?	Multimeric, α: W, [4Fe–4S]	Eqn (1.29), Figure 1.12, see text for details
P. acetylenicus acetylene hydratase	PGD, –S(Cys), –OH_2	α, α: W, [4Fe–4S]	Eqn (1.28), Figure 1.12, see text for details

[a]PCD, pyranopterin cytidine dinucleotide; PGD, pyranopterin guanosine dinucleotide; PMN, pyranopterin mononucleotide (see Figure 1.3a).
[b]Here, the chemical symbols Mo and W stand for molybdenum and tungsten centre, respectively.

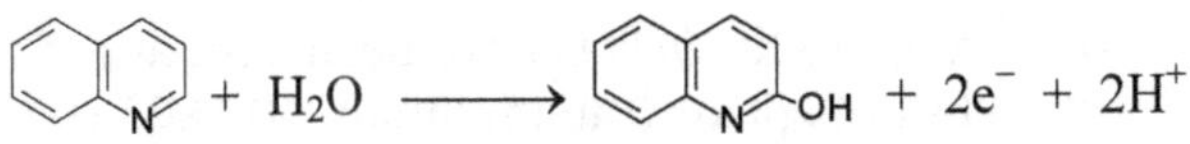

quinoline 2-hydroxyquinoline (1.9)

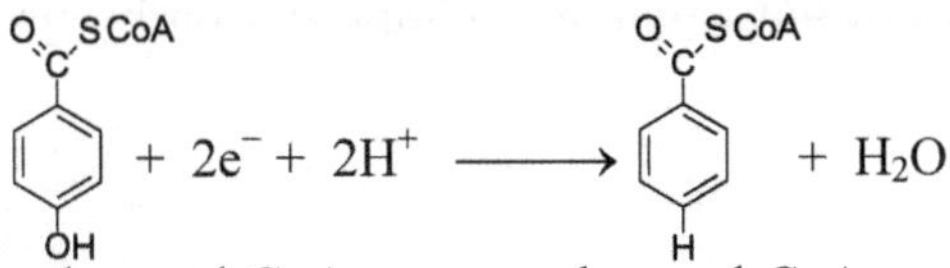

hydroxybenzoyl-CoA benzoyl-CoA (1.10)

$$CO + H_2O \rightarrow OCO + 2e^- + 2H^+ \quad (1.11)$$

a) The history of bovine milk XO, the prototype enzyme of this family, began more than a century ago, with the Schardinger observation (in 1902) of the decolourization (reduction) of methylene blue by formaldehyde in the presence of fresh milk.[234] This was followed by the discovery that milk could also oxidize hypoxanthine and xanthine to urate[235] and the partial purification of XO in 1924.[213] Being one of the most studied enzymes, our present understanding of its structure and function allows us to discuss its reaction mechanism with atomic and electronic details for each catalytic intermediate, as well as the complex mechanism by which posttranslational conformational modifications promote an "activity switch".

Mammalian XO enzymes are cytoplasmatic proteins synthesized as NAD^+-dependent dehydrogenases (denominated xanthine dehydrogenase (XD), eqn (1.12)), and are believed to exist under normal physiological conditions mostly in this form *in vivo*.[154–159] XD can be rapidly converted into a "strict" oxidase form, however, that reduces dioxygen rather than NAD^+ – the commonly studied and very well-documented XO (eqn (1.13)). This conversion can be reversible, through oxidation of key cysteine residues, or irreversible, by limited proteolysis.***[157,236–252] However, the extent and

***It is to be emphasized that mammalian XO and XD are two forms of the same protein (same gene product) that arise as the result of posttranslational conformational modifications (oxidation of key cysteine residues or limited proteolysis);[157,236–252] to designate the two forms, without regard to the oxidizing substrate, it is preferable to use the denomination "xanthine oxidoreductase". The only functional distinction between XD and XO lies in the electron acceptor used by each form: while XD transfers the electrons preferentially to NAD^+ (but it can also reduce dioxygen), XO fails to react with NAD^+ and uses exclusively dioxygen. During the XD into XO conversion process, the protein conformation at the FAD centre is modified and this conformational alteration is responsible for the differentiated oxidizing substrate specificity displayed by XO and XD (note that both dioxygen and NAD^+ react at the FAD centre). On the other hand, the protein structure at the Fe/S and molybdenum centres is not changed during the conversion and, in accordance, the two enzyme forms, XO and XD, are virtually identical with respect to the binding and catalysis of substrates at the molybdenum centre.[154–159] Hence, XO and XD can be considered as one unique entity for the discussion of the overall structural organization and molybdenum reactivity (reaction mechanism). Noteworthy, the enzyme from other sources, *e.g.* turkey and chicken,[253–255] insects[256] and bacteria[257] exists solely as a dehydrogenase form.

rate of conversion of XD into XO *in vivo* have been a matter of great controversy.[237,240,241,258–266] Physiologically, mammalian XD (XO) is a key enzyme in purine catabolism, where it catalyzes the hydroxylation of both hypoxanthine and xanthine to the terminal metabolite, urate (eqn (1.6)).[154–159] It has also been suggested to be involved in the xenobiotic compounds metabolism and in ROS-mediated signalling pathways and ROS-mediated diseases.[265–292]

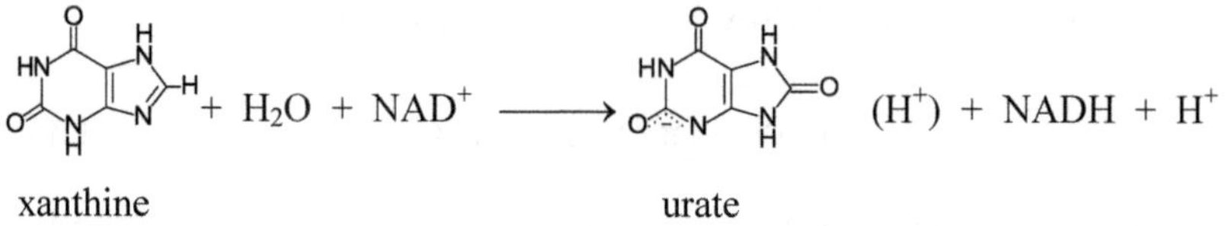

(1.12)

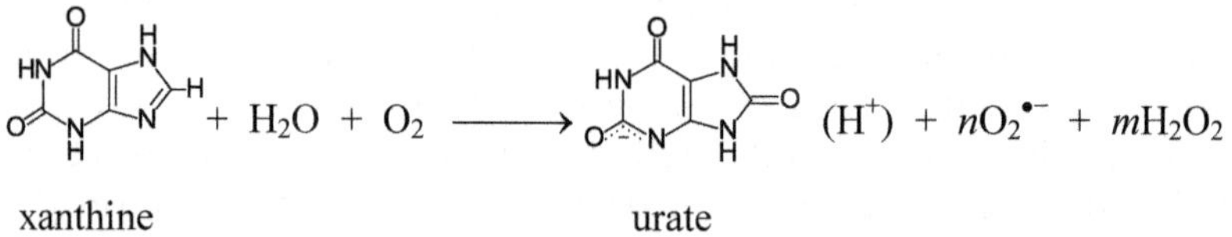

(1.13)

In its oxidized form, the molybdenum centre of mammalian XO has the characteristic catalytically labile –OH group and an essential terminal sulfo group (Mo=S). Interestingly, some prokaryotic enzymes have instead a terminal seleno group (Mo=Se) that is also essential for the activity[293,294] (*e.g.* nicotinate dehydrogenase, another member of the XO family[295–297]). Besides the molybdenum centre, mammalian XO possesses three additional redox-active centres: two [2Fe–2S] clusters and one FAD. Structurally, mammalian XOs are homodimeric enzymes (α_2) with each monomer folded into three domains, which can be thought of as pseudo $\alpha'\beta'\gamma'$ subunits, each holding one type of redox-active centre (N-terminal domain holds the Fe/S centres, central domain, the FAD site, and, C-terminal, the molybdenum centre) (Figure 1.5).[248,298] The four redox-active centres are aligned in an almost linear fashion that defines an intramolecular electron transfer pathway delivering electrons from the molybdenum centre to the FAD, the sites where the hydroxylation and dioxygen reduction take place, respectively (with Fe/S centres intermediating the electron transfer; discussed below).

Mammalian AO enzymes are structurally very similar to XO, holding one identical molybdenum centre, as well as two Fe/S centres and one FAD centre, in the same α_2 structure. They exist exclusively as an oxidase form (reducing dioxygen, not NAD^+; eqn (1.14)). While humans have only one AO, some

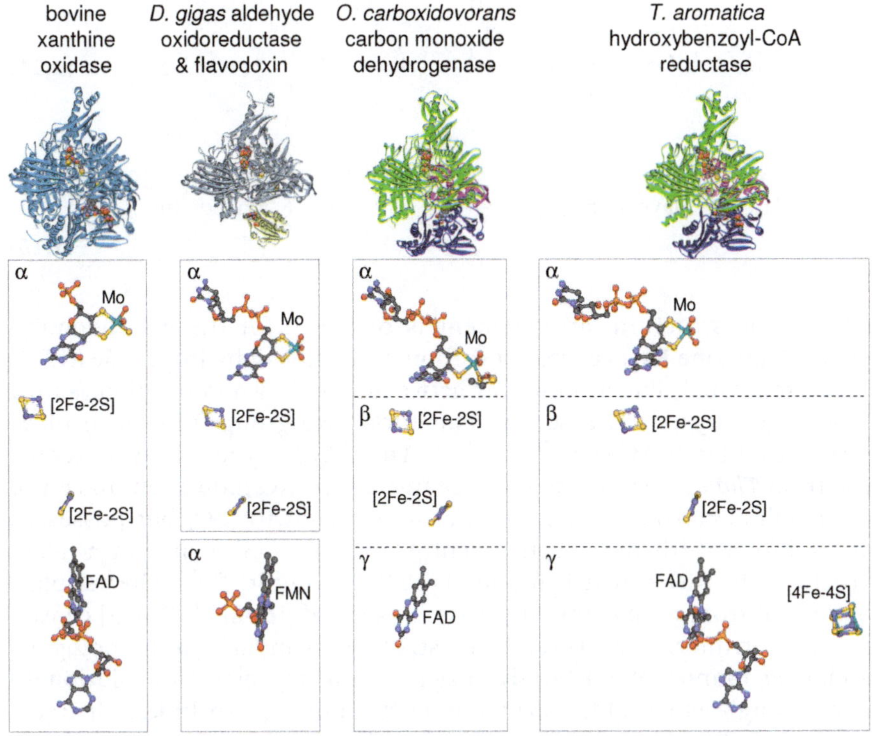

Figure 1.5 Three-dimensional structure view of the bovine xanthine oxidase, *D. gigas* aldehyde oxidoreductase plus flavodoxin, *O. carboxidovorans* carbon monoxide dehydrogenase and *T. aromatica* hydroxybenzoyl-CoA reductase (top) and of the arrangement of their redox cofactors (bottom). Only one subunit in XO and AOR (α_2) or one αβγ group in the other two enzymes ($(\alpha\beta\gamma)_2$) is represented. The number of colours reflects the number of subunits of each enzyme and its cofactor composition: XO α – blue; AOR α – light blue plus flavodoxin α – yellow; carbon monoxide dehydrogenase and hydroxybenzoyl-CoA reductase αβγ – green, pink, dark blue. The structures shown are based on the PDB files 1FO4 (XO), 1VLB (AOR), 1FX1 (flavodoxin), 1N5W (carbon monoxide dehydrogenase) and 1RM6 (hydroxybenzoyl-CoA reductase), and were produced with Accelrys DS Visualizer, Accelrys Software Inc.

mammals encode multiple tissue-specific isoforms (the mouse genome, for example, encodes four AOs).[75,299–302] Bacterial AOR from *Desulfovibrio* species holds a slightly different molybdenum centre, with a second equatorial oxo group in place of the more frequent sulfo group, and it harbours only two Fe/S centres besides the molybdenum centre (no FAD centre), in an α_2 structure.[223,303,304] In spite of this, when the AOR structure is represented with its putative physiological partner, the flavin-containing flavodoxin (eqn (1.15)), which can be regarded as a pseudo γ′ subunit, it becomes apparent that the structural homology with XO is preserved (Figure 1.5).[305]

$$\mathrm{R{-}C({=}O){-}H} + \mathrm{H_2O} + \mathrm{O_2} \longrightarrow \mathrm{R{-}C({=}O){-}O^-}\ (\mathrm{H^+}) + n\mathrm{O_2}^{\bullet-} + m\mathrm{H_2O_2} \tag{1.14}$$

$$\mathrm{R{-}C({=}O){-}H} + \mathrm{H_2O} + \mathrm{Flavodoxin_{oxidized}} \longrightarrow \mathrm{R{-}C({=}O){-}O^-}\ (\mathrm{H^+}) + \mathrm{Flavodoxin_{reduced}} \tag{1.15}$$

The same structural arrangement is observed for the molybdenum, two Fe/S and one FAD centres of carbon monoxide dehydrogenase ($(\alpha\beta\gamma)_2$ structure), which has a binuclear active site with a non-terminal sulfur (Mo–S–Cu–S(Cys)) and a second equatorial oxo group (Mo=O) in place of the usual labile Mo–OH.[229–233,306,307] The 4-hydroxybenzoyl-CoA reductase from *Thauera aromatica* has a characteristic XO molybdenum centre and displays considerable structural homology with XO, but possesses an additional fifth redox active centre, one [4Fe–4S] centre in a $(\alpha\beta\gamma)_2$ structure (the fifth centre is found in the FAD subunit).[308,309] The periplasmatic AOR from *E. coli* similarly possesses an additional [4Fe–4S] centre, in a heterotrimeric structure.[310] The structural similarity of homologous domains/subunits of XO family enzymes thus suggests an evolutionary path through which older αβγ trimeric structures (with linked or free γ subunit) evolved into the newer eukaryotic monomeric structures with pseudo α′β′γ′ subunits: three "building blocks" (polypeptides/proteins) fused into a single polypeptide.[305]

b) XO family enzymes typically catalyze, at their molybdenum centres, the hydroxylation of a C–H bond in aromatic heterocyclic compounds and aldehydes.[154–159,311] Such oxidative hydroxylations are seen in the XO-catalyzed conversion of xanthine to urate (eqn (1.6)), as well as in the reactions catalyzed by AO and AOR (eqn (1.7)), nicotinate dehydrogenase (eqn (1.8)) and quinoline 2-oxidoreductase (eqn (1.9)). There are, however, at least two important exceptions: the carbon monoxide dehydrogenase-catalyzed oxidation of carbon monoxide to carbon dioxide, which does not involve hydrolysis of a C–H bond (eqn (1.11)),[229–233,306,307] and the reaction of hydroxybenzoyl-CoA reductase, which involves the irreversible dehydroxylation of the phenol ring, an oxygen atom abstraction/reduction reaction (eqn (1.10)).[312,313]

The molecular mechanism of XO hydroxylation reaction is presently well established and believed to be essentially similar in other members of this family, namely in AO and AOR enzymes (Figure 1.6):[154–159,248,298,314,315] (1) the catalysis is initiated with the activation of the molybdenum catalytically labile hydroxyl group (Mo–OH) by a neighbouring conserved deprotonated glutamate residue, to form an Mo^{6+}–O– core (base-assisted catalysis); (2) the now deprotonated oxygen undertakes nucleophilic attack on the carbon atom to be hydroxylated, with the simultaneous transfer of hydride from

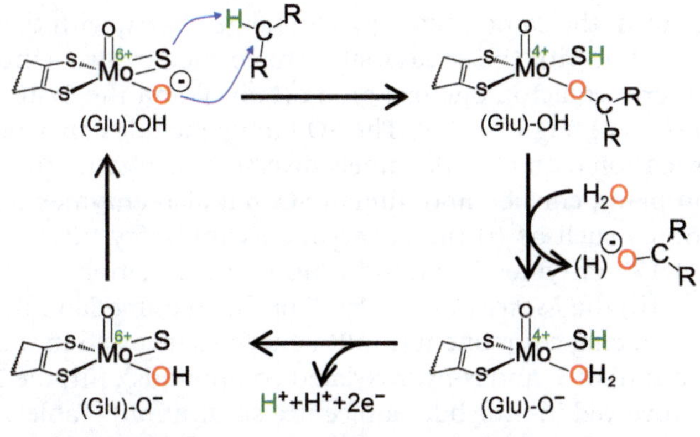

Figure 1.6 Simplified mechanistic proposal for the reaction catalyzed by mammalian xanthine oxidase. See text for details.

substrate to the essential sulfo group (Mo=S → Mo–SH), resulting in the formation of a covalent intermediate, Mo^{4+}–O–C–R(–SH) (where R represents the remainder of the substrate molecule); (3) hydroxide then displaces the hydroxylated product from the molybdenum coordination sphere to yield a Mo^{4+}–$OH_{(2)}$(–SH) core (oxidation half-reaction); (4) the two electrons transferred from the substrate to the molybdenum during the reductive half-reaction (Mo^{6+} → Mo^{4+}) are rapidly transferred, *via* the Fe/S centres, to the FAD (Mo → Fe/S → FAD), where the reduction of dioxygen (or NAD^+, in the case of the XD form) takes place (oxidative half-reaction); and (5) in the now oxidized molybdenum centre (Mo^{4+} → Mo^{6+}), the sulfo group is deprotonated and the initial Mo^{6+}–OH(=S) core is regenerated. Intramolecular electron transfer through XO (Mo → Fe/S → FAD) is, therefore, an integral aspect of the XO catalysis. Of note, (i) water is the ultimate source of the oxygen atom incorporated into the hydroxylated product, as is characteristic of molybdoenzymes (Section 1.3) (the molybdenum labile hydroxyl group (Mo–OH) that ends up in the product of the reaction (urate) is regenerated from a solvent water molecule when the catalytic cycle is closed)[316,317] and (ii) dioxygen is only the oxidizing substrate. The hydroxylation reactions catalyzed by these molybdoenzymes is, in this way, quite different from the reaction catalyzed by monooxygenases as it generates, rather than consumes, reducing equivalents and uses dioxygen as an oxidant rather than as the source of the oxygen atom incorporated into product.

1.4.2 The Sulfite Oxidase Family

The molybdenum centre of the enzymes of the sulfite oxidase (SO) family is closely related to that of the XO family, but with the distinctive feature of having the polypeptide chain, through a cysteine residue, directly coordinated to the molybdenum. In these enzymes, the molybdenum centre displays (in its oxidized form) the same square-pyramidal geometry, with the apical oxo group (Mo=O), but with the equatorial plane formed by two sulfur atoms of the pyranopterin cofactor, one oxo group (Mo=O) and the cysteine thiolate ligand (Mo–S(Cys)) (Figure 1.3b). The SO family includes enzymes involved in sulfite oxidation (eqn (1.16)), namely diverse prokaryotic sulfite dehydrogenases and plant, chicken and human SO, but also enzymes that catalyze other reactions, such as (i) the eukaryotic assimilatory nitrate reductases (NaR; eqn (1.17); enzymes involved in nitrate assimilation in plants, algae and fungi[†††]), (ii) the *Escherichia coli* YedY or the mammalian mitochondrial amidoxime reducing component (mARC; enzymes involved in the reduction (dehydroxylation) of *S*- and *N*-hydroxylated compounds), (iii) the MOSC protein family (involved in molybdenum centre sulfuration) (Table 1.2) and (iv) other archaeal and bacterial enzymes of unknown function, identified on the basis of genomic analyses.

$$SO_2{}^{2-} + H_2O \rightarrow OSO_2{}^{2-} + 2e^- + 2H^+ \tag{1.16}$$

$$ONO_2{}^- + 2e^- + 2H^+ \rightarrow NO_2{}^- + H_2O \tag{1.17}$$

a) Chicken liver SO,[‡‡‡] the benchmark enzyme of this family, as well as the SO from humans and other vertebrates, is a homodimeric enzyme (α_2), with each monomer folded into two domains, one comprising one *b*-type haem and the other the molybdenum centre, both of which can (once more) be thought of as pseudo $\alpha'\beta'$ subunits (Figure 1.7).[318] Physiologically, the vertebrate SO, located in the mitochondrial intermembrane space, is a key enzyme in the catabolism of sulfur-containing amino acids and other compounds, catalyzing the oxidation of the toxic sulfite to sulfate (at the molybdenum centre), with the simultaneous reduction of cytochrome *c* (at the *b*-type haem) (eqn (1.18)).[319] Interestingly, SO contributes to the generation of cellular energy, through its cytochrome *c* reduction activity[320,321] – a relic of a prokaryotic respiratory process? Remarkably, the molybdenum and haem centres of chicken SO have been found to be more than 30 Å apart in the crystal structure.[318] Hence, during catalysis, a conformational alteration would have to take place, through the flexible polypeptide that links the two domains, to bring the two centres into sufficiently close proximity as to allow rapid intramolecular electron transfer observed, (sulfite)Mo → Fe(cytochrome).[322–324]

[†††]It should be noted that the eukaryotic assimilatory NaR is distinct from any type of prokaryotic NaR enzymes, which are classified as members of the dimethylsulfoxide reductase family.

[‡‡‡]Although this enzyme can reduce both cytochrome *c* and dioxygen, its physiological (and preferred) partner is thought to be the cytochrome *c*. Hence, instead of sulfite oxidase (oxidase means an enzyme that reduces dioxygen), a more appropriate name according to the Enzyme Nomenclature Committee (IUBMB) would be sulfite oxidoreductase. A similar reasoning would apply to the prokaryotic sulfite dehydrogenases, whose physiological partners are most often cytochromes. Only the plant enzymes are true oxidases.

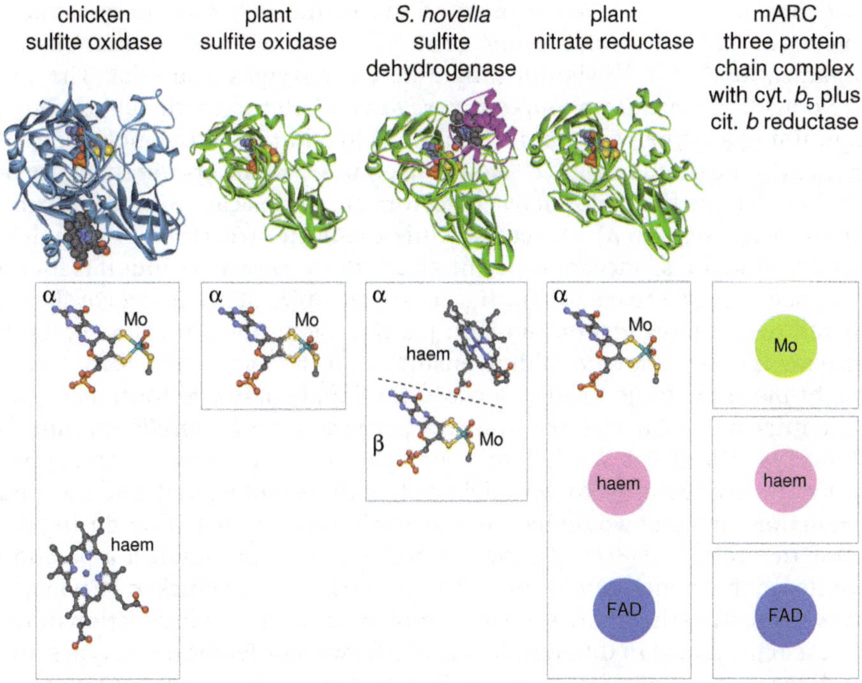

Figure 1.7 Three-dimensional structure view of chicken and plant (*A. thaliana*) sulfite oxidase, *S. novella* sulfite dehydrogenase and plant (*P. angusta*) nitrate reductase (top) and of the arrangement of their redox cofactors (bottom). Only one subunit in chicken and plant SO and plant NaR (α_2) is represented. The number of colours reflects the number of subunits of each enzyme and its cofactor composition: chicken SO α – blue; plant SO and NaR α – green; bacterial sulfite dehydrogenase αβ – green, pink. In the case of plant NaR, only the 3D structure of the domain holding the molybdenum centre is represented. The three protein chain complex with mARC is represented only schematically (its structure is not known yet). The structures shown are based on the PDB files 1SOX (chicken SO), 1OGP (*A. thaliana* SO), 2BPB (*S. novella* sulfite dehydrogenase) and 2BIH (NaR molybdenum domain), and were produced with Accelrys DS Visualizer, Accelrys Software Inc.

Contrary to vertebrate SO, the plant enzymes are peroxisomal and use (reduce) dioxygen as the physiological partner (eqn (1.19)). Most significantly, these enzymes hold only the molybdenum centre (no haem) in an α_2 structure that nevertheless shows the same basic architecture and fold structure as the molybdenum domain of vertebrate SO.[325–328] The "non-uniformity" of sulfite-oxidizing enzymes continues with the prokaryotic counterparts (sulfite dehydrogenases): these are typically periplasmatic enzymes that display different structural organizations and redox cofactor compositions, and whose physiological electron acceptor is most often a cytochrome.[87–90,329–331] While most of the sulfite dehydrogenases appear to contain only the molybdenum centre in a homodimeric or monomeric organization,[88,90,320,330] significantly different enzymes also exist. The sulfite dehydrogenase from *Starkeya novella* is a heterodimeric (αβ) enzyme, containing a *c*-type haem (in its β subunit) in addition to the molybdenum centre (in the α subunit).[332–334] However, contrary to the crystal structure of chicken SO, the two redox-active cofactors are in this case in close proximity to each other (16 Å). Moreover, while the basic structure of the molybdenum domain is, once more, maintained, the *S. novella* β subunit is found wrapped around the α subunit. In this way, in spite of not being covalently linked to the molybdenum domain (as the haem domain of the chicken enzyme is), the *S. novella* folding ensures that the haem is "locked" in the "right place", with no need of a conformational change to form a competent intramolecular electron transfer pathway. The *S. novella* enzyme is thus crystallized as a stable dimer, with the cofactors in the "correct positions", in clear contrast to the chicken SO with its mobile molybdenum and haem domains that would have to move relatively to each other during the catalytic cycle. Interestingly, the *S. novella c* haem is found in a position similar to that which can be modelled for the *b* haem of chicken enzyme,[334] thus suggesting that there is an ideal molybdenum/haem interaction mode that is reproduced in different forms of life, with different haem types and in different physiological contexts. The question now is what other structural variations in sulfite oxidizing enzymes are yet to be discovered? And, more intriguing, why have vertebrate SO evolved so as to require the flexible linker movement to reposition the two cofactors? This is particularly interesting (and significant) in light of the fact that the structure and fold of the molybdenum domain is maintained in all sulfite oxidizing enzymes whose structure is presently known.

$$SO_2^{2-} + H_2O + 2\text{cyt. } c\ (Fe^{3+}) \rightarrow OSO_2^{2-} + 2\text{cyt. } c\ (Fe^{2+}) + 2H^+ \quad (1.18)$$

$$SO_2^{2-} + H_2O + O_2 \rightarrow OSO_2^{2-} + 2O_2^{\bullet -} + 2H^+ \quad (1.19)$$

Eukaryotic assimilatory nitrate reducing enzymes**** also belong to the SO family. These enzymes are involved in the first and rate-limiting step of nitrate assimilation in plants, algae and fungi (Figure 1.1, grey arrow), where they catalyze the nitrate reduction to nitrite, at their molybdenum centre, with the simultaneous oxidation of NAD(P)H (eqn (1.20)).[158,335–343] They are cytoplasmatic homodimeric enzymes (α_2), containing, besides the characteristic molybdenum centre, one *b*-type haem and one FAD centre that is responsible for the NAD(P)H binding and oxidation (Figure 1.7).[335,344,345] Consistent with their classification as SO family members, the NaR molybdenum domain is strikingly similar to one of chicken and plant SO. The FAD domain has been suggested to contribute to the correct positioning of the haem domain for intramolecular electron transfer. Interestingly, the phosphorylation of a serine residue in the linker region between the molybdenum and haem domains,[346,347] followed by the binding of a specific regulatory protein, effectively inhibits the enzyme through the inhibition of the intramolecular electron transfer.[§§§,353–356]

$$ONO_2^- + NADH + H^+ \rightarrow NO_2^- + H_2O + NAD^+ \qquad (1.20)$$

Another member of the SO family is the recently described mARC (the fourth mammalian molybdoenzyme, after XO, AO and SO).[357] Its physiological function is not known, but it is probably involved in the detoxification of mutagenic and toxic aromatic hydroxyl-amines, such as *N*-hydroxylated DNA base derivates,[358,359] through the catalysis of the reduction of *N*- and *S*-hydroxylated compounds at the molybdenum centre (eqn (1.21)),[360–363] among other possible roles.[67,364,365] This mitochondrial enzyme (outer membrane-bounded, facing the cytoplasm) is monomeric and contains only the molybdenum centre (with no additional redox-active centres)[361,364,366,367] that coordinates the conserved cysteine residue[368] and an apical oxo group (Mo=O).[369] However, mARC functions in concert with cytochrome b_5 and a NADH-dependent cytochrome b_5 reductase, which are involved in electron transfer from NADH to the mARC (Figure 1.7), thus forming a three-protein electron transfer chain resembling the arrangement of redox-active cofactors seen in NaR enzymes: NAD(P)H → FAD → haem → Mo → *N*- and *S*-hydroxylated compounds.[357,360,366,367,370,371]

$$R_2N\text{–}OH + 2e^- + 2H^+ \longrightarrow R_2N\text{–}H + H_2O \qquad (1.21)$$

b) The members of the SO family are thought to be proper oxotransferases, as exemplified by SO and NaR, that catalyze the "simple" transfer of one oxygen atom to, or from, a lone electron pair of the substrate, with their molybdenum atom mediating the transfer: (i) SO-catalyzed sulfite oxidation to sulfate (eqn (1.16)), where the $Mo^{6+}{=}O_{(equatorial)}$ centre is the direct oxygen donor (oxygen insertion/oxidation reaction), and (ii) NaR-catalyzed nitrate reduction to nitrite (eqn (1.17)), where the Mo^{4+} is the direct oxygen acceptor (oxygen abstraction/reduction reaction). As in the XO family, water is the ultimate oxygen atom donor or acceptor (the oxygen atom from nitrate ends up eventually in solvent). Also, as seen in XO, the two half-reactions of chicken/human SO and plant NaR are physically separated, with one of the half-reactions being carried out at the molybdenum centre and the other elsewhere

[§§§]As expected from its key role on the nitrogen metabolism, NaR is highly regulated by complex transcriptional, translational and posttranslational mechanisms that respond to nitrogen, carbon dioxide and dioxygen availabilities, pH, temperature and light.[337,339–341,343,348] The posttranslational regulation involves the phosphorylation of a serine residue in the linker region between the molybdenum and haem domains.[349,350] The phosphorylation is catalyzed by protein kinases, including AMP-activated[351] and calcium-dependent kinases.[352] This phosphorylation creates a recognition site that recruits a specific regulatory protein (one member of the 14-3-3 family), whose binding effectively inhibits the enzyme.[353–356] (*i.e.* at the haem of SO or the FAD of NaR); therefore, intramolecular electron transfer (Mo → haem or FAD → haem → Mo) is, once again, an integral aspect of catalysis.

Chicken/human SO-catalyzed sulfite oxidation is presently well understood (Figure 1.8):[195,318,372–379] (1) catalysis is initiated at the oxidized molybdenum centre, initiated by nucleophilic attack of the sulfite lone-pair of electrons on the catalytically labile equatorial oxo group of the molybdenum (Mo^{6+}=O), which can be thought of as an electrophilic oxygen; [¶¶¶] this results in the formation of a covalent Mo^{4+}–O–SO_3 intermediate, where the molybdenum atom has become reduced by two electrons with a formal triple bond to the apical oxo group (Mo=O_{apical});[190–196] (2) the presence of this apical "spectator oxo" (so named because it is not directly involved in the oxygen atom transfer) facilitates the subsequent cleavage of the Mo–O(substrate)$_{equatorial}$ bond (weakens it); product (sulfate) is then released to yield a Mo^{4+}–$OH_{(2)}$ core (the reductive half-reaction); (3) finally, the two electrons transferred from the substrate to the molybdenum are intramolecularly transferred, one at a time, to the haem, where cytochrome *c* (the physiological partner) will be reduced, and the initial Mo^{6+}=O core is regenerated (oxidative half-reaction). In these enzymes with an obligatory one-electron cofactor (haem), a transient Mo^{5+} species is an obligate intermediate formed as the two electrons of Mo^{4+} are sequentially transferred to the haem, which must wait for the physiological oxidizing partner to be re-oxidized.

In the plant SO that is devoid of additional redox cofactors, the oxidative half-reaction is also carried out at molybdenum centre, with the one-electron reduction of dioxygen to superoxide radical anion, which, then, spontaneously dismutates to hydrogen peroxide,[387] again implicating the formation of a Mo^{5+} intermediate. It should be emphasized that this reactivity of a molybdenum centre towards dioxygen is not common, although haem-deficient variants of mammalian SO exhibit a dioxygen reduction/sulfite oxidation activity similar to that exhibited by the plant enzyme,[388] suggesting that the absence of a second redox-active centre (haem) allows the enzyme to transfer the electrons to the dioxygen *via* the molybdenum centre. However, this is not the case of the sulfite dehydrogenase from *Sinorhizobium meliloti* (SorT), which, in spite of having only the molybdenum centre in a similar active site, does not react with dioxygen[389] – which leaves the intriguing reactivity of molybdenum centres toward dioxygen yet to be fully understood.

The molecular mechanism of NaR-catalyzed nitrate reduction to nitrite is believed to be the reverse of that seen with SO, with the abstraction of an oxygen atom and the intramolecular electron flow occurring in the opposite direction. Hence (Figure 1.8):[337,340,343,344,390–393] (1) catalysis is initiated with the reduction of FAD by NAD(P)H (the reductive half-reaction); (2) the two

---

¶¶¶This labile oxo group is doubled bonded to the molybdenum in the 6+ oxidation state, which is an electron sink, and, thus, it can be thought of as electrophilic. Supporting the proposal of the nucleophilic attack to this electrophilic oxygen, there are several experimental evidences with model complexes[142,380–386] and theoretical data.[374,376]

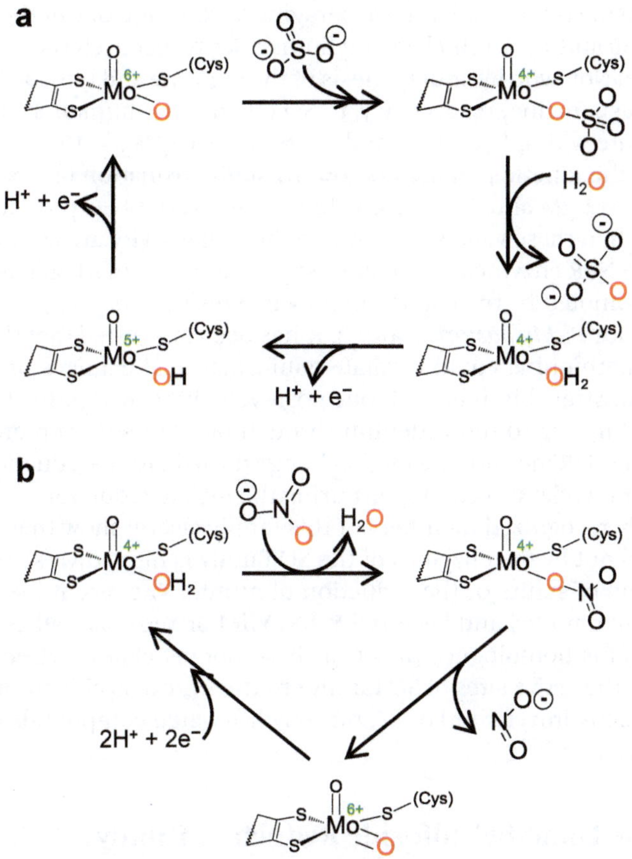

Figure 1.8 Simplified mechanistic proposal for the reaction catalyzed by vertebrate sulfite oxidase (a) and plant nitrate reductase (b) See text for details.

electrons are transferred intramolecularly, through the haem, to the molybdenum centre (FAD → haem → Mo); in the now reduced molybdenum centre, the equatorial labile oxo group is protonated (Mo^{6+}=O → Mo^{4+}–OH); (3) nitrate binds to the molybdenum, displacing the labile group; this results in the formation of the covalent intermediate Mo^{4+}–O–NO_2; (4) the subsequent O–N bond cleavage releases the product (nitrite) and regenerates the Mo^{6+}=O core (the oxidative half-reaction).

The similarities between the molybdenum-binding domains and molybdenum centres of SO and NaR, together with the complementarity of their reaction mechanisms, raise the intriguing question of whether a SO enzyme might efficiently catalyze the nitrate reduction (and *vice versa*)? The answer seems to reside in two amino acid residues that are found in the SO active site, but not in NaR: the SO substrate binding pocket comprises five residues that are conserved in all known eukaryotic SO, three arginines, one tyrosine and one tryptophan; three of these residues are also conserved in NaR enzymes, but the tyrosine is replaced by an asparagine in NaR, and one of the arginines (arginine 450 and 472 in chicken and human SO, respectively) is replaced by a methionine. Noteworthy, mutagenesis of the arginine and tyrosone to methionine and asparagine, respectively, in SO enables the human SO enzyme to reduce nitrate with a $k_{cat} \approx 1\ s^{-1}$ and $K_m \approx 25\ \mu M$ ($k_{cat}/K_m \approx 10^4\ M^{-1}s^{-1}$), while decreasing the kinetic parameters toward sulfite oxidation ($k_{cat} \approx 25\ s^{-1}$ and $1\ s^{-1}$ and $K_m \approx 3\ \mu M$ and 5 mM, for wild-type and variant SO proteins, respectively).[394] When these values of nitrate reduction activity are compared with ones of true NaR enzymes, it becomes apparent how remarkable is the modification promoted by the change of only two residues (*e.g.* $k_{cat} \approx 159\ s^{-1}$ and $K_m \approx 30\ \mu M$ for *Pichia pastoris* NaR[395]). It has been suggested that the SO arginine/NaR methionine could mediate conformational changes of the active site upon substrate binding, with one positively charged arginine hampering the nitrate binding to molybdenum through one of its oxygen atoms.[394] In addition, the NaR methionine might change the oxidation–reduction properties of the molybdenum centre, favouring the nitrate reduction.

The newly recognized members of this family clearly show that the chemistry carried out by the enzymes of the SO family is more diverse than "just" the oxidation of sulfite or the reduction of nitrate. The recent identification of mammalian mARC and bacterial YcbX, YiiM or YedY, as well as of several MOSC proteins homologues (most of these not yet characterized), demonstrated that the active sites of SO family enzymes are also able to catalyze the reduction (dehydroxylation) of *S*- and *N*-hydroxylated compounds and sulfuration reactions.[368,396–403]

1.4.3 The Dimethylsulfoxide Reductase Family

The enzymes in this family are characterized by having four sulfur atoms of two pyranopterin cofactor molecules[‖‖‖] coordinated to the molybdenum atom (in its oxidized form) in a trigonal prismatic geometry completed by oxygen and/or sulfur and/or selenium atoms in a diversity of arrangements (Figure 1.3b); the molybdenum atom is most often directly coordinated by the polypeptide chain *via* aspartate, serine, cysteine or selenocysteine residue side chains.**** The dimethylsulfoxide reductase (DMSOR) family consists only of prokaryotic enzymes, but having remarkably different activities, including DMSOR (eqn (1.22)), arsenite oxidase (eqn (1.23)) and arsenate reductase (eqn (1.24)), dissimilatory NaR (eqn (1.17); respiratory

‖‖‖The pyranopterin cofactor molecule is apparently always found esterified with a guanosine monophosphate (Figure 1.3a).

****There is at least one exception: the arsenite oxidase (eqn (1.23)) from *Alcaligenes faecalis*, whose crystal structure showed a molybdenum centre coordinated by the two characteristic pyranopterin molecules plus an apical oxo group (Mo=O), in a square-pyramidal coordination geometry, with no coordination to the polypeptide chain.[404–406] X-ray absorption data showed the presence of two oxo groups (MoO_2) or of one oxo plus a hydroxyl group (Mo=O(–OH)),[407] suggesting that the molybdenum centre in the crystal could have been reduced in the synchrotron beam.

membrane-bound and periplasmatic enzymes associated with the generation of a proton motive force (Figure 1.1) or acting as an electron sink to eliminate excess of reducing equivalents) and assimilatory NaR (eqn (1.17); prokaryotic cytoplasmatic enzymes involved in nitrogen assimilation (Figure 1.1)), as well as enzymes that do not catalyze oxygen atom transfer reactions, such as formate dehydrogenase (FDH; eqn (1.25)) and polysulfide reductase (eqn (1.26)), among many others (Table 1.2).

$$(H_3C)_2S{=}O + 2e^- + 2H^+ \longrightarrow (H_3C)_2S{-}H + H_2O \qquad (1.22)$$

$$AsO_3^{3-} + H_2O \rightarrow OAsO_3^{3-} + 2e^- + 2H^+ \qquad (1.23)$$

$$OAsO_3^{3-} + 2e^- + 2H^+ \rightarrow AsO_3^{3-} + H_2O \qquad (1.24)$$

$$HCOO^- \rightarrow CO_2 + 2e^- + H^+ \qquad (1.25)$$

$$(S_n)^{2-} + 2e^- + 2H^+ \rightarrow (2H^+)S^{2-} + (S_{n-1})^{2-} \qquad (1.26)$$

a) The DMSOR family is the most diverse of the three families, comprising enzymes with widely varying subunit composition and makeup of additional redox-active cofactors.[228,408,409] To restrict the information presented to a manageable size, only a limited number of enzymes will be described, namely two DMSOR, one polysulfide reductase and several NaR and FDH enzymes.

Rhodobacter sphaeroides DMSOR, the benchmark enzyme of this family, as well as its homologue from *R. capsulatus*, is a periplasmatic monomeric protein containing only the molybdenum centre as a redox-active group, with the molybdenum atom coordinated (in its oxidized form) by the characteristic two pyranopterin molecules (in the form of the dinucleotide of guanine) plus one catalytically labile oxo group (Mo=O) and one serine side chain oxygen atom (Mo–O(Ser)) (Figure 1.9).[224,410–412] Physiologically, this DMSOR is a dissimilatory enzyme that catalyzes the reduction of DMSO to dimethylsulfide (DMS) (eqn (1.22)), but without contributing to the transmembrane proton gradient (thus acting as an electron sink to eliminate excess of reducing equivalents).[413–415]

Contrary to the "simple" enzyme from *Rhodobacter*, the "complex" DMSOR from *E. coli* contributes to the transmembrane proton gradient, through the oxidation of menaquinol coupled to DMSO reduction.[416–420] This *E. coli* respiratory enzyme is a membrane-bound periplasm-facing heterotrimeric (αβγ) enzyme, which is suggested to be comprised of (with no crystal structure available yet): (i) a periplasmatic DMSO-reducing subunit that holds the molybdenum centre, similar to the *Rhodobacter* one, and one [4Fe–4S] centre; the amino acid residue sequence of the molybdenum-binding domain of this subunit is similar to the corresponding part of the *Rhodobacter* enzyme, with the small domain containing the Fe/S centre inserted in an N-terminal domain of the subunit; (ii) a second periplasmatic subunit containing four [4Fe–4S] centres that is thought to be involved in intramolecular electron transfer (from the menaquinol to the DMSO-reducing molybdenum centre); and (iii) a membrane-integral subunit that has no

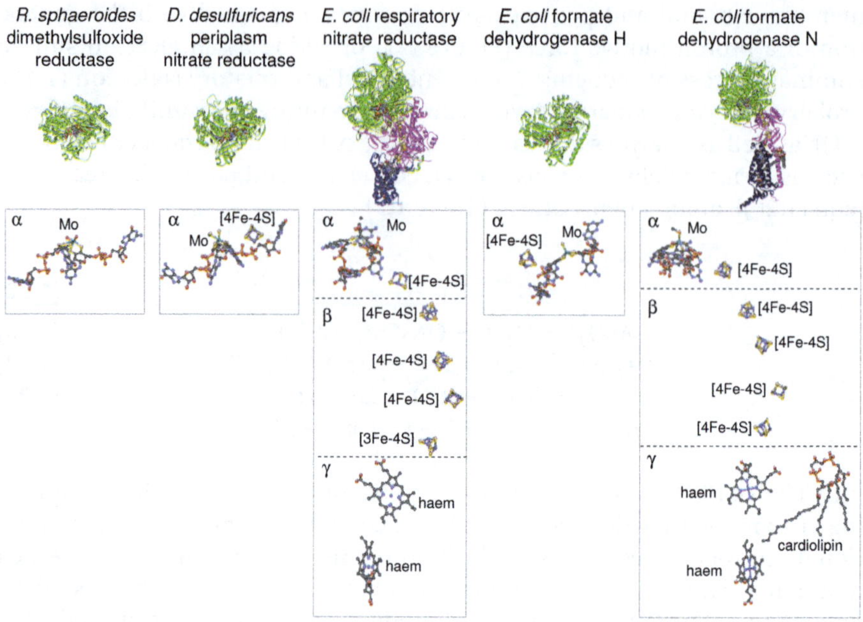

Figure 1.9 Three-dimensional structure view of *R. sphaeroides* dimethylsulfoxide reductase, *D. desulfuricans* periplasmatic nitrate reductase, *E. coli* respiratory nitrate reductase, *E. coli* formate dehydrogenase H and *E. coli* formate dehydrogenase N (top) and of the arrangement of their redox cofactors (bottom). Only one αβγ group of the *E. coli* respiratory NaR ($(\alpha\beta\gamma)_2$) and *E. coli* FDH N ($(\alpha\beta\gamma)_3$) is represented. The number of colours reflects the number of subunits of each enzyme and its cofactor composition: DMSOR, *D. desulfuricans* periplasmatic NaR and *E. coli* FDH H α – green; *E. coli* respiratory NaR and *E. coli* FDH N αβγ – green, pink, dark blue. The structures shown are based on the PDB files 1EU1 (DMSOR), 2NAP (*D. desulfuricans* periplasmatic NaR), 1Q16 (*E. coli* respiratory NaR), 2IV2 (*E. coli* FDH H) and 1KQF (*E. coli* FDH N), and were produced with Accelrys DS Visualizer, Accelrys Software Inc.

redox-active cofactors and that presumably binds the menaquinol necessary for enzyme reduction.

The respiratory, membrane-bound periplasm-faced, polysulfide reductase from *Thermus thermophilus*, involved in the reduction of inorganic sulfur to sulfide, a sulfur atom transfer reaction (eqn (1.26)), has a similar subunit organization and cofactor composition, despite being organized as a dimer of trimers ($(\alpha\beta\gamma)_2$).[421–423] Its crystal structure shows that the cofactorless, membrane-integral subunit does bind the menaquinol and in a position near one of the Fe/S centres (named FS3) of the electron transfer β subunit, as expected (and suggested for the *E. coli* DMSOR). The molybdenum centre of the polysulfide reductase, however, is coordinated by a cysteine residue (Mo–S(Cys)) rather than the serine found in the DMSORs and probably by a catalytically labile hydroxyl group (Mo–OH) (the pyranopterin cofactors are also in the form of a guanine dinucleotide).

The NaR and FDH enzymes can have similarly "simple" and "complex" subunit organizations and cofactor compositions, most often correlated to their physiological function. For example, the "simple" monomeric periplasmatic *Desulfovibrio desulfuricans* NaR and monomeric cytoplasmatic *E. coli* FDH-H, both of which contain only the molybdenum centre and one [4Fe–4S] centre, contrast with the "complex" $(\alpha\beta\gamma)_2$ membrane-bound respiratory *E. coli* NaR and $(\alpha\beta\gamma)_3$ membrane-bound respiratory *E. coli* FDH-N, both of which contain seven additional redox-active cofactors (Fe/S and haems) in addition to their molybdenum centres (all described below).

Prokaryotes use nitrate for dissimilatory and assimilatory processes (Figure 1.1) and encode three distinct NaR enzymes[††††] present in different subcellular locations:[174,175,228,424–434] (a) membrane-bound cytoplasm-faced respiratory NaR, associated with the generation of a proton motive force across the cytoplasmatic membrane; [‡‡‡‡] (b) periplasmatic NaR, involved in the generation of a proton motive force or acting as an electron sink to eliminate excess of reducing equivalents; and (c) cytoplasmatic assimilatory NaR, involved in nitrogen assimilation.[§§§§] In response to their different biological roles/subcellular locations, these enzymes have different subunit organizations and cofactor compositions (Figure 1.9). The respiratory enzyme from *E. coli* NaRGHI (product of the *narG*, *H* and *I* genes) is a dimer of heterotrimers $(\alpha\beta\gamma)_2$, made up of:[175,435] (a) a cytoplasmatic nitrate-reducing NaRG subunit that holds the molybdenum centre and one [4Fe–4S] centre; (b) an electron transfer NaRH subunit that holds one [3Fe–4S] and three [4Fe–4S] centres, responsible for the intramolecular electron transfer from the membrane quinol pool to the nitrate-reducing molybdenum centre; and (c) a membrane-bound quinol-oxidizing NaRI subunit that holds two *b*-type haems. On the other hand, the periplasmatic NaR from *D. desulfuricans* (product of the *napA* gene) is a monomeric enzyme with only one [4Fe–4S] centre in addition to the molybdenum centre,[436,437] while the enzyme from *C. necator* (*napA* and *napB* genes) is a dimer harbouring two haems in addition to the molybdenum centre.[438]

All these prokaryotic NaR catalyze the two-electron reduction of nitrate to nitrite (eqn (1.17)) at their molybdenum centre. Interestingly, in spite of catalyzing the same reaction and having the molybdenum atom coordinated

[††††] In the "molybdenum-community" or the "prokaryotic field", these three enzymes are commonly referred to as Nar (membrane-bound, respiratory), Nap (periplasmatic) and Nas (cytoplasmatic, assimilatory). However, for readers outside those fields, this denomination is not intuitive, because the letters "r", "p" and "s" do not specify the differences between the three types of enzymes (differences in function, subcellular location or even in active site structure). As such, herein the abbreviation NaR, from nitrate reductase, will be used, specifying the substrate and the type of reaction catalyzed, as is recommended by the Enzyme Nomenclature Committee (IUBMB) for the trivial names of enzymes.

[‡‡‡‡] Interestingly, evidence has been recently described for the occurrence of both archaeal and bacterial periplasm-faced NaRGHI-type nitrate reductases, *i.e.* enzymes similar to the ones described above, but with the active site on the outside of the cytoplasmatic membrane.[428,430,432] This exciting hypothesis awaits confirmation.

[§§§§] As indicated above, the eukaryotic assimilatory cytoplasmatic NaR, belonging to the SO family, is a distinct enzyme.

by the characteristic four sulfur atoms from two pyranopterin cofactor molecules (in the form of the guanine dinucleotide), the three types of NaR enzymes display significant differences in the remainder of the molybdenum coordination sphere. In the respiratory membrane-bound NaR, the molybdenum atom is further coordinated by an aspartate residue that is coordinated in a bidentate fashion,[174,435] or alternatively by one terminal oxo group plus one oxygen atom from the aspartate residue coordinated in a monodentate mode; the monodentate and bidentate aspartate binding modes correspond, possibly, to oxidized and reduced (by the synchrotron beam) proteins, respectively. However, in the periplasmatic NaR from *Desulfovibrio desulfuricans* or *Cupriavidus necator*, the molybdenum atom is coordinated instead by a cysteine sulfur atom plus one terminal sulfo group, forming a partial disulfide bond within each other.[436–439] The *E. coli* and *Rhodobacter sphaeroides* periplasmatic NaR, on their turn, complete the molybdenum coordination with the cysteine sulfur atom plus a terminal hydroxyl group.[440,441] The cytoplasmatic assimilatory NaR is the least studied one of the NaRs, but it is likely that a cysteine residue is coordinated to the molybdenum atom in this enzyme.[442]

As with the NaRs, the prokaryotic FDH enzymes (eqn (1.25)) are also involved in different biochemical pathways, operate in different subcellular locations and use diverse physiological redox partners (membrane quinols, cytoplasmatic and periplasmatic cytochromes, ferredoxins, NAD or coenzyme F_{420} (an obligate two-electron acceptor, flavin derivative)).[228,443,444] As a result, FDHs are also structurally heterogeneous proteins, displaying diverse subunit organizations and cofactor compositions,[¶¶¶¶] and again only a few examples will be here described.

E. coli expresses three FDHs: (a) a cytoplasmatic enzyme designated formate dehydrogenase H (FDH-H)[445] that is part of the formate-hydrogen lyase system and involved in the oxidation of formate and generation of molecular hydrogen formation under fermentative (anaerobic) growth conditions;[228,446] (b) a membrane-bound periplasm-facing FDH designated formate dehydrogenase N (FDH-N) that is part of the anaerobic nitrate-formate respiratory pathway (catalyzed by a supermolecular formate : nitrate oxidoreductase system formed with the NaRGHI) involved with the generation of a proton motive force;[423,447–453] and (c) a second membrane-bound periplasm-faced FDH, designated formate dehydrogenase O (FDH-O), that operates under aerobic conditions in another nitrate-formate respiratory pathway (this with the NaRZWV enzyme).[454–458]

The *E. coli* FDH-H is a monomeric enzyme with only one [4Fe–4S] centre in addition to the molybdenum centre (Figure 1.9).[459–462] The molybdenum centre is the site of formate oxidation, and the Fe/S centre is responsible for the subsequent intramolecular electron transfer to the physiological acceptor,

[¶¶¶¶] Of note, besides molybdenum- and tungsten-containing enzymes (the latter ones described in Section 1.4.4.), there are also metal-independent FDH enzymes, dependent on NAD^+, that belong to the D-specific dehydrogenases of the 2-oxyacid family.[444]

probably a ferredoxin protein; the crystal structure shows the Fe/S centre in the N-terminal domain, as is suggested for the subunit of *E. coli* DMSOR. The molybdenum is coordinated by four sulfur atoms of the two pyranopterins (in the form of guanine dinucleotide), characteristic of this family, and by a conserved essential selenocysteine residue (Mo–Se(Cys)) and a labile hydroxyl group.[462] However, a subsequent reinterpretation of the original electron density map has revealed a different coordination sphere for the formate-reduced FDH-H, with a terminal sulfur atom rather than oxygen, and with the selenocysteine residue not bound to the molybdenum, but shifted away (9 Å).[463] Accordingly, the oxidized molybdenum centre of FDH-H is presently being thought of as a trigonal prismatic structure coordinated by the selenocysteine residue and one terminal sulfur atom, in addition to the four sulfur atoms from the two pyranopterins.[||||||][463]

The *E. coli* FDH-N is a complex trimer of trimers ($(\alpha\beta\gamma)_3$), with eight redox-active cofactors in an overall organization of the $\alpha\beta\gamma$ unit that is quite similar to that seen in the respiratory NaRGHI (Figure 1.9).[449,450,465–468] The FDH-N $\alpha\beta\gamma$ unit consists of:[450] (i) a periplasmatic formate-oxidizing subunit that holds one molybdenum centre and one [4Fe–4S] centre, with the domains involved in binding the Fe/S and molybdenum centres structurally similar to the FDH-H monomer; (ii) a periplasmatic electron transfer subunit that harbours four [4Fe–4S] centres and is responsible for the intramolecular electron transfer from the formate-oxidizing molybdenum centre to the membrane quinone pool; and (iii) a membrane-bound quinone-reducing subunit that has two *b*-type haems. The trimer of $\alpha\beta\gamma$ units, $(\alpha\beta\gamma)_3$, is very tightly packed and a cardiolipin molecule is maintained at the trimer interface, thus suggesting that this complex arrangement is physiologically meaningful. The molybdenum of FDH-N is coordinated by selenocysteine and a terminal sulfur atom, as seen in its FDH-H counterpart, in addition to the characteristic four sulfur atoms of the two pyranopterins (in the form of guanine dinucleotide).[464]

Noteworthy, the *E. coli* FDH-O is believed to be structurally similar to the homologous FDH-N (for which no crystal structure is yet available).[455,469,470] The FDH from *D. desulfuricans* is also a complex periplasmatic enzyme, with a heterotrimeric structure ($\alpha\beta\gamma$), but possesses only two [4Fe–4S] centres (one in each of the α and β subunits) with four *c* haems (γ subunit) and the molybdenum centre (the last coordinated by a selenocysteine and, probably, a terminal sulfur; no crystal structure is yet available).[471,472] The enzymes from *Ralstonia eutropha*[473–475] and *Rhodobacter capsulatus*[476] are even more complex, holding seven Fe/S centres plus one FMN per $\alpha\beta\gamma$ protomer in addition to the molybdenum centre (which, in these two enzymes, is expected to be coordinated by a cysteine residue, and a terminal sulfur atom; again, no crystal structure is as yet available).

[||||||]Besides the crystallographic data, also the recent identification of a sulfurtransferase that would insert the sulfur atom into the FDH molybdenum centre[464] supports that all *E. coli* FDH could have a terminal sulfur atom.

To summarize, the DMSOR family is the most diverse of the molybdenum-containing enzymes families, consisting of enzymes with different structural organization and composition of redox-active cofactors (Fe/S centres, haems and flavins), ranging from, *e.g.*, (i) the monomeric *Rhodobacter* DMSOR with only the molybdenum centre, to (ii) the monomeric *Desulfovibrio desulfuricans* periplasmatic NaR that contains the molybdenum centre plus one Fe/S centre, and ending in (iii) the multimeric respiratory *E. coli* FDH-N, with several Fe/S and haems, besides the molybdenum centre, or in other FDH enzymes that further hold one FMN or FAD.

b) Matching its structural diversity, the DMSOR family is also catalytically extremely versatile, including enzymes that catalyze a variety of reaction types, including several not involving oxygen atom transfer, *e.g.*: (i) cleavage of a C–H bond as carried out by different FDHs that catalyze the reversible formate oxidation to carbon dioxide (eqn (1.25)); (ii) sulfur atom transfer as seen in the polysulfide reductase-catalyzed inorganic sulfur reduction to sulfide (eqn (1.26)); (iii) hydroxyl transfer with *simultaneous* oxidation and reduction (where water is not the oxygen atom acceptor or donor), such as the reductive dehydroxylation and concomitant oxidative hydroxylation catalyzed by pyrogallol : phloroglucinol transhydroxylase (eqn (1.27)); (iv) and even a hydration reaction, as observed in the acetylene hydratase-catalyzed hydration of acetylene to acetaldehyde, a non-redox reaction (eqn (1.28))); (v) besides the more common oxygen atom transfer to, or from, a lone electron pair of the substrate, as can be observed in arsenite oxidase-catalyzed oxidation of arsenite to arsenate (eqn (1.23)) or in dimethylsulfide dehydrogenase-catalyzed oxidation of DMS to DMSO (reverse of eqn (1.22)) (oxygen insertion/oxidation reaction) and in NaR (eqn (1.17)), arsenate reductase (eqn (1.24)) or DMSOR reactions (eqn (1.22)) (oxygen abstraction/reduction reaction).[161,409,423,477]

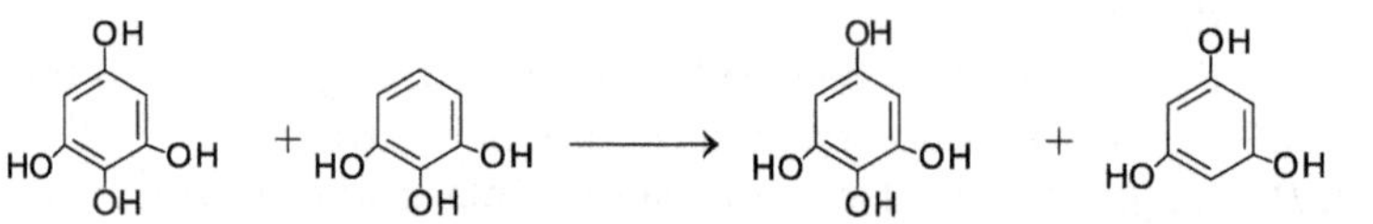

pyrogallol phloroglucinol (1.27)

$$H–C \equiv C–H + H_2O \rightarrow H_3C–COH \qquad (1.28)$$

The electrons derived from (or necessary to carry out) these reactions are intramolecularly transferred to the electron acceptor (or from the electron donor) through the different redox-active centres of each enzyme (Fe/S centres, haems and flavins); therefore, the oxidative and reductive half-reactions are physically separated and the intramolecular electron transfer is an integral aspect of catalysis – the obvious exception is the monomeric, molybdenum-only, *Rhodobacter* DMSOR and related enzymes (Table 1.2). The reaction mechanism of *Rhodobacter* DMSOR is now well established and believed to be essentially similar in other DMSOR enzymes (Figure 1.10a):[478–481] (1) catalysis is initiated with enzyme reduction by the pentaheme DorC protein (the reductive half-reaction); upon reduction of the molybdenum, the catalytically labile oxo group is protonated and released in the form of a water molecule, resulting in the formation of a penta-coordinate, square-pyramidal core, with the serine residue occupying the apical position ((Ser)O–Mo^{6+}=O → (Ser)O–Mo^{4+}), *via* a (Ser)O–Mo^{5+}–OH intermediate; (2) DMSO binding to the molybdenum yields a hexa-coordinate (Ser)O–Mo^{4+}–O–S(dimethyl) intermediate; (3) the subsequent O–S bond cleavage releases the product (DMS) and regenerates the initial (Ser)O–Mo^{6+}=O core (the oxidative half-reaction).

The outline of the mechanism of the oxygen abstraction reaction catalyzed by DMSOR shares some aspects with the eukaryotic NaR reaction (from the SO family): in both enzymes a catalytically labile oxo ligand of the molybdenum is lost to allow the binding of the substrate, *via* the oxygen atom to be transferred, resulting in the formation of a Mo^{4+}–O(substrate) complex; following

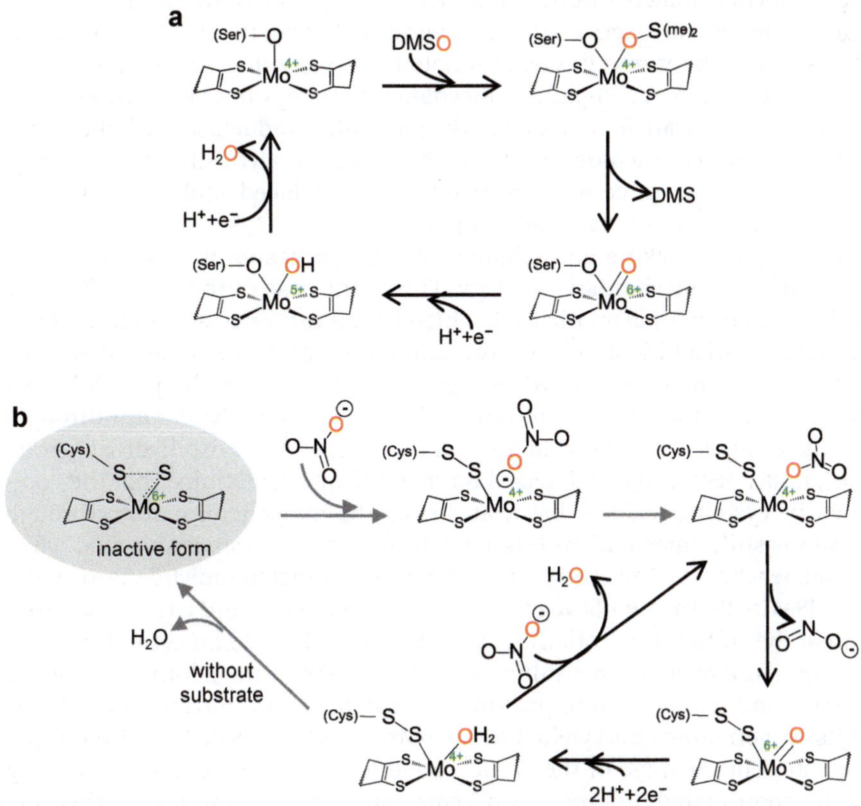

Figure 1.10 Simplified mechanistic proposal for the reaction catalyzed by dimethylsulfoxide reductase (a) and *D. desulfuricans* periplasmatic nitrate reductase (b). See text for details.

the O-substrate bond cleavage, the product is released and the initial molybdenum core, with a "fresh" labile oxo group, is regenerated. Moreover, in both cases water is the ultimate destination of the oxygen atom abstracted from the substrate. However, to catalyze these oxygen atom abstraction reactions, the DMSOR enzyme cycles between a hexa-coordinated mono-oxo Mo^{6+} core and a penta-coordinated *des*oxo Mo^{4+} core, while the eukaryotic NaR does it between a penta-coordinated di-oxo Mo^{6+} core and a mono-oxo Mo^{4+} core. Therefore, in spite of using different "enzymatic machineries", both enzymes carry out a similar chemical outcome: direct oxygen atom transfer from the substrate to the metal, followed by "recycling" with the solvent.

A similar mechanism has been suggested for the periplasmatic NaR from *E. coli*, but with nitrate binding to the Mo^{5+} oxidation state (rather than Mo^{4+}):[440] (1) reduction of the molybdenum centre by one electron results in a penta-coordinated *des*oxo Mo^{5+} core ((Cys)S–Mo^{6+}–OH → (Cys)S–Mo^{5+}; this is parallel to the DMSOR penta-coordinated desoxo (Ser)O–Mo^{4+} core); nitrate binds to this complex, through the oxygen atom to be abstracted, to form the hexa-coordinated intermediate with bound substrate; (2) after reduction by the second electron, the O–N bond is cleaved, the product (nitrite) is released and the initial hexa-coordinated mono-oxo Mo^{6+} core regenerated upon protonation. A comparable mechanistic strategy has been proposed for the sulfur atom transfer reaction of the polysulfide reductase, with the terminal sulfur atom of the polysulfide displacing the catalytically labile hydroxyl group to become directly coordinated to the reduced molybdenum atom ((Cys)S–Mo^{6+}–OH → (Cys)S–Mo^{4+}–S(polysulfide)).[482,483]

The enzymes lacking an exchangeable oxygen ligand, however, must follow a different mechanistic strategy. This is the case of the *D. desulfuricans* and *C. necator* periplasmatic NaR, which have a terminal sulfur atom, forming a partial disulfide bond with the sulfur atom of the cysteine residue side chain, rather than an oxo/hydroxyl group. Starting from the principle that the sulfo ligand would have to remain bound to the molybdenum throughout the catalytic cycle (*i.e.* that it is not be released as an oxo/hydroxyl group can), it has been proposed, based on theoretical calculations, that the coordinating cysteine residue dissociates from the molybdenum atom through a "sulfur shift" mechanism (Figure 1.10b):[484,485] catalysis is initiated when nitrate reaches the (inactive) oxidized hexa-coordinated molybdenum centre ((Cys)S–Mo=S); the repulsive environment generated would trigger the insertion of the sulfur atom into the (Cys)S–Mo bond, to yield an active (Cys)S–S–Mo core. As a result of this sulfur shift, the molybdenum is formally reduced to Mo^{4+} and a *new* binding position is created to bind nitrate through one of its oxygen atoms and yield a hexa-coordinated (Cys)S–S–Mo–ONO_2 intermediate. The cleavage of the O–NO_2 bond leads to product release, leaving a hexa-coordinated mono-oxo Mo^{6+} core, with a "fresh" oxo group; this core can then be reduced and, after displacement of the oxo labile group by a new molecule of nitrate, a new catalytic cycle can start. This mechanism can be rationalized using the same generalization introduced above with the comparison between the eukaryotic NaR and DMSOR reactions: direct oxygen atom transfer from the substrate to the metal, followed by "recycling" with the solvent, involving, in this case, hexa-coordinated intermediates created by an activating initial sulfur-shifting.

The sulfur shift is an elegant mechanism to explain how a metal with no available coordination position can bind a substrate, converting an "unready enzyme" (with a complete coordination sphere) into a "ready enzyme". The complete coordination, with a ligand that is not exchangeable (Mo=S) plus another ligand that could be exchanged (Mo–S(Cys)), could provide a mechanism to protect the active site from "unwanted" compounds (inhibition) or may be a regulatory strategy to control the enzyme's activity (mechanism "on/off") to respond to the cellular needs. Still, high-resolution structures coupled with spectroscopic and kinetic data are needed to support the existence of the two polypeptide alternating conformations and the identity of the different catalytic intermediates.

The FDH enzymes also have molybdenum coordinated by a terminal sulfur atom plus a selenocysteine or cysteine residue and a similar sulfur shift mechanism has been proposed to create a vacant position for formate binding, with formate triggering the formation of the active hexa-coordinated (Cys)Se–S–Mo–OCO(H) or (Cys)S–S–Mo–OCO(H) intermediate (in selenocysteine or cysteine-containing FDH, respectively) (Figure 1.11a).[486,487] In this case, however, the substrate (formate) has one proton to be abstracted (formate oxidation; **** eqn (1.25)). Therefore, based on theoretical calculations, it has been proposed that:[486,487] (1) the (Cys)Se–S (or (Cys)S–S) bond is cleaved; (2) the selenol anion formed ($(Cys)Se^-$), stabilized by a hydrogen bond with a key histidine residue, abstracts the formate Cα proton to yield a carbon dioxide moiety coordinated in a bidentate mode to molybdenum ($C_{carbon\ dioxide}$–S–Mo–O–$C_{carbon\ dioxide}$); (3) carbon dioxide is eventually released, leaving a penta-coordinated Mo^{4+}=S centre; (4) the catalytic cycle would be closed with the oxidation of Mo^{4+} to Mo^{6+}, *via* intramolecular electron transfer, and deprotonation of the selenocysteine/cysteine residue. The molybdenum centre can, then, bind a new formate molecule and start a new catalytic cycle; in the absence of formate, the selenocysteine/cysteine-containing loop is reoriented, the (Cys)Se–Mo=S (or (Cys)S–Mo=S) bond reformed and the enzyme returns to the inactive hexa-coordinate form.

Other reaction mechanisms have been suggested for FDH,[463,490,491] but all rely on the dissociation of the selenocysteine/cysteine residue and assume the mandatory requirement of formate to bind directly to the molybdenum atom. However, there is no clear-cut experimental evidence for formate binding to the molybdenum atom. Moreover, because the formate oxidation is not an oxygen atom transfer reaction, the direct formate binding to the molybdenum might not even be necessary. In accordance, a very recent proposal pointed to a different chemical strategy, through which formate oxidation occurs *via* hydride transfer and not proton plus two electrons transfer

****The FDH-catalyzed reaction is not an oxygen-atom transfer reaction; the reaction product is carbon dioxide ($HCOO^- \rightarrow CO_2 + 2e^- + H^+$) and not hydrogencarbonate ($HCOO^- + H_2O$ -X $\rightarrow HOCOO^- + 2e^- + 2H^+$).[488] In fact, oxidation of ^{13}C-labelled formate in ^{18}O-enriched water demonstrated clearly the formation of $^{13}CO_2$ gas with no ^{18}O atoms.[488,489]

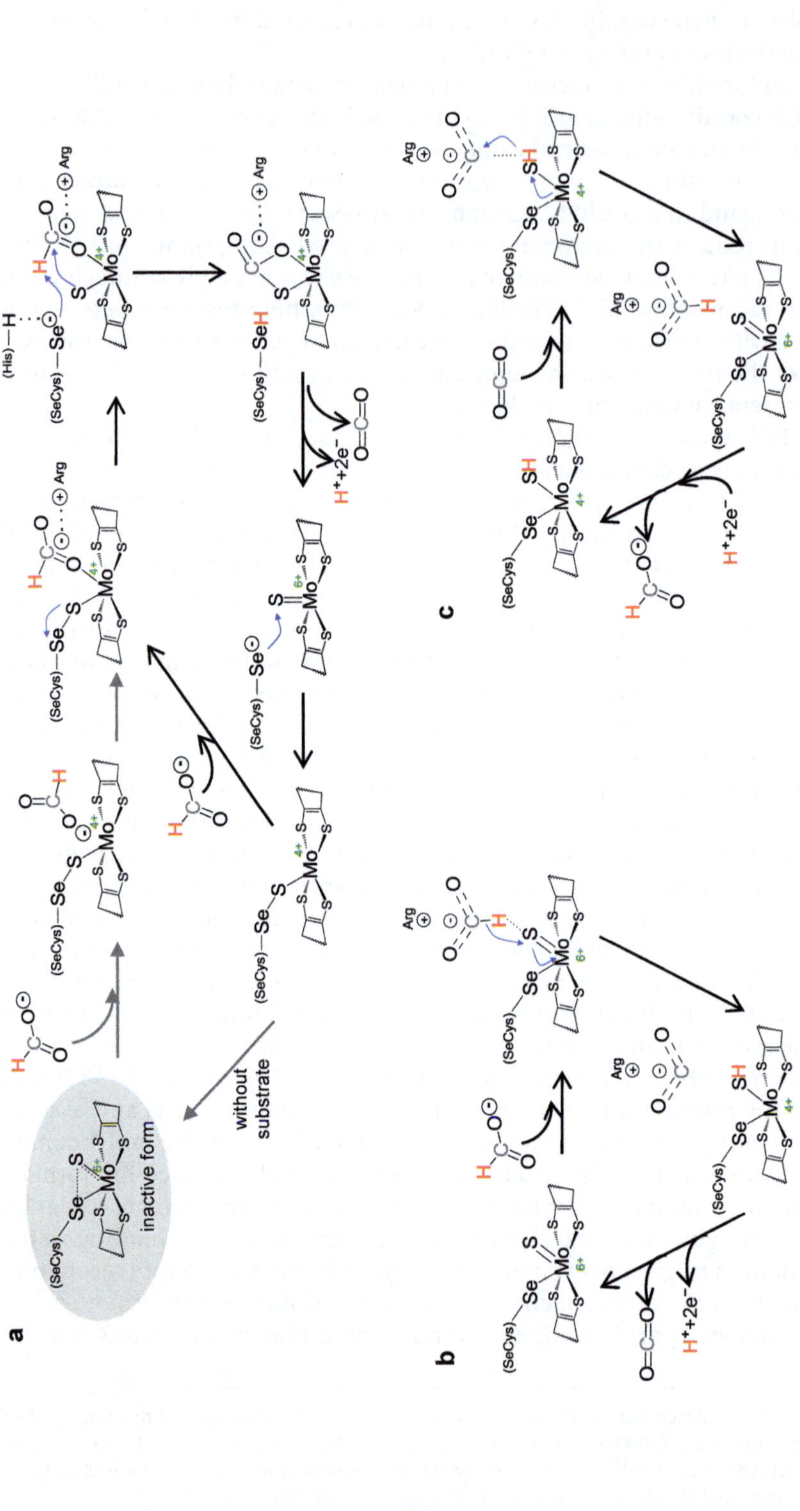

Figure 1.11 Simplified mechanistic proposals for the reaction catalyzed by formate dehydrogenase. Formate dehydrogenase-catalyzed formate oxidation (a, b) and carbon dioxide reduction (c). See text for details.

(Figure 1.11b).[492] The direct hydride acceptor is the terminal sulfo group of the oxidized molybdenum centre (Mo^{6+}=S) and formate does not bind directly to the molybdenum atom; the selenocysteine/cysteine residue was suggested to remain bound to the molybdenum atom throughout the all catalytic cycle, even though it is not a requirement in this mechanistic hypothesis.[492] The same chemical strategy has also been suggested to explain the reverse reaction of carbon dioxide reduction to formate, with the terminal sulfo group of the reduced molybdenum centre (Mo^{4+}-SH) being thought of as the hydride donor (Figure 1.11c).[493] Certainly, much more research is essential to critically evaluate the diverse mechanistic proposals.

To conclude, it should be emphasized that the DMSOR family comprises many other enzymes in addition to those described here; the array of reactions catalyzed is extensive (Table 1.2) and well beyond the oxygen, hydrogen and sulfur atom transfer reactions considered here.

1.4.4 The Tungstoenzymes Family

The active site of tungstoenzymes harbours one tungsten atom coordinated by four sulfur atoms of two pyranopterin cofactor molecules, plus oxygen and/or sulfur and/or selenium atoms in a trigonal prismatic geometry (as observed in the DMSOR family of molybdoenzymes) or in a distorted octahedral coordination geometry (Figure 1.3b). This family consists of prokaryotic enzymes of diverse functions, including (i) aldehyde oxidoreductases (eqn (1.7)), as the enzymes from *Pyrococcus furiosus* aldehyde : ferredoxin oxidoreductases (W-AOR), formaldehyde : ferredoxin oxidoreductase and glyceraldehyde-3-phosphate : ferredoxin oxidoreductase, or the *Clostridium formicoaceticum* carboxylic acid reductase, (ii) FDHs (eqn (1.25)) from, *e.g.*, *Moorella thermoacetica* (*Clostridium thermoaceticum*), *Desulfovibrio gigas* and others, (iii) *Pelobacter acetylenicus* acetylene hydratase (eqn (1.28)), (iv) *Geobacter metallireducens* benzoyl-CoA reductase (eqn (1.29)) and (v) *M. thermoautotrophicum N*-formylmethanofuran dehydrogenase (eqn (1.30)) (Table 1.2). Interestingly, except for the last two, all these enzymes have a known molybdenum-containing counterpart (which seems to have some evolutionary significance, as noted in Section 1.2).

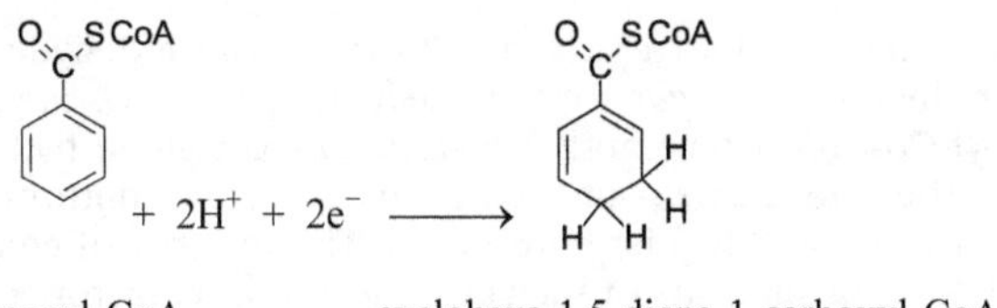

(1.29)

benzoyl-CoA cyclohexa-1,5-diene-1-carboxyl-CoA

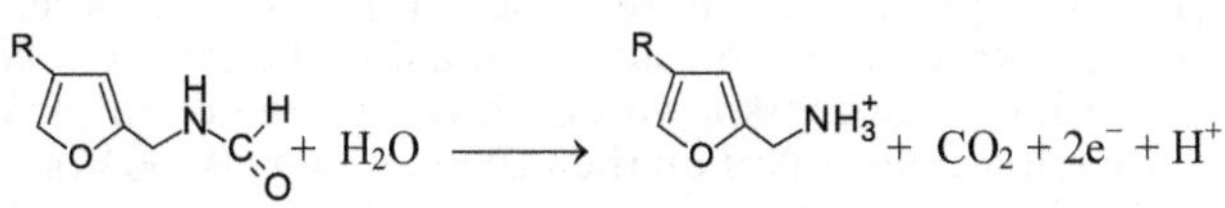

(1.30)

formyl-methanofuran methanofuran

a) Compared to the molybdenum-containing enzymes, our knowledge about the structural and mechanistic properties of the tungstoenzymes is more limited. Nevertheless, the first crystal structure of a pyranopterin-containing enzyme was that of the W-AOR from the hyperthermophilic archaeon *Pyrococcus furiosus*[222] – one of the benchmarks of tungstoenzymes.

W-AOR enzymes are involved in handling aldehydes, catalyzing the reversible oxidation of aldehydes to the respective carboxylic acids, with the simultaneous reduction of ferredoxin. The *P. furiosus* W-AOR is a homodimeric enzyme (α_2) with one tungsten centre, the site of substrate oxidation, and one [4Fe–4S] centre that is involved in intramolecular electron transfer to the physiological partner (a [4Fe–4S] ferredoxin) (Figure 1.12).[222] The tungsten centre is coordinated by the characteristic two pyranopterin cofactor molecules, in this case in the mononucleotide form; their phosphates groups interact with a Mg^{2+} ion that intervenes between them. The first coordination sphere of tungsten, in its oxidized form, is completed with two oxo groups (W=O) or, possibly, with one oxo and one hydroxyl groups (W=O(–OH); as

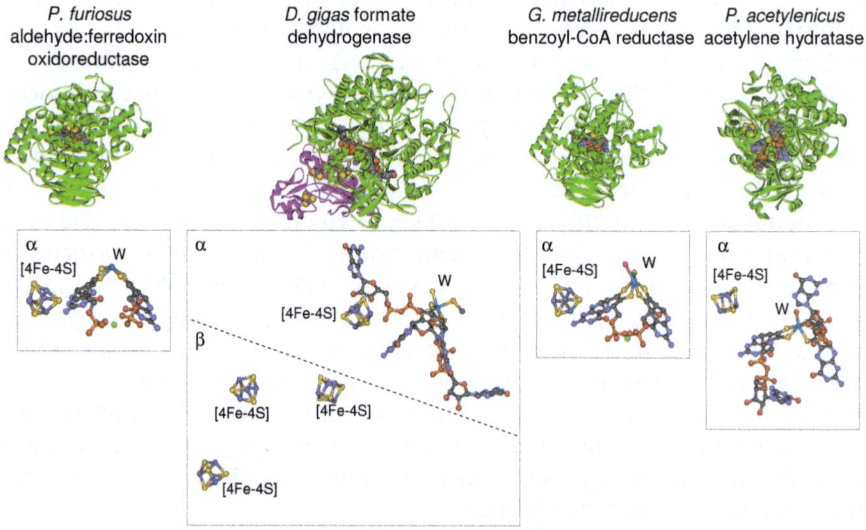

Figure 1.12 Three-dimensional structure view of *P. furiosus* aldehyde : ferredoxin oxidoreductase, *D. gigas* formate dehydrogenase, *G. metallireducens* benzoyl-CoA reductase and *P. acetylenicus* acetylene hydratase (top) and of the arrangement of their redox cofactors (bottom). Only one subunit in W-AOR (α_2) is represented. The number of colours reflect the number of subunits of each enzyme and its cofactor composition: W-AOR, benzoyl-CoA reductase and acetylene hydratase α – green; FDH αβ – green, pink. In the case of the benzoyl-CoA reductase, only the 3D structure of the subunit holding the tungsten centre is presented; the sixth unknown ligand is represented in pink. The structures shown are based on the PDB files 1AOR (W-AOR), 1H0H (FDH), 4Z3Z (benzoyl-CoA reductase) and 2E7Z (acetylene hydratase), and were produced with Accelrys DS Visualizer, Accelrys Software Inc.

observed in the molybdenum centre of arsenite oxidized), but with no ligand from the polypeptide chain. The reduced form is probably penta-coordinate, with only one terminal oxygen atom. The closely related formaldehyde : ferredoxin oxidoreductase displays a homotetrameric structure, but with each monomer comprising a similar active site architecture and also one [4Fe–4S] centre.[494–496]

Several FDH enzymes contain tungsten and, in spite of the structural/functional similarities with the molybdenum-containing FDH, they belong to the tungstoenzymes family, according to our organization. This is also the case of the *N*-formylmethanofuran dehydrogenase, which together with the FDH contributes to carbon dioxide fixation (reduction) into *N*-formylmethanofuran and formate (acetate), respectively. The FDH from *D. gigas* is a periplasmatic heterodimer (αβ) that harbours four [4Fe–4S] centres and the tungsten centre (Figure 1.12).[497–499] As anticipated, the tungsten centre is the active site, where formate oxidation to carbon dioxide takes place (eqn (1.25)); the Fe/S centres are responsible for the subsequent intramolecular electron transfer to the physiological acceptor, a periplasmatic *c*-type cytochrome. Noteworthy, the α subunit (folded into four domains) is homologous to the *E. coli* molybdenum-containing FDH-H and contains one Fe/S centre (bound by the N-terminal domain) and the tungsten centre (bound mainly through the other three domains). Also the β subunit, folded into two domains, one containing one Fe/S and the other two Fe/S centres (total of three), displays a similar arrangement to the one found in the *E. coli* FDH-N β subunit (although in FDH-N both domains contain two Fe/S centres each (total of four)). The tungsten centre is coordinated by the characteristic pair of pyranopterin molecules, in this case in the form of dinucleotide of guanine, plus one selenocysteine (Mo–Se(Cys)) and one terminal sulfur atom (Mo=S), again similar to the *E. coli* molybdenum-containing FDH enzymes (also the conserved histidine and arginine residues are present in the active site pocket).

Other bacteria from the *Desulfovibrio* genus encode both molybdenum- and tungsten-containing FDH enzymes (but with no crystal structure available yet). *D. alaskensis* expresses three FDH,[500] two of which have been characterized and one found to be a tungsten-containing FDH (the product of the *W-fdh* genes). The second enzyme has been shown to incorporate either molybdenum or tungsten (product of *Mo/W-fdh* genes), thus being the first FDH known to be able to do this. However, this enzyme should preferentially harbour molybdenum, as suggested by the slight upregulation of the *Mo/W-fdh* genes when the bacterium is grown under molybdenum supplementation and strong downregulation under tungsten supplementation.[500] Both enzymes have the characteristic tungsten centre coordinated by two pyranopterin molecules, in the form of dinucleotide of guanine, and several Fe/S centres.[500,501] Analysis of the *D. vulgaris* genome has also suggested the presence of three periplasmatic FDHs,[502,503] one of which is a dimeric FDH that can also incorporate either molybdenum or tungsten.[503,504]

Noteworthy, there are also tungsten-containing NAD(P)-dependent FDHs, including enzymes from the acetogenic *Moorella thermoacetica* (*Clostridium thermoaceticum*),[218,505] *Clostridium formicoaceticum*[215,506] or *C. carboxidivorans*,[507] but also of the methylotrophic *Methylobacterium extorquens*.[18] One of the FDH of the aerobic *M. extorquens*[19] is a heterodimeric (αβ) tungsten-containing/NAD(P)-dependent FDH.[18] The α subunit displays ≈35% identity with *E. coli* molybdenum-containing FDH-H and has been suggested to harbour at least one Fe/S centre in addition to a tungsten centre coordinated by a cysteine residue. The β subunit contains the putative binding motifs for an Fe/S centre and the FMN (responsible for the NAD(P) binding). The NADP-dependent tungsten-containing FDH from *M. thermoacetica* – the first enzyme shown to have tungsten – is an $(\alpha\beta)_2$ heterotetramer, containing several Fe/S centres (at least two [2Fe–2Fe] and two [4Fe–4S]) in addition to the selenocysteine-coordinated tungsten centre.[13–15,218] The presence of a tungsten-containing FDH in strictly aerobic bacteria may indicate that tungstoenzymes are not restricted to anaerobic organisms and are probably more widespread than previously thought.

Tungsten is also essential for the benzoyl-CoA reductase of the "Fe^{3+}-respiring" *Geobacter metallireducens*, a key enzyme in the microbial anaerobic degradation of monocyclic aromatic compounds. This complex enzyme, suggested to consist of eight subunits,[508] catalyzes the reduction of benzoyl-CoA to cyclohexa-1,5-diene-1-carboxyl-CoA by ferredoxin (eqn (1.31)), a remarkable reaction in terms of its extremely low reduction potential ($E^{\circ\prime}$ = −622 mV) and in terms of the chemistry carried out, because it does not involve the "classic" oxygen atom abstraction/insertion.[509,510] The benzoyl-CoA reductase subunit containing the active site was recently crystallized.[511] Like W-AOR, with which it shares 33% sequence identity,[222,512] it has one [4Fe–4S] centre, involved in intramolecular electron transfer from the other redox centres of the enzyme, and the tungsten centre, which in benzoyl-CoA reductase is accommodated in a highly hydrophobic pocket (Figure 1.12). The tungsten atom is coordinated by the two characteristic pyranopterin cofactor molecules, in the mononucleotide form, plus a cysteine sulfur atom (W–S(Cys)) and a six ligand whose nature has not been not established. While analysis of the electron density excludes hydroxyl or oxo ligands and suggests an electron-rich species, such as a sulfur atom or a diatomic group, X-ray absorption fine structure spectroscopy (EXAFS) has excluded the scenario of a coordination sphere with six sulfur ligands and suggested instead a linear diatomic ligand such as a CN or CO. As cyanide incubation results in activation of the enzyme, whereas CO has no effect, the unambiguous identification of the sixth ligand must wait for crystal structures with resolutions below 1.5 Å. Of note, the tungsten centre of benzoyl-CoA reductase displays a distorted octahedral coordination geometry, rather than the characteristic trigonal prismatic geometry. Another noteworthy feature of the benzoyl-Co A reductase is the presence of zinc[508] and of a Zn^{2+}-binding site, where the metal is tetrahedrally coordinated by two histidine and two glutamate residues, 11.5 Å away from the tungsten atom.[511] The invariance of these four amino acid residues and the proximity to the tungsten centre suggest that the zinc may have a key role in enzyme activity, possibly to protect the tungsten centre from bulk solvent and to stabilize the region surrounding the cofactor molecules.[511]

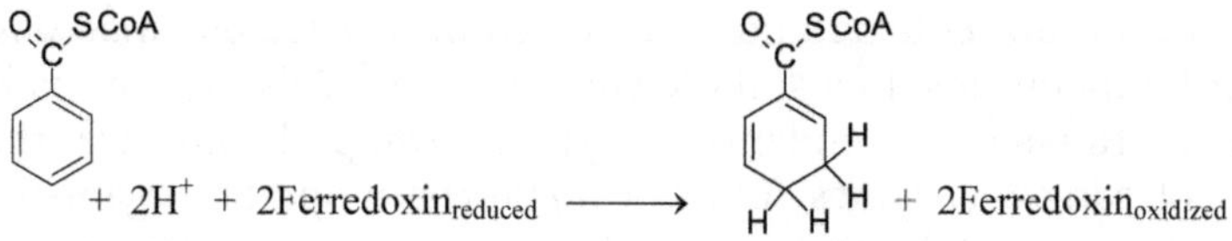

(1.31)

benzoyl-CoA cyclohexa-1,5-diene-1-carboxyl-CoA

The tungsten-containing acetylene hydratase from the anaerobe *Pelobacter acetylenicus* also catalyzes a remarkable reaction in terms of the chemistry carried out: the non-redox hydration of acetylene to acetaldehyde (eqn (1.28)) as part of an anaerobic degradation pathway of unsaturated hydrocarbons.[477,513,514] Acetylene hydratase is a monomeric enzyme, harbouring a [4Fe–4S] centre in addition to the tungsten centre, which is adjacent to a hydrophobic acetylene-binding pocket (Figure 1.12).[515,516] The tungsten is coordinated by the two characteristic pyranopterin cofactor molecules, in the form of dinucleotide of guanine, plus a cysteine sulfur atom (W–S(Cys)) and a tightly coordinated water molecule as the sixth ligand. Of note, although the acetylene hydratase-catalyzed reaction is non-redox (its tungsten and Fe/S centres do not change their oxidation states during catalysis), the tungsten centre must be reduced to W^{4+} for the enzyme to be active.[517] As with the benzoyl-CoA reductase, the tungsten centre of acetylene hydratase has a geometry closer to the octahedral (or trigonal antiprismatic) due to a slight rotation of one of the pyranopterin cofactor molecules.[516]

b) Contrary to the reaction mechanisms of XO, SO or DMSOR, which are presently well understood, the tungstoenzymes were less thoroughly studied; still, there are several recent remarkable breakthroughs.[508,511,514,516,518–520] On the other hand, the formate oxidation mechanism is believed to be, overall, similar to the molybdenum-containing enzymes.

The presently known tungstoenzymes raise some puzzling questions: (i) why are the vast majority of these molybdo- plus tungstoenzymes active only with molybdenum (in other words, why do the vast majority of organisms use only molybdenum) and (ii) why do so few organisms have enzymes that function only with tungsten (*i.e.* what turns those few tungsten-dependent organisms into "exceptions" to the "rule" of molybdenum-users)?

Tungsten was, until the 1980s, regarded as an antagonist and inhibitor of the molybdoenzymes (see note §, in Section 1.2),[97–105] which delayed the recognition of its own biological relevance.[13,216] Subsequently, after the identification of the first tungstoenzymes, it became apparent that for almost all of the known tungstoenzymes there is a homologous molybdoenzyme, either in the same or in different organisms.[12] While this general observation makes sense in terms of the chemical similarities between tungsten and molybdenum, and given that both metals are coordinated by the same cofactor molecule, it did not help to provide a clear answer to the above questions. A plausible explanation can, however, be made if one takes into account the chemical singularities of tungsten, relatively to molybdenum, and its different bioavailability, as was discussed in Section 1.2, suggesting that tungsten is a better choice for low reduction potential reactions carried out anaerobically at higher temperatures, in a sulfidic environment.[3] This hypothesis is supported by the tungsten-dependence of hyperthermophilic archaea, the large number of tungstoenzymes found in hyperthermophilic organisms and by the fact that most of the prokaryotes that contain tungstoenzymes are obligate anaerobes[521,522] (although notably there are exceptions, in some aerobic methylotrophic organisms). In addition, this hypothesis is consistent with the reduction potentials of the reactions catalyzed by the tungsten-dependent benzoyl-CoA reductase, W-AOR/carboxylic acid reductase, *N*-formylmethanofuran dehydrogenase and FDH, that have values of −622 mV, −580 mV, −500 mV and −430 mV, respectively.[523,524] Also the non-existence, as far as is presently known, of a tungsten-containing NaR (catalyzing a reaction with a reduction potential of +420 mV) is in full agreement with this hypothesis. Still, the *Pyrobaculum aerophilum* NaR can incorporate and be active with tungsten, when grown in the absence of molybdenum (although in the presence of both metals, the enzyme was reported to contain mostly molybdenum).[525,526] A tungsten-containing form of the *E. coli* trimethylamine *N*-oxide reductase, which catalyzes a reaction with a reduction potential of +130 mV, is formed when this metal is more available in the medium.[527] On the other hand, the above hypothesis dictates that molybdenum would be a good choice for reactions with higher reduction potentials, under either anaerobic or aerobic conditions, but not at high temperatures.[3,12,92–96] In agreement, it seems that prokaryotes that grow at low or moderate temperatures are able to express tungstoenzymes and homologous and/or different molybdoenzymes. This is the case, *e.g.*, of the sulfate reducer bacterium *Desulfovibrio alaskensis* that expresses both tungsten- and molybdenum-containing FDH and one molybdenum-containing AOR. Yet, as this example illustrates, molybdoenzymes are also able to catalyze reactions with low reduction potential values (namely the aldehyde/carboxylic acid reaction). Therefore, to reconcile the obligatory tungsten-dependence of the different aldehyde oxidoreductases of *P. furiosus*,[519,528,529] it may be that it is only under high-temperature conditions that the unique chemical properties of tungsten are most advantageous – future work will determine if this is universally correct.

A third intriguing question regards why some – indeed very few and only prokaryotic – enzymes can function with either molybdenum or tungsten. This is the case of one of the FDH enzymes from *D. alaskensis* and from *D. vulgaris*, which are active with either molybdenum or tungsten. It is also the case of the molybdenum-containing *R. capsulatus* DMSOR[530] and *E. coli* trimethylamine *N*-oxide reductase,[527] whose tungsten-substituted forms are active. Obviously, in these cases, one is tempted to suggest that the chemical similarities between molybdenum and tungsten and the fact that both metals are coordinated by the same cofactor is the basis for the utilization of both metals. However, as far as is presently known, these cases are "rare exceptions" and not the "rule". In fact, the substitution of molybdenum by tungsten,[97–105] and *vice versa*,[3,12,528,529] typically results in metal-free and tungsten-/molybdenum-substituted enzymes, all with no enzymatic activity. In this scenario of "metal selectivity", the question arises as to why the above-mentioned enzymes are not selective. It can be argued that it would be an advantage to incorporate both metals. This could be particularly relevant in sulfate-reducing bacteria, for which molybdenum bioavailability can be limited as a result of sulfide production by these organisms (whose environmental accumulation would lead to the molybdenum precipitation). But, if this were the case, why would other molybdo- and tungstoenzymes from these organisms be selective for one or the other metal? *D. gigas*, *e.g.*, has a tungsten-containing FDH and a molybdenum-containing AOR, both of which are active only with the respective metal. Clearly, the incorporation of both or of only one of the two metals in active prokaryotic enzymes is far from being understood.

1.4.5 The Nitrogenases

The active site of the molybdenum-containing nitrogenases holds a complex high-nuclearity centre, comprising one molybdenum atom, several iron and sulfur atoms plus a carbon atom, [$MoFe_7S_9C$] (Figure 1.3b). (The molybdenum-independent nitrogenases will not be here discussed (see note †, in Section 1.2) and the term "nitrogenase" will be here used to refer only to the molybdenum-dependent enzyme.) Nitrogenase enzymes have been purified from a variety of bacteria, including *Azotobacter vinelandii*, *Klebsiella pneumoniae* and *Clostridium pasteurianum*.[531–533] The enzymes from these diazotrophs catalyze the remarkable reaction of dinitrogen reduction (fixation) to ammonium, with the cleavage of the exceptionally stable N≡N triple bond (eqn (1.32)), a key step in the biogeochemical cycle of nitrogen (Figure 1.1, blue arrow; see Section 1.2). Of note, nitrogenases are also able to catalyze the reduction of the isoelectronic carbon monoxide to hydrocarbon,[47,49,50,534–536] cyanide or acetylene, as well as of carbon dioxide,[537,538] ethylene, nitrous oxide, nitrite, azide and hydrazine.[45,538–548]

$$N{\equiv}N + 8H^+ + 8e^- + 16MgATP \rightarrow 2NH_3 + H_2 + 16MgADP + 16P_i \qquad (1.32)$$

a) The history of nitrogenase is quite old, with the first mention of the organism's ability to fix dinitrogen occurring in the middle of the 19th century;[549] the term "nitrogenase" was introduced in 1934 by Burk and collaborators.[550,551] Since then, the progress of our knowledge about nitrogenases has been extraordinary.

The nitrogenase from *Azotobacter vinelandii* is comprised by of two proteins (Figure 1.13):[44,54,531,532,540,552–579] (i) the "iron protein" (NifH), a homodimer containing one ATP-binding site per monomer and one [4Fe–4S] centre that bridges the two monomers of the dimer; and (ii) the "molybdenum–iron protein"

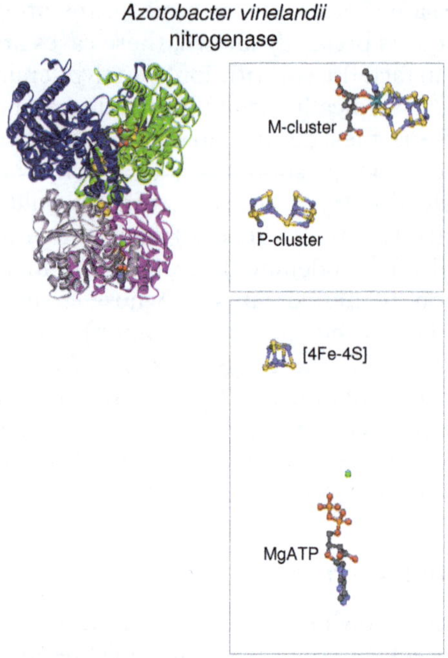

Figure 1.13 Three-dimensional structure view of *A. vinelandii* nitrogenase (top) and of the arrangement of its redox cofactors (bottom). Top – Only one αβ unit of the "molybdenum–iron protein" ($(\alpha\beta)_2$) is represented, in green and dark blue; both α subunits of the "iron protein" (α_2) are represented, in pink and light pink. Bottom – For the "molybdenum–iron protein", the M-cluster of one α subunit and the P-cluster that is found in the α/β subunit interface are represented. For the "iron protein", the MgATP of one monomer and the [4Fe–4S] centre that bridges the two monomers of the dimer are represented. The structures shown are based on the PDB files 1N2C, and were produced with Accelrys DS Visualizer, Accelrys Software Inc.

(NifDK), an $(\alpha\beta)_2$ heterotetramer consisting of one [Fe_8S_7] "P cluster" that bridges the αβ pair of each dimer, and one [$MoFe_7S_9C$–homocitrate] "M-cluster", in each α subunit. Both P- and M-clusters, with their high nuclearity and unusual structures, are unique to the nitrogenase enzymes. The P-cluster can be viewed as two [4Fe–3S] cubanes bridged by a μ_6-sulfide, with the six sulfur atoms coming from six cysteine residues;[553,554,566,577,580–583] it can exist in a fully reduced, all-ferrous, form and in a two-electron oxidized form.[555,558,559,572,576,584–586] The M-cluster can be viewed as a [4Fe–3S] and [Mo–3Fe–3S] cubane bridged by three μ_2-sulfides, with only one sulfur coming from a cysteine residue, the one that coordinates the terminal iron ion.[554,566,576,577,587,588] A histidine residue coordinates the molybdenum, whose coordination sphere is completed by the 2-hydroxyl and 2-carboxyl groups of homocitrate and the three bridging sulfurs. While it has been assumed that molybdenum is in the 4+ oxidation state (Mo^{4+}),[589–596] a recent reassignment suggests an unprecedented 3+ state

(Mo^{3+}) instead.[597,598] The central cavity of the M-cluster holds a recently identified μ_6-coordinated carbide ion (C^{4-});[566,576,577,599–604] it is thought to play a role in stabilizing the M-cluster structure[601,602] and/or in modulating the reactivity of the M-cluster, or even in the interaction with the substrate.[605,606] The M-cluster can undergo a reversible one-electron oxidation–reduction process.[540]

The "iron protein" acts as an ATP-dependent reductase for the molybdenum–iron protein (eqn (1.32)). It has been proposed that upon ATP binding/hydrolysis, a conformational rearrangement takes place that allows the "iron protein" to bind to the "molybdenum–iron protein", forming a pseudo γ_2 electron transfer subunit that would transfer electrons from its [4Fe–4S] centre to the P-cluster of the "molybdenum–iron protein".[44,559,563,567,572,607–612] The electrons are subsequently transferred from the P-cluster to the M-cluster, the active site where reduction of dinitrogen takes place. Nitrogenase-catalyzed substrate reduction thus requires nucleotide-dependent electron delivery from the "iron protein" to the "molybdenum–iron protein". After substrate reduction, products are released and the complex between the "iron protein" and the "molybdenum–iron protein" dissociates.

b) Unravelling the mechanism by which the exceptionally stable triple bond of dinitrogen, N≡N, is cleaved by nitrogenase is an exciting and highly challenging task. This is particularly obvious when one compares the biological process to the harsh industrial conditions used for ammonia production from dinitrogen in the Haber–Bosch process.[613–615] For our society, the relevance of the dinitrogen reduction can be appreciated when one realizes that the factor that most frequently limits agricultural production is the nitrogen availability and that nearly half of the existing human population could not be fed without the use of industrially produced nitrogen fertilizers.[30] Understanding of biological dinitrogen reduction could thus help in the development of improved "greener" synthetic catalysts or of genetically modified organisms with the capacity to fix their own nitrogen. However, while the structure of the nitrogenase components and of their metal centres is presently well known, there are still many uncertainties regarding the metals' oxidation states, the electronic structure of the M-cluster and how this controls the centre's reactivity; this knowledge is essential to fully understand the mechanism of dinitrogen reduction.

Reflecting the interest in this enzyme, numerous experimental[538,545,558,562,574,616–646] and theoretical[593,596,647–663] studies have been carried out to shed light on several aspects of the nitrogenase reaction mechanism. A recent mechanistic proposal suggests that the activation and reduction of dinitrogen depends on the prior reductive elimination of iron-based hydrides in the form of dihydrogen from the M-cluster, it having long been recognized that evolution of dihydrogen is obligatory in dinitrogen reduction.[53,643,664–666] Molybdenum, as anticipated, plays a critical role in the mechanism of dinitrogen reduction, as is demonstrated by enzymes formed in molybdenum-deficient environments and by tungsten-substituted inactive enzymes.[667] Although it is not the substrate binding site, which is thought to be one of the irons,[575,668] molybdenum may be essential to tune the reduction potential of the cofactor or of the neighbouring iron, or alternatively be involved in modifying the electronic structure of the cofactor to facilitate the substrate binding or protonation.

1.4.6 A Novel Heterometallic Cluster Containing Molybdenum Found in Biology

Besides the molybdenum and tungsten centres that form the active site of several enzymes (described in the previous sections), there are also a few other heteronuclear centres whose physiological function is not yet fully understood.[669–673] X-ray absorption fine structure (EXAFS) studies have revealed the presence of a novel linear cluster formulated as $[S_2MoS_2CuS_2MoS_2]$ (Figure 1.3b), a unique heterometallic cluster in biological systems.[669,670] The cluster is contained in a small monomeric "orange protein" isolated from sulfate reducing bacteria, and is non-covalently bound to the protein. Native "orange protein" has only been isolated from *Desulfovibrio gigas*, although all sequenced *Desulfovibrio* complete genomes have putative open reading frames encoding similar proteins. The metal cluster reconstitution of recombinant apo-"orange protein" has been obtained for *D. gigas* protein through a protein metal cluster assisted synthesis and NMR data sustained that the cluster is stabilized inside the protein by both electrostatic and hydrophobic interactions.[672,674] In different sulfate reducing bacteria the genes encoding putative "orange proteins" are located in operons that also encode other proteins, such as putative iron–sulfur ATPases.

1.5 Outlook

Our knowledge of molybdenum- and tungsten-containing enzymes has progressed considerably since the observation of Schardinger in 1902 of the decolourization (reduction) of methylene blue by formaldehyde in presence of fresh milk (through the action of the presently well-known molybdoenzyme xanthine oxidase). During the last years, the structural, spectroscopic and kinetic data, complemented by theoretical information (computational chemistry), about model compounds and enzymes, have provided an extraordinarily detailed picture of molybdenum- and tungsten-containing enzymes. The metabolic pathways responsible for the pyranopterin cofactor synthesis, metal insertion and subsequent modification (when this is the case) and their regulation are also presently well understood.

For many enzymes, crystallographic and spectroscopic studies have provided atomic and electronic details about the active site centres, the other redox-active centres present and the overall structure. Studies carried out with model compounds, together with the computational chemistry and kinetics, has enabled us to discuss enzyme reaction mechanism in atomic and electronic detail for each catalytic intermediary, and establish key structure-activity relationships for several enzymes. Genome analysis and microbiological work have revealed the phylogenetic relationships between different enzymes and anticipated the discovery of new enzymes.

Still, there are numerous open questions for the future. The deeper we go into a topic, the more questions arise and, particularly for the newly discovered enzymes, much work remains to be done regarding their structural characterization and reactivity and mechanistic probing. This introductory chapter on the general topic of molybdenum and tungsten in biology is an "appetizer" for the overall book, where more detailed information will be discussed in the different chapters. Relevant aspects of the chemistry of molybdenum and tungsten in terms of biochemistry are described and an overview of the different families (xanthine oxidase, sulfite oxidase, dimethylsulfoxide reductase, tungstoenzymes families, including nitrogenases) given with emphasis on structures and mechanistic aspects. We hope that we have managed to transmit that, because of its impact on the biogeochemical cycles, in particular of nitrogen and carbon, and on the present life existence and survival, our planet would be quite different without molybdenum!

Abbreviations

AO, aldehyde oxidase;
AOR, bacterial molybdenum-containing aldehyde oxidoreductases
DMS, dimethylsulfide
DMSO, dimethylsulfoxide
DMSOR, dimethylsulfoxide reductase
FDH, formate dehydrogenase (all types of formate dehydrogenase enzymes)
FDH-H, *E. coli* formate dehydrogenase H, from the formate-hydrogen lyase system
FDH-N, *E. coli* formate dehydrogenase N, from the anaerobic nitrate-formate respiratory pathway
FDH-O, *E. coli* formate dehydrogenase O, from the aerobic respiratory pathways
Fe/S, iron–sulfur centre
M, metal, molybdenum or tungsten
mARC, mitochondrial amidoxime reducing component
MOSC, from molybdenum cofactor sulfurase C-terminal domain (proteins involved in pyranopterin cofactor biosynthesis)
Mo/W-bis PGD, molybdenum/tungsten-bis pyranopterin guanosine dinucleotide-containing enzymes
Myr, million years
NaR, nitrate reductase (all types of nitrate reductase enzymes, prokaryotic and eukaryotic ones)
NaRGHI, respiratory nitrate reductase (prokaryotic), after the name of the encoding genes, *narG*, *H* and *I*
NaRZWV, respiratory nitrate reductase (prokaryotic), after the name of the encoding genes, *narZ*, *W* and *V*
PCD, pyranopterin cytidine dinucleotide
PGD, pyranopterin guanosine dinucleotide
PMN, pyranopterin mononucleotide

RO, Sreactive oxygen species
SO, sulfite oxidase
XD, xanthine dehydrogenase
XO, xanthine oxidase
W-AOR, tungsten-containing aldehyde : ferredoxin oxidoreductases (to distinguish from the molybdenum-containing aldehyde oxidoreductases).

Acknowledgements

This work was supported by the Unidade de Ciências Biomoleculares Aplicadas-UCIBIO, which is financed by national funds from FCT/MEC (UID/Multi/04378/2013) and co-financed by the ERDF under the PT2020 Partnership Agreement (POCI-01-0145-FEDER-007728). LBM wishes to thank Fundacão para a Ciência e a Tecnologia, MEC, for a fellowship grant (SFRH/BPD/111404/2015, which is financed by national funds and co-financed by FSE).

References

1. Y. Zhang and V. N. Gladyshev, *J. Mol. Biol.*, 2008, **379**, 881.
2. Y. Zhang, S. Rump and V. N. Gladyshev, *Coord. Chem. Rev.*, 2011, **255**, 1206.
3. M. K. Johnson, D. C. Rees and M. W. W. Adams, *Chem. Rev.*, 1996, **96**, 2817.
4. R. Hille, *Chem. Rev.*, 1996, **96**, 2757.
5. R. Hille, *Trends Biochem. Sci.*, 2002, **27**, 360.
6. G. Schwarz, R. R. Mendel and M. W. Ribbe, *Nature*, 2009, **460**, 839.
7. R. Hille and R. Mendel, *Coord. Chem. Rev.*, 2011, **255**, 991.
8. R. Mendel and T. Kruse, *Biochim. Biophys. Acta*, 2012, **1823**, 1568.
9. R. Hille, *Dalton Trans.*, 2013, **42**, 3029.
10. R. Hille, J. Hall and P. Basu, *Chem. Rev.*, 2014, **114**, 3963.
11. A. D. Anbar, *Science*, 2008, **322**, 1481.
12. A. Kletzin and M. W. Adams, *FEMS Microbiol. Rev.*, 1996, **18**, 5.
13. J. R. Andreesen and M. Makdessi, *Ann. N. Y. Acad. Sci.*, 2008, **1125**, 215.
14. J. R. Andreesen and L. G. Ljungdahl, *J. Bacteriol.*, 1973, **116**, 867.
15. J. R. Andreesen and L. G. Ljungdahl, *J. Bacteriol.*, 1975, **120**, 6.
16. F. M. Gírio, J. C. Marcos and M. T. Amaral-Colaço, *FEMS Microbiol. Lett.*, 1992, **97**, 161.
17. F. M. Gírio, M. T. Amaral-Colaço and M. M. Attwood, *Appl. Microbiol. Biotechnol.*, 1994, **40**, 898.
18. M. Laukel, L. Chistoserdova, M. E. Lidstrom and J. A. Vorholt, *Eur. J. Biochem.*, 2003, **270**, 325.
19. L. Chistoserdova, G. J. Crowther, J. A. Vorholt, E. Skovran, J. C. Portais and M. E. Lidstrom, *J. Bacteriol.*, 2007, **189**, 9076.
20. N. N. Greenwood and A. Earnshaw, *Chemistry of the elements*, Pergamon Press, Oxford, 1984.

21. F. Yang, E. Troncy, M. Francoeur, B. Vinet, P. Vinay, G. Czaika and G. Blaise, *Clin. Chem.*, 1997, **43**, 657.
22. S. Lal and R. S. Patil, *Environ. Monit. Assess.*, 2001, **68**, 37.
23. A. Lennartson, *Nat. Chem.*, 2014, **6**, 746.
24. S. J. Ferguson, *Curr. Opin. Chem. Biol.*, 1998, **2**, 182.
25. D. J. Richardson and N. J. Watmough, *Curr. Opin. Chem. Biol.*, 1999, **3**, 207.
26. J. J. R. Fraústo da Silva and R. J. P. Williams, *The Biological Chemistry of the Elements, The Inorganic Chemistry of Life*, Oxford University Press, Oxford, 2001.
27. J. N. Galloway, F. J. Dentener, D. G. Capone, E. W. Boyer, R. W. Howarth, S. P. Seitzinger, G. P. Asner, C. C. Cleveland, P. A. Green, E. A. Holland, D. M. Karl, A. F. Michaels, J. H. Porter, A. R. Townsend and C. J. Vorosmarty, *Biogeochemistry*, 2004, **70**, 153.
28. H. Bothe, S. J. Ferguson and W. E. Newton, *Biology of the nitrogen cycle*, ed. H. Bothe, S. J. Ferguson and W. E. Newton, Elsevier, Amsterdam, 2007, ch. 17, p. 263.
29. M. S. M. Jetten, *Environ. Microbiol.*, 2008, **10**, 2903.
30. D. E. Canfield, A. N. Glazer and P. G. Falkowski, *Science*, 2010, **330**, 192.
31. R. M. Martínez-Espinosa, J. A. Cole, D. J. Richardson and N. J. Watmough, *Biochem. Soc. Trans.*, 2011, **39**, 175.
32. D. Fowler, M. Coyle, U. Skiba, M. A. Sutton, J. N. Cape, S. Reis, L. J. Sheppard, A. Jenkins, B. Grizzetti, J. N. Galloway, P. Vitousek, A. Leach, A. F. Bouwman, K. Butterbach-Bahl, F. Dentener, D. Stevenson, M. Amann and M. Voss, *Philos. Trans. R. Soc., B*, 2013, **368**, 164.
33. J. N. Galloway, A. M. Leach, A. Bleeker and J. W. Erisman, *Philos. Trans. R. Soc., B*, 2013, **368**, 120.
34. J. Raymond, J. L. Siefert, C. R. Staples and R. E. Blankenship, *Mol. Biol. Evol.*, 2004, **21**, 541.
35. N. Gruber and J. N. Galloway, *Nature*, 2008, **451**, 293.
36. B. Thamdrup, *Annu. Rev. Ecol. Evol. Syst.*, 2012, **43**, 407.
37. R. D. Joerger and P. E. Bishop, *Crit. Rev. Microbiol.*, 1988, **16**, 1.
38. R. R. Eady, *Chem. Rev.*, 1996, **96**, 3013.
39. P. S. Kessler, J. Mclarnan and J. A. Leigh, *J. Bacteriol.*, 1997, **179**, 541.
40. D. Rehder, *J. Inorg. Biochem.*, 2000, **80**, 133.
41. R. R. Eady, *Coord. Chem. Rev.*, 2003, **237**, 23.
42. C. Rüttimann-Johnson, L. M. Rubio, D. R. Dean and P. W. Ludden, *J. Bacteriol.*, 2003, **185**, 2383.
43. D. C. Crans, J. J. Smee, E. Gaidamauskas and L. Yang, *Chem. Rev.*, 2004, **104**, 849.
44. D. C. Rees, A. F. Tezcan, C. A. Haynes, M. Y. Walton, S. Andrade, O. Einsle and J. B. Howard, *Philos. Trans. R. Soc., A*, 2005, **363**, 971.
45. K. Fisher, M. J. Dilworth and W. E. Newton, *Biochemistry*, 2006, **45**, 4190.
46. C. C. Lee, Y. Hu, M. W. Ribbe and R. H. Holm, *Proc. Natl. Acad. Sci. U. S. A.*, 2009, **106**, 9209.
47. C. C. Lee, Y. Hu and M. W. Ribbe, *Science*, 2010, **329**, 642.
48. I. Dance, *Dalton Trans.*, 2011, **40**, 5516.

49. C. C. Lee, Y. Hu and M. W. Ribbe, *Angew. Chem., Int. Ed.*, 2011, **50**, 5545.
50. Y. Hu, C. C. Lee and M. W. Ribbe, *Dalton Trans.*, 2012, **41**, 1118.
51. E. S. Boyd and W. J. Peters, *Front. Microbiol.*, 2013, **4**, 201.
52. S. E. McGlynn, E. S. Boyd, J. W. Peters and V. J. Orphan, *Front. Microbiol.*, 2013, **3**, 419.
53. B. M. Hoffman, D. Lukoyanov, Z. Y. Yang, D. R. Dean and L. C. Seefeldt, *Chem. Rev.*, 2014, **114**, 4041.
54. Y. Hu and M. W. Ribbe, *JBIC, J. Biol. Inorg. Chem.*, 2015, **20**, 435.
55. See the chapters dedicated to nitrogenase in this Book for details on this enzyme.
56. E. S. Boyd, A. D. Anbar, S. Miller, T. L. Hamilton, M. Lavin and J. W. Peters, *Geobiology*, 2011, **9**, 221.
57. D. Santos, P. C. , Z. Fang, S. W. Mason, J. C. Setubal and R. Dixon, *BMC Genomics*, 2012, **13**, 162.
58. S. E. McGlynn, E. S. Boyd, J. W. Peters and V. J. Orphan, *Front. Microbiol.*, 2013, **3**, 491.
59. B. A. MacKay and M. D. Fryzuk, *Chem. Rev.*, 2004, **104**, 385.
60. Q. Cheng, *J. Integr. Plant Biol.*, 2008, **50**, 786.
61. H. P. Jia and E. A. Quadrelli, *Chem. Soc. Rev.*, 2014, **43**, 547.
62. L. B. Maia and J. J. G. Moura, *Chem. Rev.*, 2014, **114**, 5273.
63. L. B. Maia and J. J. G. Moura, *JBIC, J. Biol. Inorg. Chem.*, 2015, **20**, 403.
64. L. B. Maia and J. J. G. Moura, *JBIC, J. Biol. Inorg. Chem.*, 2011, **16**, 443.
65. L. B. Maia, V. Pereira, L. Mira and J. J. G. Moura, *Biochemistry*, 2015, **54**, 685.
66. J. Wang, S. Krizowski, K. Fischer, D. Niks, J. Tejero, C. Sparacino-Watkins, L. Wang, P. Ragireddy, S. Frizzell, E. E. Kelley, Y. Zhang, P. Basu, R. Hille, G. Schwarz and M. T. Gladwin, *Antioxid. Redox Signal*, 2015, **23**, 283.
67. C. E. Sparacino-Watkins, J. Tejero, B. Sun, M. C. Gauthier, J. Thomas, V. Ragireddy, B. A. Merchant, J. Wang, I. Azarov, P. Basu and M. T. Gladwin, *J. Biol. Chem.*, 2014, **289**, 10345.
68. R. J. P. Williams and J. J. R. Fraúto da Silva, *Biochem. Biophys. Res. Commun.*, 2002, **292**, 293.
69. M. Seo, H. Koiwai, S. Akaba, T. Komano, T. Oritani, Y. Kamiya and T. Koshiba, *Plant J.*, 2000, **23**, 481.
70. M. Seo, A. J. Peeters, H. Koiwai, T. Oritani, A. Marion-Poll, J. A. Zeevaart, M. Koornneef, Y. Kamiya and T. Koshiba, *Proc. Natl. Acad. Sci. U. S. A*, 2000, **97**, 12908.
71. S. Akaba, M. Seo, N. Dohmae, K. Takio, H. Sekimoto, Y. Kamiya, N. Furuya, T. Komano and T. Koshiba, *J. Biochem.*, 1999, **126**, 395.
72. D. Y. Huang, A. Furukawa and Y. Ichikawa, *Arch. Biochem. Biophys.*, 1999, **364**, 264.
73. E. Garattini, M. Fratelli and M. Terao, *Cell. Mol. Life Sci.*, 2008, **65**, 1019.
74. D. C. Pryde, D. Dalvie, Q. Hu, P. Jones, R. S. Obach and T. D. Tran, *J. Med. Chem.*, 2010, **53**, 8441.
75. E. Garattini and M. Terao, *Drug Metab. Rev.*, 2011, **43**, 374.

76. T. L. Swenson and J. E. Casida, *Toxicol. Sci.*, 2013, **133**, 22.
77. C. Hesberg, R. Haensch, R. R. Mendel and F. Bittner, *J. Biol. Chem.*, 2004, **279**, 13547.
78. Z. Yesbergenova, G. Yang, E. Oron, D. Soffer, R. Flur and M. Sagi, *Plant J.*, 2005, **42**, 862.
79. M. Zarepour, K. Kaspari, S. Stagge, R. Rethmeier, R. R. Mendel and F. Bittner, *Plant Mol. Biol.*, 2010, **72**, 301.
80. G. M. Montero-Moran, M. Li, E. Rendon-Huerta, F. Jourdan, D. J. Lowe, A. W. Stumpff-Kane, M. Feig, C. Scazzocchio and R. P. Hausinger, *Biochemistry*, 2007, **46**, 5293.
81. J. L. Johnson, *Prenatal Diagn.*, 2003, **23**, 6.
82. J. O. Sass, A. Gunduz, C. A. R. Funayama, B. Korkmaz, K. G. D. Pinto, B. Tuysuz, L. Y. Santos, E. Taskiran, M. F. Turcato, C. W. Lam, J. Reiss, M. Walter, C. Yalcinkaya and J. S. Camelo, *Brain Dev.*, 2010, **32**, 544.
83. N. Carmi-Nawi, G. Malinger, H. Mandel, K. Ichida, T. Lerman-Sagie and D. Lev, *J. Child Neurol.*, 2011, **26**, 460.
84. K. Vijayakumar, R. Gunny, S. Grunewald, L. Carr, K. W. Chong, C. DeVile, R. Robinson, N. McSweeney and P. Prabhakar, *Pediatr. Neurol.*, 2001, **45**, 246.
85. G. Schwarz and A. Belaidi, *Met. Ions Life Sci.*, 2013, **13**, 415.
86. U. Kappler and J. H. Enemark, *JBIC, J. Biol. Inorg. Chem.*, 2015, **20**, 253.
87. U. Kappler, B. Bennett, J. Rethmeier, G. Schwarz, R. Deutzmann, A. G. McEwan and C. Dahl, *J. Biol. Chem.*, 2000, **275**, 13202.
88. A. Di Salle, G. D'Errico, F. La Cara, R. Cannio and M. Rossi, *Extremophiles*, 2006, **10**, 587.
89. K. Denger, S. Weinitschke, T. H. M. Smits, D. Schleheck and A. M. Cook, *Microbiology*, 2008, **154**, 256.
90. J. J. Wilson and U. Kappler, *Biochim. Biophys. Acta*, 2009, **1787**, 1516.
91. K. Kishida, Y. Sohrin, K. Okamura and J. Ishibashi, *Earth Planet. Sci. Lett.*, 2004, **222**, 819.
92. I. Dellien, F. M. Hall and L. Hepler, *Chem. Rev.*, 1976, **76**, 283.
93. G. E. Callis and R. A. Wentworth, *Bioinorg. Chem.*, 1977, **7**, 57.
94. N. Ueyama, H. Oku and A. Nakamura, *J. Am. Chem. Soc.*, 1992, **114**, 7310.
95. S. K. Das, D. Biswas, R. Maiti and S. Sarkar, *J. Am. Chem. Soc.*, 1996, **118**, 1387.
96. J. Wiegel and M. W. W. Adams, *Protein Sci.*, 2008, **8**, 1564.
97. E. S. Higgins, D. A. Richert and W. W. Westerfeld, *Proc. Soc. Exp. Biol. Med.*, 1956, **92**, 509.
98. Y. M. Heimer, J. L. Wray and P. Filner, *Plant Physiol.*, 1969, **44**, 1197.
99. P. J. Aparicio, J. Cardenas, W. G. Zumft, J. M. Vega, J. Herrera, A. Paneque and M. Losada, *Phytochemistry*, 1971, **10**, 1487.
100. J. R. Benemann, G. M. Smith, P. J. Kostel and C. E. McKenna, *FEBS Lett.*, 1973, **29**, 219.
101. J. L. Johnson, H. J. Cohen and K. V. Rajagopalan, *J. Biol. Chem.*, 1974, **249**, 5046.

102. J. L. Johnson and K. V. Rajagopalan, *J. Biol. Chem.*, 1976, **251**, 5505.
103. R. H. Scott, G. T. Sperl and J. A. DeMoss, *J. Bacteriol.*, 1979, **137**, 719.
104. C. Chauret and R. Knowles, *Can. J. Microbiol.*, 1991, **37**, 744.
105. C. D. Garner and L. J. Stewart, *Met. Ions Biol. Syst.*, 2002, **39**, 699.
106. B. Schoepp-Cothenet, R. Lis, P. Philippot, A. Magalon, M. J. Russell and W. Nitschke, *Sci. Rep.*, 2012, **2**, 263.
107. E. Lebrun, M. Brugna, F. Baymann, D. Muller, D. Lievremot, M. C. Lett and W. Nitschke, *Mol. Biol. Evol.*, 2003, **20**, 686.
108. H. D. Holland, *Geochim. Cosmochim. Acta*, 2002, **66**, 3811.
109. A. Bekker, H. D. Holland, P. L. Wang, D. Rumble, H. J. Stein, J. L. Hannah, L. L. Coetzee and N. J. Beukes, *Nature*, 2004, **427**, 117.
110. D. E. Canfield, *Annu. Rev. Earth Planet. Sci.*, 2005, **33**, 1.
111. R. E. Kopp, J. L. Kirschvink, I. A. Hilburn and C. Z. Nash, *Proc. Natl. Acad. Sci. U. S. A.*, 2005, **102**, 11131.
112. A. D. Anbar, Y. Duan, T. W. Lyons, G. L. Arnold, B. Kendall, R. A. Creaser, A. J. Kaufman, G. W. Gordon, C. Scott, J. Garvin and R. Buick, *Science*, 2007, **317**, 1903.
113. A. J. Kaufman, D. T. Johnston, J. Farquhar, A. L. Masterson, T. W. Lyons, S. Bates, A. D. Anbar, G. L. Arnold, J. Garvin and R. Buick, *Science*, 2007, **317**, 1900.
114. J. J. Brocks, G. A. Logan, R. Buick and R. E. Summons, *Science*, 1999, **285**, 1033.
115. J. A. Karhu and H. D. Holland, *Geology*, 1996, **24**, 867.
116. D. E. Canfield, *Nature*, 1998, **396**, 450.
117. J. Farquhar and B. A. Wing, *Earth Planet. Sci. Lett.*, 2003, **213**, 1.
118. O. J. Rouxel, A. Bekker and K. J. Edwards, *Science*, 2005, **307**, 1088.
119. D. A. Fike, J. P. Grotzinger, L. M. Pratt and R. E. Summons, *Nature*, 2006, **444**, 744.
120. D. E. Canfield, S. W. Poulton and G. M. Narbonne, *Science*, 2007, **315**, 92.
121. C. Scott, T. W. Lyons, A. Bekker, Y. Shen, S. W. Poulton, X. Chu and A. D. Anbar, *Nature*, 2008, **452**, 456.
122. S. W. Poulton, P. W. Fralick and D. E. Canfield, *Nature*, 2004, **431**, 173.
123. A. D. Anbar and A. H. Knoll, *Science*, 2002, **297**, 1137.
124. H. D. Holland, *The Chemical Evolution of the Atmosphere and Oceans*, Princeton Univ. Press, Princeton NJ, 1984.
125. H. Holland, *Philos. Trans. R. Soc., B*, 2006, **361**, 903.
126. T. W. Lyons, C. T. Reinhard and N. J. Planavsky, *Nature*, 2014, **506**, 307.
127. T. W. Lyons, C. T. Reinhard and N. J. Planavsky, *Curr. Biol.*, 2014, **24**, R276.
128. B. Rasmussen, I. R. Fletcher, J. J. Brocks and M. R. Kilburn, *Nature*, 2008, **455**, 1101.
129. J. B. Glass, F. Wolfe-Simon and A. D. Anbar, *Geobiology*, 2009, **7**, 100.
130. R. Navarro-González, C. P. McKay and D. N. Mvondo, *Nature*, 2001, **412**, 61.
131. X. Zhang, D. M. Sigman, F. M. M. Morel and A. M. L. Kraepiel, *Proc. Natl. Acad. Sci. U. S. A.*, 2014, **111**, 4782.

132. A. L. Zerkle, C. H. House, R. P. Cox and D. E. Canfield, *Geobiology*, 2006, **4**, 285.
133. E. S. Boyd, A. M. Costas, T. L. Hamilton, F. Mus and J. W. Peters, *J. Bacteriol.*, 2015, **197**, 1690.
134. D. J. Beerling, *The Emerald Planet. How Plants Changed Earth's History*, Oxford University Press, Oxford, 2007.
135. D. J. Beerling, C. N. Hewitt, J. A. Pyle and J. A. Raven, *Philos. Trans. R. Soc., A*, 2007, **365**, 1627.
136. P. Falkowski and A. H. Knoll, *Evolution of Primary Production in the Sea*, Academic Press, London, 2007.
137. S. J. N. Burgmayer and E. I. Stiefel, *J. Chem. Educ.*, 1985, **62**, 943.
138. R. H. Holm and J. M. Berg, *Acc. Chem. Res.*, 1986, **19**, 363.
139. E. E. Harlan, J. M. Berg and R. H. Holm, *J. Am. Chem. Soc.*, 1986, **108**, 6992.
140. R. H. Holm, *Chem. Rev.*, 1987, **87**, 1401.
141. R. H. Holm, *Coord. Chem. Rev.*, 1990, **100**, 183.
142. Z. Xiao, C. G. Young, J. H. Enemark and A. G. Wedd, *J. Am. Chem. Soc.*, 1992, **114**, 9194.
143. R. H. Holm and J. P. Donahue, *Polyhedron*, 1993, **12**, 571.
144. B. E. Scbultz, S. F. Gbeller, M. C. Muetterties, M. J. Scott and R. H. Holm, *J. Am. Chem. Soc.*, 1993, **115**, 2714.
145. B. E. Schultz and R. H. Holm, *Inorg. Chem.*, 1993, **32**, 4244.
146. S. K. Das, P. K. Chaudhury, D. Biswas and S. Sarkar, *J. Am. Chem. Soc.*, 1994, **116**, 9061.
147. J. H. Enemark and C. G. Young, *Adv. Inorg. Chem.*, 1994, **40**, 1.
148. H. Oku, N. Ueyama, M. Kondo and A. Nakamura, *Inorg. Chem.*, 1994, **33**, 209.
149. R. H. Holm, P. Kennepohl and E. I. Solomon, *Chem. Rev.*, 1996, **96**, 2239.
150. L. J. Laughlin and C. G. Young, *Inorg. Chem.*, 1996, **35**, 1050.
151. J. H. Enemark, J. A. Cooney, J. J. Wang and R. H. Holm, *Chem. Rev.*, 2004, **104**, 1175.
152. J. P. Donahue, *Chem. Rev.*, 2006, **106**, 4747.
153. H. Sugimoto and H. Tsukube, *Chem. Soc. Rev.*, 2008, **37**, 2609.
154. A. Majumdar, *Dalton Trans.*, 2014, **43**, 8990.
155. Y. Hasenaka, T. Okamura and K. Onitsuka, *Dalton Trans.*, 2015, **44**, 6260.
156. P. Basu and S. J. N. Burgmayer, *JBIC, J. Biol. Inorg. Chem.*, 2015, **20**, 373.
157. T. Nishino, K. Okamoto, B. T. Eger, E. F. Pai and T. Nishino, *FEBS J.*, 2008, **275**, 3278.
158. R. Hille, T. Nishino and F. Bittner, *Coord. Chem. Rev.*, 2011, **255**, 1179.
159. K. Okamoto, T. Kusano and T. Nishino, *Curr. Pharm. Des.*, 2013, **19**, 2606.
160. W. Reichenbecher and B. Schink, *Biochim. Biophys. Acta*, 1999, **1430**, 245.
161. A. Messerschmidt, H. Niessen, D. Abt, O. Einsle, B. Schink and P. M. H. Kroneck, *Proc. Natl. Acad. Sci. U. S. A.*, 2004, **101**, 11571.

162. K. G. Matz, R. P. Mtei, B. Leung, S. J. N. Burgmayer and M. L. Kirk, *J. Am. Chem. Soc.*, 2010, **132**, 7830.
163. K. G. Matz, R. P. Mtei, R. Rothstein, M. L. Kirk and S. J. N. Burgmayer, *Inorg. Chem.*, 2011, **50**, 9804.
164. B. R. Williams, Y. Fu, G. P. A. Yap and S. J. N. Burgmayer, *J. Am. Chem. Soc.*, 2012, **134**, 1958.
165. C. Dong, J. Yang, S. Leimkühler and M. L. Kirk, *Inorg. Chem.*, 2014, **53**, 7077.
166. R. A. Rothery and J. H. Weiner, *JBIC, J. Biol. Inorg. Chem.*, 2015, **20**, 349.
167. B. Fischer, J. H. Enemark and P. Basu, *J. Inorg. Biochem.*, 1998, **72**, 13.
168. L. Marbella, B. Serli-Mitasev and P. Basu, *Angew. Chem., Int. Ed.*, 2009, **48**, 3996.
169. F. J. Hine, A. J. Taylor and C. D. Garner, *Coord. Chem. Rev.*, 2010, **254**, 1570.
170. P. Basu and S. N. J. Burgmayer, *Coord. Chem. Rev.*, 2011, **255**, 1016.
171. C. Schulzke, *Eur. J. Inorg. Chem.*, 2011, 1189.
172. B. R. Williams, Y. Fu, G. P. A. Yap and S. J. N. Burgmayer, *J. Am. Chem. Soc.*, 2012, **134**, 19584.
173. R. A. Rothery, B. Stein, M. Solomonson, M. L. Kirk and J. H. Weiner, *Proc. Natl. Acad. Sci. U. S. A.*, 2012, **109**, 14773.
174. M. G. Bertero, R. A. Rothery, M. Palak, C. Hou, D. Lim, F. Blasco, J. H. Weiner and N. C. J. Strynadka, *Nat. Struct. Biol.*, 2003, **10**, 681.
175. M. Jormakka, D. Richardson, B. Byrne and S. Iwata, *Structure*, 2004, **12**, 95.
176. D. P. Kloer, C. Hagel, J. Heider and G. E. Schulz, *Structure*, 2006, **14**, 1377.
177. J. H. Enemark and C. D. Garner, *J. Biol. Inorg. Chem.*, 1997, **2**, 817.
178. B. Bradshaw, D. Collison, C. D. Garner and J. A. Joule, *Chem. Commun.*, 2001, 123.
179. J. G. J. Jacques, V. Fourmond, P. Arnoux, M. Sabaty, E. Etienne, S. Grosse, F. Biaso, P. Bertrand, D. Pignol, C. Léger, B. Guigliarelli and B. Burlat, *Biochim. Biophys. Acta*, 2014, **1837**, 277.
180. B. S. Lim and R. H. Holm, *J. Am. Chem. Soc.*, 2001, **123**, 1920.
181. C. Lorber, J. P. Donahue, C. A. Goddard, E. Nordlander and R. H. Holm, *J. Am. Chem. Soc.*, 1998, **120**, 8102.
182. B. S. Lim, K. M. Sung and R. H. Holm, *J. Am. Chem. Soc.*, 2000, **122**, 7410.
183. A. Majumdar, K. Pal, K. Nagarajan and S. Sarkar, *Inorg. Chem.*, 2007, **46**, 6136.
184. A. Majumdar, K. Pal and S. Sarkar, *Dalton Trans.*, 2009, 1927.
185. R. L. McNaughton, R. S. Lim, S. Z. Knottenbelt, R. H. Holm and M. L. Kirk, *J. Am. Chem. Soc.*, 2008, **130**, 4628.
186. K. M. Sung and R. H. Holm, *J. Am. Chem. Soc.*, 2001, **123**, 1931.
187. K. M. Sung and R. H. Holm, *Inorg. Chem.*, 2001, **40**, 4518.
188. J. J. Wang, O. P. Kryatova, E. V. Rybak-Akimova and R. H. Holm, *Inorg. Chem.*, 2004, **43**, 8092.

189. R. P. Mtei, G. Lyashenko, B. Stein, N. Rubie, R. Hille and M. L. Kirk, *J. Am. Chem. Soc.*, 2011, **133**, 9762.
190. A. K. Rappe and W. A. Goddard, *Nature*, 1980, **285**, 311.
191. A. K. Rappe and W. A. Goddard, *J. Am. Chem. Soc.*, 1982, **104**, 3287.
192. F. E. Inscore, R. McNaughton, L. Westcott, M. E. Helton, R. Jones, I. K. Dhawan, J. H. Enemark and M. L. Kirk, *Inorg. Chem.*, 1999, **38**, 1401.
193. Y. Izumi, T. Glaser, K. Rose, J. McMaster, P. Basu, J. H. Enemark, K. O. Hodgson and E. I. Solomon, *J. Am. Chem. Soc.*, 1999, **121**, 10035.
194. R. L. McNaughton, M. E. Helton, N. Rubie and M. L. Kirk, *Inorg. Chem.*, 2000, **39**, 4386.
195. K. Peariso, R. L. McNaughton and M. L. Kirk, *J. Am. Chem. Soc.*, 2002, **124**, 9006.
196. K. Peariso, M. E. Helton, E. N. Duesler, S. E. Shadle and M. L. Kirk, *Inorg. Chem.*, 2007, **46**, 1259.
197. D. E. Green and H. Beinert, *Biochim. Biophys. Acta*, 1953, **11**, 599.
198. E. C. Renzo, E. Kaleita, P. Heytler, J. J. Oleson, B. L. Hutchings and J. H. Williams, *J. Am. Chem. Soc.*, 1953, **75**, 753.
199. D. A. Richert and W. W. Westerfeld, *J. Biol. Chem.*, 1953, **203**, 915.
200. J. R. Totter, W. T. Buenett, R. Monroe, I. B. Whitney and C. Comar, *Science*, 1953, **118**, 555.
201. P. G. Avis, F. Bergel, R. C. Bray and K. V. Shooter, *Nature*, 1954, **173**, 1230.
202. D. J. Nicholas and A. Nason, *J. Biol. Chem.*, 1954, **207**, 353.
203. R. C. Bray, B. G. Malmstrom and T. Vanngard, *Biochem. J.*, 1959, **73**, 193.
204. R. C. Bray, *Biochem. J.*, 1961, **81**, 196.
205. R. C. Bray, G. Palmer and H. Beinert, *J. Biol. Chem.*, 1964, **239**, 2667.
206. T. Horio and A. S. Pietro, *Proc. Natl. Acad. Sci. U. S. A.*, 1964, **51**, 1226.
207. G. Palmer, R. C. Bray and H. Beinert, *J. Biol. Chem.*, 1964, **239**, 2657.
208. G. Palmer and R. H. Sands, *J. Biol. Chem.*, 1966, **241**, 253.
209. S. G. Mayhew, D. Petering, G. Palmer and G. P. Foust, *J. Biol. Chem.*, 1969, **244**, 2830.
210. G. Palmer and V. Massey, *J. Biol. Chem.*, 1969, **244**, 2641.
211. R. C. Bray and J. C. Swann, *Struct. Bonding*, 1972, **11**, 107.
212. R. C. Bray, *Biol. Magn. Reson.*, 1980, **2**, 45.
213. M. Dixon and S. Thurlow, *Biochem. J.*, 1924, **18**, 971.
214. J. R. Andreesen, E. Ghazzawi and G. Gottschalk, *Arch. Microbiol.*, 1974, **96**, 103.
215. L. G. Ljungdahl and J. R. Andreesen, *FEBS Lett.*, 1975, **54**, 279.
216. L. G. Ljungdahl, *Trends Biochem. Sci.*, 1976, **1**, 63.
217. J. B. Jones and T. C. Stadtman, *J. Bacteriol.*, 1977, **130**, 1404.
218. I. Yamamoto, T. Saiki, S.-M. Liu and L. G. Ljungdahl, *J. Biol. Chem.*, 1983, **258**, 1826.
219. H. White, G. Strobl, R. Feicht and H. Simon, *Eur. J. Biochem.*, 1989, **184**, 89.
220. S. Mukund and M. W. W. Adams, *J. Biol. Chem.*, 1990, **265**, 11508.
221. G. N. George, R. C. Prince, S. Mukund and M. W. W. Adams, *J. Am. Chem. Soc.*, 1992, **114**, 3521.

222. M. K. Chan, S. Mukund, A. Kletzin, M. W. W. Adams and D. C. Rees, *Science*, 1995, **267**, 1463.
223. M. J. Romao, M. Archer, I. Moura, J. J. G. Moura, J. LeGall, R. Engh, M. Schneider, P. Hof and R. Huber, *Science*, 1995, **270**, 1170.
224. H. Schindelin, C. Kisker, J. Hilton, K. V. Rajagopalan and D. C. Rees, *Science*, 1996, **272**, 1615.
225. A. Cvetkovic, A. L. Menon, M. P. Thorgersen, J. W. Scott, F. L. Poole II, F. E. Jenney Jr, W. A. Lancaster, J. L. Praissman, S. Shanmukh, B. J. Vaccaro, S. A. Trauger, E. Kalisiak, J. V. Apon, G. Siuzdak, S. M. Yannone, J. A. Tainer and M. W. W. Adams, *Nature*, 2010, **466**, 779.
226. Y. Zhang and V. N. Gladyshev, *J. Biol. Chem.*, 2011, **286**, 23623.
227. R. Roy and M. W. Adams, *Met. Ions Biol. Syst.*, 2002, **39**, 673.
228. S. Grimaldi, B. Schoepp-Cothenet, P. Ceccaldi, B. Guigliarelli and A. Magalon, *Biochim. Biophys. Acta*, 2013, **1827**, 1048.
229. P. Hanzelmann, H. Dobbek, L. Gremer, R. Huber and O. Meyer, *J. Mol. Biol.*, 2000, **301**, 1221.
230. H. Dobbek, L. Gremer, R. Kiefersauer, R. Huber and O. Meyer, *Proc. Natl. Acad. Sci. U. S. A.*, 2002, **99**, 15971.
231. M. Resch, H. Dobbek and O. Meyer, *JBIC, J. Biol. Inorg. Chem.*, 2005, **10**, 518.
232. B. W. Stein and M. L. Kirk, *Chem. Commun.*, 2014, **50**, 1104.
233. R. Hille, S. Dingwall and J. Wilcoxen, *JBIC, J. Biol. Inorg. Chem.*, 2015, **20**, 243.
234. F. Schardinger, *Z. Unters. Nahr. Genussm.*, 1902, **5**, 1113.
235. E. J. Morgan, C. P. Stewart and F. G. Hopkins, *Proc. R. Soc. London, Ser. B*, 1992, **94**, 109.
236. E. Della Corte and F. Stirpe, *Biochem. J.*, 1968, **108**, 349.
237. F. Stirpe and E. Della Corte, *J. Biol. Chem.*, 1969, **244**, 3855.
238. E. Della Corte and F. Stirpe, *Biochem. J.*, 1972, **126**, 739.
239. M. G. Battelli, E. Lorenzoni and F. Stripe, *Biochem. J.*, 1973, **131**, 191.
240. W. R. Waud and K. V. Rajagopalan, *Arch. Biochem. Biophys.*, 1976, **172**, 354.
241. W. R. Waud and K. V. Rajagopalan, *Arch. Biochem. Biophys.*, 1976, **172**, 365.
242. Z. W. Kaminski and M. M. Jezewska, *Biochem. J.*, 1982, **207**, 341.
243. M. Nakamura and I. Yamazaki, *J. Biochem.*, 1982, **92**, 1279.
244. T. Saito and T. Nishino, *J. Biol. Chem.*, 1989, **264**, 10015.
245. Y. Amaya, K. Yamazaki, M. Sato, K. Noda and T. Nishino, *J. Biol. Chem.*, 1990, **265**, 14170.
246. T. Nishino and T. Nishino, *J. Biol. Chem.*, 1997, **272**, 29859.
247. Y. Kuwabara, T. Nishino, K. Okamoto, T. Matsumura, B. T. Eger, E. F. Pai and T. Nishino, *Proc. Natl. Acad. Sci. U. S. A.*, 2003, **100**, 8170.
248. C. Enroth, B. T. Eger, K. Okamoto, T. Nishino, T. Nishino and E. F. Pai, *Proc. Natl. Acad. Sci. U. S. A.*, 2000, **97**, 10723.
249. T. Nishino, K. Okamoto, Y. Kawaguchi, H. Hori, T. Matsumura, B. T. Eger, E. F. Pai and T. Nishino, *J. Biol. Chem.*, 2005, **280**, 24888.

250. A. Tsujii and T. Nishino, *Nucleosides, Nucleotides Nucleic Acids*, 2008, **27**, 881.
251. H. Ishikita, B. T. Eger, K. Okamoto, T. Nishino and E. F. Pai, *J. Am. Chem. Soc.*, 2012, **134**, 999.
252. T. Nishino, K. Okamoto, Y. Kawaguchi, T. Matsumura, B. T. Eger, E. F. Pai and T. Nishino, *FEBS J.*, 2015, **282**, 3075.
253. K. V. Rajagopalan and P. Handler, *J. Biol. Chem.*, 1967, **242**, 4097.
254. M. J. Barber, R. C. Bray, D. J. Lowe and M. P. Coughlan, *Biochem. J.*, 1976, **153**, 297.
255. M. J. Barber, R. C. Bray, R. Cammack and M. P. Coughlan, *Biochem. J.*, 1977, **163**, 279.
256. D. Barrett and N. A. Davidson, *J. Insect Physiol.*, 1975, **21**, 1447.
257. J. J. Truglio, K. Theis, S. Leimkühler, R. Rappa, K. V. Rajagopalan and C. Kisker, *Structure*, 2002, **10**, 115.
258. E. Della Corte, G. Gozzetti, F. Novello and F. Stirpe, *Biochim. Biophys. Acta*, 1969, **191**, 164.
259. R. S. Roy and J. M. McCord, *Fed. Proc.*, 1982, **41**, 767.
260. T. D. Engerson, T. G. McKelvey, D. B. Rhyne, E. B. Boggio, S. J. Snyder and H. P. Jones, *J. Clin. Invest.*, 1987, **79**, 1564.
261. D. A. Parks, T. K. Williams and J. S. Beckman, *Am. J. Physiol.*, 1988, **254**, G768.
262. G. Cighetti, M. Del Puppo, R. Paroni and K. M. Galli, *FEBS Lett.*, 1990, **274**, 82.
263. Y. Xia and J. L. Zweier, *J. Biol. Chem.*, 1995, **270**, 18797.
264. W. M. Frederiks and K. S. Bosch, *Hepatology*, 1996, **24**, 1179.
265. F. Stirpe, M. Ravaioli, M. G. Battelli, S. Musiani and G. L. Grazi, *Am. J. Gastroenterol.*, 2002, **97**, 2079.
266. E. E. Kelley, T. Hock, N. K. Khoo, G. R. Richardson, K. K. Johnson, P. C. Powell, G. I. Giles, A. Agarwal, J. R. Lancaster Jr. and M. M. Tarpey, *Free Radical Biol. Med.*, 2006, **40**, 952.
267. D. N. Granger, G. Rutili and J. M. McCord, *Gastroenterology*, 1981, **81**, 22.
268. J. M. McCord, *N. Engl. J. Med.*, 1985, **312**, 159.
269. C. S. Lieber, *N. Engl. J. Med.*, 1988, **319**, 1639.
270. J. L. Zweier, P. Kuppusamy and G. A. Lutty, *Proc. Natl. Acad. Sci. U. S. A.*, 1988, **85**, 4046.
271. A. I. Cederbaum, *Free Radical Biol. Med.*, 1989, **7**, 537.
272. S. Kato, T. Kawase, J. Alderman, N. Inatomi and C. Lieber, *Gastroenterology*, 1990, **98**, 203.
273. R. Nordmann, C. Ribière and H. Rouach, *Free Radical Biol. Med.*, 1992, **12**, 219.
274. L. S. Terada, J. J. Dormish, P. F. Shanley, J. A. Leff, B. O. Anderson and J. E. Repine, *Am. J. Physiol.*, 1992, **263**, L394–L401.
275. J. L. Zweier, R. Broderick, P. Kuppusamy, S. Thompson-Gorman and G. A. Lutty, *J. Biol. Chem.*, 1994, **269**, 24156.

276. L. Mira, L. Maia, L. Barreira and C. F. Manso, *Arch. Biochem. Biophys.*, 1995, **318**, 53.
277. A. Weinbroum, V. G. Nielsen, S. Tan, S. Gelman, S. Matalon, K. A. Skinner, E. Bradley Jr. and D. A. Parks, *Am. J. Physiol.*, 1995, **268**, G988.
278. T. Nishino, S. Nakanishi, K. Okamoto, J. Mizushima, H. Hori, T. Iwasaki, T. Nishino, K. Ichimori and H. Nakazawa, *Biochem. Soc. Trans.*, 1997, **25**, 783.
279. R. Harrison, *Biochem. Soc. Trans.*, 1997, **25**, 786.
280. R. M. Wright and J. E. Repine, *Biochem. Soc. Trans.*, 1997, **25**, 799.
281. K. B. Beckman and B. N. Ames, *Physiol. Rev.*, 1998, **78**, 547.
282. H. Suzuki, F. A. Delano, D. A. Parks, N. Jamshidi, D. N. Granger, H. Ishii, M. Suematsu, B. W. Zweifach and G. W. Schmid-Schonbein, *Proc. Natl. Acad. Sci. U. S. A.*, 1998, **95**, 4754.
283. R. M. Wright, J. L. McManaman and J. E. Repine, *Free Radical Biol. Med.*, 1999, **26**, 348.
284. R. Harrison, *Free Radical Biol. Med.*, 2002, **33**, 774.
285. D. Wu and A. I. Cederbaum, *Alcohol Res. Health*, 2003, **27**, 277.
286. C. E. Berry and J. M. Hare, *J. Physiol.*, 2004, **555**, 589.
287. L. Maia, A. Vala and L. Mira, *Free Radical Res.*, 2005, **39**, 979.
288. L. Maia, R. O. Duarte, A. Ponces-Freire, J. J. G. Moura and L. Mira, *JBIC, J. Biol. Inorg. Chem.*, 2007, **12**, 777.
289. D. M. Small, J. S. Coombes, N. Bennett, D. W. Johnson and G. C. Gobe, *Nephrology*, 2012, **17**, 311.
290. M. M. Bachschmid, S. Schildknecht, R. Matsui, R. Zee, D. Haeussler, R. A. Cohen, D. Pimental and B. Loo, *Ann. Med.*, 2013, **45**, 17.
291. N. R. Madamanchi and M. S. Runge, *Free Radical Biol. Med.*, 2013, **61**, 473.
292. T. Nishino and K. Okamoto, *JBIC, J. Biol. Inorg. Chem.*, 2015, **20**, 195.
293. T. Schrader, A. Rienhofer and J. R. Andreesen, *Eur. J. Biochem.*, 1999, **264**, 862.
294. W. T. Self and T. C. Stadtman, *Proc. Natl. Acad. Sci. U. S. A.*, 2000, **97**, 7208.
295. V. N. Gladyshev, S. V. Khangulov and T. C. Stadtman, *Proc. Natl. Acad. Sci. U. S. A.*, 1994, **91**, 232.
296. V. N. Gladyshev and P. Lecchi, *BioFactors*, 1995, **5**, 93.
297. N. Wagener, A. J. Pierik, A. Ibdah, R. Hille and H. Dobbek, *Proc. Natl. Acad. Sci. U. S. A.*, 2009, **106**, 11055.
298. K. Okamoto, K. Matsumoto, R. Hille, B. T. Eger, E. F. Pai and T. Nishino, *Proc. Natl. Acad. Sci. U. S. A.*, 2004, **101**, 7931.
299. N. A. Turner, W. A. Doyle, A. M. Ventom and R. C. Bray, *Eur. J. Biochem.*, 1995, **232**, 646.
300. M. Mahro, C. Coelho, J. Trincão, D. Rodrigues, M. Terao, E. Garattini, M. Saggu, F. Lendzian, P. Hildebrandt, M. J. Romão and S. Leimkühler, *Drug Metab. Dispos.*, 2011, **39**, 1939.
301. C. Coelho, M. Mahro, J. Trincão, A. T. P. Carvalho, M. J. Ramos, M. Terao, E. Garattini, S. Leimkühler and M. J. Romão, *J. Biol. Chem.*, 2012, **287**, 40690.

302. P. Mu, M. Zheng, M. Xu, Y. Zheng, X. Tang, Y. Wang, K. Wu, Q. Chen, L. Wang and Y. Deng, *Drug Metab. Dispos.*, 2014, **42**, 511.
303. R. Huber, P. Hof, R. O. Duarte, J. J. G. Moura, I. Moura, M.-Y. Liu, J. Legall, R. Hille, M. Archer and M. J. Romão, *Proc. Natl. Acad. Sci. U. S. A.*, 1996, **93**, 8846.
304. J. M. Rebelo, J. M. Dias, R. Huber, J. J. G. Moura and M. J. Romão, *JBIC, J. Biol. Inorg. Chem.*, 2001, **6**, 791.
305. L. Krippahl, N. Palma, I. Moura and J. J. G. Moura, *Eur. J. Inorg. Chem.*, 2006, **19**, 3835.
306. B. Zhang, C. F. Hemann and R. Hille, *J. Biol. Chem.*, 2010, **285**, 12571.
307. M. Gnida, R. Ferner, L. Gremer, O. Meyer and W. Meyer- Klaucke, *Biochemistry*, 2003, **42**, 222.
308. J. Gibson, M. Dispensa and C. S. Harwood, *J. Bacteriol.*, 1997, **179**, 634.
309. M. Unciuleac, E. Warkentin, C. C. Page, M. Boll and U. Ermler, *Structure*, 2004, **12**, 2249.
310. M. A. S. Correia, A. R. Otrelo-Cardoso, V. Schwuchow, T. Santos-Silva, S. Leimkühler and M. J. Romão, Structural Insights into a Unique Periplasmic Aldehyde Oxidoreductase from Escherichia coli, *Abstract presentation to the 6th European Conference Chemistry in Life Sciences*, Lisboa, Portugal, 2015.
311. H. Cao, J. Hall and R. Hille, *Biochemistry*, 2014, **53**, 533.
312. J. Johannes, M. Unciuleac, T. Friedrich, E. Warkentin, U. Ermler and M. Boll, *Biochemistry*, 2008, **47**, 4964.
313. M. Boll, *Biochim. Biophys. Acta*, 2005, **1707**, 34.
314. J. M. Pauff, J. Zhang, C. E. Bell and R. Hille, *J. Biol. Chem.*, 2008, **283**, 4818.
315. B. W. Stein and M. L. Kirk, *JBIC, J. Biol. Inorg. Chem.*, 2015, **20**, 183.
316. K. N. Murray, J. G. Watson and S. Chaykin, *J. Biol. Chem.*, 1966, **241**, 4798.
317. R. Hille and H. Sprecher, *J. Biol. Chem.*, 1987, **262**, 10914.
318. C. Kisker, H. Schindelin, A. Pacheco, W. A. Wehbi, R. M. Garrett, K. V. Rajagopalan, J. H. Enemark and D. C. Rees, *Cell*, 1997, **91**, 973.
319. O. W. Griffith, *Methods Enzymol.*, 1987, **143**, 366.
320. U. Kappler, *Biochim. Biophys. Acta*, 2011, **1807**, 1.
321. K. V. Rajagopalan and M. P. Coughlan, *Molybdenum and molybdenum-containing enzymes*, Pergamon Press, Oxford, 1980.
322. A. Pacheco, J. T. Hazzard, G. Tollin and J. H. Enemark, *JBIC, J. Biol. Inorg. Chem.*, 1999, **4**, 390.
323. C. J. Feng, R. V. Kedia, J. T. Hazzard, J. K. Hurley, G. Tollin and J. H. Enemark, *Biochemistry*, 2002, **41**, 5816.
324. K. Johnson-Winters, A. R. Nordstrom, S. Emesh, A. V. Astashkin, A. Rajapakshe, R. E. Berry, G. Tollin and J. H. Enemark, *Biochemistry*, 2010, **49**, 1290.
325. T. Leustek and K. Saito, *Plant Physiol.*, 1999, **120**, 637.
326. T. Eilers, G. Schwarz, H. Brinkmann, C. Witt, T. Richter, J. Nieder, B. Koch, R. Hille, R. Hansch and R. R. Mendel, *J. Biol. Chem.*, 2001, **276**, 46989.

327. N. Schrader, K. Fischer, K. Theis, R. R. Mendel, G. Schwarz and C. Kisker, *Structure*, 2003, **11**, 1251.
328. G. Schwarz and R. R. Mendel, *Annu. Rev. Plant Biol.*, 2006, **57**, 623.
329. J. D. Myers and D. J. Kelly, *Microbiology*, 2005, **151**, 233.
330. G. D'Errico, A. Di Salle, F. La Cara, M. Rossi and R. Cannio, *J. Bacteriol.*, 2006, **188**, 694.
331. K. Denger, S. Weinitschke, T. H. M. Smits, D. Schleheck and A. M. Cook, *Microbiology*, 2008, **154**, 256.
332. U. Kappler, C. G. Friedrich, H. G. Truper and C. Dahl, *Arch. Microbiol.*, 2001, **175**, 102.
333. U. Kappler and S. Bailey, *J. Biol. Chem.*, 2005, **280**, 24999.
334. T. Utesch and M. A. Mroginski, *J. Phys. Chem. Lett.*, 2010, **1**, 2159.
335. N. M. Crawford, M. Smith, D. Bellissimo and R. W. Davis, *Proc. Natl. Acad. Sci. U. S. A.*, 1988, **85**, 5006.
336. I. L. Johnstone, P. C. McCabe, P. Greaves, S. J. Gurr, G. E. Cole, M. A. Brow, S. E. Unkles, A. J. Clutterbuck, J. R. Kinghorn and M. A. Innis, *Gene*, 1990, **90**, 181.
337. L. P. Solomonson and M. J. Barber, *Annu. Rev. Plant Mol. Biol.*, 1990, **41**, 225.
338. N. M. Crawford, *Plant Cell*, 1995, **7**, 859.
339. S. C. Huber, M. Bachmann and J. L. Huber, *Trends Plant Sci.*, 1996, **1**, 432.
340. E. H. Campbell, *Annu. Rev. Plant Mol. Biol.*, 1999, **50**, 277.
341. W. M. Kaiser, H. Weiner and S. C. Huber, *Physiol. Plant.*, 1999, **105**, 385.
342. M. Stitt, *Curr. Opin. Plant Biol.*, 1999, **2**, 178.
343. W. H. Campbell, *Cell. Mol. Life Sci.*, 2001, **58**, 194.
344. K. Fischer, G. G. Barbier, H. J. Hecht, R. R. Mendel, W. H. Campbell and G. Schwarz, *Plant Cell*, 2005, **17**, 1167.
345. G. G. Lu, W. H. Campbell, G. Schneider and Y. Lindqvist, *Structure*, 1994, **2**, 809.
346. J. L. Huber, S. C. Huber, W. H. Campbell and M. G. Redinbaugh, *Arch. Biochem. Biophys.*, 1992, **296**, 58.
347. M. Bachmann, N. Shiraishi, W. H. Campbell, B. C. Yoo, A. C. Harmon and S. C. Huber, *Plant Cell*, 1996, **8**, 505.
348. H. C. Kelker and P. Filner, *Biochim. Biophys. Acta*, 1971, **252**, 69.
349. J. R. Magalhaes, F. L. I. M. Silva, I. Salgado, O. Ferrarese-Filho, P. Rockel and W. M. Kaiser, *Physiol. Mol. Biol. Plants*, 2002, **8**, 11.
350. H. C. Oliveira, E. E. Saviani, J. F. P. Oliveira and I. Salgado, *Tropical Plant Pathol.*, 2010, **35**, 104.
351. Y. Sakihama, S. Nakamura and H. Yasmasaki, *Plant Cell Physiol.*, 2002, **43**, 290.
352. Y. Liu, R. Wu, Q. Wan, G. Xie and Y. Bi, *Plant Cell Physiol.*, 2007, **48**, 511.
353. J. Sang, M. Jiang, F. Lin, S. Xu, A. Zhang and M. Tan, *J. Integr. Plant Biol.*, 2008, **50**, 231.

354. N. Srivastava, V. K. Gonugunta, M. R. Puli and A. S. Raghavendra, *Planta*, 2009, **229**, 757.
355. S. J. Wu, J. L. Qi, W. J. Zhang, S. H. Liu, F. H. Xiao, M. S. Zhang, G. H. Xu, W. G. Zhao, M. W. Shi, Y. J. Pang, H. G. Shen and Y. H. Yang, *Plant Cell Physiol.*, 2009, **50**, 118.
356. J. C. Chi, J. Roeper, G. Schwarz and K. Fischer-Schrader, *JBIC, J. Biol. Inorg. Chem.*, 2015, **20**, 277.
357. A. Havemeyer, F. Bittner, S. Wollers, R. Mendel, T. Kunze and B. Clement, *J. Biol. Chem.*, 2006, **281**, 34796.
358. M. Mikula, T. Rubel, J. Karczmarski, K. Goryca, M. Dadlez and J. Ostrowski, *Funct. Integr. Genomics*, 2011, **11**, 215.
359. N. Krompholz, C. Krischkowski, D. Reichmann, D. Garbe-Schönberg, R. R. Mendel, F. Bittner, B. Clement and A. Havemeyer, *Chem. Res. Toxicol.*, 2012, **25**, 2443.
360. S. Gruenewald, B. Wahl, F. Bittner, H. Hungeling, S. Kanzow, J. Kotthaus, U. Schwering, R. R. Mendel and B. Clement, *J. Med. Chem.*, 2008, **51**, 8173.
361. A. Havemeyer, S. Grünewald, B. Wahl, F. Bittner, R. Mendel, P. Erdélyi, J. Fischer and B. Clement, *Drug Metab. Dispos.*, 2010, **38**, 1917.
362. D. Froriep, B. Clement, F. Bittner, R. R. Mendel, D. Reichmann, W. Schmalix and A. Havemeyer, *Xenobiotica*, 2013, **43**, 780.
363. H. H. Jakobs, D. Danilo Froriep, A. Havemeyer, R. R. Mendel, F. Bittner and B. Clement, *ChemMedChem*, 2014, **9**, 2381.
364. J. Kotthaus, B. Wahl, A. Havemeyer, J. Kotthaus, D. Schade, D. Garbe-Schönberg, R. Mendel, F. Bittner and B. Clement, *Biochem. J.*, 2011, **433**, 383.
365. E. P. A. Neve, A. Nordling, T. B. Andersson, U. Hellman, U. Diczfalusy, I. Johansson and M. Ingelman-Sundberg, *J. Biol. Chem.*, 2012, **287**, 6307.
366. B. Wahl, D. Reichmann, D. Niks, N. Krompholz, A. Havemeyer, B. Clement, T. Messerschmidt, M. Rothkegel, H. Biester, R. Hille, R. R. Mendel and F. Bittner, *J. Biol. Chem.*, 2010, **285**, 37847.
367. J. M. Klein, J. D. Busch, C. Potting, M. J. Baker, T. Langer and G. Schwarz, *J. Biol. Chem.*, 2012, **287**, 42795.
368. V. Anantharaman and L. Aravind, *FEMS Microbiol. Lett.*, 2002, **207**, 55.
369. A. Rajapakshe, A. V. Astashkin, E. L. Klein, D. Reichmann, R. R. Mendel, F. Bittner and J. H. Enemark, *Biochemistry*, 2011, **50**, 8813.
370. A. Havemeyer, J. Lang and B. Clement, *Drug Metab. Rev.*, 2011, **43**, 524.
371. B. Plitzko, G. Ott, D. Reichmann, C. J. Henderson, E. Wolf, R. Mendel, F. Bittner, B. Clement and A. Havemeyer, *J. Biol. Chem.*, 2013, **288**, 20228.
372. R. Hille, *Biochim. Biophys. Acta*, 1994, **1184**, 143.
373. M. S. Brody and R. Hille, *Biochim. Biophys. Acta*, 1995, **1253**, 133.
374. M. A. Pietsch and M. B. Hall, *Inorg. Chem.*, 1996, **35**, 1273.
375. M. S. Brody and R. Hille, *Biochemistry*, 1999, **38**, 6668.
376. L. M. Thomson and M. B. Hall, *J. Am. Chem. Soc.*, 2001, **123**, 3995.
377. H. L. Wilson and K. V. Rajagopalan, *J. Biol. Chem.*, 2004, **279**, 15105.

378. B. W. Kail, L. M. Perez, S. D. Zaric, A. J. Millar, C. G. Young, M. B. Hall and P. Basu, *Chem.–Eur. J.*, 2006, **12**, 7501.
379. S. Bailey, T. Rapson, K. Winters-Johnson, A. V. Astashkin, J. H. Enemark and U. Kappler, *J. Biol. Chem.*, 2009, **284**, 2053.
380. J. M. Berg and R. H. Holm, *J. Am. Chem. Soc.*, 1984, **106**, 3035.
381. J. M. Berg and R. H. Holm, *J. Am. Chem. Soc.*, 1985, **107**, 917.
382. J. M. Berg and R. H. Holm, *J. Am. Chem. Soc.*, 1985, **107**, 925.
383. S. A. Roberts, C. G. Young, C. A. Kipke, W. E. Cleland, K. Yamanouchi, M. D. Carducci and J. H. Enemark, *Inorg. Chem.*, 1990, **29**, 3650.
384. V. N. Nemykin and P. Basu, *Inorg. Chem.*, 2005, **44**, 7494.
385. P. Basu, V. N. Nemykin and R. S. Sengar, *Inorg. Chem.*, 2009, **48**, 6303.
386. P. Basu, B. W. Kail and C. G. Young, *Inorg. Chem.*, 2010, **49**, 4895.
387. R. S. Byrne, R. Haensch, R. R. Mendel and R. Hille, *J. Biol. Chem.*, 2009, **284**, 35479.
388. A. A. Belaidi, J. Röper, S. Arjune, S. Krizowski, A. Trifunovic and G. Schwarz, *Biochem. J.*, 2015, **469**, 211.
389. L. Low, J. R. Kilmartin, P. V. Bernhardt and U. Kappler, *Front. Microbiol.*, 2011, **2**, 58.
390. S. Gutteridge, R. C. Bray, B. A. Notton, R. J. Fido and E. J. Hewitt, *Biochem. J.*, 1983, **213**, 137.
391. S. P. Cramer, L. P. Solomonson, M. W. W. Adams and L. E. Mortenson, *J. Am. Chem. Soc.*, 1984, **106**, 1467.
392. L. P. Solomonson, M. J. Barber, W. D. Howard, J. L. Johnson and K. V. Rajagopalan, *J. Biol. Chem.*, 1984, **259**, 849.
393. G. N. George, J. A. Mertens and W. H. Campbell, *J. Am. Chem. Soc.*, 1999, **121**, 9730.
394. J. A. Qiu, H. L. Wilson and K. V. Rajagopalan, *Biochemistry*, 2012, **51**, 1134.
395. G. G. Barbier, R. C. Joshi, E. R. Campbell and W. H. Campbell, *Protein Expression Purif.*, 2004, **37**, 61.
396. L. Loschi, S. J. Brokx, T. L. Hills, G. Zhang, M. G. Bertero, A. L. Lovering, J. H. Weiner and N. C. J. Strynadka, *J. Biol. Chem.*, 2004, **279**, 50391.
397. G. N. George, C. J. Doonan, R. A. Rothery, N. Boroumand and J. H. Weiner, *Inorg. Chem.*, 2007, **46**, 2.
398. S. G. Kozmin, P. Leroy, Y. I. Pavlov and R. M. Schaaper, *Mol. Microbiol.*, 2008, **68**, 51.
399. J. Yang, R. Rothery, J. Sempombe, J. H. Weiner and M. L. Kirk, *J. Am. Chem. Soc.*, 2009, **131**, 15612.
400. S. G. Kozmin, J. Wang and R. M. Schaaper, *J. Bacteriol.*, 2010, **192**, 2026.
401. K. G. V. Havelius, S. Reschke, S. Horn, A. Doerlng, D. Niks, R. Hille, C. Schulzke, S. Leimkuehler and M. Haumann, *Inorg. Chem.*, 2011, **50**, 741.
402. M. J. Pushie, C. J. Doonan, K. Moquin, J. H. Weiner, R. Rothery and G. N. George, *Inorg. Chem.*, 2011, **50**, 732.
403. L. J. Giles, C. Ruppelt, J. Yang, R. R. Mendel, F. Bittner and M. L. Kirk, *Inorg. Chem.*, 2014, **53**, 9460.

404. P. J. Ellis, T. Conrads, R. Hille and P. Kuhn, *Structure*, 2001, **9**, 125.
405. S. Silver and L. T. Phung, *Appl. Environ. Microbiol.*, 2005, **71**, 599.
406. T. P. Warelow, M. Oke, B. Schoepp-Cothenet, J. U. Dahl, N. Bruselat, G. N. Sivalingam, S. Leimkuhler, K. Thalassinos, U. Kappler, J. H. Naismith and J. M. Santini, *PLoS One*, 2013, **8**, e72535.
407. T. Conrads, C. Hemann, G. N. George, I. J. Pickering, R. C. Prince and R. Hille, *J. Am. Chem. Soc.*, 2002, **124**, 11276.
408. J. J. Moura, C. D. Brondino, J. Trincao and M. J. Romao, *JBIC, J. Biol. Inorg. Chem*, 2004, **9**, 791.
409. R. A. Rothery, G. J. Workun and J. H. Weiner, *Biochim. Biophys. Acta*, 2008, **1778**, 1897.
410. N. R. Bastian, C. J. Kay, M. J. Barber and K. V. Rajagopalan, *J. Biol. Chem.*, 1991, **266**, 45.
411. F. Schneider, J. Lowe, R. Huber, H. Schindelin, C. Kisker and J. Knablein, *J. Mol. Biol.*, 1996, **263**, 53.
412. H. K. Li, C. Temple, K. V. Rajagopalan and H. Schindelin, *J. Am. Chem. Soc.*, 2000, **122**, 7673.
413. T. Satoh and F. N. Kurihara, *J. Biochem.*, 1987, **102**, 191.
414. M. J. Barber, H. Vanvalkenburgh, A. J. Trimboli, V. V. Pollock, P. J. Neame and N. R. Bastian, *Arch. Biochem. Biophys.*, 1995, **320**, 266.
415. I. Yamamoto, N. Wada, T. Ujiiye, M. Tachibana, M. Matsuzaki, H. Kajiwara, Y. Watanabe, H. Hirano, A. Okubo, T. Satoh and S. Yamazaki, *Biosci., Biotechnol., Biochem.*, 1995, **59**, 1850.
416. D. Sambasivarao, D. G. Scraba, C. Trieber and J. H. Weiner, *J. Bacteriol.*, 1990, **172**, 5938.
417. J. H. Weiner, R. A. Rothery, D. Sambasivarao and C. A. Trieber, *Biochim. Biophys. Acta*, 1992, **1102**, 1.
418. J. H. Weiner, G. Shaw, R. J. Turner and C. A. Trieber, *J. Biol. Chem.*, 1993, **268**, 3238.
419. R. A. Rothery, N. Kalra, R. J. Turner and J. H. Weiner, *J. Mol. Microbiol. Biotechnol.*, 2002, **4**, 133.
420. N. R. Stanley, F. Sargent, G. Buchanan, J. R. Shi, V. Stewart, T. Palmer and B. C. Berks, *Mol. Microbiol.*, 2002, **43**, 1005.
421. R. Hedderich, O. Klimmek, A. Kroger, R. Dirmeier, M. Keller and K. O. Stetter, *FEMS Microbiol. Rev.*, 1998, **22**, 353.
422. W. Dietrich and O. Klimmek, *Eur. J. Biochem.*, 2002, **269**, 1086.
423. M. Jormakka, K. Yokoyama, T. Yano, M. Tamakoshi, S. Akimoto, T. Shimamura, P. Curmi and S. Iwata, *Nat. Struct. Mol. Biol.*, 2008, **15**, 730.
424. D. J. Richardson, *Microbiology*, 2000, **146**, 551.
425. L. Potter, H. Angove, D. Richardson and J. Cole, *Adv. Microb. Physiol.*, 2001, **45**, 51.
426. J. F. Stolz and P. Basu, *ChemBioChem*, 2002, **3**, 198.
427. P. G. González, C. Correia, I. Moura, C. D. Brondino and J. J. G. Moura, *J. Inorg. Biochem.*, 2006, **100**, 1015.
428. R. M. Martinez-Espinosa, E. J. Dridge, M. J. Bonete, J. N. Butt, C. S. Butler, F. Sargent and D. J. Richardson, *FEMS Microbiol. Lett.*, 2007, **276**, 129.

429. M. Kern and J. Simon, *Microbiology*, 2009, **155**, 2784.
430. E. C. Lowe, S. Bydder, R. S. Hartshorne, H. L. U. Tape, E. J. Dridge, C. M. Debieux, K. Paszkiewicz, I. Singleton, R. J. Lewis, J. M. Santini, D. J. Richardson and C. S. Butler, *J. Biol. Chem.*, 2010, **285**, 18433.
431. H. Dobbek, *Coord. Chem. Rev.*, 2011, **255**, 1104.
432. B. Kraft, M. Strous and H. E. Tegetmeyer, *J. Biotechnol.*, 2011, **155**, 104.
433. M. J. Pushie and G. N. George, *Coord. Chem. Rev.*, 2011, **255**, 1055.
434. C. Sparacino-Watkins, J. F. Stolzb and P. Basu, *Chem. Soc. Rev.*, 2014, **43**, 676.
435. M. G. Bertero, R. A. Rothery, N. Boroumand, M. Palak, F. Blasco, N. Ginet, J. H. Weiner and N. C. J. Strynadka, *J. Biol. Chem.*, 2005, **280**, 14836.
436. J. M. Dias, M. E. Than, A. Humm, R. Huber, G. P. Bourenkov, H. D. Bartunik, S. Bursakov, J. Calvete, J. Caldeira, C. Carneiro, J. J. G. Moura, I. Moura and M. J. Romao, *Structure*, 1999, **7**, 65.
437. S. Najmudin, P. J. Gonzalez, J. Trincao, C. Coelho, A. Mukhopadhyay, N. Cerqueira, C. C. Romao, I. Moura, J. J. G. Moura, C. D. Brondino and M. J. Romao, *JBIC, J. Biol. Inorg. Chem.*, 2008, **13**, 773.
438. C. Coelho, P. J. Gonzalez, J. J. G. Moura, I. Moura, J. Trincao and M. J. Romao, *J. Mol. Biol.*, 2011, **408**, 932.
439. C. Coelho, P. J. Gonzalez, J. Trincão, A. L. Carvalho, S. Najmudin, T. Hettman, S. Dieckman, J. J. G. Moura, I. Moura and M. J. Romão, *Acta Crystallogr., Sect. F*, 2007, **63**, 516.
440. B. J. N. Jepson, S. Mohan, T. A. Clarke, A. J. Gates, J. A. Cole, C. S. Butler, J. N. Butt, A. M. Hemmings and D. J. Richardson, *J. Biol. Chem.*, 2007, **282**, 6425.
441. P. Arnoux, M. Sabaty, J. Alric, B. Frangioni, B. Guigliarelli, J. M. Adriano and D. Pignol, *Nat. Struct. Biol.*, 2003, **10**, 928.
442. B. J. N. Jepson, L. J. Anderson, L. M. Rubio, C. J. Taylor, C. S. Butler, E. Flores, A. Herrero, J. N. Butt and D. J. Richardson, *J. Biol. Chem.*, 2004, **279**, 32212.
443. T. Hartmann, N. Schwanhold and S. Leimkühler, *Biochim. Biophys. Acta*, 2014, **1854**, 1090.
444. L. B. Maia, J. J. G. Moura and I. Moura, *JBIC, J. Biol. Inorg. Chem.*, 2015, **20**, 287.
445. K. Trchounian, A. Poladyan, A. Vassilian and A. Trchounian, *Crit. Rev. Biochem. Mol. Biol.*, 2012, **47**, 236.
446. K. Bagramyan and A. Trchounian, *Biochemistry*, 2003, **68**, 1159.
447. R. W. Jones, A. Lamont and P. B. Garland, *Biochem. J.*, 1980, **190**, 79.
448. B. L. Berg, J. Li, J. Heider and V. Stewart, *J. Biol. Chem.*, 1991, **266**, 22380.
449. F. Blasco, B. Guigliarelli, A. Magalon, M. Asso, G. Giordano and R. A. Rothery, *Cell. Mol. Life Sci.*, 2001, **58**, 179.
450. M. Jormakka, S. Tornroth, B. Byrne and S. Iwata, *Science*, 2002, **295**, 1863.
451. D. Richardson and G. Sawers, *Science*, 2002, **295**, 1842.

452. M. Jormakka, B. Byrne and S. Iwata, *Curr. Opin. Struct. Biol.*, 2003, **13**, 418.
453. M. Jormakka, B. Byrne and S. Iwata, *FEBS Lett.*, 2003, **545**, 25.
454. G. Sawers, J. Heider, E. Zehelein, A. Bock., 1991, 173, 4983.
455. J. Pommier, M. A. Mandrand, S. E. Holt, D. H. Boxer and G. Giodano, *Biochim. Biophys. Acta*, 1992, **1107**, 305.
456. G. Sawers, *Antonie van Leeuwenhoek*, 1994, **66**, 57.
457. H. Abaibou, J. Pommier, S. Benoit, G. Giordano and M. A. Mandrand-berthelot, *J. Bacteriol.*, 1995, **177**, 7141.
458. S. Benoit, H. Abaibou and M. A. Mandrand-Berthelot, *J. Bacteriol.*, 1998, **180**, 6625.
459. F. Zinoni, A. Birkmann, T. C. Stadtman and A. Böck, *Proc. Natl. Acad. Sci. U. S. A.*, 1986, **83**, 4650.
460. M. J. Axley, D. A. Grahame and T. C. Stadtman, *J. Biol. Chem.*, 1990, **265**, 18213.
461. V. N. Gladyshev, J. C. Boyington, S. V. Khangulov, D. A. Grahame, T. C. Stadtman and P. D. Sun, *J. Biol. Chem.*, 1996, **271**, 8095.
462. J. C. Boyington, V. N. Gladyshev, S. V. Khangulov, T. C. Stadtman and P. D. Sun, *Science*, 1997, **275**, 1305.
463. H. C. A. Raaijmakers and M. J. Romao, *JBIC, J. Biol. Inorg. Chem.*, 2006, **11**, 849.
464. R. Thome, A. Gust, R. Toci, R. Mendel, F. Bittner, A. Magalon and A. Walburger, *J. Biol. Chem.*, 2012, **287**, 4671.
465. B. C. Berks, *Mol. Microbiol.*, 1996, **22**, 393.
466. F. Sargent, E. G. Bogsch, N. R. Stanley, M. Wexler, C. Robinson, B. C. Berks and T. Palmer, *EMBO J.*, 1998, **17**, 3640.
467. F. Sargent, N. R. Stanley, B. C. Berks and T. Palmer, *J. Biol. Chem.*, 1999, **274**, 36073.
468. M. Jormakka, S. Tornroth, J. Abramson, B. Byrne and S. Iwata, *Acta Crystallogr., Sect. D: Biol. Crystallogr.*, 2002, **58**, 160.
469. G. Plunkett, V. Burland, D. L. Daniels and F. R. Blattner, *Nucleic Acids Res.*, 1993, **21**, 3391.
470. H. Abaibou, J. Pommier, J. P. Benoit, G. Giordano and M. Mandrand, *J. Bacteriol.*, 1995, **177**, 141.
471. C. Costa, M. Teixeira, J. LeGall, J. J. G. Moura and I. Moura, *JBIC, J. Biol. Inorg. Chem.*, 1997, **2**, 198.
472. M. Rivas, P. Gonzalez, C. D. Brondino, J. J. G. Moura and I. Moura, *J. Inorg. Biol.*, 2007, **101**, 1617.
473. J. Friedebold and B. Bowien, *J. Bacteriol.*, 1993, **175**, 4719.
474. J. Friedebold, F. Mayer, E. Bill, A. X. Trautwein and B. Bowien, *Biol. Chem. Hoppe-Seyler*, 1995, **376**, 561.
475. J. I. Oh and B. Bowien, *J. Biol. Chem.*, 1998, **273**, 26349.
476. T. Hartmann and S. Leimkuhler, *FEBS J.*, 2013, **280**, 6083.
477. B. Schink, *Arch. Microbiol.*, 1985, **142**, 295.
478. B. E. Schultz, R. Hille and R. H. Holm, *J. Am. Chem. Soc.*, 1995, **117**, 827.

479. G. N. George, J. Hilton and K. V. Rajagopalan, *J. Am. Chem. Soc.*, 1996, **118**, 1113.
480. S. D. Garton, J. Hilton, H. Oku, B. R. Crouse, K. V. Rajagopalan and M. K. Johnson, *J. Am. Chem. Soc.*, 1997, **119**, 12906.
481. G. N. George, J. Hilton, C. Temple, R. C. Prince and K. V. Rajagopalan, *J. Am. Chem. Soc.*, 1999, **121**, 1256.
482. K. Nagarajan, H. K. Joshi, P. K. Chaudhury, K. Pal, J. J. A. Cooney, J. H. Enemark and S. Sarkar, *Inorg. Chem.*, 2004, **43**, 4532.
483. T. Prisner, S. Lyubenova, Y. Atabay, F. MacMillan, A. Kroger and O. Klimmek, *J. Biol. Inorg. Chem.*, 2003, **8**, 419.
484. M. Hofmann, *J. Biol. Inorg. Chem.*, 2009, **14**, 1023.
485. N. M. Cerqueira, P. J. Gonzalez, C. D. Brondino, M. J. Romao, C. C. Romao, I. Moura and J. J. G. Moura, *J. Comput. Chem.*, 2009, **30**, 2466.
486. N. M. F. S. A. Cerqueira, P. A. Fernandes, P. J. Gonzalez, J. J. G. Moura and M. J. Ramos, *Inorg. Chem.*, 2013, **52**, 10766.
487. C. S. Mota, M. G. Rivas, C. D. Brondino, I. Moura, J. J. G. Moura, P. J. Gonzalez and N. M. F. S. A. Cerqueira, *J. Biol. Inorg. Chem.*, 2011, **16**, 1255.
488. R. K. Thauer, B. Kaufer and G. Fuchs, *Eur. J. Biochem.*, 1975, **55**, 111.
489. S. V. Khangulov, V. N. Gladyshev, G. C. Dismukes and T. C. Stadtman, *Biochemistry*, 1998, **37**, 3518.
490. M. Tiberti, E. Papaleo, N. Russo, L. Gioia and G. Zampella, *Inorg. Chem.*, 2012, **51**, 8331.
491. P. Schrapers, T. Hartmann, R. Kositzki, H. Dau, S. Reschke, C. Schulzke, S. Leimkühler and M. Haumann, *Inorg. Chem.*, 2015, **54**, 3260.
492. D. Niks, J. Duvvuru, M. Escalona and R. Hille, *J. Biol. Chem.*, 2016, **291**, 1162.
493. L. Maia, L. Fonseca, I. Moura and J. J. G. Moura, Reduction of carbon dioxide by a molybdenum-containing formate dehydrogenase: a kinetic and mechanistic study aiming to inspire a catalyst for carbon dioxide scavenging, submitted.
494. Y. Hu, S. Faham, R. Roy, M. W. W. Adams and D. C. Rees, *J. Mol. Biol.*, 1999, **286**, 899.
495. R. Roy, S. Mukund, G. J. Shut, D. M. Dunn, R. Weiss and M. W. W. Adams, *J. Bacteriol.*, 1999, **181**, 1171.
496. E. Bol, L. E. Bevers, P. L. Hagedoorn and W. R. Hagen, *J. Biol. Inorg. Chem.*, 2006, **11**, 999.
497. M. J. Almendra, C. D. Brondino, O. Gavel, A. S. Pereira, P. Tavares, S. Bursakov, R. Duarte, J. Caldeira, J. J. G. Moura and I. Moura, *Biochemistry*, 1999, **38**, 16366.
498. H. Raaijmakers, S. Teixeira, J. M. Dias, M. J. Almendra, C. D. Brondino, I. Moura, J. J. G. Moura and M. J. Romao, *J. Biol. Inorg. Chem.*, 2001, **6**, 398.
499. H. Raaijmakers, S. Macieira, J. M. Dias, S. Teixeira, S. Bursakov, R. Huber, J. J. G. Moura, I. Moura and M. J. Romao, *Structure*, 2002, **10**, 1261.

500. C. S. Mota, O. Valette, P. J. Gonzalez, C. D. Brondino, J. J. G. Moura, I. Moura, A. Dolla and M. G. Rivas, *J. Bacteriol.*, 2011, **193**, 2917.
501. C. D. Brondino, M. C. G. Passeggi, J. Caldeira, M. J. Almendra, M. J. Feio, J. J. G. Moura and I. Moura, *J. Biol. Inorg. Chem.*, 2004, **9**, 154.
502. J. F. Heidelberg, R. Seshadri, S. A. Haveman, C. L. Hemme, I. T. Paulsen, J. F. Kolonay, J. A. Eisen, N. Ward, B. Methe, L. M. Brinkac, S. C. Daugherty, R. T. Deboy, R. J. Dodson, A. S. Durkin, R. Madupu, W. C. Nelson, S. A. Sullivan, D. Fouts, D. H. Haft, J. Selengut, J. D. Peterson, T. M. Davidsen, N. Zafar, L. Zhou, D. Radune, G. Dimitrov, M. Hance, K. Tran, H. Khouri, J. Gill, T. R. Utterback, T. V. Feldblyum, J. D. Wall, G. Voordouw and C. M. Fraser, *Nat. Biotechnol.*, 2004, **22**, 554.
503. S. M. da Silva, J. Voordouw, C. Leita, M. Martins, G. Voordouw and I. A. C. Pereira, *Microbiology*, 2013, **159**, 1760.
504. S. M. Silva, C. Pimentel, F. M. A. Valente, C. Rodrigues-Pousada and I. A. C. Pereira, *J. Bacteriol.*, 2011, **193**, 2909.
505. J. C. Deaton, E. I. Solomon, G. D. Watt, P. J. Wetherbee and C. N. Durfor, *Biochem. Biophys. Res. Commun.*, 1987, 149, 424.
506. U. Leonhardt and J. R. Andreesen, *Arch. Microbiol.*, 1977, **115**, 277.
507. A. Alissandratos, H. K. Kim, H. Matthews, J. E. Hennessy, A. Philbrook and C. J. Easton, *Appl. Environ. Microbiol.*, 2013, **79**, 741.
508. J. W. Kung, C. Löffler, K. Dörner, D. Heintz, S. Gallien, A. V. Dorsselaer, T. Friedrich and M. Boll, *Proc. Natl. Acad. Sci. U. S. A.*, 2009, **106**, 17687.
509. J. W. Kung, S. Baumann, M. von Bergen, M. Müller, P. L. Hagedoorn, W. R. Hagen and M. Boll, *J. Am. Chem. Soc.*, 2010, **132**, 9850.
510. M. Boll, C. Löffler, B. E. Morris and J. W. Kung, *Environ. Microbiol.*, 2014, **16**, 612.
511. T. Weinert, S. G. Huwiler, J. W. Kung, S. Weidenweber, P. Hellwig, H. J. Stärk, T. Biskup, S. Weber, J. J. H. Cotelesage, G. N. George, U. Ermler and M. Boll, *Nat. Chem. Biol.*, 2015, **11**, 586.
512. S. Wischgoll, D. Heintz, F. Peters, A. Erxleben, E. Sarnighausen, R. Reski, A. Van Dorsselaer and M. Boll, *Mol. Microbiol.*, 2005, **58**, 1238.
513. B. M. Rosner and B. Schink, *J. Bacteriol.*, 1995, **177**, 5767.
514. F. Brink, *Met. Ions Life Sci.*, 2014, **14**, 15.
515. O. Einsle, H. Niessen, D. J. Abt, G. Seiffert, B. Schink, R. Huber, A. Messerschmidt and P. M. H. Kroneck, *Acta Crystallogr., Sect. F*, 2005, **61**, 299.
516. G. N. Seiffert, G. M. Ullmann, A. Messerschmidt, B. Schink, P. M. H. Kroneck and O. Einsle, *Proc. Natl. Acad. Sci. U. S. A.*, 2007, **104**, 3073.
517. R. U. Meckenstock, R. Krieger, S. Ensign, P. M. H. Kroneck and B. Schink, *Eur. J. Biochem.*, 1999, **264**, 176.
518. E. Bol, N. J. Broers and W. R. Hagen, *J. Biol. Inorg. Chem.*, 2008, **13**, 75.
519. R. Z. Liao, *J. Biol. Inorg. Chem.*, 2013, **18**, 175.
520. Y. F. Liu, R. Z. Liao, W. J. Ding, J. G. Yu and R. Z. Liu, *J. Biol. Inorg. Chem.*, 2011, **16**, 745.
521. M. W. W. Adams, *J. Appl. Microbiol.*, 1998, **85**, 108S.
522. A. M. Sevcenco, M. W. H. Pinkse, E. Bol, G. C. Krijger, H. T. Wolterbeek, P. Verhaert, P. L. Hagedoorn and W. R. Hagen, *Metallomics*, 2009, **1**, 395.

523. R. K. Thauer, K. Jungermann and K. Decker, *Bacteriol. Rev.*, 1977, **41**, 100.
524. R. K. Thauer, *Biochem. Biophys Acta*, 1990, **1018**, 1852.
525. S. Afshar, E. Johnson, S. de Vries and I. Schroder, *J. Bacteriol.*, 2001, **183**, 5491.
526. S. de Vries, M. Momcilovic, M. J. F. Strampraad, J. P. Whitelegge, A. Baghai and I. Schroeder, *Biochemistry*, 2010, **49**, 9911.
527. J. Buc, C. L. Santini, R. Giordani, M. Czjzek, L. F. Wu and G. Giordano, *Mol. Microbiol.*, 1999, **32**, 159.
528. L. E. Bevers, P. L. Hagedoorn and W. R. Hagen, *Coord. Chem. Rev.*, 2009, **253**, 269.
529. A. M. Sevcenco, L. E. Bevers, M. W. H. Pinkse, G. C. Krijger, H. T. Wolterbeek, P. D. E. M. Verhaert, W. R. Hagen and P. L. Hagedoorn, *J. Bacteriol.*, 2010, **192**, 4143.
530. L. J. Stewart, S. Bailey, B. Bennett, J. M. Charnock, C. D. Garner and A. S. McAlpine, *J. Mol. Biol.*, 2000, **299**, 593.
531. J. Christiansen, D. R. Dean and L. C. Seefeldt, *Annu. Rev. Plant Physiol. Plant Mol. Biol.*, 2001, **52**, 269.
532. D. M. Lawson and B. E. Smith, *Met. Ions Biol. Syst.*, 2002, **39**, 75.
533. R. R. Eady, *Met. Ions Biol. Syst.*, 1995, **31**, 363.
534. Y. Hu, C. C. Lee and M. W. Ribbe, *Science*, 2011, **333**, 753.
535. Z. Y. Yang, D. R. Dean and L. C. Seefeldt, *J. Biol. Chem.*, 2011, **286**, 19417.
536. Y. Hu and M. W. Ribbe, *Biochim. Biophys. Acta*, 2013, **1827**, 1112.
537. Z.-Y. Yang, V. R. Moure, D. R. Dean and L. C. Seefeldt, *Proc. Natl. Acad. Sci. U. S. A.*, 2012, **109**, 19644.
538. L. C. Seefeldt, Z.-Y. Yang, S. Duval and D. R. Dean, *Biochim. Biophys. Acta*, 2013, **1827**, 1102.
539. L. C. Seefeldt, M. E. Rasche and S. A. Ensign, *Biochemistry*, 1995, **34**, 5382.
540. B. K. Burgess, *Chem. Rev.*, 1996, **96**, 2983.
541. M. E. Rasche and L. C. Seefeldt, *Biochemistry*, 1997, **36**, 8574.
542. K. Fisher, M. J. Dilworth, C.-H. Kim and W. E. Newton, *Biochemistry*, 2000, **39**, 10855.
543. S. M. Mayer, W. G. Niehaus and D. R. Dean, *J. Chem. Soc., Dalton Trans.*, 2002, 802.
544. R. Y. Igarashi, D. Santos, P. C. , W. G. Niehaus, I. G. Dance, D. R. Dean and L. C. Seefeldt, *J. Biol. Chem.*, 2004, **279**, 34770.
545. D. Santos, P. C. , R. Y. Igarashi, H.-I. Lee, B. M. Hoffman, L. C. Seefeldt and D. R. Dean, *Acc. Chem. Res.*, 2005, **38**, 208.
546. B. M. Barney, J. McClead, D. Lukoyanov, M. Laryukhin, T.-C. Yang, D. R. Dean, B. M. Hoffman and L. C. Seefeldt, *Biochemistry*, 2007, **46**, 6784.
547. P. C. Dos Santos, S. M. Mayer, B. M. Barney, L. C. Seefeldt and D. R. Dean, *J. Inorg. Biochem.*, 2007, **101**, 1642.
548. S. Shaw, D. Lukoyanov, K. Danyal, D. R. Dean, B. M. Hoffman and L. C. Seefeldt, *J. Am. Chem. Soc.*, 2014, **136**, 12776.

549. C. C. Jodin, *R. Acad. Paris*, 1862, **55**, 612.
550. D. Burk, *Ergeb. Enzymforsch.*, 1934, **3**, 23.
551. D. Burk, H. Lineweaver and C. K. Horner, *J. Bacteriol.*, 1934, **27**, 325.
552. M. M. Georgiadis, H. Komiya, P. Chakrabarti, D. Woo, J. J. Kornuc and D. C. Rees, *Science*, 1992, **257**, 1653.
553. J. Kim and D. C. Rees, *Science*, 1992, **257**, 1677.
554. J. Kim and D. C. Rees, *Nature*, 1992, **360**, 553.
555. M. K. Chan, J. Kim and D. C. Rees, *Science*, 1993, **260**, 792.
556. J. Kim, D. Woo and D. C. Rees, *Biochemistry*, 1993, **32**, 7104.
557. J. B. Howard and D. C. Rees, *Chem. Rev.*, 1996, **96**, 2965.
558. J. W. Peters, M. H. B. Stowell, S. M. Soltis, M. G. Finnegan, M. K. Johnson and D. C. Rees, *Biochemistry*, 1997, **36**, 1181.
559. H. Schindelin, C. Kisker, J. L. Schlessman, J. B. Howard and D. C. Rees, *Nature*, 1997, **387**, 370.
560. S. M. Mayer, D. M. Lawson, C. A. Gormal, S. M. Roe and B. E. Smith, *J. Mol. Biol.*, 1999, **292**, 871.
561. S. B. Jang, L. C. Seefeldt and J. W. Peters, *Biochemistry*, 2000, **39**, 641.
562. S. B. Jang, L. C. Seefeldt and J. W. Peters, *Biochemistry*, 2000, **39**, 14745.
563. H.-J. Chiu, J. W. Peters, W. N. Lanzilotta, M. J. Ryle, L. C. Seefeldt, J. B. Howard and D. C. Rees, *Biochemistry*, 2001, **40**, 641.
564. M. Sørlie, J. Christiansen, B. J. Lemon, J. W. Peters, D. R. Dean and B. J. Hales, *Biochemistry*, 2001, **40**, 1540.
565. P. Strop, P. M. Takahara, H.-J. Chiu, H. C. Angove, B. K. Burgess and D. C. Rees, *Biochemistry*, 2001, **40**, 651.
566. O. Einsle, F. A. Tezcan, S. L. A. Andrade, B. Schmid, M. Yoshida, J. B. Howard and D. C. Rees, *Science*, 2002, **297**, 1696.
567. B. Schmid, O. Einsle, H.-J. Chiu, A. Willing, M. Yoshida, J. B. Howard and D. C. Rees, *Biochemistry*, 2002, **41**, 15557.
568. B. Schmid, M. W. Ribbe, O. Einsle, M. Yoshida, L. M. Thomas, D. R. Dean, D. C. Rees and B. K. Burgess, *Science*, 2002, **296**, 352.
569. S. B. Jang, M. S. Jeong, L. C. Seefeldt and J. W. Peters, *J. Biol. Inorg. Chem.*, 2004, **9**, 1028.
570. M. S. Jeong, *Mol. Cells*, 2004, **18**, 374.
571. S. Sen, R. Igarashi, A. Smith, M. K. Johnson, L. C. Seefeldt and J. W. Peters, *Biochemistry*, 2004, **43**, 1787.
572. F. A. Tezcan, J. T. Kaiser, D. Mustafi, M. Y. Walton, J. B. Howard and D. C. Rees, *Science*, 2005, **309**, 1377.
573. S. Sen, A. Krishnakumar, J. McClead, M. K. Johnson, L. C. Seefeldt, R. K. Szilagyi and J. W. Peters, *J. Inorg. Biochem.*, 2006, **100**, 1041.
574. R. Sarma, D. W. Mulder, E. Brecht, R. K. Szilagyi, L. C. Seefeldt, H. Tsuruta and J. W. Peters, *Biochemistry*, 2007, **46**, 14058.
575. R. Sarma, B. M. Barney, S. Keable, D. R. Dean, L. C. Seefeldt and J. W. Peters, *J. Inorg. Biochem.*, 2010, **104**, 385.
576. K. M. Lancaster, M. Roemelt, P. Ettenhuber, Y. Hu, M. W. Ribbe, F. Neese, B. Bergmann and S. DeBeer, *Science*, 2011, **334**, 974.

577. T. Spatzal, M. Aksoyoglu, L. Zhang, S. L. Andrade, E. Schleicher, S. Weber, D. C. Rees and O. Einsle, *Science*, 2011, **334**, 940.
578. O. Einsle, *J. Biol. Inorg. Chem.*, 2014, **19**, 737.
579. T. Spatzal, K. A. Perez, O. Einsle, J. B. Howard and D. C. Rees, *Science*, 2014, **345**, 1620.
580. D. R. Dean, J. T. Bolin and L. Zheng, *J. Bacteriol.*, 1993, **175**, 6737.
581. D. J. Lowe, K. Fisher and R. N. F. Thorneley, *Biochem. J.*, 1993, **292**, 93.
582. J. W. Peters, K. Fisher, W. E. Newton and D. R. Dean, *J. Biol. Chem.*, 1995, **270**, 27007.
583. L. Ma, M. A. Brosius and B. K. Burgess, *J. Biol. Chem.*, 1996, **271**, 10528.
584. G. D. Watt, A. Burns, S. Lough and D. L. Tennent, *Biochemistry*, 1980, **19**, 4926.
585. K. K. Surerus, M. P. Hendrich, P. D. Christie, D. Rottgardt, W. H. Orme-Johnson and E. Münck, *J. Am. Chem. Soc.*, 1992, **114**, 8579.
586. A. J. Pierik, H. Wassink, H. Haaker and W. R. Hagen, *Eur. J. Biochem.*, 1993, **212**, 51.
587. V. K. Shah and W. J. Brill, *Proc. Natl. Acad. Sci. U. S. A.*, 1977, **74**, 3249.
588. B. K. Burgess, *Chem. Rev.*, 1990, **90**, 1377.
589. S. P. Cramer, W. O. Gillum, K. O. Hodgson, L. E. Mortenson, E. I. Stiefel, J. R. Chisnell, W. J. Brill and V. K. Shah, *J. Am. Chem. Soc.*, 1978, **100**, 3814.
590. S. P. Cramer, K. O. Hodgson, W. O. Gillum and L. E. Mortenson, *J. Am. Chem. Soc.*, 1978, **100**, 3398.
591. S. J. Yoo, H. C. Angove, V. Papaefthymiou, B. K. Burgess and E. Münck, *J. Am. Chem. Soc.*, 2000, **122**, 4926.
592. T. Lovell, R. A. Torres, W. Han, T. Liu, D. A. Case and L. Noodleman, *Inorg. Chem.*, 2002, **41**, 5744.
593. T. T. Lovell, T. T. Liu, D. A. Case and L. Noodleman, *J. Am. Chem. Soc.*, 2003, **125**, 8377.
594. L. C. Seefeldt and B. M. Hoffman, *Inorg. Chem.*, 2007, **46**, 11437.
595. I. Dance, *Inorg. Chem.*, 2011, **50**, 178.
596. T. V. Harris and R. K. Szilagyi, *Inorg. Chem.*, 2011, **50**, 4811.
597. R. Bjornsson, F. A. Lima, T. Spatzal, T. Weyhermueller, P. Glatzel, O. Einsle, F. Neese and S. DeBeer, *Chem. Sci.*, 2014, **5**, 3096.
598. R. Bjornsson, F. Neese, R. R. Schrock, O. Einsle and S. DeBeer, *J. Biol. Inorg. Chem.*, 2015, **20**, 447.
599. K. M. Lancaster, Y. Hu, U. Bergmann, M. W. Ribbe and S. DeBeer, *J. Am. Chem. Soc.*, 2013, **135**, 610.
600. J. A. Wiig, Y. Hu, C. C. Lee and M. W. Ribbe, *Science*, 2012, **337**, 1672.
601. J. A. Wiig, C. C. Lee, Y. Hu and M. W. Ribbe, *J. Am. Chem. Soc.*, 2013, **135**, 4982.
602. H.-I. Lee, P. M. C. Benton, M. Laryukhin, R. Y. Igarashi, D. R. Dean, L. C. Seefeldt and B. M. Hoffman, *J. Am. Chem. Soc.*, 2003, **125**, 5604.
603. T.-C. Yang, N. K. Maeser, M. Laryukhin, H.-I. Lee, D. R. Dean, L. C. Seefeldt and B. M. Hoffman, *J. Am. Chem. Soc.*, 2005, **127**, 12804.

604. D. Lukoyanov, V. Pelmenschikov, N. Maeser, M. Laryukhin, T. C. Yang, L. Noodleman, D. R. Dean, D. A. Case, L. C. Seefeldt and B. M. Hoffman, *Inorg. Chem.*, 2007, **46**, 11437.
605. M. E. Moret and J. C. Peters, *J. Am. Chem. Soc.*, 2011, **133**, 18118.
606. J. G. Rebelein, Y. Hu and M. W. Ribbe, *Angew. Chem., Int. Ed.*, 2014, **53**, 11543.
607. L. E. Mortenson, *Biochim. Biophys. Acta*, 1964, **81**, 473.
608. L. E. Mortenson, *Proc. Natl. Acad. Sci. U. S. A.*, 1964, **52**, 272.
609. W. A. Bulen, R. C. Burns and J. R. LeComte, *Proc. Natl. Acad. Sci. U. S. A.*, 1965, **53**, 532.
610. R. C. Burns and W. A. Bulen, *Biochim. Biophys. Acta*, 1965, **105**, 437.
611. R. V. Hageman and R. H. Burris, *Proc. Natl. Acad. Sci. U. S. A.*, 1978, **75**, 2699.
612. L. Chen, N. Gavini, H. Tsuruta, D. Eliezer, B. K. Burgess, S. Doniach and K. O. Hodgson, *J. Biol. Chem.*, 1994, **269**, 3290.
613. F. Haber, *Naturwissenschaften*, 1922, **10**, 1041.
614. F. Haber, *Naturwissenschaften*, 1923, **11**, 339.
615. R. Schlögl, *Angew. Chem., Int. Ed.*, 2003, **42**, 2004.
616. M. G. Duyvis, H. Wassink and H. Haaker, *FEBS Lett.*, 1996, **380**, 233.
617. L. C. Seefeldt and D. R. Dean, *Acc. Chem. Res.*, 1997, **30**, 260.
618. H. C. Angove, S. J. Yoo, E. Münck and B. K. Burgess, *J. Biol. Chem.*, 1998, **273**, 26330.
619. M. G. Duyvis, H. Wassink and H. Haaker, *Biochemistry*, 1998, **37**, 17345.
620. W. N. Lanzilotta, J. Christiansen, D. R. Dean and L. C. Seefeldt, *Biochemistry*, 1998, **37**, 11376.
621. J. M. Chan, J. Christiansen, D. R. Dean and L. C. Seefeldt, *Biochemistry*, 1999, **38**, 5779.
622. J. M. Chan, M. J. Ryle and L. C. Seefeldt, *J. Biol. Chem.*, 1999, **274**, 17593.
623. W. N. Lanzilotta, V. D. Parker and L. C. Seefeldt, *Biochim. Biophys. Acta*, 1999, **1429**, 411.
624. A. C. Nyborg, J. L. Johnson, A. Gunn and G. D. Watt, *J. Biol. Chem.*, 2000, **275**, 39307.
625. M. J. Ryle and L. C. Seefeldt, *J. Biol. Chem.*, 2000, **275**, 6214.
626. P. E. Wilson, A. C. Nyborg and G. D. Watt, *Biophys. Chem.*, 2001, **91**, 281.
627. B. M. Barney, H.-I. Lee, D. Santos, P. C. , B. M. Hoffman, D. R. Dean and L. C. Seefeldt, *Dalton Trans.*, 2006, 2277.
628. J. B. Howard and D. C. Rees, *Proc. Natl. Acad. Sci. U. S. A.*, 2006, **103**, 17088.
629. J. W. Peters and R. K. Szilagyi, *Curr. Opin. Chem. Biol.*, 2006, **10**, 101.
630. B. M. Hoffman, D. R. Dean and L. C. Seefeldt, *Acc. Chem. Res.*, 2009, **42**, 609.
631. L. C. Seefeldt, B. M. Hoffman and D. R. Dean, *Annu. Rev. Biochem.*, 2009, **78**, 701.

632. K. Danyal, B. S. Inglet, K. A. Vincent, B. M. Barney, B. M. Hoffman, F. A. Armstrong, D. R. Dean and L. C. Seefeldt, *J. Am. Chem. Soc.*, 2010, **132**, 13197.
633. K. Danyal, D. Mayweather, D. R. Dean, L. C. Seefeldt and B. M. Hoffman, *J. Am. Chem. Soc.*, 2010, **132**, 6894.
634. L. E. Roth, J. C. Nguyen and F. A. Tezcan, *J. Am. Chem. Soc.*, 2010, **132**, 13672.
635. K. Danyal, D. R. Dean, B. M. Hoffman and L. C. Seefeldt, *Biochemistry*, 2011, **50**, 9255.
636. L. E. Roth and F. A. Tezcan, *ChemCatChem*, 2011, **3**, 1549.
637. D. Lukoyanov, Z. Y. Yang, B. M. Barney, D. R. Dean, L. C. Seefeldt and B. M. Hoffman, *Proc. Natl. Acad. Sci. U. S. A.*, 2012, **109**, 5583.
638. D. Mayweather, K. Danyal, D. R. Dean, L. C. Seefeldt and B. M. Hoffman, *Biochemistry*, 2012, **51**, 8391.
639. L. E. Roth and F. A. Tezcan, *J. Am. Chem. Soc.*, 2012, **134**, 8416.
640. K. Rupnik, Y. Hu, C. C. Lee, J. A. Wiig, M. W. Ribbe and B. J. Hales, *J. Am. Chem. Soc.*, 2012, **134**, 13749.
641. L. C. Seefeldt, B. M. Hoffman and D. R. Dean, *Curr. Opin. Chem. Biol.*, 2012, **16**, 19.
642. S. Duval, K. Danyal, S. Shaw, A. K. Lytle, D. R. Dean, B. M. Hoffman, E. Antony and L. C. Seefeldt, *Proc. Natl. Acad. Sci. U. S. A.*, 2013, **110**, 16414.
643. B. M. Hoffman, D. Lukoyanov, D. R. Dean and L. C. Seefeldt, *Acc. Chem. Res.*, 2013, **46**, 587.
644. T. Spatzal, O. Einsle and S. L. Andrade, *Angew. Chem., Int. Ed. Engl.*, 2013, **52**, 10116.
645. L. Zhang, J. T. Kaiser, G. Meloni, K. Y. Yang, T. Spatzal, S. L. Andrade, O. Einsle, J. B. Howard and D. C. Rees, *Angew Chem., Int. Ed. Engl.*, 2013, **52**, 10529.
646. A. D. Scott, V. Pelmenschikov, Y. Guo, L. Yan, H. Wang, S. J. George, C. H. Dapper, W. E. Newton, Y. Yoda, Y. Tanaka and S. P. Cramer, *J. Am. Chem. Soc.*, 2014, **136**, 15942.
647. M. C. Durrant, *Biochem. J.*, 2001, **355**, 569.
648. T. Lovell, J. Li, T. Liu, D. A. Case and L. Noodleman, *J. Am. Chem. Soc.*, 2001, **123**, 12392.
649. M. C. Durrant, *Biochemistry*, 2002, **41**, 13934.
650. M. C. Durrant, *Biochemistry*, 2002, **41**, 13946.
651. Z. Cui, A. J. Dunford, M. C. Durrant, R. A. Henderson and B. E. Smith, *Inorg. Chem.*, 2003, **42**, 6252.
652. B. Hinnemann and J. K. Nørskov, *J. Am. Chem. Soc.*, 2003, **125**, 1466.
653. R. A. Torres, T. Lovell, L. Noodleman and D. A. Case, *J. Am. Chem. Soc.*, 2003, **125**, 1923.
654. B. Hinnemann and J. K. Nørskov, *J. Am. Chem. Soc.*, 2004, **126**, 3920.
655. B. Hinnemann and J. K. Nørskov, *Phys. Chem. Chem. Phys.*, 2004, **6**, 843.

656. R. N. F. Thorneley, H. C. Angove, M. C. Durrant, S. A. Fairhurst, S. J. George, A. Sinclair, J. D. Tolland and P. C. Hallenbeck, *J. Inorg. Biochem.*, 2003, **96**, 18.
657. M. C. Durrant, A. Francis, D. J. Lowe, W. E. Newton and K. Fisher, *Biochem. J.*, 2006, **397**, 261.
658. B. Hinnemann and J. K. Nørskov, *Top. Catal.*, 2006, **37**, 55.
659. F. Neese, *Angew. Chem., Int. Ed.*, 2006, **45**, 196.
660. I. Dance, *J. Am. Chem. Soc.*, 2007, **129**, 1076.
661. J. Kästner and P. E. Blöchl, *J. Am. Chem. Soc.*, 2007, **129**, 2998.
662. V. Pelmenschikov, D. A. Case and L. Noodleman, *Inorg. Chem.*, 2008, **47**, 6162.
663. J. B. Varley and J. K. Nørskov, *ChemCatChem*, 2013, **5**, 732.
664. Z.-Y. Yang, N. Khadka, D. Lukoyanov, B. M. Hoffman, D. R. Dean and L. C. Seefeldt, *Proc. Natl. Acad. Sci. U. S. A.*, 2013, **110**, 16327.
665. Y. Hu and M. W. Ribbe, *J. Biol. Inorg. Chem.*, 2014, **19**, 731.
666. D. Lukoyanov, Z. Y. Yang, N. Khadka, D. R. Dean, L. C. Seefeldt and B. Hoffman, *J. Am. Chem. Soc.*, 2015, **137**, 3610.
667. S. Siemann, K. Schneider, M. Oley and A. Muller, *Biochemistry*, 2003, **42**, 3846.
668. B. M. Barney, R. Y. Igarashi, P. C. Dos Santos, D. R. Dean and L. C. Seefeldt, *J. Biol. Chem.*, 2004, **279**, 53621.
669. G. N. George, I. J. Pickering, E. Y. Yu, R. C. Prince, S. A. Bursakov, O. Y. Gavel, I. Moura and J. J. G. Moura, *J. Am. Chem. Soc.*, 2000, **122**, 8321.
670. S. A. Bursakov, O. Y. Gavel, G. Di Rocco, J. Lampreia, J. Calvete, A. S. Pereira, J. J. Moura and I. Moura, *J. Inorg. Biochem.*, 2004, **98**, 833.
671. M. G. Rivas, M. S. Carepo, C. S. Mota, M. Korbas, M. C. Durand, A. T. Lopes, C. D. Brondino, A. S. Pereira, G. N. George, A. Dolla, J. J. Moura and I. Moura, *Biochemistry*, 2009, **48**, 873.
672. M. S. Carepo, S. R. Pauleta, A. G. Wedd, J. J. G. Moura and I. Moura, *J. Biol. Inorg. Chem.*, 2014, **19**, 605.
673. A. Cvetkovic, A. L. Menon, M. P. Thorgersen, J. W. Scott, F. L. Poole II, F. E. Jenney, W. A. Lancaster, J. L. Praissman, S. Shanmukh, B. J. Vaccaro, S. A. Trauger, E. Kalisiak, J. V. Apon, G. Siuzdak, S. M. Yannone, J. A. Tainer and M. W. W. Adams, *Nature*, 2010, **466**, 779.
674. S. R. Pauleta, A. G. Duarte, M. S. Carepo, A. S. Pereira, P. Tavares, I. Moura and J. J. G. Moura, *Biomol. NMR Assignments*, 2007, **1**, 81.
675. Source: http://www.webelements.com.
676. N. I. Schauer and J. F. Ferry, *J. Bacteriol.*, 1982, **150**, 1.
677. M. J. Barber, L. M. Siege, N. L. Schauer, H. D. May and J. G. Ferry, *J. Biol. Chem.*, 1983, **258**, 10839.
678. N. I. Schauert and J. G. Ferry, *J. Bacteriol.*, 1986, **165**, 405.
679. A. P. Shuber, E. C. Orr, M. A. Recny, P. Schendel, H. D. May, N. L. Schauer and J. G. Ferry, *J. Biol. Chem.*, 1986, **261**, 12942.
680. J. L. Johnson, N. R. Bastian, N. L. Schauer, J. G. Ferry and K. V. Rajagopalan, *FEMS Microbiol. Lett.*, 1991, **77**, 213.

681. K. Noolling and J. N. Reeve, *J. Bacteriol.*, 1997, **179**, 899.
682. M. Szaleniec, T. Borowski, K. Schuehle, M. Witko and J. Heider, *J. Am. Chem. Soc.*, 2010, **132**, 6014.
683. M. Szaleniec, A. Salwinski, T. Borowski, J. Heider and M. Witko, *Int. J. Quantum Chem.*, 2012, **112**, 1990.
684. M. Szaleniec, A. Dudzik, B. Kozik, T. Borowski, J. Heider and M. Witko, *J. Inorg. Biochem.*, 2014, **139**, 9.
685. P. Kalimuthu, J. Heider, D. Knack and P. V. Bernhardt, *J. Phys. Chem. B*, 2015, **119**, 3456.
686. M. Czjzek, D. Santos, J. P. , J. Pommier, G. Giordano, V. Mejean and R. Haser, *J. Mol. Biol.*, 1998, **284**, 435.
687. L. Zhang, K. J. Nelson, K. V. Rajagopalan and G. N. George, *Inorg. Chem.*, 2008, **47**, 1074.

CHAPTER 2

Abundance, Ubiquity and Evolution of Molybdoenzymes

VADIM N. GLADYSHEV[a] AND YAN ZHANG*[b,c]

[a]Division of Genetics, Department of Medicine, Brigham and Women's Hospital, Harvard Medical School, Boston, MA 02115, USA; [b]College of Life Sciences and Oceanography, Shenzhen University, Shenzhen, 518060, Guangdong Province, P. R. China; [c]Institute for Nutritional Sciences, Shanghai Institutes for Biological Sciences, Chinese Academy of Sciences, Shanghai 200031, P. R. China
*E-mail: yzhang0101@hotmail.com

2.1 Introduction

Molybdenum (Mo) is an essential transition metal for organisms in all three domains of life.[1–3] It is a key component of the active site of molybdoenzymes that catalyze transfer reactions in the metabolism of carbon-, nitrogen- and sulfur-containing compounds.[2–7] With the exception of nitrogenase that contains a unique iron(Fe)–Mo cofactor,[8,9] all known molybdoenzymes use the Mo cofactor (Moco) where Mo is coordinated to an organic tricyclic pyranopterin moiety.[1–6,10–12] In many microorganisms, tungsten (W) is found to be coordinated by the pyranopterin in place of Mo giving rise to tungstoenzymes.[13–15] It has now become clear that W is an essential element for nearly all enzymes of the aldehyde : ferredoxin oxidoreductase (AOR) family.[14–16] Moreover, it can be selectively transported into some prokaryotic cells by certain ABC-types of transporters.[17,18] However, due to the similar ionic radii

RSC Metallobiology Series No. 5
Molybdenum and Tungsten Enzymes: Biochemistry
Edited by Russ Hille, Carola Schulzke, and Martin L. Kirk

Published by the Royal Society of Chemistry, www.rsc.org

and chemical properties, it is often difficult to distinguish the use of Mo and W by certain molybdoenzyme families. In this chapter, the term Moco refers to the utilization of both elements if not mentioned specifically.

More than 50 molybdenum- and tungsten-containing enzymes have been identified in bacteria, archaea, algae, land plants, fungi and animals.[2] On the basis of sequence similarity and structural properties, these enzymes can be categorized and divided into four large and non-overlapping families: sulfite oxidase (SO), xanthine oxidase (XO), dimethylsulfoxide reductase (DMSOR) and W-containing AOR families.[2–7] Each family is further classified into a variety of subfamilies based on substrate preferences.[19,20] It has been reported that members of all four families can be found in different prokaryotes, whereas eukaryotes only contain members of the SO and XO families.[10,21] Recently, a novel molybdoenzyme family has been found in both eukaryotes (named mARC, mitochondrial amidoxime reducing component) and bacteria (YcbX and YiiM).[22–24] These most recently identified Moco-binding proteins share a significant degree of homology with the C-terminal Moco-binding domain of eukaryotic Moco sulfurase (MOSC) and show catalytic activity strictly dependent on Moco.

The proper functions of molybdoenzymes also depend on additional gene products that transport molybdate anions into cells and synthesize Moco. Thus, identification of the Mo transport systems and the Moco biosynthesis pathway is important for understanding Mo utilization in extant organisms. In prokaryotes, molybdate may be taken up through three high-affinity ATP binding-cassette (ABC) transport systems: ModABC, WtpABC and TupABC.[17,18,25,26] However, the molecular nature of Mo transporters is not well understood in eukaryotes. MOT1 is the first high-affinity molybdate transporter family identified in plant-type eukaryotic organisms such as land plants and green algae, but it is absent in animals.[27,28] Cellular localization of MOT1 is not well defined as different cellular compartments (plasma membrane, mitochondria and vacuole) have been reported for this protein.[28–30] Recently, a second type of molybdate transporter, MOT2, has been identified in algae and animals including humans, which might be related to the uptake of Mo from extracellular milieu.[31,32] On the other hand, Moco is synthesized in an ancient and ubiquitous multi-step pathway, which is similar in prokaryotes and eukaryotes. Details of the biosynthetic steps and genes involved in Moco synthesis have been described in many review articles[1–7,10–12] and other chapters in this book.

Recent advances in next-generation sequencing have led to a dramatic increase in the production of genomic sequences, which are now available for a large number of organisms from all three domains of life. These data provide an opportunity to examine occurrence and evolutionary patterns of genes, pathways and biological processes, including Mo utilization and molybdoenzymes. Comprehensive analyses of Mo transport systems, Moco biosynthesis pathway and different molybdoenzyme families can greatly help in our understanding of how Mo is used by different organisms and how the distribution and biological functions of molybdoenzymes have been shaped by evolutionary processes. Here, we focus on recent developments involving comparative genomic analyses of Mo utilization and discuss the occurrence and abundance of molybdoenzymes as well as the macroevolutionary trends of Mo utilization.

2.2 Molybdate Uptake and Molybdenum Cofactor Biosynthesis

Moco biosynthesis and its insertion into molybdoenzymes depend on transport of molybdate anion, activation of molybdate and incorporation of Mo into molybdopterin (MPT). A general scheme of Mo uptake and Moco biosynthesis pathway in bacteria and eukaryotes is shown in Figure 2.1.

A common theme in Mo uptake is that this element is transported in the form of the oxyanion molybdate (MoO_4^{2-}). In bacteria, molybdate is mainly taken up through the high-affinity ModABC system, which is the first identified Mo transporter in this group of organisms. In *Escherichia coli*, the ABC transporter consists of a periplasmic molybdate-binding protein ModA, two integral membrane proteins of the ModB family and two hydrophilic peripheral membrane proteins of the ModC family that are ATPases.[25,33,34] ModA may bind both molybdate and tungstate in the periplasm *via* several conserved amino acid residues.[35] Besides, two additional ABC transporters, WtpABC (Mo/W) and TupABC (W-specific), have been reported in different organisms.[17,18] WtpA is homologous to ModA, which contains the same domain (*i.e.* COG0725), whereas TupA contains a different domain (COG4662) with low similarity to ModB. Using site-directed mutagenesis in combination with ligand affinity analyses, a recent study has shown that a single amino acid (aspartate) might be involved in tungstate selectivity in WtpA proteins.[36]

Studies on Mo uptake in eukaryotes are quite limited. As mentioned above, only two types of eukaryotic Mo transporters have been characterized, MOT1 and MOT2. MOT1, which belongs to the sulfate transporter superfamily, was first described in *Arabidopsis thaliana* and *Chlamydomonas reinhardtii*.[27,28] Members of the MOT1 family are present in higher plants, algae and fungi and are related to molybdate transport/accumulation.[27,32] MOT2 was first identified in *C. reinhardtii*. Homologs of MOT2 have also been detected in some plants, algae and animals.[31,32] Further studies are needed to examine its role in Mo transport/homeostasis.

Many reviews have been published which broadly cover the process of Moco biosynthesis in bacteria and eukaryotes,[1–7,10–12] and we only give a brief description here. The Moco biosynthetic pathway can be divided into three highly conserved steps that are present in all Mo-utilizing organisms: (1) formation of the cyclic pyranopterin monophosphate (cPMP) from GTP; (2) transformation of cPMP into MPT; and (3) insertion of Mo into MPT to form Moco.[37] In some bacteria and archaea, additional modifications of Moco are needed, such as the attachment of a nucleotide (GMP or CMP) to MPT to form the dinucleotide form of Moco (MPT-guanine dinucleotide, MGD;

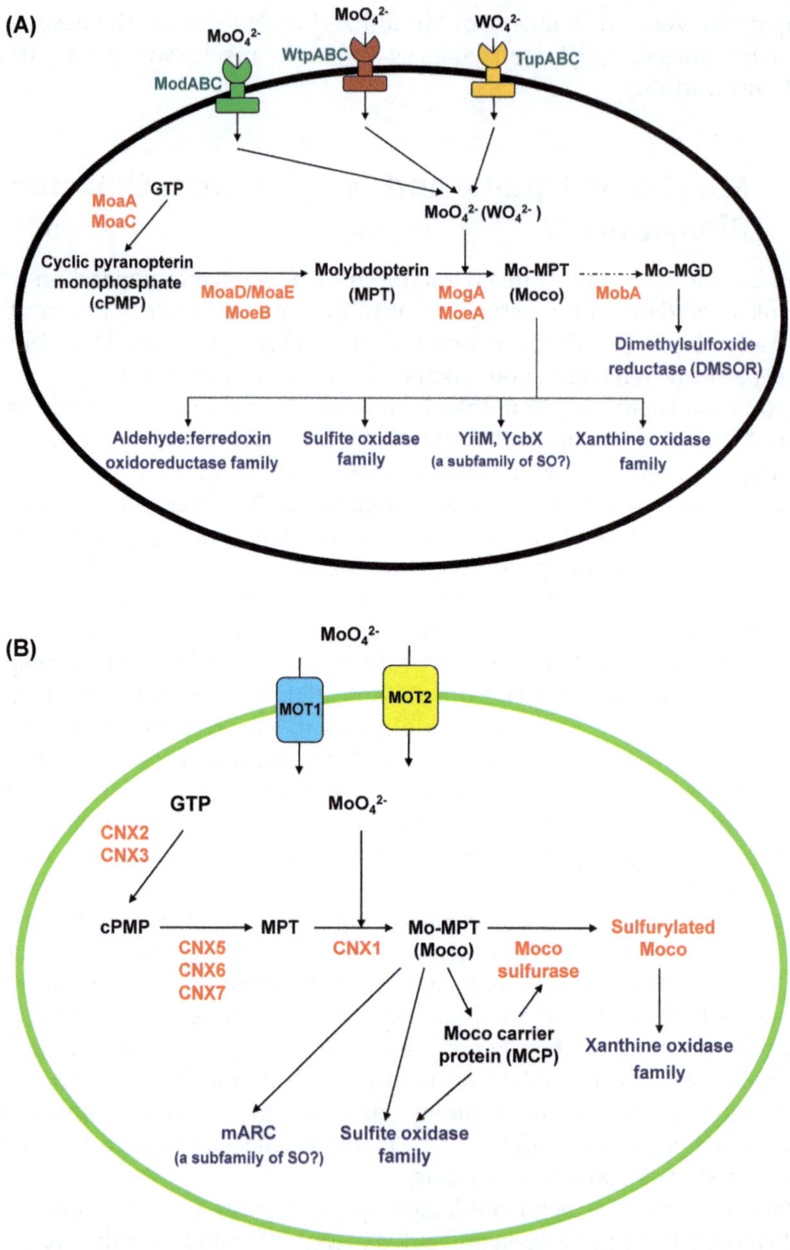

Figure 2.1 Overview of Moco biosynthetic pathways. (A) Biosynthesis of Moco in bacteria. (B) Biosynthesis of Moco in eukaryotes. The proteins from *E. coli* and *A. thaliana* that catalyze the respective steps in the pathway are indicated. MGD, MPT-guanine dinucleotide.

or MPT-cytosine dinucleotide, MCD), or substitution of a terminal oxygen ligand of Moco with a sulfur ligand.[37,38] In *E. coli*, six loci have been implicated in this process: moa, mob, moc, mod, moe and mog, comprising 16 genes.[38] With the exception of mod that encodes the ModABC transporters, these genes code for proteins involved in the biosynthesis of Moco (Figure 2.1A). In eukaryotes, Moco biosynthesis involves two cellular compartments, mitochondria and cytosol.[39] It should be mentioned that different nomenclature has been used for the genes involved in Moco biosynthesis in plants and humans.[40,41] In plants, the Moco synthetic genes are named according to the CNX nomenclature (*C*ofactor for *N*itrate reductase and *X*anthine dehydrogenase). At least six gene products (named CNX1–3 and CNX5–7, Figure 2.1B) are involved in Moco biosynthesis, which are homologous to their counterparts in bacteria.[32,41,42] In addition, the Moco sulfurase is required for post-translational activation of members of the XO family including xanthine dehydrogenase (XDH) and aldehyde oxidase (AO) by transferring a sulfur atom to the desulfo-Moco of these enzymes.[5,43,44] In addition, a Moco carrier protein (MCP) and homologous Moco-binding proteins (MoBP) were recently identified in *C. reinhardtii* and land plants, respectively.[45,46] Both groups of proteins show significant structural similarity and might be involved in cellular distribution of Moco.[46]

2.3 Classification of Molybdoenzymes

Based on cofactor composition and catalytic function, distinct groups of molybdoenzymes can be defined: (i) nitrogenase that has a unique Mo–Fe cofactor in the active site; (ii) all other molybdoenzymes that bind Moco. A list of the majority of known molybdoenzyme families and proteins are shown in Table 2.1.

Nitrogenase is required for biological nitrogen fixation, which is a key step in the global nitrogen cycle.[8,9] This enzyme catalyzes the nucleotide-dependent reduction of atmospheric dinitrogen to bioavailable ammonia, concomitant with the reduction of protons to yield hydrogen gas.[47,48] In the Mo–Fe protein component of the nitrogenase system, the Mo–Fe cofactor (the M-cluster) serves as the active site, and is assembled through the formation of an 8 Fe core prior to the insertion of Mo and homocitrate into this core.[49]

The active site of all other well-characterized molybdoenzymes utilizes Moco as cofactor. These molybdoenzymes are categorized into several families: SO, XO, DMSOR, AOR and the newly identified MOSC-containing (mARC/YcbX/YiiM) proteins. Each family includes different subfamilies based on sequence similarity, spectroscopic properties and substrate preferences (Table 2.1). It should be noted that compared to prokaryotes, which contain diverse members of all families, eukaryotes only possess relatively few molybdoenzymes, including SO and assimilatory nitrate reductase (SO family), XDH and AO (XO family), and mARC protein (MOSC-containing family).[5,10,23]

Table 2.1 Classification of molybdoenzymes.

Group	Family	Major subfamily
Moco-binding	Sulfite oxidase	Sulfite oxidase
		Sulfite dehydrogenase
		Nitrate reductase (assimilatory)
	Xanthine oxidase	Xanthine oxidase
		Xanthine dehydrogenase
		Aldehyde oxidase
		Aldehyde oxidoreductase
		4-Hydroxybenzoyl-CoA reductase
		CO dehydrogenase
		Quinoline 2-oxidoreductase
		Isoquinoline 1-oxidoreductase
		Quinoline 4-carboxylate-2-oxidoreductase
		Quinaldine 4-oxidoreductase
		Quinaldic acid 4-oxidoreductase
		Nicotinic acid hydroxylase
		6-Hydroxynicotinate hydroxylase
		Nicotine dehydrogenase
		Nicotinate hydroxylase
		Picolinate hydroxylase
		Pyridoxal oxidase
	Dimethylsulfoxide reductase	Dimethylsulfoxide reductase
		Biotin sulfoxide reductase
		Trimethylamine-*N*-oxide reductase
		Nitrate reductase (dissimilatory)
		Formate dehydrogenase
		Formylmethanofuran dehydrogenase
		Polysulfide/thiosulfate/arsenate reductase
		Arsenite oxidase
		Pyrogallol-phloroglucinol transhydroxylase
	Aldehyde : ferredoxin oxidoreductase (W-specific)	Aldehyde : ferredoxin oxidoreductase
		Formaldehyde ferredoxin oxidoreductase
		Glyceraldehyde-3-phosphate ferredoxin oxidoreductase
		Carboxylic acid reductase
		Hydroxycarboxylate viologen oxidoreductase
		Aldehyde dehydrogenase
	MOSC-containing protein (a subfamily of SO?)	mARC/YcbX/YiiM
Fe–Mo-binding	Nitrogenase	Nitrogenase

Members of the SO family generally catalyze the transfer of an oxygen atom either to or from the substrate.[50] They usually contain a conserved cysteine residue that links Mo with the apo-protein. Enzymes belonging to this family include sulfite oxidases and dehydrogenases from both prokaryotes and eukaryotes, and the assimilatory nitrate reductase from eukaryotes.[51] In eukaryotes, SO is located in the mitochondrial intermembrane space and catalyzes the oxidation of sulfite to sulfate, which is the terminal reaction in the oxidative degradation of the sulfur-containing amino acids, cysteine and methionine.[20,51] Assimilatory nitrate reductase is only detected in autotrophic organisms such as plants and fungi, which catalyzes the reduction of nitrate to nitrite.[52]

The XO family catalyzes oxidative hydroxylation of a wide range of substrates (most notably aldehydes and aromatic heterocycles).[2,53] It contains the largest and most diverse molybdoenzyme set present in all three domains of life (Table 2.1). All members of this family have redox-active centers, in addition to the Mo center. Typical enzymes of this family include AO, XDH and many other bacterial enzymes such as aldehyde oxidoreductase, 4-hydroxybenzoyl-CoA reductase and nicotinic acid hydroxylase.[54–58]

The DMSOR family is diverse in both structure and function, but all these enzymes have two equivalents of MGD bound to Mo at the active site.[2,13] Members of the DMSOR family are exclusively found in prokaryotes and include DMSORs, biotin sulfoxide reductase, dissimilatory nitrate reductases, formate dehydrogenases, trimethylamine-*N*-oxide reductase and arsenite oxidase (Table 2.1). In the absence of oxygen and in the presence of their respective substrates, these enzymes mainly function as terminal reductases where their respective cofactors serve as terminal electron acceptors.[59] The widespread enzyme DMSOR is either membrane-associated or located in the periplasm, which catalyzes reductive deoxygenation of dimethyl sulfoxide to dimethyl sulfide.[38] Biotin sulfoxide reductase reduces biotin sulfoxide to biotin and has an essential role in assimilation of free oxidized methionines.[60] Formate dehydrogenase, another important member of this family, catalyzes the oxidation of formate to bicarbonate and is also a selenium-containing enzyme in many organisms.[61]

AOR is a W-specific family that catalyzes the interconversion of aldehydes and carboxylates using ferredoxin as the physiological electron acceptor. The AOR protein is the first member of this family to be structurally characterized as a protein containing a Wco cofactor (the W atom complexed with two pyranopterin dithiolenes).[14,62] Other enzymes include formaldehyde ferredoxin oxidoreductase, glyceraldehyde-3-phosphate ferredoxin oxidoreductase, carboxylic acid reductase and hydroxycarboxylate viologen oxidoreductase.[62–65]

A novel MOSC-containing protein family has been discovered in both eukaryotes and bacteria.[23,24] The eukaryotic mARC protein was first identified in the outer mitochondrial membrane in pig cells. It consists of two conserved domains: N-terminal MOSC_N and C-terminal MOSC domains, which are also present in Moco sulfurases. Human genome harbors two mARC genes (referred to as hmARC-1/MOSC-1 and hmARC-2/MOSC-2).[23,66] Both proteins catalyze the N-reduction of a great variety of N-hydroxylated substrates such as N-hydroxy-cytosine and *N*(ω)-hydroxy-L-arginine with different specificities, implying that they may play an important role in drug metabolism and regulation of nitric oxide biosynthesis.[66–69] Two other members of the MOSC-containing protein family, YcbX and YiiM, could reduce 6-*N*-hydroxylaminopurine to adenine.[24,70] Both proteins contain MOSC domain; however, the *E. coli* YcbX contains the MOSC_N, MOSC and an additional iron–sulfur cluster binding domains, whereas YiiM has MOSC and a C-terminal 3-alpha domains. A comparative and phylogenetic analysis has suggested that mARC/YcbX and YiiM might belong to different subfamilies of the MOSC-containing family.[71] Recently, human mARC proteins were reported to be able to catalyze reduction of nitrite to nitric oxide (NO).[72] The mARC proteins possess a strictly conserved Cys residue at the C-terminal region that coordinates a bound Mo center having the same structure as seen in SO proteins.[2] Electron paramagnetic resonance (EPR) spectroscopy analyses confirm that mARC proteins are remarkably similar to the low-pH form of SO.[66] Thus, an alternative interpretation has now been suggested that mARC/MOSC-containing proteins might constitute a subfamily of the SO family.[2,53] On the other hand, sequences of mARC/YcbX/YiiM proteins lack significant similarity to members of the SO family. Future studies are needed to verify structural and functional similarities between mARC and SO proteins. In this chapter, mARC and SO proteins are considered as different molybdoenzyme families, while the alternative classification approach is also suggested (Figure 2.1 and Table 2.1).

2.4 Occurrence and Evolution of Molybdenum Utilization and Molybdoenzymes

Several studies have focused on the identification of genes involved in Mo/W uptake and Moco biosynthesis and genes encoding novel molybdoenzymes, although this has been done for only a limited number of organisms. Thus far, systematic analyses of the phylogeny and evolutionary trends of Mo utilization have been rare. However, some of these bioinformatics studies have been conducted to investigate the occurrence and evolutionary dynamics of Mo utilization at the level of Mo transport, Moco biosynthesis and molybdoenzymes in a wide range of sequenced prokaryotes and eukaryotes.[71,73–75] These studies have provided more detailed information regarding Mo utilization across the three domains of life and have demonstrated its widespread occurrence, yet limited use of this element in different taxa and species.

2.4.1 Occurrence of Molybdenum Transport and Moco Biosynthesis Pathway Genes

Comparative studies on Mo utilization and evolution have shown a wide distribution of genes encoding Mo/W transport systems and enzymes involved in the Moco biosynthesis pathway.[73,74] Very recently, we analyzed these genes in more than 5000 organisms (unpublished data). In bacteria, ModABC is found in approximately 85% of sequenced Mo-utilizing bacteria, and is the most frequently used Mo transporter (Figure 2.2). The distribution of WtpABC and TupABC is much more restricted, especially WtpABC, which is

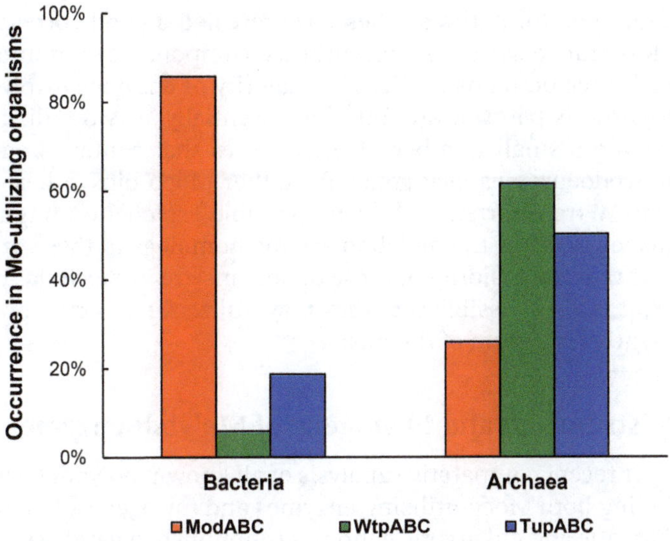

Figure 2.2 Distribution of Mo/W transporters in bacteria and archaea. Three types of high-affinity Mo/W ABC transporters are known: ModABC, WtpABC and TupABC.

only detected in less than 6% of Mo-utilizing bacteria. In contrast, WtpABC is present in the majority of Mo-utilizing archaea, whereas the ModABC system showed a very limited distribution in these organisms (Figure 2.2). Thus, it appears that ModABC and WtpABC serve as the major Mo/W transporters in bacteria and archaea, respectively. The third W-specific transporter, TupABC, is present in approximately half of the Mo-utilizing archaea, where it is much more frequently observed than in bacteria (less than 20%). This observation suggests a more important role of W in archaea than in bacteria (partially due to the presence of W-specific AOR members in archaea). Besides the three known Mo/W transporters, a distant ModABC-like system has been predicted in several *Pyrobaculum* species which are hyperthermophilic archaea.[73] Phylogenetic analysis of these components has revealed that they form a distant group of ModABC transporters and might be also involved in Mo/W uptake.

As discussed above, MOT1 and MOT2 are the only well-characterized Mo transporters in eukaryotes. Previous analyses have shown that MOT1 mainly occurs in land plants, green algae, certain fungi (Fungi/Ascomycota/Pezizomycotina) and stramenopiles.[74] We have analyzed the occurrence of MOT2 in more than 200 sequenced eukaryotes. A much wider distribution of MOT2 is observed, which is present in stramenopiles, green algae, land plants and the majority of vertebrates including mammals (unpublished data). However, the lack of both transporters in some other Mo-utilizing organisms (such as all arthropoda) implies that currently unknown Mo transport systems may be present in these organisms.

In general, bioinformatics studies have revealed a good correspondence between Mo transport/Moco biosynthesis components and molybdoenzymes in all three domains of life. The majority of enzymes involved in the Moco biosynthesis pathway are found in essentially all Mo-utilizing organisms. However, a small number of prokaryotes that contain homologs of known molybdoenzymes lack genes for either Moco biosynthesis components or Mo/W transporters.[71,73,74] One possible explanation might be that Moco is dispensable for the molybdoenzyme homologs in these organisms. Considering that the majority of these organisms are intracellular parasites/symbionts, it is also possible that they may utilize the Mo transport and/or Moco biosynthesis system of the host.

2.4.2 Distribution and Phylogeny of Molybdoenzymes

Based on our recent comparative analysis of all known molybdoenzyme families (including both Moco-utilizing enzymes and nitrogenase) in more than 5000 prokaryotic and eukaryotic genomes (unpublished data), we have characterized the occurrence of each molybdoenzyme family in different groups of organisms (Figure 2.3). The distribution patterns of the Moco biosynthesis pathway and different molybdoenzyme families in various phyla or clades of archaea, bacteria and eukaryotes are shown in Figure 2.4.

First, members of the DMSOR family, such as DMSOR, nitrate reductase (dissimilatory) and formate dehydrogenase, are present in more than 90% Mo-utilizing organisms in both bacteria and archaea, which is consistent with previous observations.[71,73,74] In bacteria, SO, XO and YcbX/YiiM families are also widespread in the majority of bacterial species (>60%) and

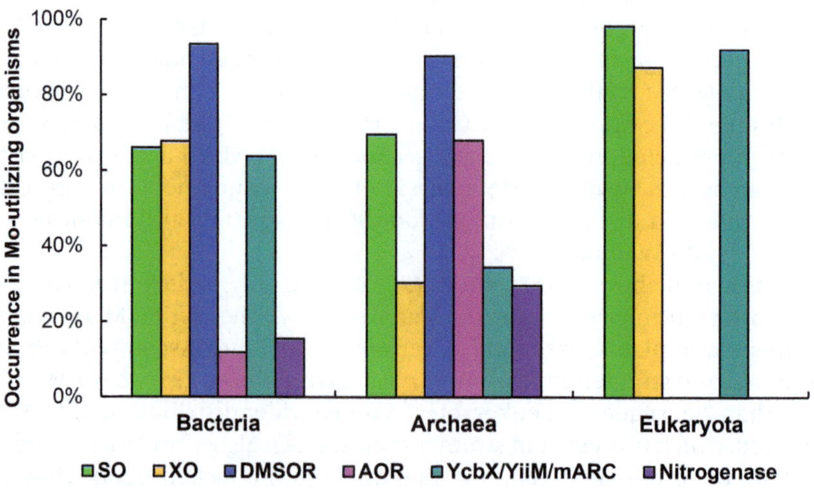

Figure 2.3 Distribution of molybdoenzymes in all Mo-utilizing organisms in the three domains of life. SO, sulfite oxidase; XO, xanthine oxidase; DMSOR, dimethylsulfoxide reductase, AOR, aldehyde:ferredoxin oxidoreductase.

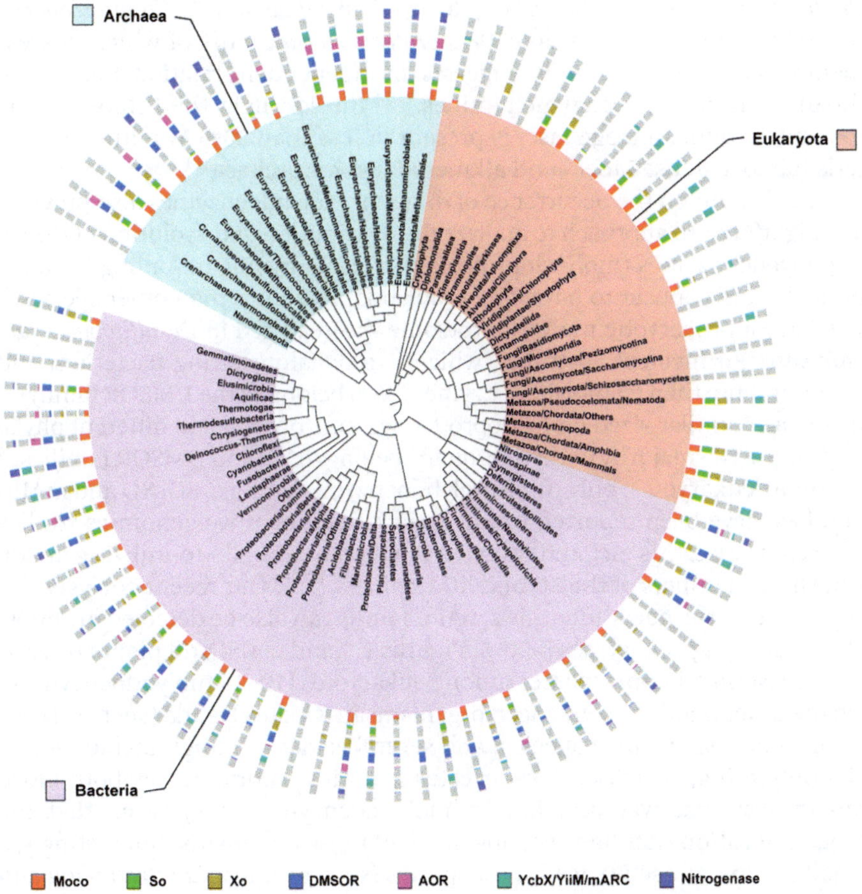

Figure 2.4 Distribution of the Moco biosynthesis pathway and molybdoenzyme families in prokaryotes and eukaryotes. The phylogenetic tree is simplified to only show major bacterial taxa and branches. The seven tracks (circles) around the tree (from inside to outside) represent the distribution patterns of the Moco biosynthesis pathway, and SO, XO, DMSOR, AOR, YcbX/YiiM/mARC and nitrogenase families, respectively. The length of the colored section of each column represents the percentage of organisms that have either Moco pathway or the corresponding molybdoenzyme family among all sequenced organisms in this phylum.

most clades, whereas the W-containing AOR family are only found in ~10% Mo/W-utilizing organisms (Figures 2.3 and 2.4). Members of the AOR family have a much wider occurrence in archaea than in bacteria (~70%), implying the importance of this family for the former organisms. XO and YcbX/YiiM families show a much more limited occurrence (less than 35%) and are absent in most archaeal phyla (including all methanogenic archaea). Many prokaryotic species possess two or three molybdoenzyme families, and often several subfamilies within these families. However, the very low occurrence or even absence of SO, XO and YcbX/YiiM families in certain phyla (*e.g.* SO in

Chlorobi and *Firmicutes/Clostridia*, XO in *Chlamydiae* and *Epsilonproteobacteria*, YcbX/YiiM in *Aquificae* and *Deltaproteobacteria*), most of which possess members the DMSOR family, suggests an independent relationship among the different molybdoenzyme families.[73,74] Interestingly, the FeMo-utilizing molybdoenzyme, nitrogenase, is present in less than 20% Mo-utilizing bacteria (almost all use Moco) and all methanogenic archaea.[73]

Investigation on the occurrence of different molybdoenzymes also provides a straightforward approach to analyze the distribution and evolution of the set of molybdoenzymes (molybdoproteome) of each organism. Among bacteria, proteobacteria appear to have larger molybdoproteomes than other species.[74] The largest prokaryotic molybdoproteome was reported in *Desulfitobacterium hafniense* (Firmicutes/Clostridia), which is a dehalorespiring bacterium and contains more than 60 molybdoenzymes (95% belong to the DMSOR family).[74] In archaea, similar-sized molybdoproteomes were observed in different phyla/clades, and the majority of these proteins belong to XO and DMSOR families.[74]

Among eukaryotes, only three molybdoenzyme families, SO, XO and mARC families, have been reported so far. Previous comparative genomics studies have confirmed this fact and revealed that essentially all Mo-utilizing organisms have members of the SO and XO families.[71,73,74] Our recent analysis also suggests that the newly identified mARC family can also be detected in almost all Mo-utilizing eukaryotes (>90%, Figures 2.3 and 2.4). Land plants possess the largest molybdoproteomes among eukaryotes (10–14 molybdoenzymes). On the other hand, all sequenced organisms in Kinetoplastida (such as *Leishmania* and *Trypanosoma* species) and several clades of Fungi (including saccharomycotina, schizosaccharomycetes and microsporidia) lack both Moco biosynthesis pathway and known molybdoenzymes, suggesting that the Moco utilization trait has been lost in all of these organisms. Interestingly, a small group of unsequenced yeast species (such as *Candida nitratophila* and *Pichia pastoris*) have been previously reported to utilize Moco-containing nitrate reductase, which catalyzes the reduction of nitrate to nitrite in autotrophic organisms such as plants, algae and fungi.[76,77] However, the fact that both homologs of this molybdoenzyme and the Moco biosynthesis pathway are absent in sister species of these Mo-utilizing yeasts (such as *C. albicans* and *C. glabrata*) and many other yeasts points to the loss of Mo utilization in these species. It is possible that either Mo-dependent reduction of nitrate is unnecessary for these organisms or alternative Mo-independent mechanisms may have evolved.

2.4.3 Factors that May Affect Evolution of Mo Utilization and Molybdoenzymes

The requirement for the Moco biosynthesis machinery and different molybdoenzymes in the three domains of life as well as their scattered occurrence in a great many phyla/clades illustrate a highly dynamic nature of Mo utilization. Thus, it is possible that certain factors (such as environmental conditions) play a role in acquisition/loss of Mo utilization in the course of evolution.

Previously, we have examined this in prokaryotes by analyzing the possible role of several factors, including environmental conditions (*e.g.* oxygen requirement, temperature, optimal pH and habitat), genome size, Gram strain and GC content in the evolution of Mo utilization and molybdoenzymes.[73,74] The majority of bacteria that do not use Moco appear to exhibit host-associated lifestyles, implying that such a habitat might lead to the loss of Mo utilization due to limited space and resources or the availability of the Moco biosynthesis pathway of the host. Further analysis of different types of host-associated conditions (such as intra- or extracellular and symbiotic or parasitic) has revealed that the majority of obligate intracellular symbionts and parasites have lost the ability to utilize Mo, whereas most extracellular symbionts have retained use of the metal. It is possible that, due to limited bioavailability of Mo and the evolutionary pressure on genome size (most intracellular parasites and symbionts have more condensed genomes than extracellular organisms), Mo is not essential for intracellular parasitic life. However, Mo is still needed for extracellular symbionts, which may have larger space and less constrained free lifestyles. In eukaryotes, most parasitic organisms (such as the majority of apicomplexa and all kinetoplastids) also do not use Mo. Similar trends are observed for some other metals, including nickel, cobalt and selenium.[74,78,79] Other factors, including oxygen utilization, Gram strain, optimal temperature and pH, appear to have no significant influence on the evolution of Mo utilization.

Analysis of the relationship between environmental factors and molybdoenzyme families has demonstrated additional features for individual molybdoenzymes. The majority of AOR-containing organisms favor anaerobic environments, whereas organisms possessing other molybdoenzymes (SO, XO or DMSOR proteins) favor aerobic conditions.[71,73] In addition, organisms containing nitrogenase (such as all methanogenic archaea) favor anaerobic and relatively warm conditions. Thus, different molybdoenzyme families show dynamic and relatively independent evolutionary histories in spite of the fact that they all require Mo uptake and Moco biosynthesis processes.

Several previous studies on other trace elements such as copper and selenium have suggested that the size of cuproproteomes or selenoproteomes correlates with oxygen utilization and/or aquatic/terrestrial environment.[80–82] However, no significant correlation is observed between these factors and molybdoproteomes.[74] In the future, it would be interesting to identify additional factors that may influence the composition and evolution of molybdoproteomes.

2.5 Concluding Remarks

In recent years, much effort has been devoted to studying the metabolism and biological functions of Mo. Both Moco biosynthesis components and molybdoenzymes have been characterized in numerous organisms. However, understanding of evolution, abundance and ubiquity of Mo utilization is somewhat limited. This chapter is written to provide an overview of the general features and evolutionary trends in Mo utilization from the point of view of bioinformatics and comparative genomics, which should broaden the scope of Mo research by comprehensive analyses of homologs of all known molybdoenzymes, Mo/W uptake systems and Moco biosynthesis genes. These observations may not only lead to significant advances in our understanding of Mo utilization across the three domains of life, but may also offer new insights and ideas for further experimental studies. In the future, additional computational and comparative analyses on the identity and functions of Mo metabolism and molybdoenzyme genes should help explain biological and biomedical effects of Mo and identify new Mo-dependent biological processes.

Acknowledgements

This work was supported by the National Natural Science Foundation of China under No. 31171233, the Natural Science Foundation of Guangdong Province (No. 2015A030313555), Chinese Academy of Sciences (2012OHTP10), and NIH GM061603.

References

1. K. V. Rajagopalan and J. L. Johnson, The pterin molybdenum cofactors, *J. Biol. Chem.*, 1992, **267**(15), 10199–10202.
2. R. Hille, J. Hall and P. Basu, The mononuclear molybdenum enzymes, *Chem. Rev.*, 2014, **114**(7), 3963–4038.
3. G. Schwarz and R. R. Mendel, Molybdenum cofactor biosynthesis and molybdenum enzymes, *Annu. Rev. Plant Biol.*, 2006, **57**, 623–647.
4. R. R. Mendel, Molybdenum: biological activity and metabolism, *Dalton Trans.*, 2005, **21**, 3404–3409.
5. R. R. Mendel and F. Bittner, Cell biology of molybdenum, *Biochim. Biophys. Acta*, 2006, **1763**(7), 621–635.
6. R. R. Mendel, Biology of the molybdenum cofactor, *J. Exp. Bot.*, 2007, **58**(9), 2289–2296.
7. R. Hille, Molybdenum and tungsten in biology, *Trends Biochem. Sci.*, 2002, **27**(7), 360–367.
8. L. M. Rubio and P. W. Ludden, Biosynthesis of the iron-molybdenum cofactor of nitrogenase, *Annu. Rev. Microbiol.*, 2008, **62**, 93–111.
9. P. C. Dos Santos, D. R. Dean, Y. Hu and M. W. Ribbe, Formation and insertion of the nitrogenase iron-molybdenum cofactor, *Chem. Rev.*, 2004, **104**(2), 1159–1173.
10. G. Schwarz, R. R. Mendel and M. W. Ribbe, Molybdenum cofactors, enzymes and pathways, *Nature*, 2009, **460**(7257), 839–847.
11. G. Schwarz, Molybdenum cofactor biosynthesis and deficiency, *Cell. Mol. Life Sci.*, 2005, **62**(23), 2792–2810.
12. R. R. Mendel, The molybdenum cofactor, *J. Biol. Chem.*, 2013, **288**(19), 13165–13172.

13. M. J. Romão, Molybdenum and tungsten enzymes: a crystallographic and mechanistic overview, *Dalton Trans.*, 2009, (21), 4053–4068.
14. M. K. Chan, S. Mukund, A. Kletzin, M. W. Adams and D. C. Rees, Structure of a hyperthermophilic tungstopterin enzyme, aldehyde ferredoxin oxidoreductase, *Science*, 1995, **267**(5203), 1463–1469.
15. J. R. Andreesen and K. Makdessi, Tungsten, the surprisingly positively acting heavy metal element for prokaryotes, *Ann. N. Y. Acad. Sci.*, 2008, **1125**, 215–229.
16. R. Roy and M. W. Adams, Tungsten-dependent aldehyde oxidoreductase: a new family of enzymes containing the pterin cofactor, *Met. Ions Biol. Syst.*, 2002, **39**, 673–697.
17. L. E. Bevers, P. L. Hagedoorn, G. C. Krijger and W. R. Hagen, Tungsten transport protein A (WtpA) in Pyrococcus furiosus: the first member of a new class of tungstate and molybdate transporters, *J. Bacteriol.*, 2006, **188**(18), 6498–6505.
18. K. Makdessi, J. R. Andreesen and A. Pich, Tungstate Uptake by a highly specific ABC transporter in Eubacterium acidaminophilum, *J. Biol. Chem.*, 2001, **276**(27), 24557–24564.
19. C. Kisker, H. Schindelin and D. C. Rees, Molybdenum-cofactor-containing enzymes: structure and mechanism, *Annu. Rev. Biochem.*, 1997, **66**, 233–267.
20. C. Kisker, H. Schindelin, D. Baas, J. Rétey, R. U. Meckenstock and P. M. Kroneck, A structural comparison of molybdenum cofactor-containing enzymes, *FEMS Microbiol. Rev.*, 1998, **22**(5), 503–521.
21. R. R. Mendel, Cell biology of molybdenum, *Biofactors*, 2009, **35**(5), 429–434.
22. A. Havemeyer, F. Bittner, S. Wollers, R. Mendel, T. Kunze and B. Clement, Identification of the missing component in the mitochondrial benzamidoxime prodrug-converting system as a novel molybdenum enzyme, *J. Biol. Chem.*, 2006, **281**(46), 34796–34802.
23. G. Ott, A. Havemeyer and B. Clement, The mammalian molybdenum enzymes of mARC, *JBIC, J. Biol. Inorg. Chem.*, 2015, **20**(2), 265–275.
24. S. G. Kozmin, P. Leroy, Y. I. Pavlov and R. M. Schaaper, YcbX and yiiM, two novel determinants for resistance of Escherichia coli to N-hydroxylated base analogues, *Mol. Microbiol.*, 2008, **68**(1), 51–65.
25. A. M. Grunden and K. T. Shanmugam, Molybdate transport and regulation in bacteria, *Arch. Microbiol.*, 1997, **168**(5), 345–354.
26. E. Aguilar-Barajas, C. Díaz-Pérez, M. I. Ramírez-Díaz, H. Riveros-Rosas and C. Cervantes, Bacterial transport of sulfate, molybdate, and related oxyanions, *Biometals*, 2011, **24**(4), 687–707.
27. M. Tejada-Jiménez, A. Llamas, E. Sanz-Luque, A. Galván and E. Fernández, A high-affinity molybdate transporter in eukaryotes, *Proc. Natl. Acad. Sci. U. S. A.*, 2007, **104**(50), 20126–20130.
28. H. Tomatsu, J. Takano, H. Takahashi, A. Watanabe-Takahashi, N. Shibagaki and T. Fujiwara, An Arabidopsis thaliana high-affinity molybdate transporter required for efficient uptake of molybdate from soil, *Proc. Natl. Acad. Sci. U. S. A.*, 2007, **104**(47), 18807–18812.

29. I. Baxter, B. Muthukumar, H. C. Park, P. Buchner, B. Lahner, J. Danku, K. Zhao, J. Lee, M. J. Hawkesford, M. L. Guerinot and D. E. Salt, Variation in molybdenum content across broadly distributed populations of Arabidopsis thaliana is controlled by a mitochondrial molybdenum transporter (MOT1), *PLoS Genet.*, 2008, **4**(2), e1000004.
30. A. Gasber, S. Klaumann, O. Trentmann, A. Trampczynska, S. Clemens, S. Schneider, N. Sauer, I. Feifer, F. Bittner, R. R. Mendel and H. E. Neuhaus, Identification of an Arabidopsis solute carrier critical for intracellular transport and inter-organ allocation of molybdate, *Plant Biol. (Stuttgart, Ger.)*, 2011, **13**(5), 710–718.
31. M. Tejada-Jiménez, A. Galván and E. Fernández, Algae and humans share a molybdate transporter, *Proc. Natl. Acad. Sci. U. S. A.*, 2011, **108**(16), 6420–6425.
32. M. Tejada-Jiménez, A. Chamizo-Ampudia, A. Galván and E. Fernández, Llamas Á. Molybdenum metabolism in plants, *Metallomics*, 2013, **5**(9), 1191–1203.
33. G. Wang, S. Angermüller and W. Klipp, Characterization of Rhodobacter capsulatus genes encoding a molybdenum transport system and putative molybdenum-pterin-binding proteins, *J. Bacteriol.*, 1993, **175**(10), 3031–3042.
34. J. A. Maupin-Furlow, J. K. Rosentel, J. H. Lee, U. Deppenmeier, R. P. Gunsalus and K. T. Shanmugam, Genetic analysis of the modABCD (molybdate transport) operon of Escherichia coli, *J. Bacteriol.*, 1995, **177**(17), 4851–4856.
35. Y. Hu, S. Rech, R. P. Gunsalus and D. C. Rees, Crystal structure of the molybdate binding protein ModA, *Nat. Struct. Biol.*, 1997, **4**(9), 703–707.
36. L. E. Bevers, G. Schwarz and W. R. Hagen, A molecular basis for tungstate selectivity in prokaryotic ABC transport systems, *J. Bacteriol.*, 2011, **193**(18), 4999–5001.
37. R. R. Mendel and S. Leimkühler, The biosynthesis of the molybdenum cofactors, *JBIC, J. Biol. Inorg. Chem.*, 2015, **20**(2), 337–347.
38. C. Iobbi-Nivol and S. Leimkühler, Molybdenum enzymes, their maturation and molybdenum cofactor biosynthesis in Escherichia coli, *Biochim. Biophys. Acta*, 2013, **1827**(8–9), 1086–1101.
39. J. Teschner, N. Lachmann, J. Schulze, M. Geisler, K. Selbach, J. Santamaria-Araujo, J. Balk, R. R. Mendel and F. Bittner, A novel role for Arabidopsis mitochondrial ABC transporter ATM3 in molybdenum cofactor biosynthesis, *Plant Cell*, 2010, **22**(2), 468–480.
40. J. Reiss and J. L. Johnson, Mutations in the molybdenum cofactor biosynthetic genes MOCS1, MOCS2, and GEPH, *Hum. Mutat.*, 2003, **21**(6), 569–576.
41. R. R. Mendel and R. Hänsch, Molybdoenzymes and molybdenum cofactor in plants, *J. Exp. Bot.*, 2002, **53**(375), 1689–1698.
42. R. R. Mendel and T. Kruse, Cell biology of molybdenum in plants and humans, *Biochim. Biophys. Acta*, 2012, **1823**(9), 1568–1579.

43. F. Bittner, M. Oreb and R. R. Mendel, ABA3 is a molybdenum cofactor sulfurase required for activation of aldehyde oxidase and xanthine dehydrogenase in Arabidopsis thaliana, *J. Biol. Chem.*, 2001, **276**(44), 40381–40384.
44. S. Wollers, T. Heidenreich, M. Zarepour, D. Zachmann, C. Kraft, Y. Zhao, R. R. Mendel and F. Bittner, Binding of sulfurated molybdenum cofactor to the C-terminal domain of ABA3 from Arabidopsis thaliana provides insight into the mechanism of molybdenum cofactor sulfuration, *J. Biol. Chem.*, 2008, **283**(15), 9642–9650.
45. K. Fischer, A. Llamas, M. Tejada-Jimenez, N. Schrader, J. Kuper, F. S. Ataya, A. Galvan, R. R. Mendel, E. Fernandez and G. Schwarz, Function and structure of the molybdenum cofactor carrier protein from Chlamydomonas reinhardtii, *J. Biol. Chem.*, 2006, **281**(40), 30186–30194.
46. T. Kruse, C. Gehl, M. Geisler, M. Lehrke, P. Ringel, S. Hallier, R. Hänsch and R. R. Mendel, Identification and biochemical characterization of molybdenum cofactor-binding proteins from Arabidopsis thaliana, *J. Biol. Chem.*, 2010, **285**(9), 6623–6635.
47. Y. Hu and M. W. Ribbe, Biosynthesis of the metalloclusters of molybdenum nitrogenase, *Microbiol. Mol. Biol. Rev.*, 2011, **75**(4), 664–677.
48. O. Einsle, Nitrogenase FeMo cofactor: an atomic structure in three simple steps, *JBIC, J. Biol. Inorg. Chem.*, 2014, **19**(6), 737–745.
49. Y. Hu and M. W. Ribbe, Biosynthesis of the iron-molybdenum cofactor of nitrogenase, *J. Biol. Chem.*, 2013, **288**(19), 13173–13177.
50. G. J. Workun, K. Moquin, R. A. Rothery and J. H. Weiner, Evolutionary persistence of the molybdopyranopterin-containing sulfite oxidase protein fold, *Microbiol. Mol. Biol. Rev.*, 2008, **72**(2), 228–248.
51. C. Feng, G. Tollin and J. H. Enemark, Sulfite oxidizing enzymes, *Biochim. Biophys. Acta*, 2007, **1774**(5), 527–539.
52. W. H. Campbell, Structure and function of eukaryotic NAD(P)H:nitrate reductase, *Cell. Mol. Life Sci.*, 2001, **58**(2), 194–204.
53. R. Hille, The molybdenum oxotransferases and related enzymes, *Dalton Trans.*, 2013, **42**(9), 3029–3042.
54. E. Garattini, M. Fratelli and M. Terao, Mammalian aldehyde oxidases: genetics, evolution and biochemistry, *Cell. Mol. Life Sci.*, 2008, **65**(7–8), 1019–1048.
55. R. Hille, Molybdenum-containing hydroxylases, *Arch. Biochem. Biophys.*, 2005, **433**(1), 107–116.
56. M. J. Romão, M. Archer, I. Moura, J. J. Moura, J. LeGall, R. Engh, M. Schneider, P. Hof and R. Huber, Crystal structure of the xanthine oxidase-related aldehyde oxido-reductase from D. gigas, *Science*, 1995, **270**(5239), 1170–1176.
57. M. Boll, G. Fuchs, C. Meier, A. Trautwein, A. El Kasmi, S. W. Ragsdale, G. Buchanan and D. J. Lowe, Redox centers of 4-hydroxybenzoyl-CoA reductase, a member of the xanthine oxidase family of molybdenum-containing enzymes, *J. Biol. Chem.*, 2001, **276**(51), 47853–47862.

58. T. Schräder, A. Rienhöfer and J. R. Andreesen, Selenium-containing xanthine dehydrogenase from Eubacterium barkeri, *Eur. J. Biochem.*, 1999, **264**(3), 862–871.
59. R. A. Rothery, G. J. Workun and J. H. Weiner, The prokaryotic complex iron-sulfur molybdoenzyme family, *Biochim. Biophys. Acta*, 2008, **1778**(9), 1897–1929.
60. T. Shi and J. Xie, Molybdenum enzymes and molybdenum cofactor in mycobacteria, *J. Cell. Biochem.*, 2011, **112**(10), 2721–2728.
61. T. Hartmann, N. Schwanhold and S. Leimkühler, Assembly and catalysis of molybdenum or tungsten-containing formate dehydrogenases from bacteria, *Biochim. Biophys. Acta*, 2014, (14), 00322–00327pii: S1570-9639.
62. R. Roy, A. L. Menon and M. W. Adams, Aldehyde oxidoreductases from Pyrococcus furiosus, *Methods Enzymol.*, 2001, **331**, 132–144.
63. A. Kletzin and M. W. Adams, Tungsten in biological systems, *FEMS Microbiol. Rev.*, 1996, **18**(1), 5–63.
64. S. Mukund and M. W. Adams, Glyceraldehyde-3-phosphate ferredoxin oxidoreductase, a novel tungsten-containing enzyme with a potential glycolytic role in the hyperthermophilic archaeon Pyrococcus furiosus, *J. Biol. Chem.*, 1995, **270**(15), 8389–8392.
65. H. White, G. Strobl, R. Feicht and H. Simon, Carboxylic acid reductase: a new tungsten enzyme catalyses the reduction of non-activated carboxylic acids to aldehydes, *Eur. J. Biochem.*, 1989, **184**(1), 89–96.
66. B. Wahl, D. Reichmann, D. Niks, N. Krompholz, A. Havemeyer, B. Clement, T. Messerschmidt, M. Rothkegel, H. Biester, R. Hille, R. R. Mendel and F. Bittner, Biochemical and spectroscopic characterization of the human mitochondrial amidoxime reducing components hmARC-1 and hmARC-2 suggests the existence of a new molybdenum enzyme family in eukaryotes, *J. Biol. Chem.*, 2010, **285**(48), 37847–37859.
67. J. Kotthaus, B. Wahl, A. Havemeyer, J. Kotthaus, D. Schade, D. Garbe-Schönberg, R. Mendel, F. Bittner and B. Clement, Reduction of N(ω)-hydroxy-L-arginine by the mitochondrial amidoxime reducing component (mARC), *Biochem. J.*, 2011, **433**(2), 383–391.
68. S. Gruenewald, B. Wahl, F. Bittner, H. Hungeling, S. Kanzow, J. Kotthaus, U. Schwering, R. R. Mendel and B. Clement, The fourth molybdenum containing enzyme mARC: cloning and involvement in the activation of N-hydroxylated prodrugs, *J. Med. Chem.*, 2008, **51**(24), 8173–8177.
69. H. H. Jakobs, D. Froriep, A. Havemeyer, R. R. Mendel, F. Bittner and B. Clement, The mitochondrial amidoxime reducing component (mARC): involvement in metabolic reduction of N-oxides, oximes and N-hydroxyamidinohydrazones, *ChemMedChem*, 2014, **9**(10), 2381–2387.
70. S. G. Kozmin, E. I. Stepchenkova and R. M. Schaaper, TusA (YhhP) and IscS are required for molybdenum cofactor-dependent base-analog detoxification, *Microbiologyopen*, 2013, **2**(5), 743–755.
71. Y. Zhang, S. Rump and V. N. Gladyshev, Comparative genomics and evolution of molybdenum utilization, *Coord. Chem. Rev.*, 2011, **255**(9–10), 1206–1217.

72. C. E. Sparacino-Watkins, J. Tejero, B. Sun, M. C. Gauthier, J. Thomas, V. Ragireddy, B. A. Merchant, J. Wang, I. Azarov, P. Basu and M. T. Gladwin, Nitrite reductase and nitric-oxide synthase activity of the mitochondrial molybdopterin enzymes mARC1 and mARC2, *J. Biol. Chem.*, 2014, **289**(15), 10345–10358.
73. Y. Zhang and V. N. Gladyshev, Molybdoproteomes and evolution of molybdenum utilization, *J. Mol. Biol.*, 2008, **379**(4), 881–899.
74. Y. Zhang and V. N. Gladyshev, General trends in trace element utilization revealed by comparative genomic analyses of Co, Cu, Mo, Ni, and Se, *J. Biol. Chem.*, 2010, **285**(5), 3393–3405.
75. V. N. Gladyshev and Y. Zhang, Comparative genomics analysis of the metallomes, *Met. Ions Life Sci.*, 2013, **12**, 529–580.
76. C. R. Hipkin, D. A. Kau and A. C. Cannons, Further characterization of the assimilatory nitrate reductase from the yeast Candida nitratophila, *J. Gen. Microbiol.*, 1993, **139**(3), 473–478.
77. G. G. Barbier, R. C. Joshi, E. R. Campbell and W. H. Campbell, Purification and biochemical characterization of simplified eukaryotic nitrate reductase expressed in Pichia pastoris, *Protein Expression Purif.*, 2004, **37**(1), 61–71.
78. Y. Zhang, D. A. Rodionov, M. S. Gelfand and V. N. Gladyshev, Comparative genomic analyses of nickel, cobalt and vitamin B12 utilization, *BMC Genomics*, 2009, **10**, 78, DOI: 10.1186/1471-2164-10-78.
79. L. Jiang, H. Z. Zhu, Y. Z. Xu, J. Z. Ni, Y. Zhang and Q. Liu, Comparative selenoproteome analysis reveals a reduced utilization of selenium in parasitic platyhelminthes, *PeerJ*, 2013, **1**, e202.
80. Y. Zhang, H. Romero, G. Salinas and V. N. Gladyshev, Dynamic evolution of selenocysteine utilization in bacteria: a balance between selenoprotein loss and evolution of selenocysteine from redox active cysteine residues, *Genome Biol.*, 2006, **7**(10), R94. Epub 2006 Oct 20.
81. A. V. Lobanov, D. E. Fomenko, Y. Zhang, A. Sengupta, D. L. Hatfield and V. N. Gladyshev, Evolutionary dynamics of eukaryotic selenoproteomes: large selenoproteomes may associate with aquatic life and small with terrestrial life, *Genome Biol.*, 2007, **8**(9), R198.
82. P. G. Ridge, Y. Zhang and V. N. Gladyshev, Comparative genomic analyses of copper transporters and cuproproteomes reveal evolutionary dynamics of copper utilization and its link to oxygen, *PLoS One*, 2008, **3**(1), e1378.

CHAPTER 3

Molybdenum Cofactor Biosynthesis

SILKE LEIMKÜHLER[a] AND RALF R. MENDEL*[b]

[a]Department of Molecular Enzymology, Institute of Biochemistry and Biology, University of Potsdam, Karl-Liebknecht-Str. 24–25, 14476 Potsdam, Germany; [b]Institute of Plant Biology, Braunschweig University of Technology, Humboldtstr. 1, 38106 Braunschweig, Germany
*E-mail: sleim@uni-potsdam.de, r.mendel@tu-bs.de

3.1 Introduction

Molybdenum is an essential trace element for bacteria, archaea and eukaryotes. It is biologically active in the form of molybdate and due to its low abundance in nature it enters the cell generally by active transport systems. In the cell, molybdenum is either coordinated to the dithiolene group of the molybdenum cofactor (Moco),[1] or in some bacteria to the iron-molybdenum cofactor (FeMoco) solely present in nitrogenase. Nitrogenase will be discussed in detail in Chapter 8. Here we will mainly summarize the common aspects of Moco biosynthesis in bacteria, plants and humans.[2–6]

More than 60 different enzymes binding the molybdenum cofactor were identified in the past. The vast majority of these are present in bacteria and archaea while in animals and plants only four or five different enzymes, respectively, have been characterized to date.[7] The different Moco-containing enzymes are categorized on the basis of their molybdenum ligandation sphere, dividing them into three different enzyme families: the xanthine

RSC Metallobiology Series No. 5
Molybdenum and Tungsten Enzymes: Biochemistry
Edited by Russ Hille, Carola Schulzke, and Martin L. Kirk

Published by the Royal Society of Chemistry, www.rsc.org

oxidase (XO) family, the sulfite oxidase (SO) family and the dimethylsulfoxide (DMSO) reductase family.[8] While bacteria contain enzymes of all three enzyme families, enzymes of the DMSO reductase family are not present in eukaryotes. Further exceptions occur and, somewhat surprisingly, the commonly used eukaryotic model organism *Saccharomyces cerevisiae* does not contain any active molybdoenzymes. Further genome-wide database analyses have revealed a significant number of bacteria and unicellular eukaryotes that do not require molybdenum, while all multi-cellular eukaryotes contain molybdoenzymes, which are essential for their viability.[9] In contrast, mainly anaerobic archaea and some bacteria require tungsten (W) instead of molybdenum and bind the W in a similar manner to the dithiolene group of MPT.[10] In the periodic table of elements, W lies directly below Mo and features chemical properties similar to Mo.

Nason *et al.*[11] provided the first biochemical evidence for a common cofactor to all molybdenum-containing enzymes, which was removed as a low molecular weight fraction from denatured Mo-enzymes of mammalian, plant and bacterial origin. After separation, the protein-free Moco fraction was subsequently incorporated into a cofactor-free apo-nitrate reductase thereby activating the enzyme. The chemical nature of Moco has been determined by Rajagopalan and coworkers in 1982.[1] Later crystal structures of Mo-enzymes have demonstrated the existence of a third pyrano ring between the OH-group at C3′ of the side chain and the pterin C7 atom. Once the pyrano ring is closed, a fully reduced hydrogenated pterin is formed. Because of the unique nature of the pterin in Moco, the metal-free form of the cofactor is called *m*olybdo*pt*erin or *m*etal-binding *pt*erin (MPT); the latter reflects the fact that not only Mo but also W can be coordinated by this pterin scaffold.

Genetic analyses of Moco-deficient mutants in organisms ranging from bacteria to humans have resulted in the identification of several gene loci being involved in Moco biosynthesis. These findings have provided the basis to propose an evolutionarily conserved biosynthetic pathway[12] that can be divided into three steps, according to the stable biosynthetic intermediates that can be isolated (Figure 3.1, Table 3.1):[13] the synthesis of *c*yclic *p*yranopterin *m*ono*p*hosphate (cPMP),[14] conversion of cPMP into MPT by introduction of two sulfur atoms[15] and insertion of molybdate to form Moco.[16] In bacteria, a fourth step involves the further modification of Moco by the addition of nucleotide monophosphates to the phosphate group of MPT.[17,18]

In *E. coli*, a total of five gene loci were shown to be directly involved in the biosynthesis of Moco. These gene loci were designated as *moa* (*moaAB-CDE*), *mob* (*mobAB*), *moc* (*mocA*), *moe* (*moeAB*) and *mog* (*mogA*).[19] The loci are organized in operon structures in the *E. coli* genome and code for a total of 11 different proteins required for the complex synthesis of Moco. With the exception of MoaB and MobB, an essential function in Moco biosynthesis has been demonstrated for each of these gene products (Table 3.1).[13,19] Additionally, a high-affinity molybdate transport system has been identified that is involved in the uptake of molybdate at low concentrations of molybdate.[20] The molybdate transport system is encoded by the genes *modABC*.[21] Proteins

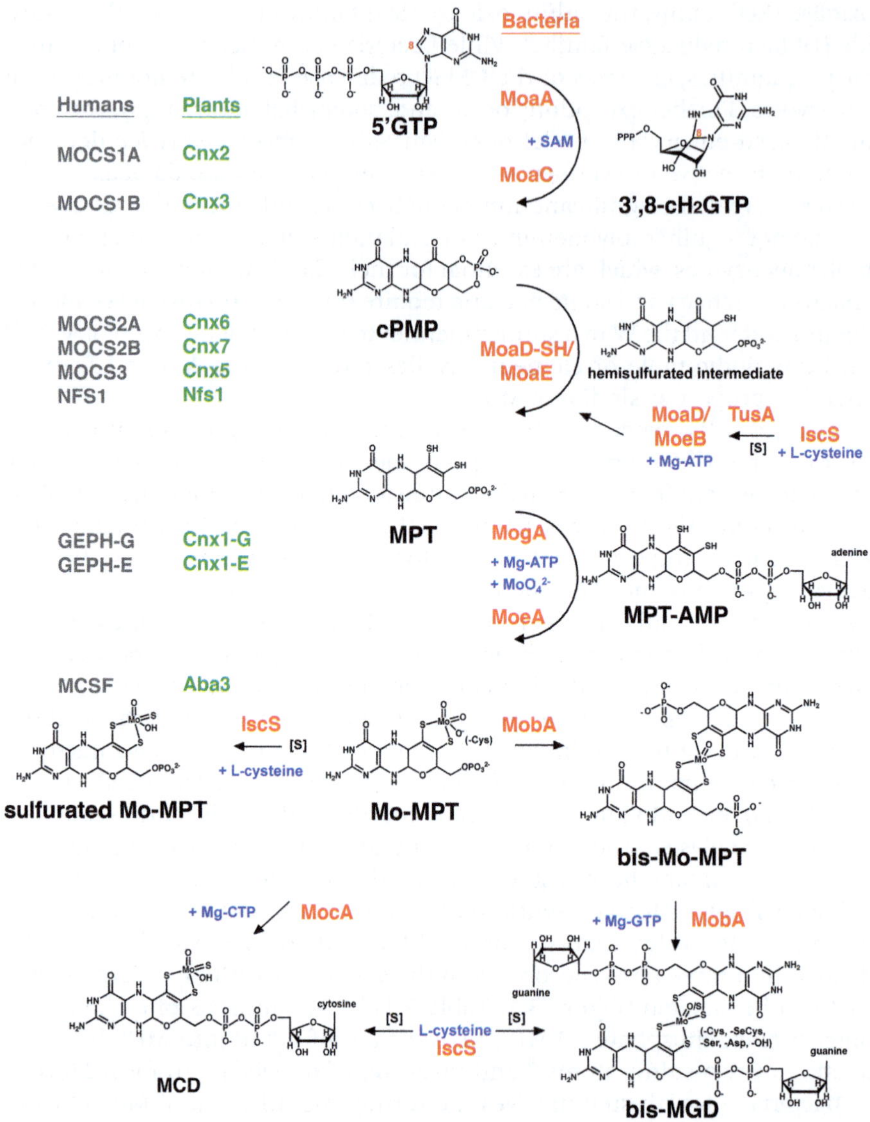

Figure 3.1 The biosynthesis of Moco. Shown is a scheme of the biosynthetic pathway for Moco biosynthesis. The central part shows the three conserved steps of Moco biosynthesis present in all organisms, the formation of cPMP, MPT and Mo-Moco. The right side shows the intermediates formed during the reactions, which are not stable. In bacteria, Mo-MPT is further modified by the addition of nucleotides, GMP or CMP. The proteins involved in the reactions are colored in red (bacteria), green (plants) and grey (humans). Moco can be further modified by the replacement of one oxo ligand by a sulfido ligand, forming the sulfurated form of Mo-MPT. Additional cosubstrates required for the reactions are colored in blue.

Table 3.1 Proteins for Moco biosynthesis in bacteria, plants and humans and the catalyzed reactions.

Bacteria	Humans	Plants	Reaction catalyzed
MoaA	MOCS1A	Cnx2	Formation of 3′,8-cH_2GTP from 5′GTP
MoaC	MOCS1B	Cnx3	Formation of cPMP from 3′,8-cH_2GTP
MoaD	MOCS2A	Cnx6	Sulfur transfer to cPMP, formation of MPT
MoaE	MOCS2B	Cnx7	Binding of cPMP, formation of MPT
MoeB	MOCS3	Cnx5	Adenylation of MPT synthase small subunit
MogA	GEPHYRIN-G	Cnx1-G	Formation of MPT–AMP
MoeA	GEPHYRIN-E	Cnx1-E	Molybdate insertion into MPT–AMP, Mo-MPT formation
MobA	—	—	Formation of bis-Mo-MPT, GMP transfer to bis-Mo-MPT, formation of bis-MGD
MocA	—	—	CMP transfer to Mo-MPT, formation of MCD
IscS	NFS1	Nfs1	L-Cysteine desulfurase, mobilizing sulfur from L-cysteine in form of a protein-bound persulfide group
	MOCS3-RLD	Cnx5-RLD	Sulfur-relay proteins involved in sulfur transfer of the persulfide-sulfur from the L-cysteine desulfurase to form the thiocarboxylate group on the small subunit of MPT synthase

with high sequence identity to the bacterial proteins involved in Moco biosynthesis are found in all organisms producing Moco; however, the gene organization differs from one organism to the next.

Some eukaryotic proteins represent a fusion of different components that are separate gene products in bacterial systems. In humans, genes and the encoded proteins have been designated MOCS (*mo*lybdenum *c*ofactor *s*ynthesis).[5,22,23] Six proteins that directly catalyze Moco biosynthesis have been identified in humans (Table 3.1): MOCS1A (MoaA), MOCS1B (MoaC), MOCS2A (MoaD), MOCS2B (MoaE), MOCS3 (MoeB) and GEPHYRIN (MogA–MoeA fusion).[24] A total of five enzymes binding Moco were identified in humans, namely sulfite oxidase (SO), xanthine oxidoreductase (XOR), aldehyde oxidase (AOX1) and the two mitochondrial amidoxime reducing components, mARC1 and mARC2.[24,25] Of these five enzymes SO is the only one shown to be essential for humans.[26,27]

In plants, the corresponding genes are designated cnx (*c*ofactor for *n*itrate reductase and *x*anthine dehydrogenase). In analogy to humans, six proteins that directly catalyze Moco biosynthesis have been identified in *Arabidopsis thaliana* (Table 3.1): Cnx2 (MoaA), Cnx3 (MoaC), Cnx6 (MoaD), Cnx7 (MoaE), Cnx5 (MoeB) and Cnx1 (MoeA–MogA fusion). In addition to the Moco-containing enzymes present in humans, plants additionally possess a nitrate reductase that is involved in nitrate assimilation.[24] For comparison, Figure 3.1

and Table 3.1 show the names of the corresponding bacterial, human and plant proteins. In the following we will briefly characterize the individual steps of Moco biosynthesis.

3.2 Formation of cPMP from 5′GTP

The biosynthesis of Moco starts from 5′GTP, and a complex rearrangement reaction results in the formation of cPMP, presenting the first stable intermediate of Moco biosynthesis[14] (Figure 3.1).[28,29] cPMP is the most stable intermediate of Moco biosynthesis with an estimated half-life of several hours at a low pH.[30] It has, thus, been possible to solve its structure (purified from *E. coli* cells) using ^{1}H-NMR, while structural elucidation of other intermediates (including MPT, MPT–AMP and Moco) has required crystallographic characterization of protein–ligand complexes.[31] cPMP is still sulfur-free but it already has the tricyclic pyranopterin structure of the mature cofactor. The synthesis of cPMP is catalyzed by the two enzymes MoaA and MoaC (Figure 3.1).[32,33] MoaA has been classified as a member of the superfamily of *S*-adenosylmethionine (SAM)-dependent radical enzymes.[34,35] The X-ray structure of MoaA was solved from the *Staphylococcus aureus* enzyme and presented a homodimer with two [4Fe–4S] clusters bound to each monomer.[36] The N-terminal [4Fe–4S] cluster thereby binds the SAM molecule and generates the 5′-deoxyadenosyl radical, which subsequently initiates the transformation of 5′GTP bound by the C-terminal [4Fe–4S] cluster.[35,37] In the course of a complex rearrangement reaction, the C8 of GTP is inserted between the C2′ and C3′ carbons of the ribose,[32,33] resulting in formation of an (8*S*)-3′,8-cyclo-7,8-dihydroguanosine 5′triphosphate (3′,8-cH_2GTP) intermediate (Figure 3.1). This intermediate formed by MoaA is then converted by MoaC to cPMP, a reaction which involves pyrophosphate cleavage and formation of the cyclic phosphate group.[32] MoaC subsequently catalyzes the complex rearrangement of 3′,8-cH_2GTP into cPMP *via* a mechanism involving dynamic conformational changes of substrate and enzyme.[38] The crystal structures of MoaC from *E. coli* and *Thermus thermophilus* have demonstrated that the protein forms an $(\alpha_2)_3$ complex.[39,40]

In humans, the similar reactions are catalyzed by the two proteins MOCS1A and MOCS1B, which are encoded by the *MOCS1* locus and present the homologs to the bacterial MoaA and MoaC proteins, respectively.[41] The corresponding homologs in plants are Cnx2 and Cnx3. Since the human and plant homologs are able to complement the *E. coli* mutant strains with a deletion in *moaA* and *moaC*, respectively, the reaction catalyzed by the MOCS1A/1B or Cnx2/3 pairs is expected to be similar. In eukaryotes, conversion of 5′GTP to cPMP takes place in the mitochondrion. In addition, an unusual bicistronic transcript of the *MOCS1* locus has been identified, with open reading frames for both *MOCS1A* and *MOCS1B* in a single transcript, separated by a stop codon.[22,42] Different splice variants of the *MOCS1* locus are synthesized and a fusion protein combining MOCS1A and MOCS1B was identified.[43] While the MOCS1A protein is also expressed as a separate protein, so far no evidence exists for the separate expression of *MOCS1B*.[44] Both transcripts of MOCS1A contain a mitochondrial targeting sequence for the translocation to the mitochondria. After its synthesis in mitochondria by the MOCS1A and MOCS1A-1B proteins, cPMP is exported to the cytosol where cPMP is further modified. This export task involves the transporter protein Atm3 in plants,[45] which is localized in the inner membrane of mitochondria and belongs to the family of ATP-binding cassette (ABC) transporters. Surprisingly, Atm3 functions also in exporting another – however, as yet unidentified – species essential for Fe–S cluster synthesis in the cytosol. However, when the Atm3 exporter is mutated, cPMP accumulates within mitochondria and the cytosol becomes depleted, with the consequence that Moco levels and Mo enzyme activities decrease in the cell. The precise role of Atm3 is still unknown. The human analog to Atm3 is designated ABCB7.

3.3 Formation of MPT by Sulfur Insertion into cPMP

In the second step of Moco biosynthesis, two sulfur atoms are inserted into cPMP by MPT synthase thereby converting it to MPT.[15,46–49] The two sulfur atoms are incorporated at the C1′ and C2′ positions of the cPMP backbone.[50] MPT synthase is composed of two MoaD and two MoaE subunits in bacteria and forms an $(\alpha\beta)_2$ heterodimer.[51] MoaD was shown to act as direct sulfur donor in this reaction with the sulfur atom being provided by the C-terminal thiocarboxylate group.[52,53] The X-ray structures were solved of the *E. coli* and *S. aureus* MoaD/MoaE protein complexes and have shown that the central dimer is formed by two MoaE subunits and that the MoaD subunits are located at each end of the MoaE dimer with the C-terminal thiocarboxylate group reaching into the cPMP binding pocket of MoaE.[50,51] Each MoaE subunit binds one cPMP molecule and both MoaE–MoaD dimers were shown to act independently during the reaction.[54] A reaction mechanism has been proposed suggesting that the first sulfur is added by one MoaD molecule first at the C2′ position of cPMP, and a hemisulfurated intermediate is formed which could be isolated (Figure 3.1). This reaction is coupled to the hydrolysis cyclic phosphate group of cPMP.[50] After the transfer of the first sulfur from the thiocarboxylate group of the first MoaD to cPMP, MoaD dissociates from the MPT synthase complex[50,54] and a new MoaD molecule binds to MoaE. The second part of the reaction involves the attack of the C1′ atom of the hemisulfurated intermediate, the sulfurtransfer to this position and the formation of MPT (Figure 3.1).

After the reaction, MPT synthase is present in an inactive form and for a turnover, the thiocarboxylate group has to be rebuilt on MoaD. This regeneration of MoaD is catalyzed by the MoeB protein.[55,56] After the completion of MPT synthesis, MoaD dissociates from MoaE and reassociates with the MoeB protein. The X-ray structure of the *E. coli* $(MoaD–MoeB)_2$ protein complex was solved and intermediates of the reaction revealed that the C-terminus of MoaD is first activated by the formation of an acyl–adenylate group at its terminal glycine under ATP consumption.[55,57,58] In the activated MoaD–AMP complex MoaD then receives the sulfur from the pyridoxal-dependent IscS sulfurtransferase to form the MoaD–thiocarboxylate group and additionally releasing $(MoeB)_2$ and AMP. The released MoaD–SH subsequently reassociates with MoaE to form the active MPT synthase heterotetramer catalyzing another round of MPT formation (Figure 3.1).[58–61]

Recently, a role of TusA for Moco biosynthesis has been identified,[62] implying that IscS does not directly interact with MoaD, but rather transfers the sulfur to TusA in a sulfur relay system, which then transfers the sulfur to MoaD. It has been shown that deletion of *tusA* in *E. coli* affects the activity of molybdoenzymes under both aerobic and anaerobic conditions. Under aerobic conditions, the Δ*tusA* strain has an overall low MPT content and accumulates cPMP. Under anaerobic conditions, nitrate reductase activity is reduced by only 50%, suggesting that TusA is not essential for Moco biosynthesis and can be replaced by other proteins. The rhodanese-like protein YnjE or the L-cysteine desulfurase SufS are able to replace IscS and TusA in its role as sulfur donor for MoaD, and under anaerobic conditions the expression of *ynjE* is known to increase in the absence of TusA.[62,63] Additionally, SufS, an additional L-cysteine desulfurase in *E. coli*, can replace IscS in its role in providing the sulfur under certain conditions. SufS is thought to be involved principally in the synthesis of FeS clusters which have been oxidatively damaged. It thus appears that redundant sulfur transfer systems exist in the cell.

The TusA protein has been implicated in the formation of s^2U of the modified wobble base 5-methylaminomethyl $(mnm)^5s^2U$ in tRNA of $tRNA^{Lys}$, $tRNA^{Glu}$ and $tRNA^{Gln}$.[64] This thiomodification is crucial for precise decoding of the mRNA. TusA has been shown to transfer the sulfur from IscS on to the TusBCD complex and finally, *via* TusE, to MnmA. The crystal structure of MnmA has revealed a P-loop in its active site, which is involved in binding of tRNA and directly activates the C2 position of the uracil ring at position 34 under ATP consumption by forming an adenylylate intermediate.[65] TusA is thus involved in at least two distinct sulfur transfer pathways, Moco biosynthesis and the biosynthesis of s^2U in tRNA. It has been suggested that TusA might provide a mechanism for the balanced distribution of IscS-labilized sulfur among the several sulfur-requiring metabolic pathways in *E. coli*.[62]

In humans, the homologs to the bacterial MoaD and MoaE proteins are MOCS2A and MOCS2B, respectively. The two genes are located in the MOCS2 locus on the human genome and by analogy to the *MOCS1* locus, *MOCS2A* and *MOCS2B* are also transcribed in a bicistronic transcript with overlapping reading frames.[66] *In vitro* translation and mutagenesis experiments have shown that independent transcripts of the two genes are formed. Two alternative *MOCS2* splice forms were identified, resulting in different first exons and leading to the separate *MOCS2A* and *MOCS2B* transcripts.[66,67]

The human MPT synthase reaction has been studied using the separately expressed MOCS2A and MOCS2B subunits in *E. coli*, and a similar mechanism has been described to the mechanism catalyzed by the *E. coli* MoaD and MoaE proteins.[68] It has been shown that both separately purified MOCS2A and MOCS2B subunits readily assemble into a functional MPT synthase, which converts cPMP to MPT in an *in vitro* assay system. Further, it was shown that also active hybrid proteins with the *E. coli* proteins were formed and that the MoaD/MOCS2B and MOCS2A/MoaE hybrid proteins were active, showing that the MPT synthase reaction is highly conserved in pro- and eukaryotes. The corresponding homologs in plants are Cnx6 and Cnx7.[52]

The human MOCS3 protein is the homolog of *E. coli* MoeB (the corresponding protein is Cnx5 in plants). Sequence alignments of MOCS3 revealed that the protein is composed of two domains and that the N-terminal domain shares amino acid sequence identities with *E. coli* MoeB and that an additional C-terminal domain is present in MOCS3, which shares amino acid sequence identities with rhodaneses (MOCS3-rhodanese like domain, RLD).[69,70] MOCS3-RLD was thereby shown to act as direct sulfur donor for the formation of the thiocarboxylate group on MOCS2A.[69] A persulfide group has been identified on the highly conserved cysteine residue of the six amino acid active site loop of the MOCS3 C-terminal rhodanese domain.[71] In total, MOCS3 can be regarded as a multi-functional and multi-domain protein combining the adenylylation activity of MOCS2A and the subsequent sulfur transfer activity of MOCS2B. By analogy to the *E. coli* system, it has been shown that the sulfur for the MOCS3-RLD persulfide group is likely provided from the L-cysteine desulfurase NFS1 present in the cytosol in humans cells. Here, the sulfur is transferred in a relay system from L-cysteine involving the proteins NFS1, MOCS3 and MOCS2A.[72,73]

MOCS3 has also been shown to play a role in tRNA thiolation in humans. For tRNA thiolation, MOCS3 directly interacts with the protein URM1, activating and sulfurating it in a similar manner to that described above for MOCS2A. For tRNA thiolation in humans it has been shown that the proteins MOCS3, URM1, CTU1 and CTU2 are essential for the biogenesis of the 2-thiouridine group in cytoplasmic tRNAs Lys, Gln and Glu.[74] The URM1 protein (ubiquitin-related modifier) has a ubiquitin-like β-grasp-fold and contains a conserved C-terminal double glycine-motif on which a thiocarboxylate group is formed for direct sulfurtransfer to mcm^5U_{34} in tRNA.[75–78] The sulfur is transferred to URM1 *via* MOCS3.[79] MOCS3 activates URM1 in the presence of ATP by formation of an acyl–adenylylate bond. URM1 is further transferred to a persulfide group on Cys412 of MOCS3-RLD, forming a disulfide bond. The disulfide bond can be cleaved by C239 of MOCS3, which releases thiocarboxylated URM1. URM1-thiocarboxylate further transfers the sulfur onto uridine 34 of the tRNA, aided by the CTU1 and CTU2 proteins and ATP consumption.[79] As described above, NFS1 supplies sulfur to MOCS3, which links tRNA modification to the sulfur supplied by L-cysteine desulfurases.[80] Additionally, and in contrast to MOCS2A, URM1 conjugates to its target *via* a lysine–isopeptide bond, revealing that it is a dual-function protein.[74] Among the targets for urmylation are the components of the tRNA thiolation machinery itself (CTU1, CTU2) and a deubiquitinating-enzyme (USP15). Other targets include proteins involved in nuclear transport such as the cellular apoptosis susceptibility protein, CAS, which promotes the shuttling of proteins between cytosol and nucleus.[74] However, the role of the protein urmylation is not as yet fully elucidated and may only occur under certain conditions in the cell such as oxidative stress.

3.4 Insertion of Molybdate into MPT

After the formation of the MPT dithiolene group, the molybdate ion can be coordinated to MPT. In *E. coli*, the proteins MoeA and MogA are involved in the molybdate insertion process. MogA has been well characterized and the crystal structure of the proteins revealed exhibits a four-domain structure, with one domain being structurally related to MoeA.[81–83] It therefore has been suggested that MoeA and MogA originated from a common ancestor and it was shown by biochemical studies that both proteins are essential for the incorporation of molybdenum into the cofactor by a multistep process.[84] Both proteins nevertheless have distinct functions during this process: MogA activated the MPT molecule for the subsequent molybdate insertion process and forms an MPT–adenylylate intermediate under ATP consumption, while MoeA helps to insert the molybdate atom and cleaves the MPT–AMP bond.[85] In higher eukaryotes, in contrast, the two separate proteins present in bacteria are fused to a single two-domain protein (Cnx1 in plants[86] and GEPHYRIN in mammals[87]). The two domains of these proteins are, thus, referred to as the G-domain (homologous to MogA) and the E-domain (homologous to MoeA). The novel intermediate MPT–AMP (Figure 3.1) was first identified in the crystal structure derived of the Cnx1-G domain from *Arabidopsis thaliana*.[88–90] Apparently, the coupling of the two functions into a single protein has been proven as beneficial during evolution for the direct channeling of product-substrate of the labile intermediate MPT–AMP from the G-domain to the E-domain.[91] However, studies in bacteria showed that active Moco can alternatively be synthesized *in vitro* by coordination of molybdate to MPT without the requirement of any proteins.[92] Therefore, the ATP-dependent activation of MPT might only be required under physiological molybdate concentrations.

In humans, the protein GEPHYRIN catalyzes the molybdate insertion step in a similar manner to that described above for the plant and *E. coli* proteins.[87,91] In GEPHYRIN, however, the order of the E- and G-domains is reversed as compared to Cnx1, with the G domain being located on its N-terminus and the E-domain on its C-terminus.[93,94] GEPHYRIN was shown to be a moonlighting protein with a dual function additionally acting in synaptic anchoring of inhibitory ligand-gated ion channels.[95] Alternative splicing has been proposed to contribute to the functional diversity of GEPHYRIN.[96]

3.5 Further Modification of Moco

3.5.1 Formation of bis-MGD

While the first three steps of Moco biosynthesis are conserved from bacteria to humans, a fourth Moco modification step exists solely in bacteria and archaea. Here, the formed Moco is further modified by the attachment of an additional nucleotide to the phosphate group of MPT. The common characteristic of proteins of the DMSO reductase family is that all proteins contain the bis-MGD cofactor in which two molybdopterin guanine dinucleotide molecules are ligated to the molybdenum atom thereby forming the so-called bis-MGD structure.[7] By studies on the *E. coli* MobA protein, synthesis of bis-MGD has been shown to occur in a two-step reaction involving only the preformed Mo-MPT molecule, MobA and Mg-GTP.[18] In the first step, MobA forms the bis-Mo-MPT intermediate from two Mo-MPT molecules as substrate, which are directly delivered from MoeA after molybdate insertion into MPT (Figure 3.1).[97] In the second step, two equivalents of GMP (from GTP) are added to the C4′ phosphate of each MPT present in the bis-Mo-MPT structure and the bis-MGD cofactor is formed.[98,99] While MobA apparently catalyzed both steps of the reaction, details of the molecular mechanism of bis-Mo-MPT formation and its binding mode to MobA have not been completely resolved to date.[18] The crystal structure of MobA showed that the N-terminal domain adopts a nucleotide (GTP)-binding Rossmann fold, and a possible MPT binding site is located in the C-terminal half of the protein.[99,100] Since two Mo-MPT moieties have to be bound to monomeric MobA for bis-Mo-MPT formation, it has been suggested that bis-MPT binding occurs on one MobA monomer by using both the MPT and GTP binding site of the protein (Figure 3.1).[18] During the formation of bis-MPT from two Mo-MPT molecules, one molecule of molybdate has to be released. After the attachment of two GMP molecules to the bis-Mo-MPT intermediate, the bis-MGD cofactor is formed and released from MobA. For the insertion of bis-MGD into apoenzymes specific molecular chaperones are required (discussed in detail in Chapter 4).

Further, it has also been shown that the bis-MGD cofactor can be modified by the insertion of a sulfido-group. For proteins like formate dehydrogenase (*E. coli* FdhF or *R. capsulatus* FdsGBA) or periplasmic nitrate reductase (*Cupriavidus necator* NapAB) it has been shown that the molybdenum center ligates a Mo=S ligand rather than Mo=O atom.[7,101–103] Sulfuration of bis-MGD was shown to be catalyzed by the involvement of specific molecular chaperones (*e.g. E. coli* FdhD for FdhF maturation) in conjunction with a sulfur transferase (*e.g.* the *E. coli* L-cysteine desulfurase IscS).[101,104] During this reaction, a persulfide sulfur is transferred from the L-cysteine desulfurase to the bound bis-MGD cofactor on the specific chaperone (discussed in detail in Chapter 4).

3.5.2 The Formation of the MCD Cofactor

Exclusively in enzymes of the xanthine oxidase family in prokaryotes, formed Moco can be further modified by the attachment of a CMP nucleotide to the phosphate group of Mo-MPT, forming the molybdopterin cytosine dinucleotide (MCD) cofactor.[61,105] In contrast to the GMP modification of Mo-MPT, the formation of a bis-MCD cofactor has not been described so far. The CMP attachment to Mo-MPT is catalyzed by the *E. coli* MocA protein, which shares a high amino acid sequence identity to MobA.[17] Due to the similarities of both proteins, the overall reaction of nucleotide attachment is expected to be similar for both proteins, however, the first part of the reaction, namely the formation of bis-Mo-MPT, is not necessarily formed on MocA.[105] Both MobA and MocA were shown to be specific for their respective nucleotides, GTP and CTP, respectively. After MCD formation, the MCD cofactor is further transferred to specific chaperones for additional modification by incorporation of a sulfido ligand that is essential for enzyme activity for all enzymes of the xanthine oxidase family.

The specific Moco binding chaperones in the case of the xanthine oxidase family belong to the XdhC family of molecular chaperones (including XdhC from *Rhodobacter capsulatus* or PaoD and YqeB from *E. coli*).[61,106] The *R. capsulatus* XdhC protein is the best-studied of the family and has been shown to bind Moco (Mo-MPT) and thereby protects it from oxidation. Again, an L-cysteine desulfurase provides the sulfur originating from L-cysteine, and further replacement of the equatorial Mo=O of Mo-MPT by a Mo=S group occurs on the XdhC-like chaperone[107] (Figure 3.1). After sulfuration, the mature cofactor is inserted into the target apoenzyme.[107,108] Since Moco is deeply buried in these prefolded proteins, the XdhC-type chaperones may additionally be directly involved in the cofactor insertion process.

3.5.3 Moco Sulfuration in Eukaryotes

In eukaryotes, sulfuration of Mo-MPT occurs prior to incorporation into members of the xanthine oxidase family of enzymes and is catalyzed by the Moco sulfurase, named ABA3 in plants, and HMCS in humans. These proteins comprise a homodimeric two-domain protein[109] with their N-terminal domains sharing structural and functional homologies to the bacterial L-cysteine desulfurases IscS. The N-terminal domain of ABA3 was shown to decompose L-cysteine to yield L-alanine and persulfide-sulfur in a pyridoxal phosphate-dependent manner.[110] The C-terminal domain of ABA3 binds Moco, and the Moco–sulfido group is directly formed by sulfurtransfer from the N-terminal domain.[111,112]

3.6 Trafficking of Moco in the Cell

After the biosynthesis of Moco, Moco has to be inserted into the target proteins. For Moco trafficking within the cell, a *M*oco *c*arrier *p*rotein (MCP) has been identified in the green algae *Chlamydomonas reinhardtii*.[113] The protein is a homotetramer capable of binding four equivalents of Moco, thereby protecting it against oxidation.[114] It has been demonstrated that Moco from MCP can be transferred directly to apo-nitrate reductase from *Neurospora crassa* to produce an active protein. In *A. thaliana*, eight different MCP-related proteins have been identified, which can be grouped into one family, and each of these proteins was shown to bind Moco with low affinity.[115] These proteins seem to be unique to eukaryotes, since homologous proteins do not exist in bacteria. However, due to the labile nature of Moco, it is not expected that Moco can exist in a "free" protein-unbound form in the cell. In bacteria, molecular chaperones specific for each molybdoenzyme exist instead, which bind Moco, perform final modification of Moco and specifically insert the correct form of Moco into the respective target proteins. The mechanism underlying Moco insertion into bacterial molybdoenzymes is described in detail in Chapter 4.

3.7 Conclusions

In general, the biosynthesis of Moco is ubiquitous and highly conserved in all organisms from bacteria to humans. However, small differences exist in Moco formation among the different phyla. For instance, in higher eukaryotes many individual Moco biosynthesis proteins appear to have several cellular roles and some proteins are shared between different biosynthetic pathways. Further, while a wide range of variations on the basic Moco structure exists in bacteria, no dinucleotide forms of Moco have been identified in eukaryotes to date. Additionally, bacteria contain a large variety of more than 60 different molybdoenzymes being involved in specific, however, usually non-essential redox-reactions. In contrast, in humans only four different molybdoenzymes have been identified, and a defect in Moco biosynthesis is lethal due to the loss of sulfite oxidase activity (see Chapter 7). A therapy for Moco deficiency of group I patients has been developed,[116] however, to date no effective therapy exists for the treatment of isolated sulfite oxidase deficiency. In the future, a broader view is necessary to elucidate the multiple roles of the proteins involved in Moco biosynthesis and their link to other cellular pathways.

Acknowledgements

The authors thank all current and former members of their research groups in addition to collaboration partners who were involved in this work over the past years and decades. The work was mainly supported by continuous grants of the Deutsche Forschungsgemeinschaft to S.L. and to R.R.M.

References

1. K. V. Rajagopalan and J. L. Johnson, *J. Biol. Chem.*, 1992, **267**, 10199.
2. S. Leimkühler, M. M. Wuebbens and K. V. Rajagopalan, *Coord. Chem. Rev.*, 2011, **255**, 1129.
3. S. Leimkühler, *Adv. Biol.*, 2014, **2014**, 808569.

4. R. R. Mendel, *J. Biol. Chem.*, 2013, **288**, 13165.
5. R. R. Mendel and G. Schwarz, *Coord. Chem. Rev.*, 2011, **255**, 1145.
6. R. R. Mendel and S. Leimkuhler, *JBIC, J. Biol. Inorg. Chem.*, 2015, **20**, 337.
7. R. Hille, J. Hall and P. Basu, *Chem. Rev.*, 2014, **114**, 3963–4038.
8. R. Hille, *Chem. Rev.*, 1996, **96**, 2757.
9. Y. Zhang, S. Rump and V. N. Gladyshev, *Coord. Chem. Rev.*, 2011, **255**, 1206.
10. F. Hagen, *Coord. Chem. Rev.*, 2011, **255**, 1117.
11. A. Nason, K. Y. Lee, S. S. Pan, P. A. Ketchum, A. Lamberti and J. DeVries, *Proc. Natl. Acad. Sci. U. S. A.*, 1971, **68**, 3242.
12. R. R. Mendel, in *Plant Biotechnology and Development – Current Topics in Plant Molecular Biology*, ed. P. M. Gresshoff, CRC Press, Boca Raton and London, 1992, vol. 1, p. 11.
13. K. V. Rajagopalan, in *Escherichia coli and Salmonella. Cellular and Molecular Biology*, ed. F. C. Neidhardt, ASM Press, Washington, DC, 1996, vol. I, p. 674.
14. M. M. Wuebbens and K. V. Rajagopalan, *J. Biol. Chem.*, 1993, **268**, 13493.
15. D. M. Pitterle, J. L. Johnson and K. V. Rajagopalan, *J. Biol. Chem.*, 1993, **268**, 13506.
16. M. S. Joshi, J. L. Johnson and K. V. Rajagopalan, *J. Bacteriol.*, 1996, **178**, 4310.
17. M. Neumann, F. Seduk, C. Iobbi-Nivol and S. Leimkühler, *J. Biol. Chem.*, 2011, **286**, 1400.
18. S. Reschke, K. G. Sigfridsson, P. Kaufmann, N. Leidel, S. Horn, K. Gast, C. Schulzke, M. Haumann and S. Leimkühler, *J. Biol. Chem.*, 2013, **288**, 29736.
19. K. T. Shanmugam, V. Stewart, R. P. Gunsalus, D. H. Boxer, J. A. Cole, M. Chippaux, J. A. DeMoss, G. Giordano, E. C. C. Lin and K. V. Rajagopalan, *Mol. Microbiol.*, 1992, **6**, 3452.
20. S. Gerber, M. Comellas-Bigler, B. A. Goetz and K. P. Locher, *Science*, 2008, **321**, 246.
21. J. A. Maupin-Furlow, J. K. Rosentel, J. H. Lee, U. Deppenmeier, R. P. Gunsalus and K. T. Shanmugam, *J. Bacteriol.*, 1995, **177**, 4851.
22. J. Reiss, N. Cohen, C. Dorche, H. Mandel, R. R. Mendel, B. Stallmeyer, M. T. Zabot and T. Dierks, *Nat. Genet.*, 1998, **20**, 51.
23. J. Reiss, *Hum. Genet.*, 2000, **106**, 157.
24. R. R. Mendel and T. Kruse, *Biochim. Biophys. Acta*, 2012, **1823**, 1568.
25. R. Hille, T. Nishino and F. Bittner, *Coord. Chem. Rev.*, 2011, **255**, 1179.
26. M. Duran, F. A. Beemer, C. van der Heiden, J. Korteland, P. K. de Bree, M. Brink, S. K. Wadman and I. Lombeck, *J. Inherited Metab. Dis.*, 1978, **1**, 175.
27. M. Duran, P. K. de Bree, J. B. C. de Klerk, L. Dorland and R. Berger, *Int. J. Pediatr.*, 1996, **11**, 334.
28. S. Leimkühler, M. Wuebbens and K. V. Rajagopalan, *Coord. Chem. Rev.*, 2011, **255**, 1129.
29. J. A. Santamaria-Araujo, B. Fischer, T. Otte, M. Nimtz, R. R. Mendel, V. Wray and G. Schwarz, *J. Biol. Chem.*, 2004, **279**, 15994.

30. M. M. Wuebbens and K. V. Rajagopalan, *J. Biol. Chem.*, 1993, **268**, 13493.
31. J. Kuper, A. Llamas, H. J. Hecht, R. R. Mendel and G. Schwarz, *Nature*, 2004, **430**, 806.
32. B. M. Hover, A. Loksztejn, A. A. Ribeiro and K. Yokoyama, *J. Am. Chem. Soc.*, 2013, **135**, 7019.
33. A. P. Mehta, S. H. Abdelwahed and T. P. Begley, *J. Am. Chem. Soc.*, 2013, **135**, 10883.
34. H. J. Sofia, G. Chen, B. G. Hetzler, J. F. Reyes-Spindola and N. E. Miller, *Nucleic Acids Res.*, 2001, **29**, 1097.
35. P. Hänzelmann, H. L. Hernandez, C. Menzel, R. Garcia-Serres, B. H. Huynh, M. K. Johnson, R. R. Mendel and H. Schindelin, *J. Biol. Chem.*, 2004, **279**, 34721.
36. P. Hänzelmann and H. Schindelin, *Proc. Natl. Acad. Sci. U. S. A.*, 2004, **101**, 12870.
37. P. Hänzelmann and H. Schindelin, *Proc. Natl. Acad. Sci. U. S. A.*, 2006, **103**, 6829.
38. B. M. Hover, N. K. Tonthat, M. A. Schumacher and K. Yokoyama, *Proc. Natl. Acad. Sci. U. S. A.*, 2015, **112**, 6347.
39. M. M. Wuebbens, M. T. Liu, K. Rajagopalan and H. Schindelin, *Structure*, 2000, **8**, 709.
40. S. P. Kanaujia, J. Jeyakanthan, N. Nakagawa, S. Balasubramaniam, A. Shinkai, S. Kuramitsu, S. Yokoyama and K. Sekar, *Acta Crystallogr., Sect. D: Biol. Crystallogr.*, 2010, **66**, 821.
41. J. Reiss, E. Christensen, G. Kurlemann, M.-T. Zabot and C. Dorche, *Hum. Genet.*, 1998, **103**, 639.
42. S. Gross-Hard and J. Reiss, *Mol. Genet. Metab.*, 2002, **76**, 340.
43. T. A. Gray and R. D. Nicholls, *RNA*, 2000, **6**, 928.
44. P. Hänzelmann, G. Schwarz and R. R. Mendel, *J. Biol. Chem.*, 2002, **277**, 18303.
45. J. Teschner, N. Lachmann, J. Schulze, M. Geisler, K. Selbach, J. Santamaria-Araujo, J. Balk, R. R. Mendel and F. Bittner, *Plant Cell*, 2010, **22**, 468.
46. D. M. Pitterle, J. L. Johnson and K. V. Rajagopalan, *FASEB J.*, 1990, **4**, A1957.
47. D. M. Pitterle and K. V. Rajagopalan, *J. Bacteriol.*, 1989, **171**, 3373.
48. D. M. Pitterle and K. V. Rajagopalan, *FASEB J.*, 1991, **5**, A468.
49. D. M. Pitterle and K. V. Rajagopalan, *J. Biol. Chem.*, 1993, **268**, 13499.
50. J. N. Daniels, M. M. Wuebbens, K. V. Rajagopalan and H. Schindelin, *Biochemistry*, 2008, **47**, 615–626.
51. M. J. Rudolph, M. M. Wuebbens, K. V. Rajagopalan and H. Schindelin, *Nat. Struct. Biol.*, 2001, **8**, 42.
52. G. Gutzke, B. Fischer, R. R. Mendel and G. Schwarz, *J. Biol. Chem.*, 2001, **276**, 36268.
53. S. Leimkühler, A. Freuer, J. A. Araujo, K. V. Rajagopalan and R. R. Mendel, *J. Biol. Chem.*, 2003, **278**, 26127.
54. M. M. Wuebbens and K. V. Rajagopalan, *J. Biol. Chem.*, 2003, **278**, 14523.
55. S. Leimkühler, M. M. Wuebbens and K. V. Rajagopalan, *J. Biol. Chem.*, 2001, **276**, 34695.

56. H. Schindelin, in *Protein Degradation: Ubiquitin and the Chemistry of Life*, ed. R. J. Mayer, A. Ciechanover and M. Rechsteiner, WILEY-VCH, Weinheim, 2005, vol. I, p. 21.
57. M. W. Lake, M. M. Wuebbens, K. V. Rajagopalan and H. Schindelin, *Nature*, 2001, **414**, 325.
58. J. Schmitz, M. M. Wuebbens, K. V. Rajagopalan and S. Leimkühler, *Biochemistry*, 2007, **46**, 909.
59. Y. Tong, M. M. Wuebbens, K. V. Rajagopalan and M. C. Fitzgerald, *Biochemistry*, 2005, **44**, 2595.
60. S. Leimkühler and K. V. Rajagopalan, *J. Biol. Chem.*, 2001, **276**, 22024.
61. C. Iobbi-Nivol and S. Leimkühler, *Biochim. Biophys. Acta*, 2013, **1827**, 1086.
62. J. U. Dahl, C. Radon, M. Bühning, M. Nimtz, L. I. Leichert, Y. Denis, C. Jourlin-Castelli, C. Iobbi-Nivol, V. Mejean and S. Leimkühler, *J. Biol. Chem.*, 2013, **288**, 5426.
63. J. U. Dahl, A. Urban, A. Bolte, P. Sriyabhaya, J. L. Donahue, M. Nimtz, T. J. Larson and S. Leimkühler, *J. Biol. Chem.*, 2011, **286**, 35801.
64. Y. Ikeuchi, N. Shigi, J. Kato, A. Nishimura and T. Suzuki, *Mol. Cell*, 2006, **21**, 97.
65. T. Numata, Y. Ikeuchi, S. Fukai, T. Suzuki and O. Nureki, *Nature*, 2006, **442**, 419.
66. B. Stallmeyer, G. Drugeon, J. Reiss, A. L. Haenni and R. R. Mendel, *Am. J. Hum. Genet.*, 1999, **64**, 698.
67. R. Hahnewald, S. Leimkühler, A. Vilaseca, C. Acquaviva-Bourdain, U. Lenz and J. Reiss, *Mol. Genet. Metab.*, 2006, **89**, 210–213.
68. S. Leimkühler, A. Freuer, J. A. Santamaria Araujo, K. V. Rajagopalan and R. R. Mendel, *J. Biol. Chem.*, 2003, **278**, 26127.
69. A. Matthies, K. V. Rajagopalan, R. R. Mendel and S. Leimkühler, *Proc. Natl. Acad. Sci. U. S. A.*, 2004, **101**, 5946.
70. D. Bordo and P. Bork, *EMBO Rep.*, 2002, **3**, 741.
71. A. Matthies, M. Nimtz and S. Leimkühler, *Biochemistry*, 2005, **44**, 7912.
72. K. Krepinsky and S. Leimkühler, *FEBS J.*, 2007, **274**, 2778.
73. Z. Marelja, W. Stöcklein, M. Nimtz and S. Leimkühler, *J. Biol. Chem.*, 2008, **283**, 25178.
74. A. G. Van der Veen, K. Schorpp, C. Schlieker, L. Buti, J. R. Damon, E. Spooner, H. L. Ploegh and S. Jentsch, *Proc. Natl. Acad. Sci. U. S. A.*, 2011, **108**, 1763.
75. S. Leidel, P. G. Pedrioli, T. Bucher, R. Brost, M. Costanzo, A. Schmidt, R. Aebersold, C. Boone, K. Hofmann and M. Peter, *Nature*, 2009, **458**, 228.
76. P. G. Pedrioli, S. Leidel and K. Hofmann, *EMBO Rep.*, 2008, **9**, 1196.
77. J. Xu, J. Zhang, L. Wang, J. Zhou, H. Huang, J. Wu, Y. Zhong and Y. Shi, *Proc. Natl. Acad. Sci. U. S. A.*, 2006, **103**, 11625.
78. J. Schmitz, M. Mullick Chowdhury, P. Hänzelmann, M. Nimtz, E. Y. Lee, H. Schindelin and S. Leimkühler, *Biochemistry*, 2008, **47**, 6479.
79. M. M. Chowdhury, C. Dosche, H. G. Löhmannsröben and S. Leimkühler, *J. Biol. Chem.*, 2012, **287**, 17297.
80. Z. Marelja, M. Mullick Chowdhury, C. Dosche, C. Hille, O. Baumann, H. G. Löhmannsröben and S. Leimkühler, *PLoS One*, 2013, **8**, e60869.

81. M. T. Liu, M. M. Wuebbens, K. V. Rajagopalan and H. Schindelin, *J. Biol. Chem.*, 2000, **275**, 1814.
82. S. Xiang, J. Nichols, K. V. Rajagopalan and H. Schindelin, *Structure*, 2001, **9**, 299.
83. J. D. Schrag, W. Huang, J. Sivaraman, C. Smith, J. Plamondon, R. Larocque, A. Matte and M. Cygler, *J. Mol. Biol.*, 2001, **310**, 419.
84. J. Nichols and K. V. Rajagopalan, *J. Biol. Chem.*, 2002, **277**, 24995.
85. J. D. Nichols and K. V. Rajagopalan, *J. Biol. Chem.*, 2005, **280**, 7817.
86. B. Stallmeyer, A. Nerlich, J. Schiemann, H. Brinkmann and R. R. Mendel, *Plant J.*, 1995, **8**, 751.
87. B. Stallmeyer, G. Schwarz, J. Schulze, A. Nerlich, J. Reiss, J. Kirsch and R. R. Mendel, *Proc. Natl. Acad. Sci. U. S. A.*, 1999, **96**, 1333.
88. J. Kuper, A. Llamas, H. J. Hecht, R. R. Mendel and G. Schwarz, *Nature*, 2004, **430**, 803.
89. A. Llamas, R. R. Mendel and G. Schwarz, *J. Biol. Chem.*, 2004, **279**, 55241.
90. A. Llamas, T. Otte, G. Multhaup, R. R. Mendel and G. Schwarz, *J. Biol. Chem.*, 2006, **281**, 18343.
91. A. A. Belaidi and G. Schwarz, *Biochem. J.*, 2013, **450**, 149.
92. S. Leimkühler and K. V. Rajagopalan, *J. Biol. Chem.*, 2001, **276**, 1837.
93. R. R. Mendel and G. Schwarz, *Coord. Chem. Rev.*, 2011, **255**, 1145.
94. B. Stallmeyer, G. Schwarz, J. Schulze, A. Nerlich, J. Reiss, J. Kirsch and R. R. Mendel, *Proc. Natl. Acad. Sci. U. S. A.*, 1999, **96**, 1333.
95. G. Schwarz, *Cell. Mol. Life Sci.*, 2005, **62**, 2792.
96. J. Herweg and G. Schwarz, *J. Biol. Chem.*, 2012, **287**, 12645.
97. C. A. Temple and K. V. Rajagopalan, *J. Biol. Chem.*, 2000, **275**, 40202.
98. T. Palmer, C.-L. Santini, C. Iobbi-Nivol, D. J. Eaves, D. H. Boxer and G. Giordano, *Mol. Microbiol.*, 1996, **20**, 875.
99. M. W. Lake, C. A. Temple, K. V. Rajagopalan and H. Schindelin, *J. Biol. Chem.*, 2000, **275**, 40211.
100. C. E. Stevenson, F. Sargent, G. Buchanan, T. Palmer and D. M. Lawson, *Structure*, 2000, **8**, 1115.
101. R. Thome, A. Gust, R. Toci, R. Mendel, F. Bittner, A. Magalon and A. Walburger, *J. Biol. Chem.*, 2012, **287**, 4671.
102. H. C. Raaijmakers and M. J. Romao, *JBIC, J. Biol. Inorg. Chem.*, 2006, **11**, 849.
103. C. Coelho, P. J. Gonzalez, J. G. Moura, I. Moura, J. Trincao and M. Joao Romao, *J. Mol. Biol.*, 2011, **408**, 932.
104. P. Arnoux, C. Ruppelt, F. Oudouhou, J. Lavergne, M. I. Siponen, R. Toci, R. R. Mendel, F. Bittner, D. Pignol, A. Magalon and A. Walburger, *Nat. Commun.*, 2015, **6**, 6148.
105. M. Neumann, G. Mittelstädt, F. Seduk, C. Iobbi-Nivol and S. Leimkühler, *J. Biol. Chem.*, 2009, **284**, 21891.
106. M. Neumann and S. Leimkühler, *Biochem. Res. Int.*, 2011, **2011**, 850924.
107. M. Neumann, W. Stöcklein, A. Walburger, A. Magalon and S. Leimkühler, *Biochemistry*, 2007, **46**, 9586.
108. M. Neumann, W. Stöcklein and S. Leimkühler, *J. Biol. Chem.*, 2007, **282**, 28493.

109. F. Bittner, M. Oreb and R. R. Mendel, *J. Biol. Chem.*, 2001, **276**, 40381.
110. T. Heidenreich, S. Wollers, R. R. Mendel and F. Bittner, *J. Biol. Chem.*, 2005, **280**, 4213.
111. S. Wollers, T. Heidenreich, M. Zarepour, D. Zachmann, C. Kraft, Y. Zhao, R. R. Mendel and F. Bittner, *J. Biol. Chem.*, 2008, **283**, 9642.
112. M. Lehrke, S. Rump, T. Heidenreich, J. Wissing, R. R. Mendel and F. Bittner, *Biochem. J.*, 2012, **441**, 823.
113. C. P. Witte, M. I. Igeno, R. Mendel, G. Schwarz and E. Fernandez, *FEBS Lett.*, 1998, **431**, 205.
114. K. Fischer, A. Llamas, M. Tejada-Jimenez, N. Schrader, J. Kuper, F. S. Ataya, A. Galvan, R. R. Mendel, E. Fernandez and G. Schwarz, *J. Biol. Chem.*, 2006, **281**, 30186.
115. T. Kruse, C. Gehl, M. Geisler, M. Lehrke, P. Ringel, S. Hallier, R. Hansch and R. R. Mendel, *J. Biol. Chem.*, 2010, **285**, 6623.
116. A. Veldman, J. A. Santamaria-Araujo, S. Sollazzo, J. Pitt, R. Gianello, J. Yaplito-Lee, F. Wong, C. A. Ramsden, J. Reiss, I. Cook, J. Fairweather and G. Schwarz, *Pediatrics*, 2010, **125**, e1249.

CHAPTER 4

Bacterial Molybdoenzymes: Chaperones, Assembly and Insertion

SILKE LEIMKÜHLER[a], OLIVIER N. LEMAIRE[b], AND CHANTAL IOBBI-NIVOL*[b]

[a]Institute of Biochemistry and Biology, Department of Molecular Enzymology, University of Potsdam, 14476 Potsdam, Germany; [b]Laboratoire de Bioénergétique et Ingénierie des Protéines, Institut de Microbiologie de la Méditerranée, CNRS, Aix Marseille Université, 31 Chemin Joseph Aiguier, 13402 Marseille Cedex 20, France
*E-mail: iobbi@imm.cnrs.fr

4.1 Introduction

Molybdoenzymes are widespread in all domains of life catalyzing key steps in carbon, sulfur and nitrogen metabolism (see Volume 1, Chapter 2) and from an evolutionary point of view they are ancient enzymes already present in LUCA including also recently evolved enzymes.[1–3] Moreover, it has been shown that in distinct bacteria the absence of genes for molybdoenzymes additionally implies a loss of molybdenum utilization.[4,5] With the exception of nitrogenase, containing a molybdenum–iron–sulfur center (described in detail in Volume I, Chapter 8), molybdoenzymes bind the molybdopterin (MPT)-based cofactor, which coordinates the molybdenum atom.[1] The

RSC Metallobiology Series No. 5
Molybdenum and Tungsten Enzymes: Biochemistry
Edited by Russ Hille, Carola Schulzke, and Martin L. Kirk

Published by the Royal Society of Chemistry, www.rsc.org

biosynthesis of the molybdenum cofactor is described in detail in Volume 1, Chapter 3. Apart from the overall chemical structure of the molybdenum cofactor (Moco), consisting of a tricyclic pyranopterin with a unique dithiolene group coordinating the molybdenum atom, differences in the ligand coordination sphere of the molybdenum ion and modifications at the MPT phosphate group by the addition of AMP, GMP or CMP moieties have been identified. According to the molybdenum ligands and the overall form of modified Moco present in the final holo-enzyme, molybdoenzymes are categorized into three families: the xanthine oxidase (XO) family, the sulfite oxidase (SO) family and the dimethyl sulfoxide reductase (DMSOR) family.[6] Enzymes of thc XO family usually present a Moco with a MPT (or MCD)–$Mo^{VI}OS(OH)$ ligand coordination sphere, as bacterial xanthine dehydrogenase and aldehyde oxidase. Enzymes of the SO family contain a Moco with a MPT–$Mo^{VI}O_2Cys$ core and this family includes bacterial sulfite dehydrogenase. Enzymes of the DMSO reductase family are found exclusively in prokaryotes and have a Moco with a MGD_2–$Mo^{VI}XY$ core, with X being either a sulfur or an oxygen ligand and Y being either a hydroxo or an amino acid ligand (Ser, Cys, Sec and Asp ligands have been identified so far). Typical enzymes of this family include formate dehydrogenases, nitrate and trimethylamine oxide (TMAO) reductases. While eukaryotes produce only a limited number of enzymes belonging to the sulfite oxidase and xanthine oxidase families (for instance, four molybdoenzymes are present in humans), enzymes of all three families are present in prokaryotes with enzymes of the DMSO reductase family being predominant.[7] So far, *Desulfitobacterium hafniense* (Firmicutes/Clostridia), with around 60 molybdoenzymes present in the genome, is the organism exhibiting the highest number of molybdoenzymes, most of them belonging to the DMSO reductase family.[8,9] The insertion of the molybdenum cofactor into molybdoenzymes is a highly regulated process that requires specific chaperones for each molybdoenzyme. However, so far, only chaperones for enzymes of the XO and DMSOR family have been identified. Enzymes of the sulfite oxidase family do not appear to require specific chaperones for Moco insertion.

4.2 The Essential Role of Dedicated Chaperones for Molybdoenzyme Assembly

4.2.1 Chaperones are Required for the Biogenesis of Cognate Molybdoenzymes

Beginning in the late 1980s, the sequencing of operons containing genes encoding molybdoenzymes has highlighted the presence of genes coding for cytoplasmic proteins of about 20 kDa (Figure 4.1). Although these proteins have been shown to be essential for the activity of the respective molybdoenzyme, they are not a subunit of the active enzyme. These

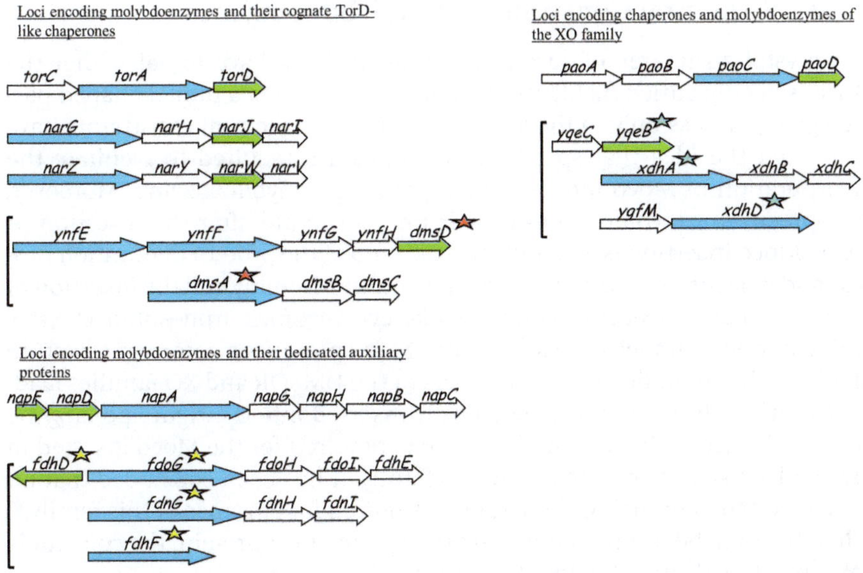

Figure 4.1 Genetic organization of loci encoding molybdoenzymes and their dedicated chaperones in *E. coli*. Genes coding for molybdoenzymes (▭) and chaperones (▭) are highlighted. In general, genes encoding the specific chaperone and the molybdoenzyme are part of the same operon. In a few examples, they are in distinct loci. ☆ indicates the genes of molybdoenzyme–chaperone couple which are not co-transcribed.

proteins accordingly have been considered to be system-specific chaperones dedicated to the respective molybdoenzyme.[10] The first study on molybdoenzyme private chaperone function was performed on the membrane-bound nitrate respiratory system of *E. coli* and in particular on the NarJ chaperone by the group of DeMoss.[11] They demonstrated that in the absence of NarJ, nitrate reductase activity was decreased in addition to the stability and assembly of the NarGHI nitrate reductase complex.[12–15] Additional studies on the trimethylamine-*N*-oxide (TMAO) respiratory system (Tor) of *E. coli* underlined the requirement for Moco-inserting chaperones and showed that in the absence of the TorD chaperone, its cognate molybdoenzyme TorA was inactive and was degraded by the Lon protease.[16,17] In *Rhodobacter capsulatus*, work with the Dor system, which is an ortholog to the *E. coli* Tor system, showed that disruption of the *dorD* gene led to the loss of both the DMSO/TMAO reductase activity of the DorA protein.[18] Further, it was shown that in the *dmsD* mutant of *E. coli* and *Halobacterium* sp. strain NRC-1, there was no bacterial growth with DMSO as supplement and the periplasmic and membrane-bound DmsABC complex was not detectable in *E. coli* crude extracts.[19–21]

4.2.2 Structural Constraints of Molybdoenzymes

The crystal structures of several molybdoenzymes have revealed that the Moco is deeply buried inside the proteins, at the end of a funnel-shaped passage giving access only to the substrate.[22] These structural constraints have reinforced the idea that specific chaperones are required to facilitate the incorporation of Moco into the corresponding molybdoenzyme. Moreover, the apo-enzymes would acquire their final structure after the insertion of Moco. Moco insertion is generally the last step of the protein maturation process and it occurs after protein folding, subunit assembly and the insertion of the additional redox-active cofactors like cytochromes, iron–sulfur clusters or flavin mono-/dinucleotides.[23] Finally, Moco presents a common structure that is further modified in the members of the DMSOR and XO families leading to various forms that are specifically inserted into the right apo-enzyme partner.[10] Since no further modification is required for the Moco inserted in the catalytic site of enzymes of the SO family, it has been concluded that no dedicated Moco-inserting chaperone is required for enzymes of this family.[24] When the molybdenum is bound to MPT by the MoeA protein, the cofactor is probably directly inserted into the enzyme.[25]

4.2.3 The Identification of Chaperones Dedicated to Specific Molybdoenzymes

For most molybdoenzymes in bacteria, especially enzymes of the DMSOR family, a specific chaperone for Moco insertion has been identified (Table 4.1).[24] One of the best characterized enzyme/chaperone systems is the TorA/TorD couple allowing the respiration of TMAO in *E. coli*. The *torA* and *torD* genes are co-transcribed and TorD has been shown to be involved in both the stabilization and maturation of the TorA apo-enzyme (Figure 4.1).[26] Due to the unique characteristics of each molybdoenzyme (cell localization, subunit composition, presence of other prosthetic groups, *etc.*) the chaperone involved in Moco insertion is specific, although there are some exceptions. One example is DmsD, which is cotranscribed with the *ynfE* and *ynfG* genes that encode selenate reductase, and has been additionally shown to play an essential role for the *E. coli* DMSO reductase DmsABC (Figure 4.1).[20,27] Other examples are the two membrane-bound nitrate reductase systems NarGHI and NarZYV. Although each possesses its own respective chaperone, NarJ for NarGHI and NarW for NarZYV, the chaperones have been shown to be interchangeable for the two systems (Figures 4.1 and 4.2).[14,28] *E. coli* contains a third nitrate reductase, NapAB, which is periplasmic and requires a specific NapD chaperone.[29] Further, it has been shown that all three formate dehydrogenases (the FdoGHI and FdnGHI complexes, as well as the FdhF protein) share the FdhD chaperone for Moco insertion (Figures 4.1 and 4.2).[30] Each of these molybdoenzyme–chaperone couples has been experimentally investigated, and crystal structures of several of these chaperones have been solved (*e.g.* TorD, DmsD, NarJ and NapD; Figure 4.2). These dedicated chaperones are also referred to as REMPs for *r*edox *e*nzyme *m*aturation *p*roteins in the

Table 4.1 Chaperones and auxiliary proteins dedicated to molybdoenzymes of the DMSOR and XO families.[a,b]

Chaperone auxiliary protein	Enzyme	Subunit composition/ gene locus/organism	Metal contained in the catalytic subunit	Ref.
	DMSO reductase family			
TorD		TorA/*torCAD*/*E. coli*		39
	TMAO reductase	TorA/*torECAD*/*S. massilia*	bisMGD	116
TorD				
—		TorZ/*torYZ*/*E. coli*		117
—	Biotine sulfoxide reductase	BisC/bis*C*/*E. coli*	bisMGD	118
NarJ		NarGHI/*narGHJI*/*E. coli*		119
NarW		NarZYV/*narZYWV*/*E. coli*	bisMGD + $[Fe_4S_4]$	
	Nitrate reductase			
NapD, NapF		NapA/*napFDAGHBC*/*E. coli*		64
DmsD (YnfI)		YnfEFGH/*ynfEFGHI*/*E. coli*		33
	Selenate reductase		bisMGD + $[Fe_4S_4]$	
SerD		SerABC/*serABCD*/*T. selenatis*		120
PcrD	Perchlorate reductase	PcrABC/*pcrABCD*/*Dechloromonas* spp.	bisMGD + $[Fe_4S_4]$	121 122
ClrD	Chlorate reductase	ClrABC/*clrABDC*/*Ideonella dechloratans*	bisMGD + $[Fe_4S_4]$	123
—	Arsenite oxidase Aio	$(AioAB)_2$/*aioAB*/*Alcaligenes faecalis*	bisMGD + $[Fe_3S_4]$	124
ArxD	Arx	?/*arxB$_2$ABCD*/*Halorhodospira halophile*	bisMGD + $[Fe_4S_4]$	125
—		ArrAB/*arrAB*/*S.* sp. ANA-3	bisMGD + $[Fe_4S_4]$	
	Arsenate reductase	ArrABC/*arrTSRCABD*/*Desulfitobacterium hafniense*		77
ArrD		DCB-2		
FdhD, FdhE		$(FdnGHI)_3$/*fdnGHI*/*E. coli*	bisMGD + $[Fe_4S_4]$	80
	Formate dehydrogenase			
FdhT		?/*so4504-15*/*S. oneidensis*	bisMGD + $[Fe_4S_4]$	126
FdsC, FdsD		$(FdsABC)_2$/*fdsGBACD*/*R. capsulatus*	bisMGD + $[Fe_4S_4]$ + $[Fe_2S_2]$	103
DmsD (YnfI)	DMSO reductase	DmsABC/*dmsABC*/*E. coli*	bisMGD + $[Fe_4S_4]$	58
DdhD	DMS dehydrogenase	DdhABC/*ddhABDC*/*Rhodovulum sulfidophilum*	bisMGD + $[Fe_4S_4]$	32
EbdD	Ethylbenzene dehydrogenase	EbdABC/*ebdABCDped*/*Azoarcus* EbN1	bisMGD + $[Fe_4S_4]$	127

(*continued*)

Table 4.1 (*continued*)

Chaperone auxiliary protein	Enzyme	Subunit composition/ gene locus/organism	Metal contained in the catalytic subunit	Ref.
	Xanthine oxidase family			
XdhC, NifS4	Xanthine oxidase	XdhAB/*xdhABC*/*R. capsulatus*	MoMPT	128
	Aldehyde oxidoreductase			129
PaoD		$(PaoABC)_3$/*paoABCD*/*E. coli*	MCD	
?	Aldehyde oxidoreductase	MOP/*Desulvovibrio gigas*	MCD	130
XdhC	Quinaldine dehydrogenase	QoxLMS/*qoxMLS*/*A.* sp. *Rü61a*	MCD	131
?	Quinoline-2-oxidoreductase	QorLMS/*qorMLS*/*Pseudomonas putida*	?	132 133
?	Isoquinoline 1-oxidoreductase	IorAB/*iorAB*/*Pseudomonas diminuta*	?	134
?	Quinoline-4-carboxylate-2-oxidoreductase	*Agrobacterium* sp. *Pseudomonas* sp.	MCD	135
?	Quinaldic acid 4-oxidoreductase	*Serratia marcecens*	?	136
?	Nicotinate hydroxylase	NdhFSLM/*ndhFSLM*/*Eubacterium barkeri*	MCD	137
coxF/coxI	CO dehydrogenase	CoxMSL/*coxMSLDEFG*/*Oligotropha carboxydovorans*	MCD	138

[a]– not found in the gene locus of structural subunits.
[b]? not known.

literature.[31] There are also exceptions to the general rule that members of the DMSO reductase family require chaperones. For example, no chaperone-requirement for the production of active biotin-sulfoxide reductase (BisC) or the periplasmic TMAO reductase TorZ has been identified so far.[19,31]

Among the three families of molybdoenzymes, members of the XO family also require molecular chaperones in bacteria, the best characterized example being the *Rhodobacter capsulatus* XdhC protein shown to be essential for the maturation of the xanthine dehydrogenase encoded by *xdhAB* (Figures 4.1 and 4.2).[23]

4.3 The TorD Family of Moco-Binding Chaperones

4.3.1 Subfamily Organization of the TorD-Like Chaperones

The TorD family of molecular chaperones contains hundreds of members that are mainly proteins identified in bacteria but also a few proteins present in archaea are included.[10] These chaperones carry out essential roles in the

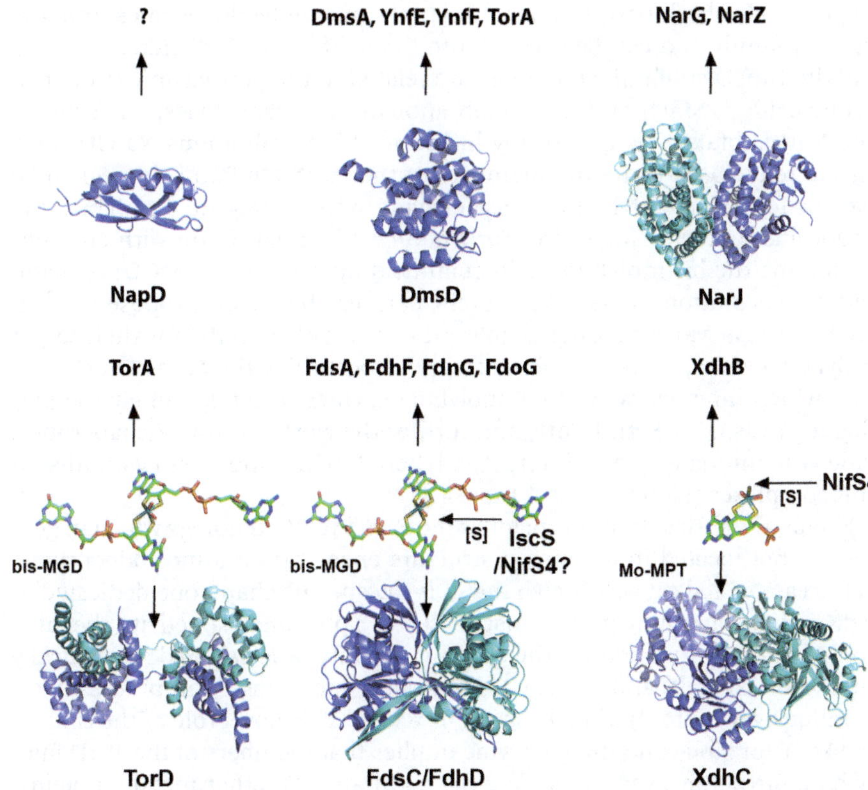

Figure 4.2 System-specific chaperone for molybdoenzymes. Shown are the crystal structures of *E. coli* NapD (PDB: 2JSX), *Salmonella typhimurium* DmsD (PDB: 1S9U), *Archaeoglobus fulgidus* NarJ (PDB: 2O9X), *Shewanella massilia* TorD (PDB: 1N1CT), *Desulfotalea psychrophila* FdhD (PDB: 2PW9) and *Bacillus halodurans* XdhC (PDB: 3ON5). While for XdhC, FdhD/FdsC and TorD Mo-MPT or bis-MGD binding was shown, respectively, for NarJ, DmsD or NapD the direct binding of bis-MGD still remains to be proven. The role of NapD is not completely clear so far; a direct binding to NapA is not clear. Additionally, NapD has a ferredoxin-like fold and thus a different structure from the other chaperones of the TorD family. An interaction of DmsD with TorA, YnjE, YnfF and DmsA was shown. NarJ interacts with NarG and NarZ. TorD was shown to be specific for TorA. *E. coli* FdhD interacts with FdnG, FdoG and FdhF, while the *R. capsulatus* homolog FdsC interacts with the FdsA subunit of *R. capsulatus* $(FdsGBA)_2$. *R. capsulatus* XdhC is the chaperone for XdhAB, the *R. capsulatus* xanthine dehydrogenase. For the activity of XDH and FDH, further modification of Moco by sulfuration occurs. The sulfur is provided by L-cysteine desulfurases, which directly interact with the chaperones FdhD and XdhC, respectively.

biogenesis for both periplasmic and cytoplasmic molybdoenzymes from the DMSOR family and can be divided into five subfamilies.[19,31] Here, the TorD and the DmsD subfamily members are related to the periplasmic trimethylamine oxide (TMAO) and the membranous DMSO reductases, respectively. The NarJ subfamily is genetically linked to the membranous NarGHI type nitrate reductases while the members of the FdhX subfamily appear to be specific for formate dehydrogenase enzyme systems (Figure 4.1). Thus, the important characteristic of the TorD family is its association with enzymes containing the bis-molybdopterin guanine dinucleotide (bis-MGD) cofactor, and a co-evolution of the two protein partners has been proposed.[19,32] In general, these Moco-binding chaperones are highly specific for their target molybdoenzyme; however, a few exceptions occur like the *E. coli* DmsD protein, which interacts with three molybdoenzymes in total, namely the paralogues DmsA, YnfE and YnfF, and further the NarJ and NarW chaperones, which can interchangeably interact with both NarGHI and NarZYV multi-subunit complexes (Figure 4.2 and Table 4.1).[33]

Finally, the fifth subfamily includes the *E. coli* YcdY homologs which in general are not located in an operon structure encoding for a molybdoenzyme. In the case of YcdY it was shown that it is the specific chaperone dedicated to YcdX, a zinc-containing protein that possesses a phosphatase activity, the function of which is required for the swarming ability of *E. coli*. YcdY specifically interacts with YcdX and thereby facilitates the metal insertion process likely by inducing a conformational change in YcdX.[34] This novel role of the chaperone YcdY for zinc-containing enzyme implies that members of the TorD family have further evolved to mediate the maturation of other metalloproteins. However, this fifth family might also imply that a so-far unknown molybdoenzyme-partner might exist for YcdY in the *E. coli* genome.

4.3.2 Structural Features of TorD-Like Chaperones

In general, proteins of the TorD family have a low identity at the amino acid level (20% or less), but share several other common characteristics including a similar secondary fold, a cytoplasmic location, a molecular weight from 18 to 25 kDa and an acidic pI.[19,31] Indeed, the TorD family members exhibit a high α-helix content when tested with predictive secondary structure programs and, for three subfamilies, this feature has been confirmed by circular dichroism studies including TorD from *E. coli*, *Shewanella oneidensis* and *S. massilia*, and DmsD and YcdY from *E. coli*.[19,35,36] Finally, the X-ray crystal structure of various members of this family (TorD, DmsD, NarJ) indicate that these specific chaperones are generally organized in 10 to 12 α-helices that account for 60–70% of the amino acid residues (Figure 4.2). In addition, they contain a long loop region separating the N- and C-terminal domains of the proteins.[10,24] At first, an "E(Q)PxDH" motif in the loop sequence was supposed to be characteristic of the TorD family, but with the availability of additional sequences it became apparent that the presence of this motif is not totally conserved and thus not as distinctive as initially thought.[19]

4.3.3 Mechanism of Molybdoenzyme Maturation Dependent on TorD-Like Chaperones

It has been shown that the chaperones of the TorD family bind to their targets in a region that encompasses either the twin-arginine signal sequence of those molybdoenzymes destined to be exported to the periplasm, or to the N-terminal sequence of cytoplasmic molybdoenzymes.[37,38] A common framework in the maturation mechanism of bis-MGD containing molybdoenzymes has been suggested according to the complexity or the location of the molybdoenzymes; however, differences also appear. Specific examples of characteristic members of subfamilies of the family of TorD-like chaperones are explained in detail below.

4.3.3.1 Maturation of TorA, a Soluble Monomeric Molybdoenzyme

The mechanism of TorA maturation by TorD has been studied in detail.[10,24] TorA is a bacterial respiratory enzyme catalyzing the reduction of TMAO into the volatile product trimethylamine (TMA), and the required electrons during respiration are delivered by the pentahemic *c*-type cytochrome TorC.[39,40] TorA is a monomeric protein located in the periplasm and contains only the bis-MGD cofactor as prosthetic group.[40] Therefore, it is a simple and well-defined model system to study the insertion of Moco without interference of subunit assembly or the requirement of the insertion of other cofactors, like iron–sulfur centers or cytochromes. TorA maturation takes place in the cytoplasm before the translocation to the periplasm (Figure 4.3).[41] It is now established that TorD is involved in the stabilization and maturation of TorA.[17] Two interaction sites of TorD with the Moco-free apo-form of TorA were identified, which include the twin-arginine signal peptide at the N-terminus and a second binding site in the core of TorA.[10] SAXS analyses confirmed a 1 : 1 binding stoichiometry of the two proteins.[42] Finally, it was shown that by binding of TorD to the core of apo-TorA the protein is protected against proteolysis by the Lon protease, while TorD binding to the N-terminal leader sequence prevents its degradation directed by an unknown protease.[17,43,44] The region of TorD which is involved in the recognition of the core of apo-TorA has been well-defined and is located at the fifth alpha helix in the TorD structure; however, the part of TorD being involved in the protection of the apo-TorA leader peptide is still controversial.[10,38,45] The two residues Asp124 and His125 (*E. coli* numbering) located in the loop region have been predicted to be involved in the binding of TorA leader sequence.[37,38] The TorD chaperone is also required for the insertion of the Moco into the active site of TorA, and TorD has been shown to bind MGD directly (Figure 4.3).[16,19,45] By binding to the core of apo-TorA, it has been suggested that TorD induces a conformational change in apo-TorA that facilitates the insertion of Moco. Moreover, TorD also acts in the final step of Moco maturation by interacting with the MobA enzyme that catalyzes the bis-MGD formation.[45] Thus, the chaperone can facilitate

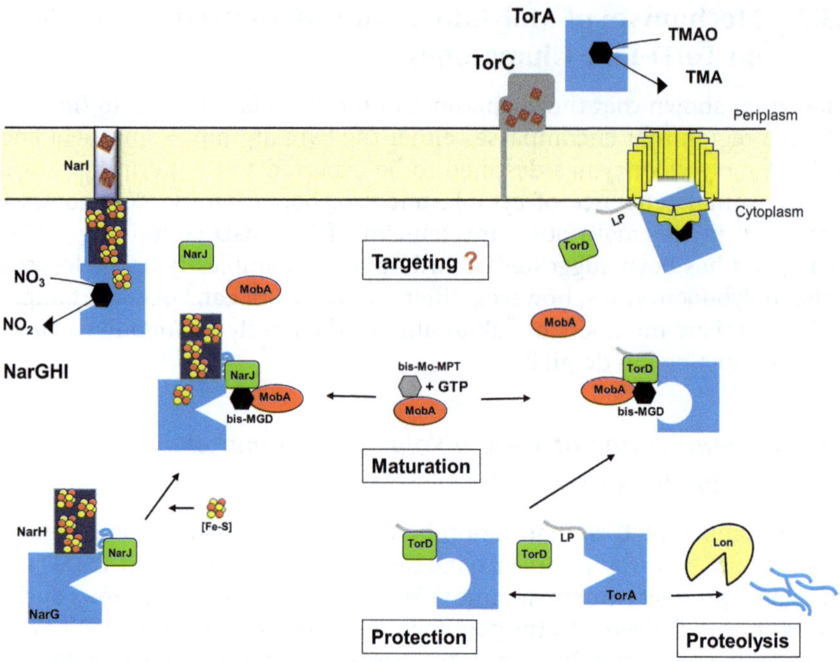

Figure 4.3 Model for TMAO reductase and nitrate reductase maturation in *E. coli*. Common features: chaperones bind the N-terminal part/leader peptide (LP) and the core of the apo-protein and also with enzymes of Moco biosynthesis. (left) NarGHI maturation: the insertion of both prosthetic groups (iron–sulfur and Moco) depends on the presence of NarJ that is also necessary for the correct membrane anchoring of the NarGHI complex. (right) TorA maturation: in an early step, TorD counteracts proteolytic attack mainly from Lon and stabilizes a competent shape of the apo-enzyme allowing the acquisition of the Moco. TorD contributes to the final step of bis-MGD synthesis by interacting with MobA and the Mo-MPT before transferring the mature form of the cofactor to apo-TorA.

the insertion of the synthesized bis-MGD cofactor into the binding site of apo-TorA and thereby might render the mature enzyme protease resistant.[17] After bis-MGD insertion into apo-TorA, final folding of TorA and release of TorD from the complex, TorA is targeted to the TAT membrane translocation machinery (Figure 4.3). Translocation is facilitated by the fact that insertion of Moco into TorA alters the affinity between the interacting region of TorA and TorD further resulting in the release of TorD.[16] This can be considered either an active proof-reading mechanism directed by TorD or a passive competition between proteins of the TAT machinery and TorD, whose affinity for the signal sequence is decreased by the conformational change in TorA upon incorporation of Moco.[10,38] Consequently, the TAT leader peptide is exposed after the release of TorD, and the mature holo-TorA can be targeted to the TAT-machinery.

4.3.3.2 Maturation of Complex Membrane-Associated Molybdoenzymes

NarGHI is a membrane-bound nitrate reductase responsible for anaerobic respiration when nitrate is available.[6,46–48] The catalytic subunit NarG is facing the cytoplasm and contains the bis-MGD cofactor and an additional iron–sulfur cluster named FS0.[49] NarH and NarI contain four iron–sulfur centers and a *b*-type cytochrome, respectively. They constitute together the electron transfer unit to the Moco-center and NarI anchors the complex to the inner membrane (Figure 4.3).[50–52] The molecular chaperone for NarGHI was identified to be the soluble protein NarJ.[53] NarJ is required for the stability and the location of the complex and the insertion of Moco.[13,54] Like TorD, NarJ interacts with two distinct regions on NarG including the N-terminal part and the core of the protein.[15] The 1–15 N-terminal amino acids of NarG adopt an α-helical conformation in solution, which is recognized by NarJ *via* hydrophobic interactions.[54] Moreover, the conformational change of NarJ was identified after its interaction with NarG.[55] In the absence of NarJ, NarG was identified to be Moco-free, further resulting in a defect in the assembly of the NarGHI complex. Here, the NarGH is not anchored to the membrane-bound NarI and an alteration of the *b*-type heme content of the membranous NarI cytochrome has been suggested.[56] Finally, it has been demonstrated that the maturation of the FS0 center must precede bis-MGD insertion into NarG. Thus, NarJ is involved in the final step of maturation and anchoring of the cytoplasmic NarGH complex to NarI.[55]

A further molybdoenzyme requiring a TorD-like chaperone is the DMSO reductase system, encoded by *dmsABC* in *E. coli* (Figures 4.1 and 4.2 and Table 4.1).[57] The DMSO reductase from *E. coli* and other species is an αβγ heterotrimer that is located in the periplasm.[58] The DmsA subunit harbors the bis-MGD cofactor and a [4Fe–4S] cluster.[59,60] The DmsB subunit contains four [4Fe–4S] clusters and DmsC is a cofactor-less, integral transmembrane subunit with eight transmembrane helices.[61] A menaquinone-binding site has been identified on DmsC, which serves as electron donor for substrate reduction.[57] *E. coli* DMSO reductase encompasses a large substrate specificity by reducing a variety of S- and N-oxides. For the activity of DmsA the chaperone DmsD is essential, which belongs to the TorD family of chaperones.[20] The role of DmsD has been studied in *E. coli* and in *Salmonella LT2*.[27] However, in contrast to other chaperones it has been shown that DmsD interacts not only with DmsA, but also with two other paralogues, YnfE and YnfF, that are the catalytic subunits of the *E. coli* selenate reductase (YnfEFGH or SerABC in *E. coli* or *Salmonella*, respectively). The overall role of DmsD has not been well characterized so far; however, studies showed that DmsD binds in a 1 : 1 stoichiometry to the 45-amino acid TAT-leader peptide of DmsA thereby resulting in minor conformational changes in DmsA.[62,63] Even if not studied yet, a role for DmsD in bis-MGD insertion into DmsA analogous to the role of other chaperones is conceivable. Indeed, it has been demonstrated that DmsD is necessary for the catalytic activity of its protein partners, but details on the mechanism of Moco insertion have not been studied yet and it is not known whether the stability or the maturation of the cognate molybdoenzymes are impaired when DmsD is missing. Further, since no interaction occurs between DmsD and the core of the DmsA, YnfE or YnfF, it is possible that DmsD follows a different mechanism for Moco insertion.[33] Thus, the precise role of DmsD for molybdoenzyme maturation has to be investigated in future studies.

4.3.3.3 bis-MGD Enzymes Can Acquire Other Accessory Proteins

No TorD-like chaperones were identified for NapA maturation, the enzyme that catalyzes the reduction of nitrate in the periplasm of bacteria.[64] The periplasmic nitrate reductase contains the NapAB dimer complexed either to the *c*-type cytochrome NapC or to the iron–sulfur-containing proteins NapGH. Depending on the bacterial species, various combinations of these subunits are possible and for example in a few *Shewanella* species both complexes are synthesized independently (Table 4.1).[65] The *nap* locus encodes an additional cytoplasmic protein named NapD, which has a ferredoxin-like fold and a molecular mass of around 9 kDa (Figures 4.1 and 4.2).[42] NapA activity in *E. coli* and *Campylobacter jejuni* depends on the presence of NapD and the absence of NapD leads to a loss of the NapA protein in *Wolinella succinogenes*.[66–68] This indicates that NapD plays a role in NapA stability. Finally, it has been shown that although an interaction occurs between NapD and the TAT leader peptide of NapA, NapD does not bind to the core of NapA.[69,70] As shown for the N-terminal part of NarG upon binding to NarJ, the interaction between NapD and the unstructured signal sequence of NapA leads to a helical conformation of the leader peptide.[71] In the purified NapAD complex of a 1 : 1 stoichiometry, the NapA protein was shown to be not completely folded with the iron–sulfur center being bound, but lacking Moco.[42] While the exact role of NapD still remains to be resolved, this indicated that NapD has a role in both Moco insertion and NapA folding.

A further protein, named NapF was shown to be involved in NapA maturation. *In vitro*, NapF was shown to play a direct role in the transfer of a [4Fe–4S] cluster to NapA and a role in the acquisition of the Moco also remains possible.[72,73] Thus, in *E. coli*, two chaperones NapF and NapD are involved in the biogenesis and the targeting of NapA, respectively. However, in some bacteria, NapF is not always present, like in several *Shewanella* species. Moreover, in *E. coli*, the lack of NapF leads only to a slower rate of NapA maturation, while NapA was identified to be active. In contrast, in *R. sphaeroides* the lack of NapF results in the absence of NapA protein.[68,73–76] A direct interaction has been identified between NapF and both the periplasmic and cytoplasmic forms of NapA.[73] This suggested that the core of the NapA is the target for NapF binding and may facilitate the iron–sulfur insertion during NapA maturation. Since binding experiments between the cofactor and the two auxiliary proteins have not been analyzed yet, the exact mechanism of Moco acquisition by apo-NapA still needs to be investigated.

4.3.3.4 Systems Without a Dedicated Chaperone

The requirement of a chaperone that is encoded by the same operon as the structural genes for the molybdoenzyme is a common feature for bis-MGD containing enzymes. However, in some cases, a gene encoding a potential accessory protein is not present in the same gene-locus. For instance, the comparison of the genomic arsenic islands of *Desulfitobacterium hafniense* DCB-2, *Shewanella* sp. ANA-3, *Geobacter lovleyi* SZ and *G. uraniireducens* Rf4 indicated that *arrD* coding for the TorD-like chaperone of ArrA is missing in *S.* ANA-3 while it is present in three other bacterial species (Table 4.1).[77] Furthermore, while in *E. coli* the major part of molybdoenzymes possesses a private chaperone involved in their maturation process, only for the molybdoenzymes TorZ and BisC no chaperone was identified so far (Table 4.1).[10,78] Thus, it is not known whether MobA directly inserts the molybdenum cofactor or whether other unknown auxiliary proteins are involved in the bis-MGD insertion process.

4.4 The Maturation of Formate Dehydrogenases

Formate dehydrogenases (FDHs) are a heterogeneous group of molybdoenzymes that catalyze the oxidation of formate to CO_2 and H^+.[79–81] Depending on the growth conditions, *E. coli* can produce three distinct FDHs, each of which contains selenocysteine coordinated to the molybdenum atom (Figure 4.1). In *E. coli*, these proteins are termed FDH-H (which is associated with hydrogenase), FDH-N (being induced in the presence of nitrate) and FDH-O (being present also during aerobic growth).[82] FDH-H is encoded by the *fdhF* gene and together with the [NiFe]-hydrogenases (Hyd3) of *E. coli* constitutes the H_2-evolving formate hydrogenlyase (FHL) enzyme complex.[83,84]

In *E. coli*, formate is generated from pyruvate during anaerobic fermentation of glucose in a radical-based, homolytic cleavage reaction catalyzed by the pyruvate–formate lyase complex (PFL).[85] To prevent accumulation of formate in the cytoplasm, formate is oxidized to CO_2 and H^+ by the FDHs. FDH-N is encoded by the *fdnGHI* operon and is expressed under anaerobic growth conditions in the presence of nitrate.[86] The enzyme has an $(\alpha\beta\gamma)_3$ mushroom-shaped structure, with the active site of the membrane-bound enzyme exposed to the periplasm.[87] The *fdnGHI* genes are co-expressed with the genes *narGHJI* encoding a membrane-bound nitrate-reductase whose active site is exposed to the cytoplasm.[88,89] The two systems comprise a formate–nitrate oxidoreductase complex, which couples formate oxidation to nitrate reduction. Protons are released in the periplasm by the oxidation of formate and are consumed in the cytosol by the reduction of nitrate, and the complex thus contributes to the transmembrane proton gradient. Both FDH-N and its associate nitrate reductase have an integral diheme-containing subunit (FdnI and NarI), and the menaquinone pool mediates electron transfer between the two.[90] The FdnH subunit of FDH-N contains four [4Fe–4S] clusters that are involved in the electron transfer reaction during substrate oxidation,[87] and have been designated FS1–4 in the structure (FS1 being proximal to the molybdenum-containing subunit). Remarkably, their arrangement resembles that seen for the iron–sulfur components of other systems, including nitrate reductase, [NiFe]-hydrogenase and NADH dehydrogenase, suggesting that they comprise an evolutionarily ancient and highly conserved arrangement of redox-active centers.[87] The FdnG subunit possesses the bis-MGD-containing Moco and additional [4Fe4S] cluster cofactor, designated as FeS0; the molybdenum ion is also coordinated by the SeCys196 of the polypeptide backbone.[87] Substrate access to the active site is provided by a predominantly positively charged funnel.

The *E. coli* FDH-O is encoded by the *fdoGHI* operon and is expressed under either aerobic or nitrate-respiring conditions.[91,92] FdoGHI is organized very similarly to FDH-N, forming an $(\alpha\beta\gamma)_3$ trimer of trimers with the active site facing the periplasm.[93] The *fdoGHI* genes are co-expressed with the *narZYWV* genes and the two systems constitute a second formate : nitrate oxidoreductase complex (although expressed under different growth conditions).[92,94] The cofactor arrangement is essentially identical to that of the FdnGHI complex; the two in fact share a 75% amino acid sequence identity between FdnGH and FdoGH, and 45% identity between FdnI and FdoI.

The third FDH in *E. coli* is the FHL-associated FDH-H enzyme (encoded by the *fdhF* gene). It is synthesized only under fermentative conditions in the absence of nitrate and is associated with the HycBCDEFG proteins that constitute the hydrogenase portion of the large multienzyme-membrane complex.[83,84,95] The electrons generated by FdhF upon oxidation of formate to CO_2 are transferred to the hydrogenase where they are used to reduce protons to molecular hydrogen. Unlike FDH-N and FDH-O, the active site of FDH-H is exposed to the cytoplasm rather than the periplasm.[95] Like FDH-N and FDH-O, however, FdhF contains the bis-MGD form of Moco, with SeCys140 being the ligand to the Mo atom.[96] Also, as with the FdnG and FdoG subunits of FDH-N and FDH-O, FdhF possesses a [4Fe–4S] cluster in close proximity to its Moco.

It was reported in 2006 by Raaijmakers and Romao that the bis-MGD cofactor of FdhF is characterized by the presence of a sulfur-ligand to the molybdenum atom at the active site.[97–100] Thus, in the oxidized state, the enzyme contains the four pterin sulfur ligands at the Mo site, a selenocysteine ligand and a –SH ligand. Since the bis-MGD cofactor in FdhF is deeply buried in the protein with the molybdenum atom being 25 Å away from the enzyme surface, it was suggested that the cofactor has to be sulfurated prior to its insertion into the FdhF-apo-enzyme.

The chaperone for all three FDHs in *E. coli* was identified to be the FdhD protein (Figure 4.2 and Table 4.1).[30] The crystal structures of two FdhD proteins from *Desulfotalea psychrophila* and *E. coli* have been solved (Figure 4.2).[101] Additional studies on the homologous protein FdsC from *Rhodobacter capsulatus*, which was copurified with bound bis-MGD (see below), suggested that FdhD-like proteins are indeed bis-MGD binding chaperones which insert the Moco into their FDH targets.[102] Additionally, the *E. coli* FdhD protein was co-crystallized in complex with GDP.[101] The FdhD-GDP co-structure showed that FdhD forms a dimer and likely binds one bis-MGD molecule per dimer with the GDP binding sites being located on each side of the monomer. Additional studies suggested that the sulfuration of the Mo site of the bis-MGD structure occurs on the FdhD protein before its insertion into the FdhF protein (Figure 4.4). An interaction of FdhD with the L-cysteine desulfurase IscS of *E. coli* was determined.[30] Additionally, two cysteine residues, Cys121 and Cys124 of FdhD were shown to be involved in the sulfur transfer reaction, predicting that Cys121 is the catalytic essential residue involved in sulfur transfer from IscS to bis-MGD, while Cys124 is rather involved in the reduction of the persulfide bond formed on Cys121 after sulfur transfer from IscS. The molecular details of the sulfur transfer reaction are not clear so far, especially since the cysteine residues are not conserved in all FdhD-like proteins and, in addition, do not seem to be essential for the sulfur transfer reaction in the *R. capsulatus* FdhD-homologue FdsC.[102]

In *R. capsulatus*, the role of two chaperones FdsC and FdsD for FDH maturation has been investigated in detail.[103] *R. capsulatus* FDH consists of an $(\alpha\beta\gamma)_2$ heterotrimer in which the large α-subunit FdsA (105 kDa) harbors the bis-MGD cofactor and a set of four [4Fe4S] clusters and one [2Fe2S] cluster (Figure 4.5). FdsA is linked to the β-subunit FdsB (52 kDa) that binds one additional [4Fe4S] cluster and the FMN cofactor.[103] The γ-subunit FdsG (15 kDa) binds a [2Fe2S] cluster. The terminal electron acceptor was shown to be NAD^+. For *R. capsulatus* FDH, the FdsC chaperone was the first chaperone to be identified to directly bind the bis-MGD cofactor.[102] Bis-MGD binding to FdsC was shown after heterologous expression and purification of FdsC in *E. coli*.[102] The bis-MGD cofactor bound to FdsC was further transferred and inserted into apo-TorA, and the production of active TorA directly showed that intact bis-MGD was bound to FdsC and that the transfer and insertion into apo-TorA did not require any additional factors.

Recent studies on the active site of purified *R. capsulatus* FdsGBA in the presence and absence of FdsC showed that in its active form *R. capsulatus* FdsA binds the bis-MGD cofactor in which the molybdenum atom is coordinated by four sulfur atoms originating from the two dithiolene moieties, one sulfur atom originating from cysteine and one additional =S ligand. The protein expressed in the absence of FdsC showed a different Mo coordination with four (dithiolene) sulfurs, one cysteine sulfur and one =O group.[104] These studies also directly proved the essential requirement of FdsC for the insertion of the sulfido group to the bis-MGD cofactor of FDH.

For *R. capsulatus* FdsC a mechanism of Moco sulfuration has been suggested, in which the sulfur is directly transferred to the bis-MGD cofactor when it is bound to FdsC (Figure 4.4).[80,102] After bis-MGD sulfuration, the cofactor is inserted into the formed $(FdsGBA)_2$ heterodimer and active FDH is formed. For the *R. capsulatus* proteins, an interaction of FdsGBA with FdsC was shown. Additionally, FdsC interacted with the FdsD protein (role not clear so far) and MobA, the protein that synthesizes the bis-MGD cofactor. Likely, all proteins form a complex, which facilitates bis-MGD formation, its sulfuration on FdsC and the effective transfer of the modified bis-MGD to the

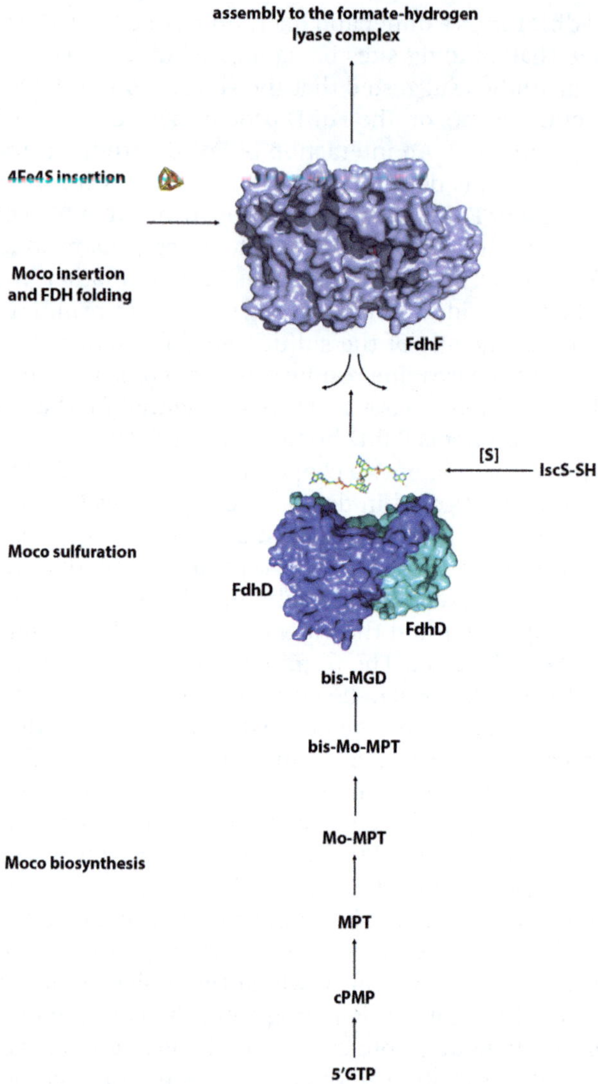

Figure 4.4 bis-MGD insertion into *E. coli* FdhF. A model of the FdhD/FdhF system from *E. coli* for insertion of sulfurated bis-MGD is shown. The FdhD dimer was shown to bind bis-MGD. It is proposed that bis-MGD is further modified by sulfuration by an interaction of FdhD with the L-cysteine desulfurase IscS of *E. coli*. It is suggested an L-cysteine desulfurase transfers the sulfur to the Mo atom by exchanging an oxo-group and adding the sulfido ligand. Afterwards, sulfurated bis-MGD is inserted into FdhF, in which a Fe_4S_4 cluster has already been inserted (here a surface representation of *E. coli* FdhF is shown, pdb 1FDO). After bis-MGD insertion and folding of FdhF, the protein is located to the inner membrane and is assembled as the formate-hydrogen lyase (FHL) complex together with HycBCDEFG. The surface representation of the structure for the FdhD-homologous protein from *Desulfotalea psychrophila* is shown (PDB: 2PW9).

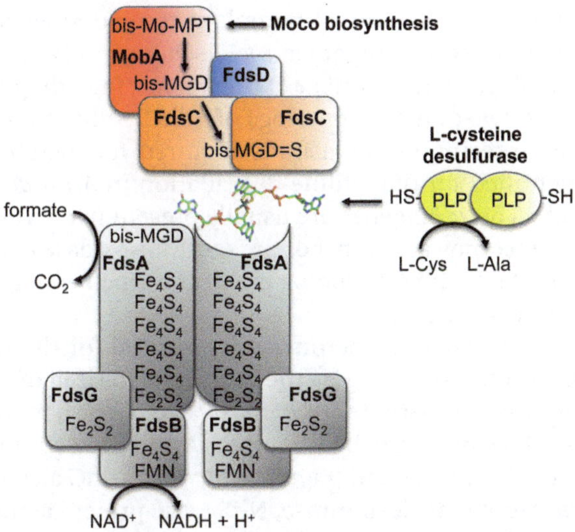

Figure 4.5 bis-MGD insertion into *R. capsulatus* FDH. A model of the FdsC/FdsA system from *R. capsulatus* for insertion of sulfurated bis-MGD is shown. FdsC and FdsD are essential for FDH activity. An interaction of FdsC, FdsD and MobA was shown. MobA synthesizes bis-MGD from Mo-MPT with bis-Mo-MPT as intermediate. FdsC was shown to bind bis-MGD, which is transferred from MobA. It is proposed that bis-MGD is further modified by sulfuration while bound to FdsC. It is suggested an L-cysteine desulfurase transfers the sulfur to the Mo atom by exchanging an oxo-group and adding the sulfido ligand. Afterwards, sulfurated bis-MGD is inserted into FdsA, which is already assembled as a $(FdsGBA)_2$ heterotrimer containing FeS clusters and FMN.

FdsA subunit of FDH. In analogy to *E. coli* FdhD, FdsC harbors two conserved Cys residues. However, substitutions of these Cys to Ala did not result in the inactivation of FDH. This implies that the role of *R. capsulatus* FdsC might be different from the one of FdhD, and a persulfide group is not formed on a Cys residue of FdsC during the reaction. Instead, bis-MGD might be directly sulfurated when it is bound to FdsC.[102]

4.5 The Step of Moco Sulfuration for Enzymes of the Xanthine Oxidase Family

The Moco of enzymes of the xanthine oxidase family has only one MPT coordinated to the ion atom to give a MPT–Mo^{VI}OS(OH) core. In contrast to the other two families of molybdoenzymes, no amino acid ligand to the metal is present. All enzymes of the XO family possess an equatorial sulfur ligand in the molybdenum coordination sphere,[6] and it is well established that sulfur incorporation into the molybdenum coordination sphere is mediated by members of the XdhC family of specific chaperones (Figure 4.2 and Table 4.1).[23]

The XdhC family contains hundreds of members that are bacterial and archaeal proteins, but no homologous proteins are present in eukaryotes

– in plants and humans, functional homologs termed Moco-sulfurases are present instead. While several organisms such as *R. capsulatus* contain only a single XdhC homolog, a large number of organisms, including *R. sphaeroides* and *E. coli*, contain two or more homologs of XdhC. XdhC homologs are also present on strain-specific megaplasmids required for metabolic pathways such as that identified for quinaldine degradation in *Arthrobacter nitroguajacolicus* Rü61a.[105] The *xdhC* genes are usually present in operons containing structural genes encoding a member of the xanthine oxidase family, although the gene for at least one XdhC homolog is cotranscribed with genes essential for Moco biosynthesis instead.[23]

The XdhC-like molecular chaperones are essential for the maturation of molybdoenzymes of the XO family,[106] the best characterized of which is the *R. capsulatus* XdhC protein.[23] Studies of *R. capsulatus* XdhC have demonstrated that it binds the Moco produced by MoeA/MogA and protects it prior to insertion of the terminal sulfur ligand (Figure 4.6).[107,108] XdhC also interacts with the *R. capsulatus* L-cysteine desulfurase, NifS4, the protein actually responsible for sulfuration in a reaction that involves transient formation of a NifS4 persulfide upon abstraction of the cysteine sulfur.[109] After Moco sulfuration, XdhC with its bound sulfurated Moco is thought to dissociate from NifS4 and insert the sulfurated Moco into the apo-XdhB subunits of the *R. capsulatus* $(\alpha\beta)_2$ XDH heterotetramer.[108,110]

It thus appears that XdhC-like proteins play several specific roles, including (1) stabilization of the newly formed Moco, (2) interaction with an L-cysteine desulfurase to ensure that Moco sulfuration occurs and (3) interaction with the specific target apo-proteins for insertion of the sulfurated Moco.[23] Because Moco is deeply buried, the XdhC proteins may act as chaperones to facilitate the proper folding of the target proteins after Moco insertion.[106] Such a model implies that enzymes of the xanthine oxidase family exist in a Moco-competent "open" apo-molybdoenzyme conformation prior to insertion of the sulfurated Moco, and after insertion adopt a final "closed" conformation that is catalytically active.[106]

The overall sequence identity of the XdhC-like proteins from different bacteria is low (~15–30%), with only very few highly conserved amino acids present in all XdhC-like proteins. However, one cysteine residue is conserved through almost all XdhC-like proteins present in the database, which corresponds to cysteine 82 in *R. capsulatus* XdhC.[23] Whether this cysteine residue is involved in the sulfuration of Moco during the maturation process remains to be established.

In eukaryotes, the sulfuration step of Moco maturation in the xanthine oxidase family is catalyzed by the Moco sulfurases ABA3 (in *Arabidopsis*) or HMCS (in humans). ABA3 has been intensively studied and is a homodimeric two-domain protein[111] with an N-terminal domain sharing structural and functional homologies to pyridoxal phosphate-dependent bacterial cysteine desulfurases such as SufS or IscS. This N-terminal domain decomposes L-cysteine to yield alanine and sulfur,[112] the latter being incorporated into a persulfide at a highly conserved cysteine residue (Cys 430) of ABA3.

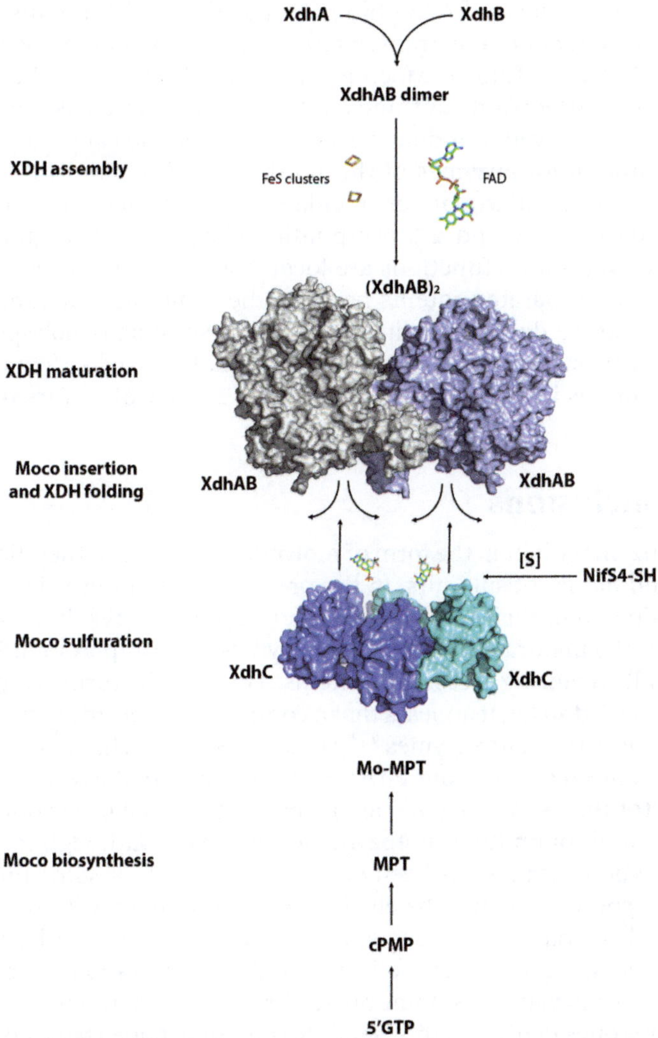

Figure 4.6 The role of XdhC-like chaperones for enzymes of the xanthine oxidase family. A model of the XdhC/XdhA system from *R. capsulatus* for insertion of sulfurated Mo-MPT is shown. It was established that XdhC binds Mo-MPT. Mo-MPT is produced by the pathway of Moco biosynthesis. The assembly of *R. capsulatus* XDH involves the synthesis of the XdhA and XdhB subunits, the dimerization of both subunits, the insertion of FeSI, FeSII and FAD into the XdhA subunit, dimerization of two (αβ) dimers *via* the XdhB subunit to form the Moco-free apo-XDH. The equatorial Mo=S ligand of Mo-MPT is inserted into Moco while bound to XdhC by the sulfurtransferase function of the NifS4 protein. After the formation of sulfurated Mo-MPT, XdhC interacts with XdhB (here a surface representation of *R. capsulatus* XDH is shown, pdb 1JRO) for final Moco insertion. After Moco insertion, the $(XdhAB)_2$ heterodimer accepts its final fold. The surface representation of the structure for the XdhC-homologous protein from *Bacillus halodurans* is depicted (pdb 3ON5).

The C-terminal domain of ABA3 binds Moco, after which the distal sulfur of the Cys 430 persulfide is transferred to the molybdenum coordination sphere.[113–115] The sulfurated Moco is subsequently transferred to the target apo-protein (the plant xanthine oxidoreductase as well as several aldehyde oxidases involved in indole-3-acetate and abscisic acid biosynthesis). Moco sulfuration for enzymes of the xanthine oxidase family thus occurs in a functionally similar manner in eukaryotes and bacteria, involving an L-cysteine desulfurase and a Moco-binding chaperone. The difference is that in eukaryotes both functions are located within a single protein while in bacteria two separate proteins perform the same reaction, and in general the L-cysteine desulfurase in the bacterial systems is not specific for Moco biosynthesis but is also involved in the sulfur transfer to a variety of acceptor proteins involved in the synthesis of a range of sulfur-containing molecules.

4.6 Conclusions

Molybdoenzymes differ in the form of molybdenum cofactor they bind, in the number and nature of subunits, in the presence of other prosthetic groups they contain and in their intracellular localization. As a result, it is not surprising that the maturation of molybdoenzymes is a complex multistep process generally requiring accessory proteins. Through the various maturation pathways studied so far, it appears that a common but specific pattern can be drawn for most molybdoenzymes.[139] Thus, the specific chaperones that are in general dedicated to only one enzyme seem to control the type of molybdenum cofactor that is inserted and are responsible for the protection and the correct assembly of multimeric enzymatic complexes. Although the maturation of molybdoenzymes has been extensively studied for some time, some important aspects remain to be elucidated. Furthermore, with the increase of released bacterial genome sequences, it appears that in at least certain organisms no specific chaperone is required for maturation of some complex molybdoenzymes. This point raises the question of the involvement of either chaperones dedicated to several distinct molybdoenzymes or general molecular chaperones.

Funding

This work was supported by grants from the CNRS and Aix-Marseille Université (to C.I.-N.) and by continuous grants of the Deutsche Forschungsgemeinschaft (to S.L.).

Acknowledgements

The authors thank all current and former members of the research group in addition to collaboration partners who were involved in the work over the past years.

References

1. K. V. Rajagopalan and J. L. Johnson, *J. Biol. Chem.*, 1992, **267**, 10199–10202.
2. B. Schoepp-Cothenet, R. van Lis, P. Philippot, A. Magalon, M. J. Russell and W. Nitschke, *Sci. Rep.*, 2012, **2**, 263.
3. F. Baymann, E. Lebrun, M. Brugna, B. Schoepp-Cothenet, M. T. Giudici-Orticoni and W. Nitschke, *Philos. Trans. R. Soc., B*, 2003, **358**, 267–274.
4. Y. Zhang, S. Rump and V. N. Gladyshev, *Coord. Chem. Rev.*, 2011, **255**, 1206–1217.
5. Y. Zhang and V. N. Gladyshev, *J. Mol. Biol.*, 2008, **379**, 881–899.
6. R. Hille, J. Hall and P. Basu, *Chem. Rev.*, 2014, **114**, 3963–4038.
7. R. Hille, J. Hall and P. Basu, *Chem. Rev.*, 2014, **114**, 3963–4038.
8. Y. Zhang and V. N. Gladyshev, *J. Biol. Chem.*, 2010, **285**, 3393–3405.
9. H. Nonaka, G. Keresztes, Y. Shinoda, Y. Ikenaga, M. Abe, K. Naito, K. Inatomi, K. Furukawa, M. Inui and H. Yukawa, *J. Bacteriol.*, 2006, **188**, 2262–2274.
10. O. Genest, V. Mejean and C. Iobbi-Nivol, *FEMS Microbiol. Lett.*, 2009, **297**, 1–9.
11. E. J. Sodergren, P. Y. Hsu and J. A. DeMoss, *J. Biol. Chem.*, 1988, **263**, 16156–16162.
12. M. Dubourdieu and J. A. DeMoss, *J. Bacteriol.*, 1992, **174**, 867–872.
13. X. Liu and J. A. DeMoss, *J. Biol. Chem.*, 1997, **272**, 24266–24271.
14. F. Blasco, J. Pommier, V. Augier, M. Chippaux and G. Giordano, *Mol. Microbiol.*, 1992, **6**, 221–230.
15. A. Vergnes, J. Pommier, R. Toci, F. Blasco, G. Giordano and A. Magalon, *J. Biol. Chem.*, 2006, **281**, 2170–2176.
16. J. Pommier, V. Mejean, G. Giordano and C. Iobbi-Nivol, *J. Biol. Chem.*, 1998, **273**, 16615–16620.
17. D. Redelberger, O. Genest, D. Arabet, V. Mejean, M. Ilbert and C. Iobbi-Nivol, *FEBS Lett.*, 2013, **587**, 3935–3942.
18. A. L. Shaw, S. Leimkühler, W. Klipp, G. R. Hanson and A. G. McEwan, *Microbiology*, 1999, **145**, 1409–1420.
19. M. Ilbert, V. Mejean and C. Iobbi-Nivol, *Microbiology*, 2004, **150**, 935–943.
20. N. Ray, J. Oates, R. J. Turner and C. Robinson, *FEBS Lett.*, 2003, **534**, 156–160.
21. J. A. Muller and S. DasSarma, *J. Bacteriol.*, 2005, **187**, 1659–1667.
22. C. Kisker, H. Schindelin and D. C. Rees, *Annu. Rev. Biochem.*, 1997, **66**, 233–267.
23. M. Neumann and S. Leimkühler, *Biochem. Res. Int.*, 2011, **2011**, 850924.
24. C. Iobbi-Nivol and S. Leimkühler, *Biochim. Biophys. Acta*, 2013, **1827**, 1086–1101.
25. M. Neumann and S. Leimkühler, *FEBS J.*, 2008, **275**, 5678–5689.
26. O. Genest, M. Ilbert, V. Mejean and C. Iobbi-Nivol, *J. Biol. Chem.*, 2005, **280**, 15644–15648.

27. D. Guymer, J. Maillard and F. Sargent, *Arch. Microbiol.*, 2009, **191**, 519–528.
28. F. Blasco, J. P. Dos Santos, A. Magalon, C. Frixon, B. Guigliarelli, C. L. Santini and G. Giordano, *Mol. Microbiol.*, 1998, **28**, 435–447.
29. J. Cole, *FEMS Microbiol. Lett.*, 1996, **136**, 1–11.
30. R. Thome, A. Gust, R. Toci, R. Mendel, F. Bittner, A. Magalon and A. Walburger, *J. Biol. Chem.*, 2012, **287**, 4671–4678.
31. R. J. Turner, A. L. Papish and F. Sargent, *Can. J. Microbiol.*, 2004, **50**, 225–238.
32. C. A. McDevitt, P. Hugenholtz, G. R. Hanson and A. G. McEwan, *Mol. Microbiol.*, 2002, **44**, 1575–1587.
33. S. P. Lubitz and J. H. Weiner, *Arch. Biochem. Biophys.*, 2003, **418**, 205–216.
34. D. Redelberger, F. Seduk, O. Genest, V. Mejean, S. Leimkühler and C. Iobbi-Nivol, *J. Bacteriol.*, 2011, **193**, 6512–6516.
35. S. Tranier, C. Iobbi-Nivol, C. Birck, M. Ilbert, I. Mortier-Barriere, V. Mejean and J. P. Samama, *Structure (Cambridge, MA, U. S.)*, 2003, **11**, 165–174.
36. K. J. Sarfo, T. L. Winstone, A. L. Papish, J. M. Howell, H. Kadir, H. J. Vogel and R. J. Turner, *Biochem. Biophys. Res. Commun.*, 2004, **315**, 397–403.
37. K. Hatzixanthis, T. A. Clarke, A. Oubrie, D. J. Richardson, R. J. Turner and F. Sargent, *Proc. Natl. Acad. Sci. U. S. A.*, 2005, **102**, 8460–8465.
38. R. L. Jack, G. Buchanan, A. Dubini, K. Hatzixanthis, T. Palmer and F. Sargent, *EMBO J.*, 2004, **23**, 3962–3972.
39. V. Mejean, C. Iobbi-Nivol, M. Lepelletier, G. Giordano, M. Chippaux and M. C. Pascal, *Mol. Microbiol.*, 1994, **11**, 1169–1179.
40. M. Czjzek, J. P. Dos Santos, J. Pommier, G. Giordano, V. Mejean and R. Haser, *J. Mol. Biol.*, 1998, **284**, 435–447.
41. C. L. Santini, B. Ize, A. Chanal, M. Muller, G. Giordano and L. F. Wu, *EMBO J.*, 1998, **17**, 101–112.
42. J. M. Dow, F. Gabel, F. Sargent and T. Palmer, *Biochem. J.*, 2013, **452**, 57–66.
43. O. Genest, F. Seduk, M. Ilbert, V. Mejean and C. Iobbi-Nivol, *Biochem. Biophys. Res. Commun.*, 2006, **339**, 991–995.
44. O. Genest, F. Seduk, L. Theraulaz, V. Mejean and C. Iobbi-Nivol, *FEMS Microbiol. Lett.*, 2006, **265**, 51–55.
45. O. Genest, M. Neumann, F. Seduk, W. Stöcklein, V. Mejean, S. Leimkühler and C. Iobbi-Nivol, *J. Biol. Chem.*, 2008, **283**, 21433–21440.
46. R. A. Rothery, A. Magalon, G. Giordano, B. Guigliarelli, F. Blasco and J. H. Weiner, *J. Biol. Chem.*, 1998, **273**, 7462–7469.
47. M. G. Bertero, R. A. Rothery, N. Boroumand, M. Palak, F. Blasco, N. Ginet, J. H. Weiner and N. C. Strynadka, *J. Biol. Chem.*, 2005, **280**, 14836–14843.
48. M. G. Bertero, R. A. Rothery, M. Palak, C. Hou, D. Lim, F. Blasco, J. H. Weiner and N. C. Strynadka, *Nat. Struct. Biol.*, 2003, **10**, 681–687.
49. R. A. Rothery, M. G. Bertero, T. Spreter, N. Bouromand, N. C. Strynadka and J. H. Weiner, *J. Biol. Chem.*, 2010, **285**, 8801–8807.

50. R. A. Rothery, F. Blasco, A. Magalon, M. Asso and J. H. Weiner, *Biochemistry*, 1999, **38**, 12747–12757.
51. R. A. Rothery, F. Blasco, A. Magalon and J. H. Weiner, *J. Mol. Microbiol. Biotechnol.*, 2001, **3**, 273–283.
52. T. Palmer, C.-L. Santini, C. Iobbi-Nivol, D. J. Eaves, D. H. Boxer and G. Giordano, *Mol. Microbiol.*, 1996, **20**, 875–884.
53. F. Blasco, J. P. Dos Santos, A. Magalon, C. Frixon, B. Guigliarelli, C. L. Santini and G. Giordano, *Mol. Microbiol.*, 1998, **28**, 435–447.
54. S. Zakian, D. Lafitte, A. Vergnes, C. Pimentel, C. Sebban-Kreuzer, R. Toci, J. B. Claude, F. Guerlesquin and A. Magalon, *FEBS J.*, 2010, **277**, 1886–1895.
55. P. Lanciano, A. Vergnes, S. Grimaldi, B. Guigliarelli and A. Magalon, *J. Biol. Chem.*, 2007, **282**, 17468–17474.
56. M. Lorenzi, L. Sylvi, G. Gerbaud, E. Mileo, F. Halgand, A. Walburger, H. Vezin, V. Belle, B. Guigliarelli and A. Magalon, *PLoS One*, 2012, **7**, e49523.
57. R. A. Rothery and J. H. Weiner, *Biochemistry*, 1993, **32**, 5855–5861.
58. P. T. Bilous and J. H. Weiner, *J. Bacteriol.*, 1988, **170**, 1511–1518.
59. C. A. Trieber, R. A. Rothery and J. H. Weiner, *J. Biol. Chem.*, 1996, **271**, 27339–27345.
60. V. W. Cheng, R. A. Rothery, M. G. Bertero, N. C. Strynadka and J. H. Weiner, *Biochemistry*, 2005, **44**, 8068–8077.
61. T. Ujiiye, I. Yamamoto, H. Nakama, A. Okubo, S. Yamazaki and T. Satoh, *Biochim. Biophys. Acta*, 1996, **1277**, 1–5.
62. C. S. Chan, T. M. Winstone, L. Chang, C. M. Stevens, M. L. Workentine, H. Li, Y. Wei, M. J. Ondrechen, M. Paetzel and R. J. Turner, *Biochemistry*, 2008, **47**, 2749–2759.
63. C. S. Chan, X. Song, S. J. Qazi, D. Setiaputra, C. K. Yip, T. Chao and R. J. Turner, *Biochem. Biophys. Res. Commun.*, 2014, **456**, 841–846.
64. G. Thomas, L. Potter and J. A. Cole, *FEMS Microbiol. Lett.*, 1999, **174**, 167–171.
65. B. J. Jepson, A. Marietou, S. Mohan, J. A. Cole, C. S. Butler and D. J. Richardson, *Biochem. Soc. Trans.*, 2006, **34**, 122–126.
66. L. Potter, H. Angove, D. Richardson and J. Cole, *Adv. Microb. Physiol.*, 2001, **45**, 51–112.
67. Y. W. Liu, A. Hitchcock, R. C. Salmon and D. J. Kelly, *Microbiology*, 2014, **160**, 2053–2066.
68. M. Kern, A. M. Mager and J. Simon, *Microbiology*, 2007, **153**, 3739–3747.
69. C. S. Chan, L. Chang, K. L. Rommens and R. J. Turner, *J. Bacteriol.*, 2009, **191**, 2091–2101.
70. C. S. Chan, L. Chang, T. M. Winstone and R. J. Turner, *FEBS Lett.*, 2010, **584**, 4553–4558.
71. S. Grahl, J. Maillard, C. A. Spronk, G. W. Vuister and F. Sargent, *Mol. Microbiol.*, 2012, **83**, 1254–1267.
72. M. F. Olmo-Mira, M. Gavira, D. J. Richardson, F. Castillo, C. Moreno-Vivian and M. D. Roldan, *J. Biol. Chem.*, 2004, **279**, 49727–49735.

73. A. Nilavongse, T. H. Brondijk, T. W. Overton, D. J. Richardson, E. R. Leach and J. A. Cole, *Microbiology*, 2006, **152**, 3227–3237.
74. F. Reyes, M. Gavira, F. Castillo and C. Moreno-Vivian, *Biochem. J.*, 1998, **331**(Pt 3), 897–904.
75. D. J. Richardson, B. C. Berks, D. A. Russell, S. Spiro and C. J. Taylor, *Cell. Mol. Life Sci.*, 2001, **58**, 165–178.
76. M. Kern and J. Simon, *Microbiology*, 2009, **155**, 2784–2794.
77. S. Yamamura and S. Amachi, *J. Biosci. Bioeng.*, 2014, **118**, 1–9.
78. S. Gon, M. T. Giudici-Orticoni, V. Mejean and C. Iobbi-Nivol, *J. Biol. Chem.*, 2001, **276**, 11545–11551.
79. C. S. Mota, M. G. Rivas, C. D. Brondino, I. Moura, J. J. Moura, P. J. Gonzalez and N. M. Cerqueira, *J. Biol. Inorg. Chem.*, 2011, **16**, 1255–1268.
80. T. Hartmann, N. Schwanhold and S. Leimkuhler, *Biochim. Biophys. Acta*, 2015, **1854**, 1090–1100.
81. J. Friedebold and B. Bowien, *J. Bacteriol.*, 1993, **175**, 4719–4728.
82. H. G. Enoch and R. L. Lester, *Methods Enzymol.*, 1982, **89 Pt D**, 537–543.
83. R. Böhm, M. Sauter and A. Bock, *Mol. Microbiol.*, 1990, **4**, 231–243.
84. M. Sauter, R. Bohm and A. Bock, *Mol. Microbiol.*, 1992, **6**, 1523–1532.
85. S. Leonhartsberger, I. Korsa and A. Bock, *J. Mol. Microbiol. Biotechnol.*, 2002, **4**, 269–276.
86. H. N. Wang and R. P. Gunsalus, *J. Bacteriol.*, 2003, **185**, 5076–5085.
87. M. Jormakka, S. Tornroth, B. Byrne and S. Iwata, *Science*, 2002, **295**, 1863–1868.
88. M. Jormakka, B. Byrne and S. Iwata, *Curr. Opin. Struct. Biol.*, 2003, **13**, 418–423.
89. H. G. Enoch and R. L. Lester, *J. Biol. Chem.*, 1975, **250**, 6693–6705.
90. M. Jormakka, B. Byrne and S. Iwata, *FEBS Lett.*, 2003, **545**, 25–30.
91. H. Abaibou, J. Pommier, S. Benoit, G. Giordano and M. A. Mandrand-Berthelot, *J. Bacteriol.*, 1995, **177**, 7141–7149.
92. J. Pommier, M. A. Mandrand, S. E. Holt, D. H. Boxer and G. Giordano, *Biochim. Biophys. Acta*, 1992, **1107**, 305–313.
93. G. Plunkett 3rd, V. Burland, D. L. Daniels and F. R. Blattner, *Nucleic Acids Res.*, 1993, **21**, 3391–3398.
94. C. Iobbi-Nivol, C. L. Santini, F. Blasco and G. Giordano, *Eur. J. Biochem.*, 1990, **188**, 679–687.
95. M. J. Axley, D. A. Grahame and T. C. Stadtman, *J. Biol. Chem.*, 1990, **265**, 18213–18218.
96. J. C. Boyington, V. N. Gladyshev, S. V. Khangulov, T. C. Stadtman and P. D. Sun, *Science*, 1997, **275**, 1305–1308.
97. H. C. Raaijmakers and M. J. Romao, *JBIC, J. Biol. Inorg. Chem.*, 2006, **11**, 849–854.
98. S. Grimaldi, B. Schoepp-Cothenet, P. Ceccaldi, B. Guigliarelli and A. Magalon, *Biochim. Biophys. Acta*, 2013, **1827**, 1048–1085.
99. C. Coelho, P. J. Gonzalez, J. G. Moura, I. Moura, J. Trincao and M. Joao Romao, *J. Mol. Biol.*, 2011, **408**, 932–948.

100. S. Najmudin, P. J. Gonzalez, J. Trincao, C. Coelho, A. Mukhopadhyay, N. M. Cerqueira, C. C. Romao, I. Moura, J. J. Moura, C. D. Brondino and M. J. Romao, *J. Biol. Inorg. Chem.*, 2008, **13**, 737–753.
101. P. Arnoux, C. Ruppelt, F. Oudouhou, J. Lavergne, M. I. Siponen, R. Toci, R. R. Mendel, F. Bittner, D. Pignol, A. Magalon and A. Walburger, *Nat. Commun.*, 2015, **6**, 6148.
102. N. Böhmer, T. Hartmann and S. Leimkühler, *FEBS Lett.*, 2014, **588**, 531–537.
103. T. Hartmann and S. Leimkühler, *FEBS J.*, 2013, **280**, 6083–6096.
104. P. Schrapers, T. Hartmann, R. Kositzki, H. Dau, S. Reschke, C. Schulzke, S. Leimkuhler and M. Haumann, *Inorg. Chem.*, 2015, **54**, 3260–3271.
105. K. Parschat, B. Hauer, R. Kappl, R. Kraft, J. Hüttermann and S. Fetzner, *J. Biol. Chem.*, 2003, **278**, 27483–27494.
106. S. Leimkühler and W. Klipp, *J. Bacteriol.*, 1999, **181**, 2745–2751.
107. S. Leimkühler, S. Angermüller, G. Schwarz, R. R. Mendel and W. Klipp, *J. Bacteriol.*, 1999, **181**, 5930–5939.
108. M. Neumann, W. Stöcklein and S. Leimkühler, *J. Biol. Chem.*, 2007, **282**, 28493–28500.
109. M. Neumann, W. Stöcklein, A. Walburger, A. Magalon and S. Leimkühler, *Biochemistry*, 2007, **46**, 9586–9595.
110. M. Neumann, M. Schulte, N. Jünemann, W. Stöcklein and S. Leimkühler, *J. Biol. Chem.*, 2006, **281**, 15701–15708.
111. F. Bittner, M. Oreb and R. R. Mendel, *J. Biol. Chem.*, 2001, **276**, 40381–40384.
112. T. Heidenreich, S. Wollers, R. R. Mendel and F. Bittner, *J. Biol. Chem.*, 2005, **280**, 4213–4218.
113. S. Wollers, T. Heidenreich, M. Zarepour, D. Zachmann, C. Kraft, Y. Zhao, R. R. Mendel and F. Bittner, *J. Biol. Chem.*, 2008, **283**, 9642–9650.
114. M. Lehrke, S. Rump, T. Heidenreich, J. Wissing, R. R. Mendel and F. Bittner, *Biochem. J.*, 2012, **441**, 823–832.
115. M. Lehrke, S. Rump, T. Heidenreich, J. Wissing, R. R. Mendel and F. Bittner, *Biochem. J.*, 2012, **441**, 823–832.
116. J. P. Dos Santos, C. Iobbi-Nivol, C. Couillault, G. Giordano and V. Mejean, *J. Mol. Biol.*, 1998, **284**, 421–433.
117. S. Gon, J. C. Patte, V. Méjean and C. Iobbi-Nivol, *J. Bacteriol.*, 2000, **182**, 5779–5786.
118. D. E. Pierson and A. Campbell, *J. Bacteriol.*, 1990, **172**, 2194–2198.
119. F. Blasco, C. Iobbi, J. Ratouchniak, V. Bonnefoy and M. Chippaux, *Mol. Gen. Genet.*, 1990, **222**, 104–111.
120. E. J. Dridge and C. S. Butler, *Biochimie*, 2010, **92**, 1268–1273.
121. K. S. Bender, C. Shang, R. Chakraborty, S. M. Belchik, J. D. Coates and L. A. Achenbach, *J. Bacteriol.*, 2005, **187**, 5090–5096.
122. N. Bardiya and J. H. Bae, *Microbiol. Res.*, 2011, **166**, 237–254.
123. H. D. Thorell, K. Stenklo, J. Karlsson and T. Nilsson, *Appl. Environ. Microbiol.*, 2003, **69**, 5585–5592.

124. T. P. Warelow, M. Oke and B. Schoepp-Cothenet, *et al.*, *PloS one*, 2013, **8**, e72535.
125. K. Zargar, A. Conrad and D. L. Bernick, *et al.*, *Environ. Microbiol.*, 2012, **14**, 1635–1645.
126. J. F. Heidelberg, I. T. Paulsen and K. E. Nelson, *et al.*, *Nat. Biotechnol.*, 2002, **20**, 1118–1123.
127. R. Rabus, M. Kube, A. Beck, F. Widdel and R. Reinhardt, *Arch. Microbiol.*, 2002, **178**, 506–516.
128. R. K. Thauer, K. Jungermann and K. Decker, *Bacteriol. Rev.*, 1977, **41**, 100–180.
129. M. Neumann, G. Mittelstädt, C. Iobbi-Nivol, M. Saggu, F. Lendzian, P. Hildebrandt and S. Leimkühler, *FEBS J.*, 2009, **276**, 2762–2774.
130. M. J. Romão, M. Archer and I. Moura, *et al., Science*, 1995, **270**, 1170–1176.
131. I. Stephan, B. Tshisuaka, S. Fetzner and F. Lingens, *Eur. J. Biochem./FEBS J.*, 1996, **236**, 155–162.
132. M. Blase, C. Bruntner, B. Tshisuaka, S. Fetzner and F. Lingens, *J. Biol. Chem.*, 1996, **271**, 23068–23079.
133. U. Frerichs-Deeken, B. Goldenstedt, R. Gahl-Janssen, R. Kappl, J. Huttermann and S. Fetzner, *Eur. J. Biochem./FEBS J.*, 2003, **270**, 1567–1577.
134. M. Lehmann, B. Tshisuaka, S. Fetzner and F. Lingens, *J. Biol. Chem.*, 1995, **270**, 14420–14429.
135. G. Bauer and F. Lingens, *Biol. Chem. Hoppe-Seyler*, 1992, **373**, 699–705.
136. S. Fetzner and F. Lingens, *Biol. Chem. Hoppe-Seyler*, 1993, **374**, 363–376.
137. N. Wagener, A. J. Pierik, A. Ibdah, R. Hille and H. Dobbek, *Proc. Natl. Acad. Sci. U. S. A.*, 2009, **106**, 11055–11060.
138. U. Schübel, M. Kraut, G. Mörsdorf and O. Meyer, *J. Bacteriol.*, 1995, **177**, 2197–2203.
139. S. Leimkühler and C. Iobbi-Nivol, *FEMS Microbiol. Rev.*, 2016, **40**, 1–18.

CHAPTER 5

The Prokaryotic Mo/W-bisPGD Enzymes Family

AXEL MAGALON*[a,b], PIERRE CECCALDI†[b,c], AND BARBARA SCHOEPP-COTHENET[b,c]

[a]CNRS, Laboratoire de Chimie Bactérienne (UMR7283), Institut de Microbiologie de la Méditerranée, CNRS, 13009 Marseille, France; [b]Aix-Marseille University, 13009 Marseille, France; [c]CNRS, Unité de Bioénergétique et Ingénierie des Protéines (UMR7281), Institut de Microbiologie de la Méditerranée, CNRS, 13009 Marseille, France
*E-mail: magalon@imm.cnrs.fr

5.1 Introduction

Molybdenum (Mo) and tungsten (W) are the only known second- and third-row transition metals required for most living organisms. In most cases, the corresponding Mo/W-enzymes catalyze important oxidation–reduction reactions in the global carbon, nitrogen and sulfur cycles. Despite the high similarity in the chemical properties of Mo and W, W-containing enzymes are rare compared to Mo-containing enzymes and essentially found in thermophilic prokaryotes.[1,2] To date, more than 50 different Mo/W-enzymes have been described in Nature.[3] However, this list is far from complete since newly isolated prokaryotic organisms and their genome sequence analysis endlessly provide novel structural and cellular organization of Mo/W-enzymes

†Present address: Department of Chemistry, Boston University, MA02215 Boston, USA.

RSC Metallobiology Series No. 5
Molybdenum and Tungsten Enzymes: Biochemistry
Edited by Russ Hille, Carola Schulzke, and Martin L. Kirk

Published by the Royal Society of Chemistry, www.rsc.org

and even novel members. The molybdoproteome moreover might be even more diverse than previously recognized in prokaryotes as suggested by the proteomic approach conducted by Cvetkovic *et al.*[4] revealing the existence of several Mo-binding proteins with unrelated sequence homology to any known molybdoenzymes.

With the exceptions of the multinuclear $MoFe_7$ cluster found in nitrogenase[5] and the binuclear MoCu center found in CO dehydrogenase,[6] molybdenum and tungsten are found in active sites in a mononuclear form coordinated by the organic cofactor pyranopterin.[7] This pterin consists of a tricyclic pyranopterin moiety synthesized through a conserved multiple-step pathway.[8] It is now well-established that all Mo/W-enzymes fall into three distinct large families, xanthine oxidase, sulfite oxidase and DMSO reductase, the latter being the focus of the present chapter. Members of the DMSO reductase family, found exclusively in prokaryotes, are distinguished by the coordination of the Mo/W atom by two pyranopterins which are themselves conjugated to nucleosides, usually either cytosine (pterin cytosine dinucleotide, PCD) or guanosine (pterin guanosine dinucleotide, PGD) (Figure 5.1). These enzymes are diverse in terms of structure and/or subunit composition (see below) but the family has been named for the first representative to have been crystallographically resolved,[9] DMSO reductase. As already discussed,[10] not only is it misleading to designate the family based on a precise enzymatic activity but, considering the extreme diversity of this superfamily, we therefore prefer the designation "Mo/W-*bis*PGD enzymes".

During the last decades, tremendous progress has been made in the genetic, structural, biochemical and mechanistic understanding of members of this enzyme superfamily through the combined action of researchers from different fields, as illustrated by several recent excellent reviews.[10–25] The quantity of sequence and structural data available to biology has increased exponentially. This approach has yielded an unexpected result: the apparent diversity in the bioenergetics of a broad range of organisms actually arises out of uniformity! Bioenergetics indeed appears today to consist of a "construction kit" (model introduced by Baymann *et al.*[26] and updated by Schoepp-Cothenet *et al.*[27]), with a very limited number of basic building blocks. Nature only modulates the assembly of these basic blocks to construct a vast range of bioenergetic enzymes that account for the multiplicity of metabolic pathways seen in Nature. The main challenge for future fundamental research is therefore to elucidate the basic and universal rules underlying the "construction kit": rules for protein and complex architecture, for assembly and for catalytic mechanisms.

The Mo/W-*bis*PGD superfamily of enzymes constitutes today one of the most prolific in the world of bioenergetics. Mo/W-*bis*PGD enzymes convert molecules ranging from the simple inorganic molecule CO_2 to molecules considered xenobiotics such as ethylbenzene, chlorate and acetylene. The Mo/W-*bis*PGD structural module serves multiple purposes in life, especially in energy harvesting, and has consequently been termed "the catalytic workhorse of bioenergetics" (illustrated in Section 5.4).[10] The most conspicuous

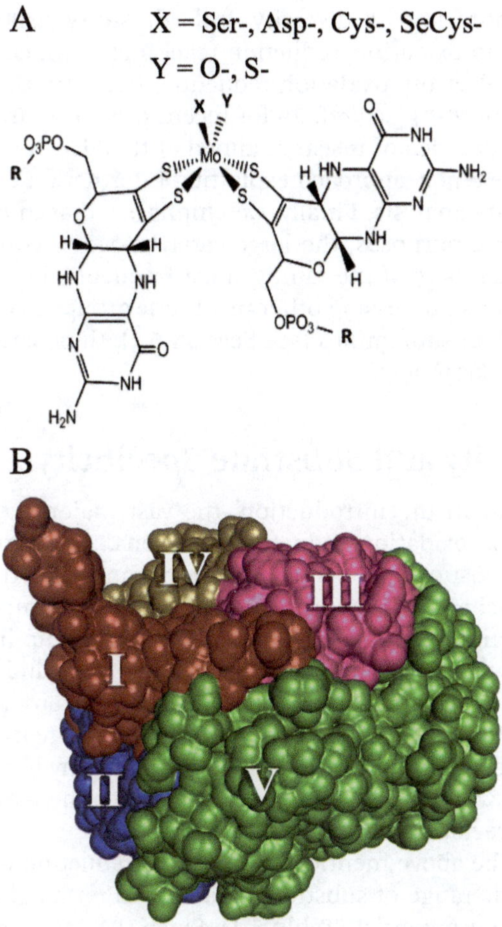

Figure 5.1 Schematic description of the catalytic subunit of Mo/W-*bis*PGD enzymes family. (A) Structure of the molybdenum cofactor showing the *bis*-pyranopterindithiolene coordination. The group R is guanosine (PGD) or cytosine (PCD). The Mo ligands X and Y can be: X = O or S; Y = Ser, Cys, SeCys, Asp. (B) Schematic representation of the domains organization found in Mo/W-*bis*PGD enzymes and illustrated with nNarGHI from *Escherichia coli*. It is to note that these domains do not correspond to consecutive sequences in the protein and are related by an internal pseudo-two-fold axis. All members of this family, except AH, have an active site funnel situated between domains II and III. Ah is the only currently known case where the funnel is situated at the intersection of domains I, II and III.

property of Mo/W-*bis*PGD proteins is their extreme substrate and oxidation–reduction versatility.[28] One of the main goals of current research on these enzymes is to elucidate the molecular details governing the substrate specificity and the reactivity of the Mo/W-*bis*PGD enzymes. In the present account, we intend to provide an alternative analysis of this Mo/W-*bis*PGD superfamily, embracing its diversity. Indeed, Mo/W-*bis*PGD enzymes are mainly involved in oxidation–reduction reactions in which the Mo/W atom undergoes two-electron oxidation/reduction between the states Mo(IV)/W(IV) and Mo(VI)/W(VI) (see ref. 29 for recent overview). In this chapter, we will summarize decades of research aimed at this deciphering (see Section 5.2) and propose a new approach exploiting information coming from evolutionary diversity analysis. Finally, we emphasize that in order to adapt to multiple energetic purposes, the large catalytic Mo/W containing-subunit, present in all members of the family, must frequently be supplemented by additional subunits possessing other prosthetic groups to generate the multitude of individual subfamilies (see Section 5.3), thus perfectly illustrating the construction kit model.

5.2 Reactivity and Substrate Specificity

As already stated in the introduction, the vast majority of Mo/W-*bis*PGD enzymes catalyze oxidation–reduction reactions. These reactions can be considered for most of them as oxygen atom transfer (OAT) or sulfur atom transfer (SAT) (Scheme 5.1, eqns (5.1) and (5.2)). Other important catalyzed oxidation–reduction reactions are conversion of formate into carbon dioxide, formally a hydrogen atom transfer (HAT) (eqn (5.3)) and hydroxylation of substrates such as resorcinol, ethylbenzene or cholesterol (eqn (5.4)). Eventually, two Mo/W-*bis*PGD enzymes catalyze non-redox reactions such as the transfer of hydroxyl group between phenols and the hydration of acetylene, reactions which will not be discussed further here. The reader is referred to more extensive reviews on this subject.[21,30]

Importantly, the above-mentioned oxidation–reduction reactions are performed on a wide range of substrates not only in terms of chemical nature but also reduction potential (Table 5.1). Since the most common reaction catalyzed by Mo/W-*bis*PGD enzymes is the activation of N–O or S–O bond, substrates can include nitrogen oxides such as nitrate and trimethylamine *N*-oxide or sulfur oxides such as dimethylsulfoxide or sulfite. Other substrates include formate, ethylbenzene, cholesterol but also toxic compounds such as arsenics (arsenate and arsenite), selenate or oxochlorates. It is noteworthy

$$R - O_n + 2\,H^+ + 2\,e^- = R - O_{n-1} + H_2O \tag{5.1}$$

$$R - S_n + 2\,H^+ + 2\,e^- = R - S_{n-1} + H_2S \tag{5.2}$$

$$HCOO^- = CO_2 + H^+ + 2\,e^- \tag{5.3}$$

$$R - CH_3 + H_2O = R - CH_2OH + 2\,H^+ + 2\,e^- \tag{5.4}$$

Scheme 5.1 Chemical equations of redox reactions catalyzed by Mo/W-*bis*PGD enzymes. (5.1) Oxygen atom transfer (OAT); (5.2) sulfur atom transfer (SAT); (5.3) hydrogen atom transfer (HAT); (5.4) hydroxyl transfer.

Table 5.1 Principal redox couples considered in this review. Non-common substrates are indicated with both chemical formula and names.

Redox couples	Redox potential at pH 7 E_0' (mV)	References
Methanofuran/formylmethanofuran $C_{34}H_{37}N_4O_{14} + CO_2/C_{35}H_{39}N_4O_{16}$	−497; −530	7 and 31
Carbon dioxide/formate $CO_2/HCOOH$	−430	32
$2H^+/H_2$	−410	32
Thiosulfate/sulfide + sulfite $S_2O_3^{2-}/HS^- + HSO_3^-$	−400	32
Sulfur/sulfide S^0/HS^-	−270	32
Pyrogallol/1,2,3,5-tetrahydroxybenzene	−84	33
Resorcinol/hydroxyhydroquinone	−33	34
Tetrathionate/thiosulfate $S_4O_6^{2-}/S_2O_3^{2-}$	+24; +170	32 and 35
1-Phenylethanol/ethylbenzene	+30	36
C25-hydroxycholesterol/cholesterol	+30	37
Arsenate/arsenite $HAsO_4^{2-}/H_3AsO_3$	+60; +139	32 and 38
TMAO/TMA	+130	39
DMSO/DMS	+160	32
Uranium trioxide/uranium dioxide UO_3/UO_2	+260; +410	40
Nitrate/nitrite NO_3^-/NO_2^-	+433	32
Selenate/selenite $SeO_4^{2-}/HSeO_3^-$	+475	32
Chlorate/chlorite ClO_3^-/ClO_2^-	+708	41
Fe(III)-citrate/Fe(II)-citrate	+385	42
Perchlorate/chlorate ClO_4^-/ClO_3^-	+788	41
O_2/H_2O	+820	32

that in contrast to the substrate diversity and the chemical reactions they support, Mo/W-*bis*PGD enzymes are relatively homogenous in terms of the architecture of the catalytic site.

The basic Mo/W active site of *bis*PGD enzymes is coordinated by four thiolate ligands arising from the pyranopterin rings. In all cases excepting the arsenite oxidase Aio, a fifth ligand comes from an amino acid side chain, the nature of which can be used to further subdivide this enzyme superfamily into three groups: a first one where Mo is coordinated by Cys or SeCys, a second one where Mo is coordinated by Asp and a third one where Mo is coordinated by Ser. Importantly, this criterion coincides with the subdivision deduced from phylogenetic analyses (see Section 5.3.). The sixth coordination site is occupied by an inorganic ion, which could be either oxygen from an oxo group or a solvent molecule, or sulfur from a sulfido group. It is of note that clear crystallographic identification of a sulfur atom has proven difficult due to the low resolutions achieved, with uncertainties that have only been disclosed recently as illustrated by the periplasmic nitrate reductase Nap, from *Cupriavidus necator* (Figure 5.2(A)).[43] It is now well accepted that formate dehydrogenases Fdh, harbor a sulfur ligand at the sixth position essential for their reactivity,[44,45] and it appears that the Nap enzymes do also. While Nap and Fdh share similar coordination of the Mo/W atom with a Cys/

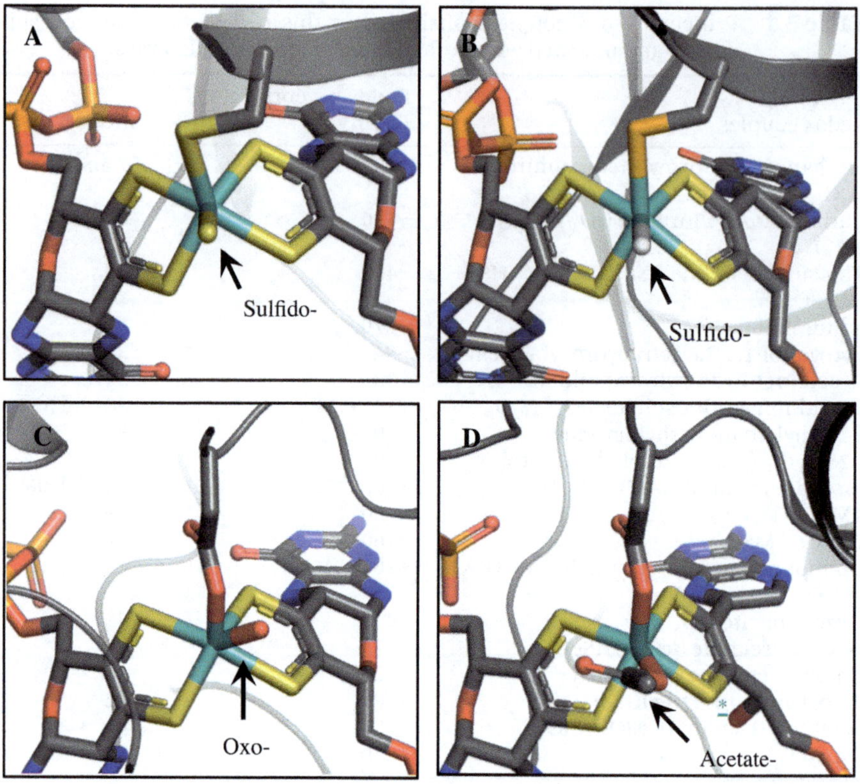

Figure 5.2 Comparison of crystal structures of four active sites of Mo-*bis*PGD enzymes. Main chain is showed as a grey cartoon and active site components (pyranopterins, proteic and exogenous ligands) as sticks. (A) Periplasmic nitrate reductase (Nap) from *C. necator* (PDB: 3ML1).[43] (B) Respiratory formate dehydrogenase (Fdh-N) from *E. coli* (PDB: 1KQF).[48] (C) Respiratory nitrate reductase (Nar) from *E. coli* (PDB: 1R27).[50] (D) Ethylbenzene dehydrogenase (Ebd) from *A. aromaticum* (PDB: 2IVF).[51]

SeCys proteic ligand in fifth position, whether sulfuration of the active site has to be systematically considered for other enzymes with Cys/SeCys ligand remains open. For instance, the polysulfide reductase Psr, from *Wolinella succinogenes* with Cys coordinated Mo displays Mo(V) EPR signals very similar to those of Nap[46] and recently interpreted as the result of a six-sulfur coordination of the metal.[47] Experimental validation of a Mo=S group is required at this stage to ascertain a complete sulfur coordination sphere of the Mo atom. In contrast, all X-ray crystal structures of enzymes with Ser/Asp coordinated Mo/W (Figure 5.3(C) and (D)) indicate an oxygen atom as sixth ligand, suggesting it would be the case for all Ser/Asp-ligated enzymes.

Overall, a conundrum exists in the Mo/W-*bis*PGD enzyme family when considering the limited chemical diversity of Mo/W active sites and a yet uncovered diversity of reactivities with both oxidation–reduction and non-oxidation–reduction reactions and large range of substrates. We intend in the

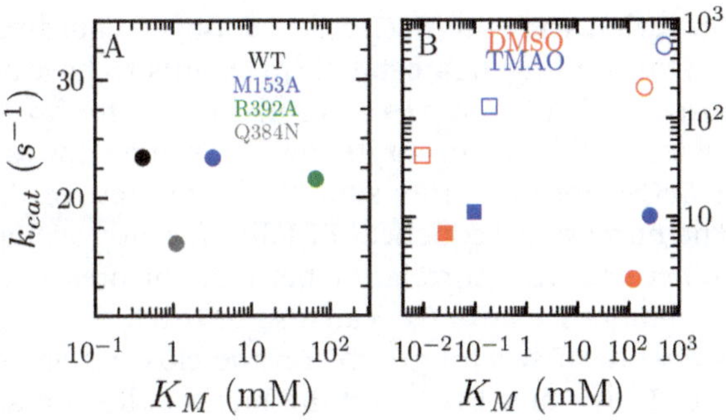

Figure 5.3 Comparative kinetic analyses. Michaelis parameters of (A) the nitrate reductase activity of wild-type and mutated NapAB,[65] and (B) the DMSO (red) and TMAO (blue) reductase activity of wild-type (open squares) and mutated Dor (closed squares, W116F)[68] and wild-type (open circles) and mutated BisC (closed circles, W90F).[69]

following account to summarize decades of research aiming at deciphering the molecular determinants of the reactivity. Several structural determinants have been screened in the Mo/W-*bis*PGD enzymes, from the ligands of the metal center, to the substrate channel, also including the pterins and their close environment.

5.2.1 Role of the Mo/W Ligands

Figure 5.2 shows the X-ray crystal structures of the active site of four enzymes, thus illustrating the structural and functional versatility of Mo/W-*bis*PGD enzymes and the difficulty of addressing the question of structure–function relationships in this family. The periplasmic nitrate reductase Nap[43] (panel A) shares exactly the same first coordination shell of the Mo ion with the formate dehydrogenase Fdh[48] (panel B), even though the type of reaction (OAT *vs.* HAT) is different. The respiratory nitrate reductase Nar[49,50] catalyzes the same reaction as Nap but with a different proteic ligation of the Mo, using Asp instead of Cys (panel C). Panel D shows the active site of ethylbenzene dehydrogenase Ebdh,[51] which has the same proteic ligand to the Mo as Nar, even though the reaction catalyzed is a hydroxylation. Despite the non-straightforward correlation between Mo/W-ligand and reactivity as illustrated above, one of the major determinants for the reactivity was originally considered as the nature of the proteic ligand.

To the best of our knowledge, the first study evaluating the impact of the substitution of the amino acid side chain coordinating the Mo atom in a Mo/W-*bis*PGD enzyme is the formate dehydrogenase, FdhF of *Escherichia coli*. Axley *et al.* substituted the SeCys by a Cys and observed a nearly 300-fold decrease in k_{cat} (9 *vs.* 2800 s^{-1}) and a lower K_m value (9 *vs.* 26 mM).[52] Sulfur was thus proposed to be less effective in formate oxidation as compared to selenium. Reexamination of the X-ray crystal structure of the formate-reduced

FdhF enzyme revealed that the SeCys is no longer coordinating the Mo atom.[53] Based on this finding, new catalytic mechanisms have been proposed whereby the dissociated SeCys serves as a general base for hydrogen abstraction from formate.[25,45,54,55] Owing to its low pK_a value ~5.2 as compared to Cys (pK_a ~ 8.2), SeCys appears more suitable for formate oxidation at physiological pH. The numerous sequences of Fdh presently available, however, suggest the absence of a strict correlation between the presence of SeCys and formate oxidation function since Cys can also be used.

The combined work of several groups on the closely related (Figure 5.4) DMSO reductases Dms[56] and Dor[57] and the biotin sulfoxide reductase Bis[58]

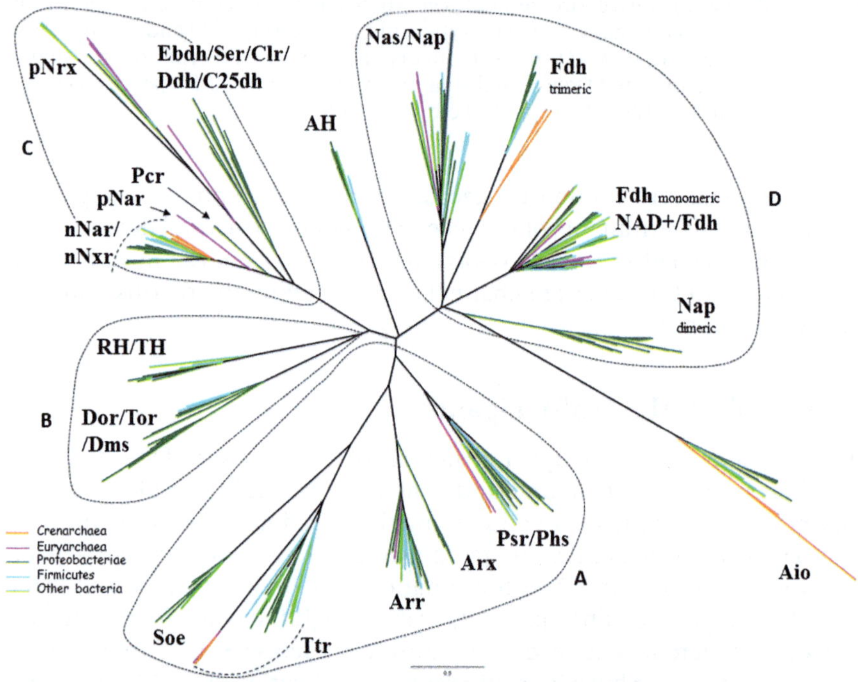

Figure 5.4 Reconstructed phylogenetic tree of the catalytic subunits of the Mo-*bis*PGD enzymes. This tree is a 3D structure-based NJ-phylogenetic tree of the Mo-*bis*PGD-enzymes: respiratory nitrate reductase (Nar), ethylbenzene dehydrogenase (Ebdh), selenite reductase (Ser), chlorate reductase (Clr), perchlorate reductase (Pcr), DMS dehydrogenase (Ddh), nitrite oxidoreductase (Nxr), periplasmic nitrate reductase (Nap), assimilatory nitrate reductase (Nas), DMSO/TMAO reductase (Dms/Dor/Tor), arsenate reductase (Arr), alternative arsenite oxidase (Arx), formate dehydrogenase (Fdh), arsenite oxidase (Aio), polysulfide reductase (Psr), tetrathionate reductase (Ttr), sulfite dehydrogenase (Soe), acetylene hydratase (AH), pyrogallol-phloroglucinol transhydroxylase (TH), resorcinol hydroxylase (RH) and sterol C25 dehydrogenase (C25dh). Violet and orange denote eury- and crenarchaeal branches, dark green, cyan and light green stand for Proteobacteria, Firmicutes and other Bacteria, respectively. This reconstructed phylogenetic tree allows subtrees (noted A to D) to be delimitated.

is very interesting, perfectly illustrating the non-straightforward correlation between metal–ligand and reactivity. Dor is able not only to convert dimethylsulfoxide (DMSO) but also, to a comparable degree, methionine sulfoxide (MetSO), trimethylamine-*N*-oxide (TMAO) and adenosyl-*N*-oxide (ANO).[59] The enzyme is clearly even more active toward *N*-oxides than *S*-oxides and the designation as a DMSO reductase has been justified on the basis of the much lower value of K_m toward DMSO than TMAO.[60] Hilton *et al.*[57] replaced the Ser coordinating the Mo atom by a Cys in Dor. The resulting S147C variant enzyme showed both decreased overall kinetic and substrate binding properties toward DMSO, demonstrating a decreased ability for both the binding and the activation of DMSO. But this mutation also increases the catalytic constant toward the ANO. The comparable studies of the equivalent residue S121 in BisC and S176 in Dms furthermore led to almost inactive enzymes for every performed mutation (S > C, S > T, S > A, S > H).[56,58]

Besides the nature of proteic metal–ligand, other tracks of exploration have been exploited to account for the functional diversity of these enzymes. As mentioned above, systematic sulfuration of the active site in Cys/SeCys has been proposed. Thome *et al.*[44] and Schrapers *et al.*[61] reported the importance for catalysis of the non-pterinic sulfido-ligands of Mo in Fdhs. We are, however, still waiting for equivalent studies with Nap. The thiolate ligands arising from the pterin rings have also been proposed to be crucial determinants for the reactivity and selectivity of the Mo/W-*bis*PGD enzymes, and the transient dissociation of one of these thiolate ligands during turnover has been proposed. This idea relies on the first high-resolution X-ray crystal structure of DorA that revealed heterogeneity of the Mo first coordination shell in the crystal.[62] Further studies on DorA and NapAB showed, however, that the Mo center is only catalytically active when coordinated to all four thiolate ligands from both pterins.[47,63]

5.2.2 Role of Amino Acids in the Immediate Environment of the Mo/W Atom

Among the molecular determinants of the reactivity of Mo/W-*bis*PGD enzymes, the surrounding amino acids in the active site vicinity, in particular those located in the substrate access channel, have been early on studied with the aim to decipher the structural basis of the selectivity. Related to this question, several publications reported the impact of single mutations of residues that shape the substrate channel and the active site cavity (the residue M153, next to C152 Mo ligand, the residue Q384 above the Mo ion and the residue R392 closing the end of the substrate channel) on the reactivity of Nap.[64,65] All three mutants show lower substrate affinity with a very similar catalytic constant (Figure 5.3(A)). Therefore, these three positions have been proposed to tune the substrate binding step, which changes independently of the overall kinetics of the reaction. Intriguingly, mutating the equivalent positions in assimilatory nitrate reductase Nas (it is to note that no crystal structure is available for any Nas) shows a greater impact on the catalytic constant than on the affinity for the substrate, suggesting different roles of these residues in either Nap or Nas.[66] This can be rationalized by the belonging of Nap and Nas to two distinct phylogenetic clades (Figure 5.4).

Equivalent studies have been conducted on Dor and Bis, studying the role of the residue W116 or W90, respectively.[67–69] These residues located in the active site vicinity (Figure 5.3(B)) were examined with the aim to decipher the structural basis of the selectivity of the whole Dor/Bis/Tor group compared to the other Mo/W-*bis*PGD enzymes. In this case, and in the same manner as in Nas, the mutated enzyme shows similar affinity toward DMSO and TMAO as for the wild-type enzyme, but a decreased catalytic constant. Together, these data obtained on Nap, Nas, Dor and Bis invite to re-examine the molecular determinants of the selectivity of specific groups of enzymes in the light of biodiversity (see Section 5.2.4).

In the case of Dor/Bis/Tor, specific residues responsible for the substrate selectivity among DMSO, TMAO and ANO have been looked for. The Y114 residue, for example, located in the active site vicinity, "above" the Mo ion, directly interacting with the oxo-ligand of the Mo *via* its hydroxyl group is only present in Dor and Bis and not in Tor. Y114F or Y114A mutants, constructed in Dor, are significantly affected for binding DMSO and ANO while the binding of TMAO is unaffected or even improved.[59,70] The symmetric introduction of a tyrosine in the equivalent position in Tor leads to changes in the ratio of the catalytic efficiency of DMSO and TMAO reduction, confirming that Tyr114 is one of the key features that distinguish Dor/Bis from Tor. This study on closely related enzymes is the first one to propose a tangible molecular determinant of selectivity.

5.2.3 Role of the Pterins

The first suggestion of involvement of the pterins in the catalytic process of Mo/W-*bis*PGD enzyme came from the resolution of the X-ray crystal structures of respiratory nitrate reductase Nar[49] and ethylbenzene dehydrogenase Ebdh[51] revealing pterins in a "ring-opened" configuration (see star in Figure 5.2(D)). But this structural feature has recently rather been proposed to be associated with the irreversible reductive activation occurring in Nap and Nar.[71,72]

Still searching for structural determinants for the reactivity of Mo/W-*bis*PGD enzymes and benefiting from increasing number of available 3D structures, Rothery *et al.* recently proposed a correlation between the geometry of pterins and their oxidation–reduction states in the enzymes, therefore tuning the electrochemical properties of their Mo center.[73] Recently, Wu *et al.* studied the role of amino acids surrounding the pterins moieties evaluating how pterins coordination influences Mo active site oxidation–reduction chemistry in the respiratory nitrate reductase Nar.[74] A clear correlation between specific activity and $Mo^{VI/IV}$ reduction potential was established. To be confirmed, this new approach has now to be applied to systems other than Nar.

It is likely that the versatility of the superfamily at least in part stems from the electrochemical plasticity of the Mo/W atom. This "metallocentric" view of enzyme catalysis, however, seems to be too limited since, as we saw above, at least an important role played by the pyranopterins[75] and the surrounding protein[23] is beginning to emerge in recent times. The former approach, the data thereof having been presented above, has concentrated on a single enzyme of the Mo/W-*bis*PGD superfamily. To date, a wealth of information contained in evolutionary diversity of both structure and function is available. To fully exploit this set of information, a new approach is to address the topic through the combined analysis of the phylogenetic relationships between enzymes and their reactivities (in terms of substrate specificities and catalytic properties).

5.2.4 What Can We Learn from Phylogenetic Analysis of the Mo-*bis*PGD Superfamily?

One of the most important messages arising from phylogenetic work is the importance of the Mo/W-*bis*PGD superfamily in the evolution of bioenergetics. This work suggests indeed that several members have been present already in the last universal common ancestor (LUCA) *i.e.* before the divergence between Archaea and Bacteria.[22,28] This is illustrated in the phylogenetic tree reconstructed from the catalytic subunits of members of the Mo/W-bisPGD superfamily showing divergence at the basis of the corresponding subtree of several members into archaeal and bacterial clades (Figure 5.4). This is the case for the respiratory nitrate reductase nNar, the formate dehydrogenase Fdh, the polysulfide reductase Psr and the arsenite oxidase Aio. The high solubility of MoS_4^{2-} salts supplied by alkaline hydrothermal vents and their abundance in the oceans likely contributed to the early adoption of Mo as a redox cofactor.[76,77] The Mo/W-*bis*PGD-containing structural module thus seems to have arisen in the earliest life forms. The second outcome of this phylogenetic work is the identification of particular structural and sequence relationships between the diverse members of this enzyme superfamily. Readers are encouraged to read the review from Rothery *et al.* for a detailed evolutionary description of each subunit,[22] whereas the aim here is to highlight the benefit of an evolutionary analysis for deciphering substrate specificities and catalytic properties in the Mo/W-*bis*PGD superfamily.

As illustrated by Figure 5.4, and as noted in previous studies, several subtrees can be defined within the Mo/W-*bis*PGD superfamily, in addition to the Aio and AH branches that have only a distant relationship with the other members of the superfamily.[22,78–80] The first group (A) contains not only the alternative arsenite oxidase Arx, the closely related arsenate reductase Arr, but also sulfur-converting enzymes such as Psr, tetrathionate reductase Ttr and sulfite oxidase Soe. The second group (B) contains Dor and Dms along with Tor, together with resorcinol hydroxylase RH and pyrogallol-phloroglucinol transhydroxylase TH. The third group (C) contains Nar, together with very diverse enzymes such as Ebdh, selenate reductase Ser, chlorate reductase Clr, perchlorate reductase Pcr, steroid C25 dehydrogenase C25dh and the nitrite oxidoreductases Nxr. The last group (D) contains the Nas and Nap but also Fdh (both NAD^+-dependent and -independent).

The nature of the coordinating amino acid can be used to subdivide Mo/W-*bis*PGD superfamily into groups more or less corresponding to the above phylogenetic analysis. Aio is the single known enzyme where no amino acid side chain coordinates the Mo ion. The two closely related but distinct groups A and D have Mo coordinated by Cys or SeCys; a coordination is also shared by the isolated AH clade. The group C is characterized by a Mo coordinated by Asp, whereas the group B is formed by enzymes with Mo coordinated by Ser. The first coordination shell of the Mo ion has long been considered crucial in determining the catalytic properties of a given enzyme. As presented above (Section 5.2.1), mutations at the position of the Mo-ligating amino acid have consequently been performed, with the aim of modifying the reactivity and selectivity of Mo-*bis*PGD enzymes. All these mutations have been shown to significantly affect the reactivity of the enzyme towards the physiological substrates. As mentioned in the previous section, except for work on the soluble DMSO reductase Dor,[57] this strategy has failed to affect the selectivity of the enzyme. This is not surprising in hindsight, since the nature of Mo coordination does not correlate with the catalyzed reaction (Figures 5.3 and 5.4). For example, both Nap and Nar reduce nitrate, with the Mo in Nap being coordinated by a Cys residue but by an Asp in Nar. Both Aio and Arx oxidize arsenite, with the Mo in Arx probably (no crystal structure is presently available) coordinated by a Cys residue but by a water molecule in Aio. Conversely, sequence alignments suggest the Mo to be coordinated by a Cys in both Arr and Ttr, whereas these enzymes perform OAT and S–S cleavage-reactions, respectively. As presented in Section 5.2.2, only a few mutations not directly targeting the Mo-ligand have been carried out.[69,74,81,82] These mutations, again neglecting information contained in evolutionary relationships between enzymes within the Mo/W-*bis*PGD family, affect overall enzymatic catalytic efficiency but do not modify the selectivity of the enzyme.

Several experimental results, however, suggest the enzymatic cross-reactivity to be correlated to phylogenetic proximity. The observation that *Pyrobaculum aerophilum*, devoid of *arr* genes, respires As(v) and concomitantly expresses *ttr* genes is perfectly in line with the close evolutionary relationship between Arr and Ttr observed in Figure 5.4.[83] In the same vein, the high enzymatic activity seen with *Aquifex aeolicus* Soe towards tetrathionate is in line with the apparent evolution of Soe from the Ttr enzyme (Figure 5.4).[84] Early studies on *E. coli* Nar showed that this enzyme is capable of reduction of chlorate or selenate, consistent with the close phylogenetic relationship of Nar with Ser and Clr (Figure 5.4).[85,86] Again the ability of the *Rhodobacter sphaeroides* Dor to convert both DMSO and TMAO is consistent with the close phylogenetic proximity of Dor and Tor as illustrated by the mixed-up sequences observed in the Tor/Dor clade in Figure 5.4.[57] We thus consider that the most promising way to modify the substrate selectivity of a given enzyme is to reproduce in the test tube what Nature has achieved in cells, *via* extracting the information on structure/function relationships contained in the global phylogenetic tree of Mo-*bis*PGD enzymes, through multiple sequence alignments and structure comparison. Experimental data on cross-reactivities cited above suggest that substrate specificity of a subgroup of enzymes resides on common amino acid and structural motifs that can be extracted from a specific phylogenetic clade analysis. Extending this idea, substrate specificity of one enzyme has to be defined by specific motifs observed in this enzyme compared to its close related homologs.

As briefly introduced above, the catalytic versatility of the Mo/W-*bis*PGD enzymes lies in the plasticity of the catalytic site, probably governed by interactions with the surrounding protein, that has yet to be decrypted and that is best addressed using information provided by phylogenetic analysis. The metabolic versatility of the Mo/W-*bis*PGD enzymes lies not only in the modularity of the catalytic subunit but also in the variation of assembly of the Mo/W-*bis*PGD subunit with multiple additional subunits.

5.3 Molecular Variation of the Mo/W-*bis*PGD Enzymes

Among the three main families of molybdoenzymes known so far,[21,23] the sulfite oxidase-type, the xanthine dehydrogenase-type and the Mo/W-*bis*PGD-type families, the last, exclusively found in prokaryotes, is the most diverse in terms of subunit composition and organization. We will briefly present the major types of elements contributing to the Mo/W-*bis*PGD complexes focusing on common functional features and highlighting eventual phylogenetic relationships thus facilitating a better description of their function. It is noteworthy that the designation of an "enzymatic complex" is not straightforward. In the case of the Nap, the formate hydrogen lyase FHL or the hydrogen-dependent carbon dioxide reductase HDCR systems, the "enzymatic entity" can be very large.[87–89] We thus admit that the selection we describe in the following presentation (Figure 5.5) may be the subject of debate.

5.3.1 The Catalytic Subunit

In the vast majority of enzymes from the Mo/W-*bis*PGD enzymes family, the catalytic subunit is typically designated A referring to the genes' order in the operons. Indeed, the gene encoding the catalytic subunit is located, in the vast majority of cases, before all other genes encoding for structural proteins. The Nar enzyme constitutes a notable but not unique exception since in the *nar* operons (there are two Nar complexes in *E. coli*), the genes encoding for the catalytic subunits are denoted *narG* and *narZ*.[90]

Although highly variable in size and sequence (from 710 residues for NasC to 1200 residues for NarG), resulting from a long evolution from the origins of life, all the known structures show very similar overall protein folds, with four core

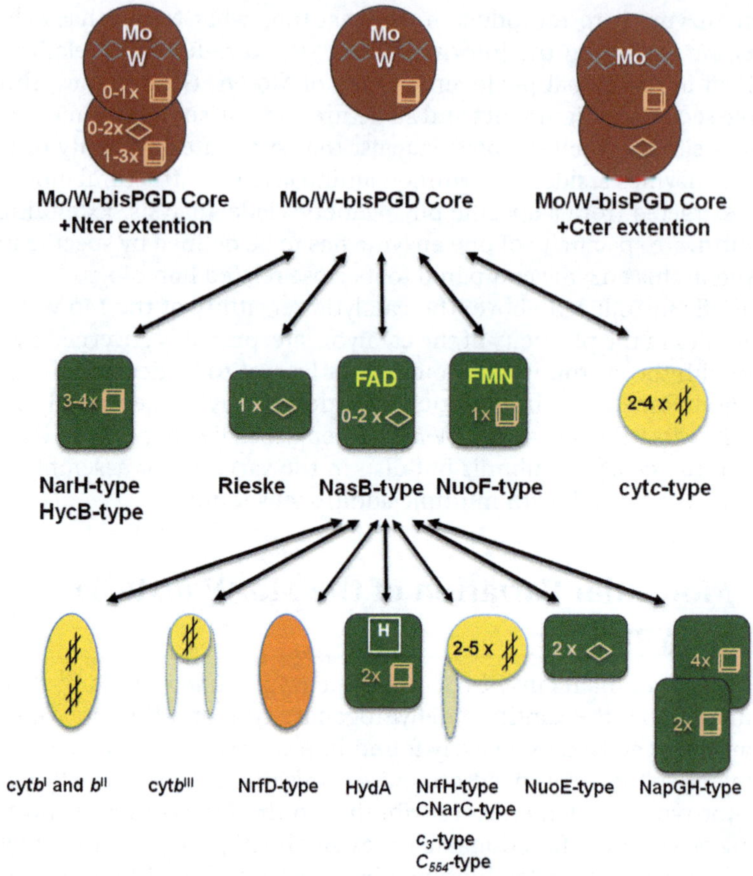

Figure 5.5 Schematic modularity cascade of the Mo/W-*bis*PGD enzymes. Several, but not all, types of Mo/W-*bis*PGD enzymes have been crystallized. We thus illustrated the modularity of the enzyme architecture by schemes rather than 3D structures in a cascade view. Indeed, whereas only three structure foldings are presently known for the α subunit, at least seven different structural types of β subunits potentially associated with an α subunit are known. Even more different types of γ subunits have been identified as electron entry/exit unit (see the text for details) combining with the β subunit to create all the variety of the Mo/W-*bis*PGD superfamily.

domains (see the work of Schoepp-Cothenet *et al.*[28] defining four different conserved domains (I–IV) in the catalytic subunit). These four domains (non-contiguously located within the primary protein sequence and with domains II and III related by an internal pseudo-two-fold axis; see Figure 5.1(B)) bury the cofactor deep inside the protein. Access to the active site is typically provided *via* a funnel-like entrance whose position at the interface between domains II and III is conserved in all crystallographically characterized family members except acetylene hydratase AH (ID: 2E7Z).[91] Much of the sequence stretches that do not align with the four core catalytic subunits are located within a fifth domain V. This domain has consequently been proposed to play an essential role in defining the substrate binding cavity.[85,86,92]

The archetypical catalytic subunit contains an Fe/S cluster known as FS0 in addition to the Mo/W cofactor. With the exception of the non-redox AH, this FS0 represents the electron transfer relay between the catalytic site and the redox-active centers in additional subunits. A first deviation from this archetype is exemplified by the Dor (ID: 1EU1),[9,93] the Tor (ID: 1TMO),[94,95] the Bis[96] and the TH (ID: 1TI2),[97] where FS0 is absent while the overall four-domain fold is conserved. Phylogenetic analyses suggest that the loss of the FS0 is a relatively late event and that the common ancestor of the whole Mo/W-*bis*PGD enzymes family almost certainly contained an FS0 cluster.[22,28] The second deviation from the canonical structure is exemplified by NasA from *Klebsiella pneumoniae*[98] or NasC from *Paracoccus denitrificans*,[99] with an additional [2Fe–2S] cluster bound in Cys-rich motifs present in an additional C-terminal region. A third deviation from the canonical structure is represented by the NAD^+-dependent Fdh catalytic subunit where an additional N-terminal extension harbors up to two [2Fe–2S] centers and up to three [4Fe–4S] clusters.[100–104]

The Cys residues responsible for the coordination of the FS0 cluster are located in the N-terminal domain I of the catalytic subunit. In most cases, a $CX_2CX_nCX_nC$ motif is found coordinating a typical [4Fe–4S] center. This is the case of Nap (ID: 1OGY, 3ML1, 2NYA and 2NAP),[43,105–107] Fdh and Fdn (ID: 1KQG, 1H0H and 1FDO),[48,108,109] Psr (ID: 2VPZ)[110] and AH (ID: 2E7Z).[91] Due to their close sequence homologies with Psr and Nap, Arr,[80] Ttr,[111] Soe,[84,112] thiosulfate reductase Phs[113] and NasA, NasC and NarB are also expected to harbor such an FS0 cluster.[98,114–116]

Besides this typical [4Fe–4S]-binding motif, several unusual ones have been found. In NarG (ID: 1Q16 and 1R27),[49] the first residue ligating FS0 is a His in an $HX_3CX_nCX_nC$ motif, giving rise to very special properties of the FS0 cluster (in particular, a high-spin ground state in the reduced form).[117,118] This FS0 binding motif is also seen in Ebdh (ID: 2IVF),[51] and proposed in Ser,[119] dimethyl sulfide dehydrogenase Ddh,[120] and Clr.[121] The trimeric Dms from *E. coli* possesses a variant $CX_3CX_nCX_nC$ binding motif of FS0,[122] and recently new binding motifs have been revealed in C25dh and pNxr,[37,78] even proposing an Asp replacing the third Cys in the binding motif of the [4Fe–4S] cluster. Aio is atypical in the Mo/W-*bis*PGD family with its FS0 being a [3Fe–4S] cluster coordinated by only three Cys, and having a high reduction potential.[123,124]

Trieber *et al.*[125] originally proposed subdividing the Mo/W-*bis*PGD family into different classes corresponding to the FS0 binding motif, and McDevitt *et al.*[120] have subsequently pointed out that this subdivision also corresponds to phylogenic relationships with the members of these different classes grouping together into separate monophyletic clades in the phylogenetic tree of the Mo/W-*bis*PGD enzymes family. Recent analyses based on increasing number of sequences argue against this idea, however.[28,78,126] It thus appears most appropriate to classify the enzymes based on the Mo/W ligand: Ser-group, Asp-group, Cys-group, water (represented by the Aio). Furthermore, such classification may be considered as functionally relevant.

Another important function of the catalytic subunit is its role in determining the final sub-cellular localization of the mature complex. In the case of periplasmic Mo-enzymes, protein assembly and cofactor insertion take place in the cytoplasm prior to translocation of the mature complex across the inner membrane by the Tat-machinery. The N-terminal, twin arginine translocation peptide signal (Tat leader sequence) responsible for the addressing of the immature enzyme to this machinery is in most cases found in the catalytic subunit, with only two exceptions presently known. The Ttr is a special case since both the catalytic and the electron transfer subunits can, in some organisms, harbor a Tat leader sequence.[127,128] Aio represents another special case since the Tat leader sequence is present in the electron transfer subunit rather than the catalytic one.[129,130] This sequence can be the archetypal RRXFLK but more often is slightly modified to RRXFIK, RRXFLR, RRXLLK, RRXWLK, RRXFLQ, RRXFLS or RRXLMQ. The Nar enzymes perfectly illustrate the importance of the leader sequence in sub-cellular location of the enzyme. The bacterial representatives, harboring a vestigial leader sequence in the immature NarG proteins are found facing the cytoplasm,[131,132] while the archaeal enzymes, with a predicted Tat leader sequence at the N-terminus of the immature pNarG, are found facing the periplasm.[133]

5.3.2 The Electron Transfer Subunit

5.3.2.1 Archetypal Subunit

A four Fe/S-harboring protein has been simplistically presented by Rothery *et al.*[22] as "THE" β subunit of Mo/W-*bis*PGD enzymes. While most of the members do possess such a subunit, the diversity of the β subunit into the Mo/W-*bis*PGD family is today well documented (Figure 5.5). Several structures of archetypal β subunits have been resolved (ID: 1TI2, 2IVF, 1H0H, 1KQG, 2VPZ, 1Q16 and 1R27). Despite the common feature of multiple Fe/S clusters, the subunits show very variable size, ranging from 191 residues in the case of the *W. succinogenes* Psr to 512 residues for nNar. Restricting our discussion to proteins of established structure, the iron–sulfur clusters can be four [4Fe–4S] clusters, as in the Fdn (ID: 1KQG)[48] or the Psr (ID: 2VPZ)[110] with typical $CX_2CX_nCX_3C$ binding motifs, or alternatively three [4Fe–4S] and one [3Fe–4S] as in nNar (ID: 1Q16,1Y4Z)[49] and Ebdh (ID: 2IVF),[51] or only three [4Fe–4S] in TH (ID: 1TI2)[97] and the W-containing Fdh (ID: 1H0H).[109] Together with the FS0 (when present), these Fe/S clusters (designated FS1–FS4 with increasing distance from the FS0) are arranged in a W shape. As in the case of the catalytic subunit, domains (more or less superimposable among the different enzymes) can be delimited in this type of β subunit. Domains I and II are bacterial ferredoxin-like domains harboring two iron–sulfur clusters, each sandwiched between a pair of α-helices on one side and a β-sheet on the other.

5.3.2.2 "Exotic" Subunits

Mo/W-*bis*PGD enzymes carrying a β subunit distinct from the consensus multiple-Fe/S protein are not rare. Seven other types of β subunits are presently known, each present in one specific system: a Rieske type protein in the AioAB heterodimer,[123,124] a diheme cytochrome (*c*-type) in the NapAB heterodimers,[105,107] a tetra-heme cytochrome (*c*-type) in the NapA monomers,[134] two different types of FAD/[2Fe–2S] proteins in the dimeric Nas,[115,135] the HycB modules in the HDCR system[89] and the NuoF-type proteins in the NAD^+-dependent Fdhs.[100–103] Only AioAB and NapAB have at present been crystallographically characterized.

NapAB enzymes are all soluble and periplasmic, with NapB more or less tightly associated to NapA. In the *R. sphaeroides* NapAB (ID: 1OGY) the tight AB association is mediated by two extended arms extending from the NapB subunit and embracing the NapA subunit.[105] In NapAB, heme II from NapB is located in the vicinity of FS0, whereas heme I is exposed to the solvent, arguing for this latter heme as the electron entry site.

The AioB Rieske protein (ID: 1 G8K and 4AAY), by its [2Fe–2S] cluster and its protein fold, is homologous to the well-characterized subunit of the complex III in mitochondria in terms of both the overall protein fold and the type of [2Fe–2S] cluster contained therein.[123,124,136] A characteristic $C–X–H–X_n–C–X_2–H$ sequence motif, observed in all Rieske proteins, binds the [2Fe–2S] cluster. This [2Fe–2S] cluster, exposed to the solvent, is the electron donor to the enzyme's physiological partner, which in proteobacterial systems (the only characterized) is a soluble *c*-type cytochrome (see ref. 137 for review). All Aio are periplasmic but some of them have been observed attached to the cytoplasmic membrane. It is not clear at present whether the Tat signal peptide, present in AioB and which would thus remain uncleaved, could be responsible for the membrane association in a manner similar to that observed in the complex III homolog.[137] However, this anchor is not responsible for the connection of the enzyme to the membranous quinone pool.

Among the assimilatory nitrate reductases, the NasBC from *Bacillus subtilis*[115] and the NasAC from *K. pneumoniae*[135] or *R. capsulatus*[138] are the only representatives characterized to a significant extent. Both enzymes are responsible for the electron transfer from NAD(P)H to the catalytic subunit in cytoplasm but are highly divergent not only at a global sequence level but also at the structural and cofactor composition levels. The NasB-type subunit is 771 residues long and therefore even larger than the catalytic subunit, whereas the NasC-type protein is only 396 residues long. Sequence and biochemical analyses suggest that NasB and NasC both harbor a FAD cofactor, but NasB also possesses clear motifs for the coordination of two [2Fe–2S] clusters.

In the NAD^+-dependent FDHs, the FdsB proteins are responsible for NAD^+ reduction linked to the oxidation of formate by the FdsA protein. Although described as a "β subunit" in this context and also in the original publication characterizing these systems, this FdsB subunit is thus in fact the "exit subunit" of the related enzymatic complex. Associated with the FdsG subunit (when present), FdsB constitutes an active diaphorase unit in the system. The sequence analysis of these FdsB proteins in the NAD^+-dependent FDHs from *Cupriavidus necator* (formerly *Ralstonia eutropha*), *Methylobacterium extorquens* or *R. capsulatus* reveal a conserved structure related to NuoF subunit from Complex I, harboring a [4Fe–4S] cluster together with an FMN cofactor.[100–103]

As mentioned in the beginning of this section, the notion of "enzymatic complex" is relatively flexible. In the case of the HDCR system, the enzymatic entity has been described as the association of a FdhF unit, one/two HycB units and one HydA.[89] In the system, HycB proteins are responsible for delivering electrons to the FdhF either from the oxidation of H_2 by the HydA unit or from the oxidation of soluble ferredoxins (preliminary reduced by a CO dehydrogenase for example). Based on sequence analyses, the HycB unit has been proposed to harbor either three [4Fe–4S] and one [2Fe–2S] clusters[139] or four [4Fe–4S] clusters.[89] This subunit is in any way not structurally related to the NarH-type proteins, but is related to the HycB component of the FHL system.

5.3.3 The Electron Entry/Exit Subunit

This component is the most variable element of the Mo/W-*bis*PGD enzymes, although each type of γ subunit is often encountered in several representatives of the Mo/W-*bis*PGD enzymes family. The cytc_3 characterized with *Desulfovibrio* sp. Fdh,[140,141] the tetrahemic c_{554}-type cytochrome characterized in *Dechloromonas* sp. Pcr[142] and the dihemic cytochrome characterized in the *Thermus thermophilus* cNar[143–145] have, however, been observed each in a single system. With or without redox-active cofactors, integrated or not into the membrane, this γ subunit is of primordial importance for integration of the enzyme into the bioenergetic chain in which it is involved. In the great majority of cases, this third subunit is responsible for the reaction with the membranous isoprenoid quinones.

5.3.3.1 b-Type Cytochromes

Two types of transmembrane diheme cytochrome *b* and one type of hydrophilic monoheme cytochrome *b*, all of which have been crystallographically characterized, are present as γ subunit in members of the Mo/W-*bis*PGD enzymes family (Figure 5.5). All represent typical "redox construction kit units" and form different enzymatic systems even out of the Mo/W-*bis*PGD enzymes family.[26,27] The first cytochrome cyt b^{I} (ID: 1KQG) usually contains four trans-membrane helices (TM) and the two hemes are coordinated by histidine residues on TM1 (1 His), TM2 (1 His) and TM4 (2 His). Cyt b^{II} (ID: 1Q16) features five TM and the heme ligands are located on TM2 (2 His) and TM5 (2 His). Cyt b^{I} constitutes the FdhN subunit.[48] Due to its sequence homology to the FdhN γ subunit, PhsC is proposed also to be a cyt b^{I} cytochrome.[22,146]

Cyt b^{II} is exemplified by the γ subunit of the bacterial nNar enzyme[49] and has been proposed as third subunit in nNxr and cNar.[147] The function of such Cyt b^{I} and Cyt b^{II} units is to connect the related enzymes to the membranous quinol pool. Why there are two structurally distinct membrane-integral *b*-type cytochromes for this purpose remains unknown. The idea that cyt b^{I} could serve as a quinone-reducing and cyt b^{II} as a quinol-oxidizing element does not match with the function of PhsC in oxidizing MK to reduce thiosulfate.

A third type of γ subunit is seen in Ebdh, as a cytochrome b^{III} with a hydrophilic antiparallel β-sheet fold. Again, a 3D-structure is available (ID: 2IVF)[51] and similar subunits have been proposed to be present in archaeal pNar,[133,148,149] Ser,[119,150,151] Ddh,[120,152] Clr[20] and C25dh.[37] This type of γ subunit does not connect the enzymes to the quinol pool, although in two cases it has been proposed as anchor of the related complex to the membrane.[133,149] All these cytochromes *b* have the unusual combination of Met and Lys residues as axial heme ligands. The wire function of this protein is achieved by exposing two distinct heme edges at the protein surface.

5.3.3.2 The NrfD-Type Subunits

The "NrfD-type" unit (named from the first crystallographically characterized representative of the group, periplasmic nitrite reductase[153]) is a membrane-integral protein devoid of any cofactor but nevertheless responsible for interacting with the intramembrane quinone pool. Sequence homologies indicate that this subunit is present in several enzymes in the Mo/W-*bis*PGD family, including Psr,[154] Ttr,[111] Soe,[84,112] Dms[155] and several Arr/Arx.[80] Although clearly related to one another, these units have a variable number of TM. The X-ray crystal structure of *T. thermophilus* PsrC has provided the sole 3D picture of this type of γ subunit. The crystal structure shows one equivalent of bound MK (ID: 2VPW), which has made it possible to visualize precisely the quinone binding site in this complex. Little information exists on the other NrfD-type representatives.

5.3.3.3 c-Type Cytochromes

c-Type cytochromes are often found in the γ subunit of the enzymes of the Mo/W-*bis*PGD superfamily, although they are structurally diverse. Over the last decade, a family of membrane-anchored *c*-type cytochromes, the NrfH/NapC/NirT family, has been recognized as being involved in the major pathway for electron transfer between the quinol pool and water-soluble periplasmic oxidoreductases, and are found even in enzymes not of the Mo/W-*bis*PGD family. These cytochromes are not strictly speaking subunits, but rather function more as dedicated physiological partners. The structural model for this family is the tetrahemic *c*-type NrfH cytochrome involved in nitrite reduction to ammonium with NrfA,[156] which is not a member of the Mo/W-*bis*PGD family. NrfH-type cytochromes demonstrated to function with

Mo/W-*bis*PGD enzymes are numerous, however, including TorC (the partner of the TorA enzyme[157]), DorC (the partner of DorA protein[158]) and NapC (the partner of the NapAB complex[159]).

In *D. vulgaris*[141] and *D. desulfuricans*,[140] a very special FdhABC$_3$ has been shown to contain a 16 kDa cytochrome. This subunit is homologous to the typical soluble cytochrome c_3 (TpIc_3) that plays the role of physiological partner for periplasmic hydrogenase, for which a 3D-structure is available (ID: 2BQ4).[160] TpIc_3 is a small globular cytochrome containing four covalently bound hemes all with *bis*-His axial coordination and quite negative reduction potentials.[161] The close proximity of the hemes results in special oxidation–reduction properties of these cytochromes that are also manifested in the C$_3$ subunit of the FdhABC$_3$.[140,141] This c_3 does not bind FdhABC$_3$ to the membrane, which is a soluble enzyme. The *pcr* operon of two *Dechloromonas* species[142] includes a *pcrC* gene coding for a protein product that more closely resembles the soluble tetrahemic cytochrome c_{554} from *Nitrosomonas europaea* (ID: 1FT5)[162] rather than the c_3 cytochrome. PcrC is not expected to be membranous and therefore may play the role of an additional electron wire for the enzyme, similarly to the c_3 subunit depicted above.

In the cNar complex from *T. thermophilus*, the diheme cNarC is required for correct assembly of an active NarGH complex and is therefore considered as a true subunit of the cNar complex.[144,145] Sequence analysis suggests that a single TM at the C-terminus part of the protein anchors the cytochrome into the cytoplasmic membrane. Such a dihemic cytochrome *c* has also been proposed, based on genome sequencing, as a partner subunit in nNxr.[163,164]

5.3.3.4 The HydA Subunit in the Hydrogen-Dependent Carbon Dioxide Reductase System

As already discussed, the HDCR system isolated from *Acetobacterium woodii*[89] contains an HydA unit responsible for the oxidation of H$_2$ linked to the reduction of CO$_2$. This subunit is proposed to be a true subunit of the complex and therefore to be distinguished from Hyd-3 in the FHL system, which is considered a physiological partner of the FdhF and not a subunit due to the fact that *fdhF* is not in the *hyc* operon. It is to note, however, that HDCR is likely encoded by separate transcriptional units in *A. woodii*. If HydA is a true subunit of the HDCR, it represents a unique case in the Mo/W-*bis*PGD family as the HydA is responsible for a catalytic function and not a simple electron transfer. Based on sequence analysis, this protein has been proposed to harbor, in addition to the H-cluster responsible for the catalytic activity, either two [4Fe–4S] clusters[89] or three [4Fe–4S] clusters and one [2Fe–2S] cluster.[139]

5.3.3.5 The NapGH Component

In the simplest Nap system, NapA requires only NapC to mediate reduction of nitrate and oxidation of membranous quinols. However, in other Nap systems, including the ones in *E. coli* and *W. succinogenes*, NapA functions together with periplasmic proteins (NapB and NapC), and membrane-integral proteins (NapG and NapH), all of which have been demonstrated to interact.[87,165] Regardless of whether a NapC is present (as in the *E. coli* system) or absent (as in *W. succinogenes*), the NapGH subcomplex is responsible for the electron delivery to NapB from the oxidation of quinols.[165,166] These two subunits are predicted to harbor four and two [4Fe–4S] clusters, respectively, but no detailed structural characterization is available.

For some *c*-type cytochromes and cytb^{III} units and NapGH component, we can wonder what their role is in their respective system. When the catalysis is for sure to be attributed to the α subunit and the electron transfer function to the β subunit, all these additional proteins could appear as functionally redundant. When available, functional studies have underpinned the hub character of these electron entry/exit subunits connecting Mo/W-*bis*PGD enzymes to multiple metabolic chains.

5.4 Metabolic Chains Involving Mo/W-*bis*PGD Enzymes

As indicated in the introduction, members of the Mo/W-*bis*PGD family are exclusively found in bacteria and archaea. With a single exception, these enzymes catalyze oxidation–reduction reactions involving the Mo(VI)/(IV) and W(VI)/W(IV) couples. As thoroughly described in the previous section, electrons are transferred to or from the active site by a variety of metal centers (Fe/S clusters, hemes *b* or *c*) that are present in associated subunits or physiological partners. Most interesting is that these architecturally complex enzymes are found in all cell compartments, including the outer surface of bacterial cells (Figure 5.6). In this section, we will highlight how members of the Mo/W-*bis*PGD family are integrated into metabolic pathways in the overwhelming majority of prokaryotes, thereby contributing to their growth, survival and virulence. Importantly, the idea is here to describe how rather limited sets of active site chemistry and structural modules have led to the generation of an extended group of enzymes that play essential roles in cell metabolism. A focus will be made on enzymes involved in the N, C and S cycles for which a plethora of information allows for comparative analysis.

Besides compounds containing nitrogen, sulfur and/or carbon that are also metabolized by eukaryotes, numerous compounds considered as toxic for the latter can be used as sources of energy by prokaryotes. There are at least eight Mo/W-*bis*PGD enzymes known today to be involved in metabolism of toxic compounds such as arsenics, selenate or oxochlorates. For an in-depth survey of the integration of these enzymes to cell metabolism, the reader is referred to more extensive reviews on this subject,[10,20,137,167] which due to space limitations are not considered in detail here. In addition, several hydrocarbon compounds, including acetylene,[30] pyrogallol,[33,168] resorcinol,[169–172] ethylbenzene[36,173,174] and cholesterol[37] are also substrates of Mo/W-*bis*PGD enzymes, although in many cases it remains unknown as to how these enzymes are integrated into cell metabolism.

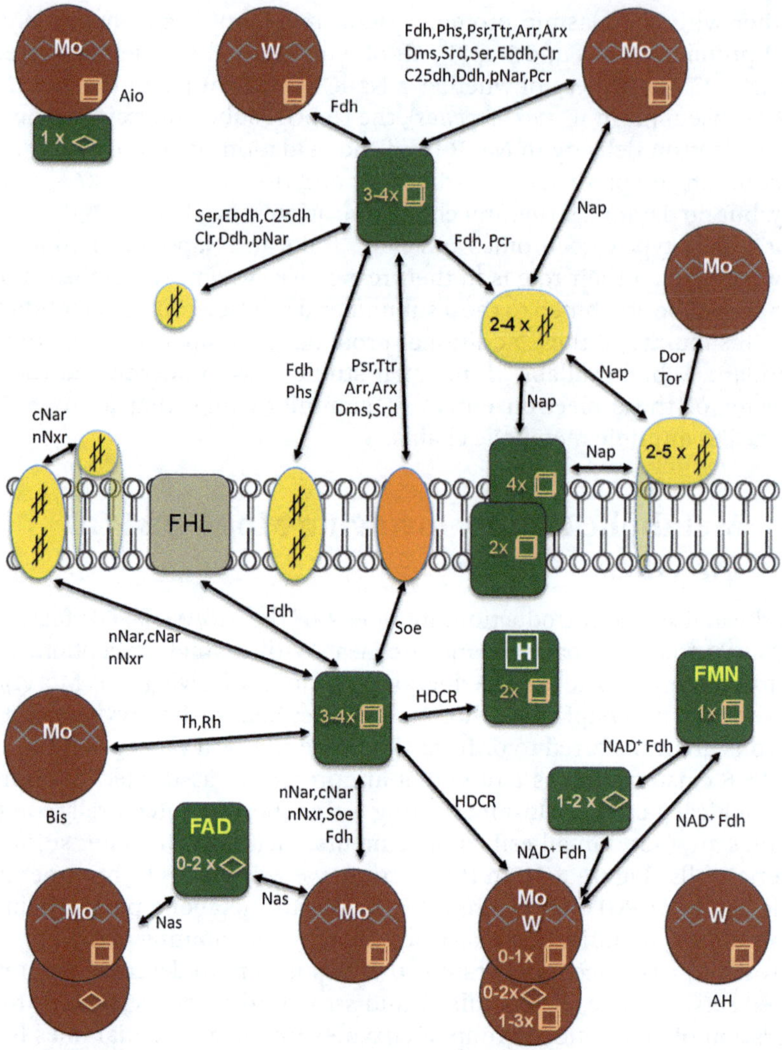

Figure 5.6 Composition and localization diversity of Mo/W-*bis*PGD enzymes. As detailed in the text, the Mo/W-*bis*PGD enzymes are textbook examples of the "redox construction kit" principle described by Nitschke and coworkers.[26,344] This representation aims at providing a clear understanding of how subunits combination (illustrated in Figure 5.5) has been used throughout evolution to build various Mo/W-*bis*PGD enzymes of various metal- and subunit-compositions, cellular localization which had in turn a direct impact on their integration to the cell metabolism.

5.4.1 Enzymes Involved in the Nitrogen Cycle

Molybdoenzymes of the Mo/W-*bis*PGD family are well known to participate in the nitrogen cycle through the conversion of nitrate to nitrite by nitrate reductases.[147] All prokaryotic nitrate reductases share a Mo/W-*bis*PGD-containing catalytic subunit where the reduction of nitrate to nitrite occurs, although they differ in their physiological function and subcellular location, which, as described below, are related issues. Three distinct classes of prokaryotic nitrate reductases have been distinguished by taking into consideration (i) localization in the cell, (ii) molecular properties of the catalytic center, (iii) structural organization and (iv) physiological function. The three types are: (1) the soluble, assimilatory nitrate reductases Nas located in the cytoplasm; (2) the membrane-associated respiratory nitrate reductases Nar; and (3) the soluble, "respiratory" nitrate reductases Nap located in the periplasm. It has to be mentioned that Nar and Nas are found in both archaea and bacteria whereas Nap enzymes are only found in a sub-group of bacteria (Figure 5.4). A fourth distinct category can be illustrated by Mo-*bis*PGD enzymes catalyzing the nitrate reduction in the reverse direction *i.e.* nitrite oxidation and named Nxr. Nxr-catalyzed nitrite oxidation is at play during the nitrification and anammox processes producing N_2 from nitrite and ammonium. All categories will be surveyed in the following sections.

5.4.1.1 Nas and Nitrogen Assimilation

Nitrate is one of the more abundant sources of inorganic nitrogen available in aquatic systems. As such, the majority of cyanobacteria possess pathways for the uptake and assimilation of nitrate. The first reduction step in nitrate assimilation, the two-electron reduction of nitrate to nitrite catalyzed by Nas, is followed by the six-electron reduction of nitrite to ammonium by a siroheme-containing nitrite reductase. The ammonium is generally used for glutamate synthesis, which plays a pivotal role in biosynthetic cell metabolism. The assimilatory reduction of nitrate to ammonium is an energetically expensive process since it requires eight electrons and involves complex prosthetic groups for the nitrate and nitrite reductase enzymes.

Extensive genetic and biochemical characterization of assimilatory nitrate-reducing systems has been performed with cyanobacteria.[175–179] For instance, the overwhelming majority of *Synechococcus* sp. can utilize nitrate or nitrite as sole nitrogen sources in oceanic surface waters and the same holds true for a number of other nitrate-assimilating bacteria.[180,181] In comparison, fewer physiological, genetic and biochemical studies have been conducted on other nitrate-assimilating bacteria.[99,115,135,182–188] Bioinformatic analyses have revealed that *nas* genes are present in many archaea[189] but, to date, physiological, genetic and biochemical characterization of nitrate assimilation have only been performed on *Haloferax mediterranei*.[190–194] A recent study has shown that nitrate assimilation promotes virulence of a plant pathogen *Ralstonia solanacearum*, the causative agent of bacterial wilt disease.[195] In particular, Nas has been shown to be essential during the early stages of plant colonization.

Although nitrate assimilation pathways are dependent on cytoplasmic nitrate and nitrite reductases, a high degree of flexibility is observed regarding the nature of the physiological electron donor to these enzymes, which has immediate consequences in terms of integration within the cell metabolism. In cyanobacteria, a ferredoxin reduced by Photosystem I constitutes the electron donor to both the nitrate and nitrite reductases NarB and NirA, respectively, thus linking nitrate assimilation to photosynthesis.[196] On the other hand, a Rieske-type ferredoxin, NasG, has been shown to be essential for coupling of the physiological electron donor, NADH, to both the nitrate and nitrite reductases NasC and NasB, respectively, in *P. denitrificans*.[99] Indeed, NADH is oxidized by the nitrite reductase NasB, a situation distinct from nitrate assimilation in plants and fungi, further requiring NasG to provide NasC with electrons. NasG, encoded by the *nasG* gene (also named *nirD*, *nasD* or *nasE*), is conserved in several bacterial *nas* gene clusters. Nitrate assimilation is thus linked to the status of the NADH pool. In haloarchaea such as *Haloferax volcanii* or *Haloferax mediterranei*, the physiological electron donor to the assimilatory nitrate is thought to be ferredoxin rather than NADH.[190]

5.4.1.2 Nar and Nitrate Respiration

A prominent example of the participation of molybdoenzymes of the Mo/W-*bis*PGD family in the nitrogen cycle is denitrification, a key process that contributes to the nitrogen cycle on Earth through the biological production of NO, N_2O and N_2 gases from nitrate under anoxic conditions (Figure 5.7(A)). Most prokaryotes are able to use nitrate as a terminal electron acceptor in energy metabolism, many of them being able to perform the entire denitrification process. While only partially described in some *Haloarchaea* species, the denitrification pathway is involved in the virulence of several bacterial species, including *Brucella*, *Pseudomonas* or the obligate human pathogens *Neisseria gonorrhoeae* and *Neisseria meningitidis*.[197–200] Conversely, *Enterobacteriaceae* are not able to denitrify nitrate or nitrite to N_2. Instead, reduction of nitrate *via* nitrite to ammonia occurs during anaerobic growth (Figure 5.7(B)). Importantly, bacterial denitrification can generate small but substantial amounts of NO as a metabolic intermediate and has been implicated in the bacterial NO production in the human gut.[201]

5.4.1.2.1 nNar and cNar. The first step of the bacterial denitrification pathway, catalyzed by the cytoplasmically oriented membrane-bound NarGHI-type nitrate reductase (*i.e.* nNar), is directly coupled to the generation of a *pmf* to sustain cell growth. The nNar system is largely conserved among all nitrate-respiring bacteria. The enzyme has been extensively studied in mesophilic nitrate-reducing bacteria such as the ammonifier *E. coli* and the denitrifiers *Paracoccus* and *Pseudomonas* species. The *E. coli* NarGHI enzyme has been crystallized and serves as the prototype in our understanding of the structure and function of respiratory nitrate reductases in other bacteria.[49,50] The overall topology of the NarGHI complex ensures that the free energy in the QH_2/NO_3^- oxidation–reduction couple is conserved as *pmf via* an electrogenic redox half-loop mechanism. In line with the nitrate/nitrite reduction potential of +420 mV, quinones of both low and high potential can serve as electron donors.[202–204] Accordingly, the reduction of nitrate can be coupled to

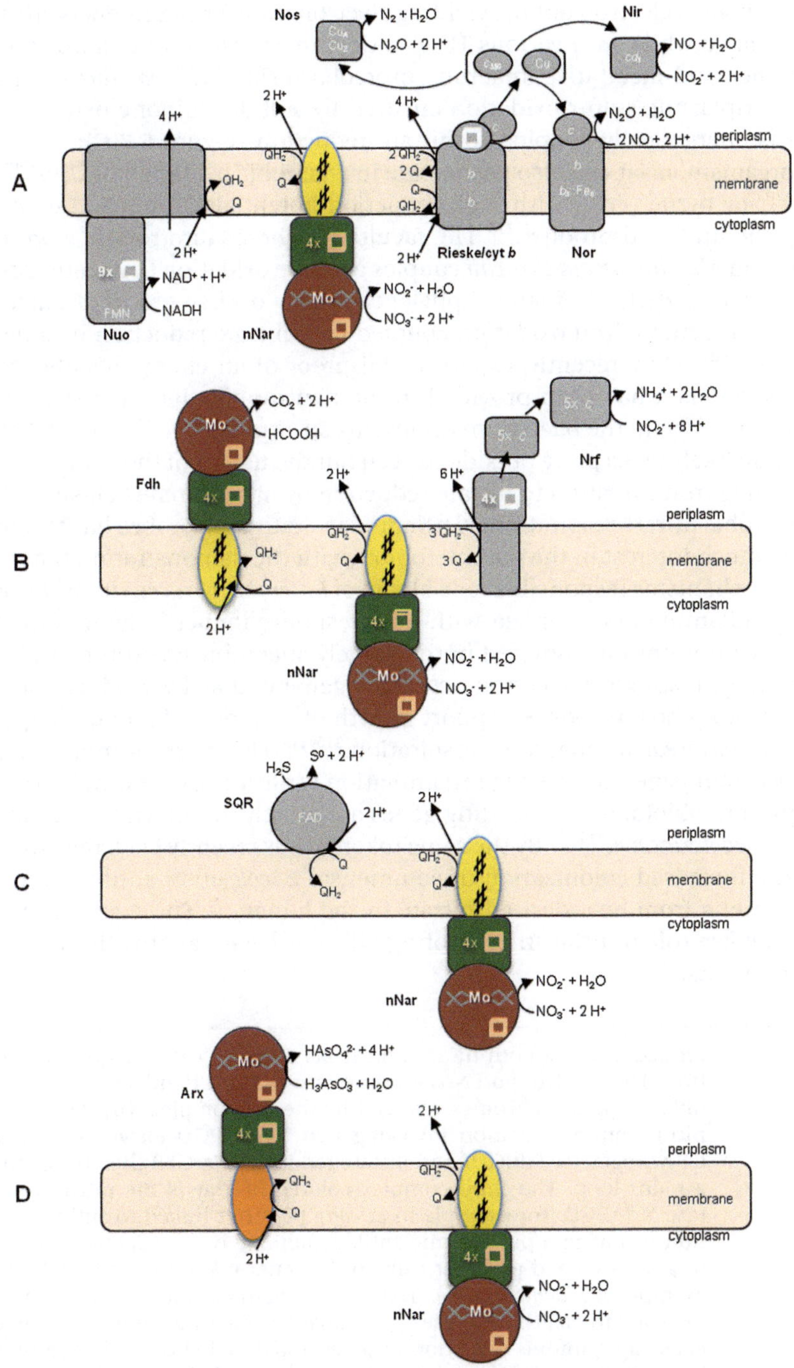

Figure 5.7 Schematic representation of a few bioenergetic chains involving nNar in prokaryotes. (A) nNar linked to the denitrification pathway in *P. denitrificans* allowing sequential conversion of nitrate to dinitrogen. Electrons are provided by complex I (colored in gray) ensuring quinone

a wide range of electron donors yielding electron transfer chains energetically conservative while with various H^+/e^- ratios. In the chemolithoautotrophic denitrifier and facultative anaerobe proteobacterium *Thiobacillus denitrificans*, periplasmic sulfur oxidation ensured by sulfide : quinone oxidoreductases is energetically coupled to nitrate reduction (Figure 5.7(C)).[205–207] In this organism, nNar can also participate in anaerobic oxidation of U(IV),[40,208] Fe(II)[209] or pyrite (FeS_2) with high reduction potentials (Table 5.1) with N_2 being the final end product.[210] The facultative chemoautotrophic γ-proteobacterium *Alkalilimnicola ehrlichii* couples arsenite oxidation to nitrate reduction (Figure 5.7(D)).[211] Finally, a phylogenetically diverse group of bacteria sustains anaerobic iron oxidation coupled with nitrate reduction by a nNar enzyme.[209,212] Only recently, experimental proof of an energy benefit from this metabolism has been provided in the neutrophilic β-proteobacterium Strain 2002[212] or in the bacterium *Acidovorax* sp. Strain 2AN.[213] To date, Fe(II) oxidation likely takes place outside the cell but the nature of the components connecting iron oxidation to nitrate reduction by nNar remains elusive.[214]

Anaerobic nitrate respiration by nNar in facultative anaerobes has recently gained much interest in the gut microbiota with the demonstration that one of the mechanisms responsible for a bloom of *Enterobacteriaceae* in the lumen during inflammation is linked with host response-induced changes of the growth conditions encountered in this largely anaerobic environment.[201,215] Specifically, respiratory electron acceptors generated as by-products of the inflammatory host response support growth of commensal and pathogenic *Enterobacteriaceae* by anaerobic respiration.[216–218] The most prominent electron acceptor generated during inflammation is nitrate. Interestingly, within the gut microbiota, nNar-encoding genes are mostly found within genomes of *Enterobacteriaceae*.[219,220] By the same token, it has recently been recognized that the increased colonization of commensal *E. coli* upon antibiotic treatment results from host-derived nitrate in the lumen.[221] These studies reinforce the key role of nNar in promoting efficient bacterial growth in anoxic environments.

---

reduction while both nNar and the bc_1 complex catalyze quinol oxidation. The Nir, Nor and Nos enzymes are reduced thanks to various soluble *c*-type cytochromes reduced by the bc_1 complex. (B) Respiratory nitrate ammonification involving two Mo-*bis*PGD enzymes, formate dehydrogenase FdnGHI and nitrate reductase NarGHI thereby forming a redox loop. The third complex colored in gray is the nitrite reductase NrfABCD (for example in *E. coli*). (C) nNar linked to sulfur oxidation involving a periplasmic sulfide : quinone oxidoreductase (colored in gray) attached peripherically to the inner membrane and nNar (for example in *T. denitrificans*). (D) nNar linked to arsenite oxidation involving two Mo-*bis*PGD enzymes, a periplasmic facing arsenite oxidase Arx ensuring quinone reduction at a yet undefined site within the membrane anchor subunit and the nNar (for example in *A. ehrlichii*). Owing to the various postulated mechanisms used by neutrophilic anaerobic nitrate-dependent iron-oxidizing bacteria, the reader should refer to ref. 214 for their extensive description.

Apart from the nitrate reductase activity displayed by nNar enzymes, several studies have reported the ability of nNar from *E. coli*[222,223] or *Salmonella typhimurium*[224] to catalyze a one-electron reduction of nitrite to nitric oxide (NO).[225] NO is a highly reactive and toxic gas that can freely diffuse into cells and attack DNA molecules or the redox-active centers of proteins.[226,227] NO formation by nNar depends on a combination of anoxic, nitrate-sufficiency and nitrite-accumulating conditions. While the importance of such activity in bioenergetic terms remains unclear, Rowley *et al.* reported that up to 20% of catabolized nitrate is converted to NO in *S. typhimurium*.[228] Such NO-producing activity needs now to be reconciled with the participation of nNar to the virulence of a number of microbial pathogens or to the reported inflammation-induced nitrate respiration.[217,229,230]

While the nNar system is principally associated with anaerobic metabolism, it appears that it is also present in a number of bacteria considered as obligate aerobes such as *B. subtilis* and in *Mycobacteria* sp.[231–234] If the physiological significance of nitrate reduction in the above-mentioned organisms is only partly understood to date, expression of an active nNar complex in *Mycobacteria* is known to be essential in protecting the hypoxic cells from environmental stresses[235] and in contributing to their virulence under hypoxic conditions by allowing cell proliferation.[236,237] Nitrate reduction is associated with ATP synthesis, maintenance of a membrane potential and thus is sufficient to restore cell survival under hypoxic acidic conditions encountered in the phagosome.[235] On the other hand, produced nitrite is actively secreted in the phagosome to avoid toxicity. Recently, the role of nNar in the anaerobic survival and virulence of *Burkholderia thailandensis*, closely related to the human pathogen *B. pseudomallei*, the causative agent of melioidosis with similar clinical features to tuberculosis such as granulomas, has been reported.[238]

Variations of the nNar organization have been observed in the hyperthermophilic bacterium *T. thermophilus* HB8, which encodes an atypical NarCGHI complex (*i.e.* cNar) and a NADH dehydrogenase (Nrc) by a self-transposable element that allows anaerobic growth with nitrate as terminal electron acceptor.[239,240] Both enzymes are considered to form a supercomplex ensuring electron transfer from NADH to nitrate using MK-8 as electron carrier.[143] The unique feature of this cNarCGHI is the presence of a fourth subunit, cNarC, a periplasmic diheme *c*-type cytochrome (see also Section 5.3). Furthermore, integrity of the NarCGHI complex is required for electron transfer towards the Nir, the Nor or the Nos reductases in denitrifying *T. thermophilus* strains.[241] Together with the diheme cytochrome *b* NarI, the cNarC subunit of the cNar enzyme is used as a branching point to deviate electrons towards enzymes of the denitrification pathway. Finally, cNar from the obligatory anaerobe *Geobacter metallireducens* is involved in degradation of aromatic compounds or U(IV) oxidation.[242,243]

5.4.1.2.2 pNar. Several archaeal denitrifiers (*e.g.* several *Haloarchaea* sp.[244–247]) possess a Nar complex that shows two main differences as compared to the cytoplasmic version found in most bacteria: a periplasmic location thanks to the presence of an intact Tat signal sequence in NarG (*i.e.* pNar) and the presence of a monohemic cytb^{III} subunit named Orf7, which may serve as membrane anchor of the NarGH complex (see also Section 5.3). Analysis of the genes of the *pnar* cluster has allowed many authors to speculate on the overall structural organization of this enzyme.[133,246,248] Accordingly, a supercomplex made up of NarGH and Rieske/cyt*b* complex has been hypothesized, which would display quinol : nitrate oxidoreductase activity coupled with an effective generation of *pmf* through an electrogenic Q-cycle-type mechanism.[133] The hydrophilic monohemic cytb^{III} subunit would then replace functionally the missing *c*-type cytochrome of the Rieske/cyt*b* complex. Apart from the *Haloarchaea*, pNar with either Mo or W ion is also encountered in hyperthermophilic denitrifier and facultative anaerobe crenarchaeon *Pyrobaculum* sp. (Figure 5.4) able to perform dissimilatory nitrate reduction using H_2, thiosulfate or organic compounds as electron donors.[148,149,249] It is presently unclear whether archaeal pNar is involved in energy conservation and only scarce information is available concerning how pNar are integrated in cell metabolism.

5.4.1.3 Nap and Respiration or Dissipation of Excess Reductant

In addition to the membrane-bound nitrate reductases are the non-electrogenic periplasmic Nap enzymes found almost exclusively in Proteobacteria and which fulfill different physiological roles depending on the organism.[147,250–252] As described above, both the membrane-bound nitrate reductases and Nap enzymes catalyze the first step of the denitrification process, which is regarded as an anaerobic or microaerobic process.[253,254] Nap can support anaerobic respiratory growth in *E. coli* under nitrate-limiting conditions using glycerol as non-fermentable carbon source owing to a higher nitrate affinity as compared to nNar.[255,256] Based on these observations, it has been considered that Nap would be required to scavenge nitrate present in low levels in the micro-oxic environment of some pathogenic bacteria as *Haemophilus influenzae* or *Campylobacter jejuni*.[165,251,257] Recently, such scavenging activity has been shown to make Nap from *S. typhimurium* effective in supporting anaerobic nitrate respiration in the inflamed gut.[258] In ε-proteobacteria, respiratory nitrate reduction sustained by Nap results in the exclusive production of either ammonium (ammonification) or dinitrogen gas (denitrification).[165,259] Respiratory nitrate ammonification has also been demonstrated in *S. oneidensis* MR-1 using Nap rather than a nNar as seen in *Enterobacteriaceae*.[260] In this particular case, CymA, an analog of NapC, serves as electron donor to Nap. Noteworthy, CymA is also the electron donor to the soluble and periplasmic cytochrome *c* nitrite reductase Nrf catalyzing the subsequent step in the ammonification process.[261] The Nap from the δ-proteobacterium *D. desulfuricans* differs from other such enzymes in two regards: the absence of the NapB subunit and the additional presence of NapM, a soluble and periplasmic tetrahemic *c*-type cytochrome.[134] How electrons enter NapA thus remains unclear and may be either *via* quinol-dependent routes involving some of the NapC,G,H proteins or by a quinol-independent route involving NapM (Figure 5.6).

Nitrate reduction by Nap enzymes can be used for alternative purposes to respiration, *e.g.* oxidation–reduction homeostasis under oxic conditions. This is typically the case during chemoheterotrophic or photoheterotrophic growth in the presence of highly reduced carbon substrates.[262–269] The excess of reducing equivalents is dissipated by Nap enzymes which are uncoupled from energy conservation. Indeed, quinols that serve as electron donors for Nap are oxidized on the periplasmic side of the cytoplasmic membrane by two distinct membrane-bound quinol-oxidizing systems (NapC, NapGH or even CymA) depending on the organism and the electrons flow towards the periplasm where they ultimately reduce nitrate to nitrite.

5.4.1.4 Nxr and Nitrification or Anammox

Nitrification consists of the sequential oxidation of ammonia to nitrate *via* nitrite as reaction intermediate and constitutes a key process in the biogeochemical nitrogen cycle in both aquatic and terrestrial ecosystems. While the first step, *i.e.* ammonia oxidation, is performed by both aerobic ammonia-oxidizing bacteria and archaea, the second step of the nitrification process is exclusively sustained by chemolithoautotrophic nitrite-oxidizing bacteria, which gain energy from the oxidation of nitrite to nitrate.[270,271] Under oxic conditions, bacterial nitrite oxidation is the principal source of nitrate, which constitutes the major fraction of the fixed nitrogen in the oceans and also the predominant inorganic nitrogen form in aerated soils.[78,126,163,164,270,272] Nitrite oxidation is catalyzed by a Mo-*bis*PGD enzyme, nNxr, a membrane-bound enzyme found in, *e.g.*, *Nitrobacter* sp. that shows high sequence similarity to nNar (Figure 5.4) while having a structural organization more closely related to cNar (Figure 5.6) (see Section 5.3).[273] *In vivo*, nNxr can catalyze the reaction in both directions, nitrate reduction being part of the dissimilatory nitrate reduction pathway using organic compounds as electron donors.[274–276] In this latter case, nNxr is supposed to have quinol oxidizing activity through the membrane anchor subunit nNxrI. Biochemical and genomic studies suggested that the reducing equivalents from nitrite oxidation by nNxr are transferred to a terminal cytochrome *c* oxidase *via* a cytochrome c_{550}.[126,277] Interestingly enough, the nNxr of *Nitrobacter* and *Nitrococcus* is oriented towards the cytoplasm while Nxr, at least in *Nitrospira*, is phylogenetically distant and oriented on the periplasmic side of the membrane (*i.e.* pNxr) (Figure 5.4).[78,126]

An alternative pathway involving Nxr is the anaerobic ammonium oxidation (anammox) process that produces N_2 by reducing nitrite and oxidizing ammonium, a unique bacterial trait performed by autotrophic anoxic ammonia-oxidizing bacteria.[278] During this anoxic process, nitrite is oxidized to nitrate *via* a pNxr with the released electrons being used for carbon dioxide reduction by the acetyl-coenzyme A pathway.[279,280] More interestingly, anammox bacteria such as *Kuenenia stuttgartiensis* and *Scalindua* sp. have been shown to perform dissimilatory nitrate reduction using organic electron donors such as formate, propionate or acetate.[281,282] As such, the reduction of nitrate to nitrite and to ammonia serves as a source of substrates for N_2 formation by the anammox process. Finally, the anammox pathway is spatially confined to an intracellular compartment termed the anammoxosome, which is bound by a single bilayer membrane.[278] It is thus considered that Nxr in *K. stuttgartiensis* is oriented towards the anammoxosome, thus participating in either nitrite oxidation or nitrate reduction.[280] The question of the exact location of the electron donor enzymes in the dissimilatory nitrate reduction pathway is opened.

5.4.2 Enzymes Involved in the Sulfur Cycle

Molybdoenzymes of the Mo/W-*bis*PGD family actively participate in the sulfur cycle in prokaryotes using a range of sulfur compounds (elemental sulfur, sulfite, thiosulfate, tetrathionate and polysulfide) as substrates. In fact, the ability to reduce sulfur compounds using H_2 or organic substrates as electron donors is widespread among bacteria and archaea, the latter being mostly hyperthermophilic.[283–287] Such anaerobic chemolithoautotrophic H_2 oxidation with sulfur reduction can be exemplified with the ε-proteobacterium *W. succinogenes* and involves alternative electron transfer chains composed of two periplasmically facing membrane-bound enzymes, a [NiFe] hydrogenase or a Fdh and a polysulfide reductase designated PsrABC.[283,288–292] Within this electron transfer chain, energy conservation is proposed to be due solely to uptake of cytoplasmic protons upon quinone reduction by the hydrogenase or Fdh complex. A similar situation has been reported in the deep-sea ε-proteobacterium *Sulforovum* sp. NBC37-1 and in the acidophilic crenarchaeon *Acidianus ambivalens* with chemolithoautrophic H_2 oxidation with S^0 reduction catalyzed by a periplasmically oriented enzymatic complex denoted PsrABC.[293,294]

Interestingly, a Mo/W-*bis*PGD enzyme has recently been shown to be involved in sulfite oxidation in the purple sulfur bacterium *Allochromatium vinosum*.[112] This cytoplasmically oriented complex (SoeABC) is found in a number of sulfite oxidizers among α,β,γ-proteobacteria where it couples sulfite oxidation to quinone reduction. Together with a periplasmic SoxYZ protein, SoeABC allows effective sulfite oxidation in the cytoplasm of these bacteria. Recently, it has been proposed that sulfite oxidation (*via* a Soe complex) could be coupled to O_2 reduction in the hyperthermophilic and microaerophilic bacterium *A. aeolicus* (Guiral M., personal communication). However, how energy arising from sulfite oxidation is conserved remains enigmatic in both organisms.

The facultative anaerobe γ-proteobacterium *Shewanella oneidensis* MR-1 is able to respire anaerobically with inorganic sulfur compounds such as S^0, SO_3^{2-}, $S_2O_3^{2-}$ as terminal electron acceptors, using lactate or formate as electron donor, thanks to the use of thiosulfate reductase encoded by a *psrABC* operon.[295,296] In this context, it is worth recalling the close phylogenetic proximity of Phs and Psr sequences (Figure 5.4). However, while the molecular basis of thiosulfate respiration has long been reported and described in *Salmonella enterica*,[113] the exact functioning of the Phs/Psr complexes remains enigmatic given the very low reduction potential of the acceptor couple [$S_2O_3^{2-}$/HS− + SO_3^{2-}] of −400 mV (Table 5.1).[146] As expected, only low-potential quinones such as MK are utilized in thiosulfate reduction in *S. typhimurium*.[297] Finally, in *S. enterica*[298] growth on S^0 or $S_2O_3^{2-}$ is only possible with fermentable carbon sources, with minor contribution to the overall yield of ATP production.[113] This suggests a role for sulfur reduction in redox-balancing of the quinone pool as in the case of Nap.[146] While evidence for chemolithoautotrophic H_2 oxidation in conjunction with sulfur reduction by hyperthermophilic archaea has long been reported,[299] evidence for the existence of a polysulfide reductase in the hyperthermophilic crenarchaeon *Pyrobaculum islandicum* supporting anaerobic respiration of S^0 or $S_2O_3^{2-}$ has only recently been provided.[300] In this context, the availability of complete genome sequences of several hyperthermophilic crenarchaea should allow a better understanding of their respiratory capabilities and identification of the involved enzymes.[301,302]

Interestingly, formate, NADH, H_2 and non-fermentable carbon substrates are used as electron donors in *S. typhimurium* for tetrathionate reduction to thiosulfate, in accordance with the more positive reduction potential of the $S_4O_6^{2-}/S_2O_3^{2-}$ couple between +24 and +170 mV (Table 5.1).[111,119,297,303] This reaction is catalyzed by tetrathionate reductase Ttr, another Mo-*bis*PGD enzyme, allowing the organism to take advantage of this abundant substrate in marine and soil sediments.[297] Importantly, anaerobic tetrathionate respiration of *S. typhimurium* has been shown to provide a significant growth advantage over strictly anaerobic bacteria in the gut microbiota.[215] Tetrathionate as well as other *S*- or *N*-oxides has been shown to be produced in the lumen during gut inflammation and used as terminal electron acceptor for anaerobic respiration.[215,304,305]

In addition to the inorganic sulfur compounds described above, several organosulfur compounds are metabolized by members of the Mo/W-*bis*PGD family. The first, dimethylsulfide (Me_2S; DMS), contributes about 50% of the global biogenic sulfur input to the atmosphere from marine environments and it has long been recognized that biological activities have a major influence on sulfur levels in marine systems. DMS is indeed the substrate for the unique Ddh enzyme in *Rhodovulum sulfidophilum* that allows the organism to grow photoautotrophically with DMS as sole electron donor.[306] In this electron transfer chain, Ddh oxidizes DMS in the periplasm, subsequently reducing a soluble cytochrome c_2, electron donor to the photosynthetic reaction center.[120] For more details, the reader is referred to the excellent review.[307]

The reaction catalyzed by the Ddh is formally the reverse of the reaction catalyzed by the more widespread dimethylsulfoxide (Me_2SO; DMSO) reductases. These enzymes are anaerobic terminal reductases frequently encountered in bacteria. In the case of heterotrimeric DmsABC, typified by the membrane-bound DMSO reductase from *E. coli*,[308] the reduction of DMSO is achieved using electrons delivered by DmsC from the MK and DMK pool.[309] In the case of the monomeric soluble DorA DMSO reductase encountered in *R. capsulatus* and *R. sphaeroides*, the electrons reducing DMSO are delivered by the membrane-bound quinol-oxidizing pentahemic cytochrome DorC in the absence of light or during photosynthetic growth.[158,310] In the case of the *S. oneidensis* Dor system, the electrons reducing DMSO are delivered *via* the multifunctional tetraheme CymA from the MK pool.[311] DMSO reductases have also been reported in *Haloarchaea* such as *Halobacterium salinarum* (formerly *Halobacterium* sp. NRC-1 [312]) where a heterotrimeric DmsABC complex is responsible for both anaerobic DMSO and TMAO respiration using organic substrates as electron donors.[313,314] Finally, DMSO reductases can also be found associated with the outer membrane of *Shewanella* sp. facing the exterior of the cell and being responsible for extracellular respiration of DMSO.[315] Herein, the multiheme CymA mediates electron transfer between the quinone pool and the DMSO reductase.[316,317] None of these DMSO reduction systems contribute directly to the proton gradient formation by themselves. The *pmf* generated by these chains comes from the involved NADH dehydrogenase.[318]

5.4.3 Enzymes Involved in the Carbon Cycle

Mo/W-*bis*PGD enzymes are known to participate in the carbon cycle through oxidoreduction of C_1 compounds. Formate dehydrogenase Fdh catalyzes the reversible conversion of formate to CO_2 while the much less studied formylmethanofuran dehydrogenase Fmd catalyzes the covalent attachment of CO_2 to methanofuran, both being either Mo or W enzymes.

Formate can be either the end product of much bacterial fermentation such as the mixed-acid fermentation of enterobacteria or an intermediate of energy metabolism in acetogenic bacteria during the reduction of CO_2 to acetic acid.[25] Thanks to a very low reduction potential, formate oxidation is readily coupled to the reduction of a number of terminal electron acceptors ranging from sulfur compounds, to nitrate, fumarate, As(V), Fe(III), Mn(IV) or oxygen. In *E. coli*, three formate dehydrogenases are present, two being involved in respiration (FdnGHI and FdoGHI), the third one (FdhF) being a component of the FHL complex involved in anaerobic fermentation. Basically, the periplasmically oriented FdnGHI (or its isoform FdoGHI) transfers electrons to the quinone pool, which ultimately reduces a number of alternative terminal electron acceptors, including nitrate, thereby forming the paradigmatic full proton motive redox loop (Figure 5.7(B)).[319] These Fdh systems may also be involved in the periplasmic nitrate ammonification pathway of enteric bacteria or *W. succinogenes* that couple formate oxidation to nitrate reduction by Nap producing nitrite further reduced to ammonium.[259] On the other hand, H_2 formed from formate oxidation by the FHL complex in the cytoplasm is recycled by the periplasmically oriented hydrogenases 1 and 2, with reducing equivalents reintroduced in the quinone pool.[320]

A slightly different situation is encountered in respiratory nitrite ammonification,[147,259] in which formate oxidation by a membrane-bound formate dehydrogenase FdhABC is coupled to nitrite reduction by a cytochrome *c* nitrite reductase (NrfAH) in the ε-proteobacterium *W. succinogenes* in a strictly periplasmic process.[259] In contrast, in the δ-proteobacteria *Desulfovibrio* sp., formate oxidation is coupled to dissimilatory sulfate reduction or nitrate reduction to ammonium with the participation of a periplasmic and soluble formate dehydrogenase (FdhAB exemplified in *D. gigas*)[109] to which a third subunit, a cytochrome c_3, can in some instances be added (exemplified in *D. desulfuricans*).[140,141,321–324] The electrons derived from formate oxidation are then transferred to a network of soluble and periplasmic *c*-type cytochromes feeding alternative respiratory pathways.[325] How energy is conserved in these systems remains an open question.

Dissimilatory reduction of Fe(III) or Mn(IV) using formate as electron donor appear to be major respiratory processes in aquatic environments and are found in both bacteria and archaea.[326,327] Herein, Fe(III) or Mn(IV) reduction involves the multiheme *c*-type cytochrome CymA connecting the quinone pool to the final reductases present in the outer membrane.[328,329] Dissimilatory arsenate reduction coupled with formate oxidation, an energy-conserving process, has been reported in the firmicutes *Desulfitobacterium* sp. where formate is the sole source of carbon as well as energy.[330]

Formate can also serve as one-carbon source for autotrophic growth of *C. necator*, a strictly respiratory facultative lithoautotrophic β-proteobacterium. Formate oxidation is directly coupled to NAD^+ reduction *via* a cytoplasmic and soluble formate dehydrogenase (S-Fdh).[331,332] This enzyme is composed of three subunits and harbors a Mo-*bis*PGD, several Fe/S clusters and a FMN as cofactors. Interestingly, the Moco-containing α subunit exhibits high similarity with the NuoG of the *E. coli* complex I and likely coordinates multiple [4Fe–4S] and [2Fe–2S] clusters.[101] A similar system exists in *R. capsulatus* with the isolation of a soluble heterotrimeric Fdh (FdsABG) coupling formate oxidation to NAD^+ reduction.[100]

Formate oxidation can also be associated with the reduction of NAD^+ and ferredoxin in the anaerobic bacterium *Clostridium acidurici* using purines as sole carbon, nitrogen and energy source.[333,334] In particular, an electron-bifurcating Fdh has been identified in this organism that shares functional properties with other flavin-based electron-bifurcating enzymes. The cytoplasmic enzyme has been purified and is composed of four subunits encoded by a single transcription unit, namely HydCBA forming a [FeFe]-hydrogenase-like complex with a FdhF predicted to be a Mo- and selenocysteine-containing Fdh.[334]

Interestingly, flavin-based electron-bifurcation, a two-electron mechanism first described in the cytoplasm of anaerobic bacteria and archaea, can involve Fdhs to perform carbon dioxide reduction.[335,336] Schematically, Fdhs are found associated with a variety of electron-bifurcating/confurcating modules, with some examples illustrated below. As such, Fdhs working in the reverse direction have recently been shown to be involved in energy conservation during methanogenesis, a pathway allowing reduction of CO_2 to methane using H_2 as reductant. Specifically, a multiprotein complex in the methanogenic euryarchaeon *Methanococcus maripaludis* composed of heterodisulfide reductase, F420-nonreducing hydrogenase, Fdh and Fmd has been isolated and characterized.[337] Importantly, when H_2 is limiting, formate can now be used as electron donor for methanogenesis, electrons being transferred to the heterodisulfide reductase. Two molybdoenzymes, formate dehydrogenase and formylmethanofuran dehydrogenase, are thus involved in this metabolism. In contrast to Fdh, much less information is available concerning the formylmethanofuran dehydrogenase. The enzyme has been purified from a number of methanogens, including *Methanothermobacter marburgensis* and *Methanosarcina barkeri*, yielding a multimeric complex containing Mo or W and having a varying number of subunits.[31,338–342] Alternatively, another type of electron-confurcating enzyme involving a Fdh has been reported in the acetogenic bacterium *Clostridium autoethanogenum* when grown autotrophically on CO.[343] The enzymatic complex couples CO_2 reduction to formate with reduced ferredoxin and NADPH as electron donors. This complex is made up of seven subunits and contains a W- and SeCys-containing Fdh (FdhA), a NADP- and ferredoxin-dependent [FeFe] hydrogenase (HytABC) together with iron–sulfur proteins ($HytDE_1E_2$). Noteworthy, the $HytE_1E_2$ subunits display strong sequence similarities with the archetypal β-subunit harboring four Fe/S clusters of Mo/W-*bis*PGD enzymes (see Section 5.4.2). It is proposed that the physiological role of this enzymatic complex is to protect cells from over-reduction of ferredoxin and NADPH during growth on CO ($E_0' = -520$ mV).[343]

Recently, a closely related hydrogen-dependent CO_2 reductase HDCR complex has been identified in the acetogenic bacterium *A. woodii* using H_2 and CO_2 as energy source.[89] The HDCR complex is involved in the first step of the Wood–Ljungdahl pathway to build new blocks for biosynthesis. The complex is composed of four subunits including a Mo-dependent Fdh (FdhF2), a [FeFe] hydrogenase (HydA2) and two iron–sulfur proteins ($HycB_2B_3$). Interestingly, CO can serve as an alternative electron donor for CO_2 reduction using ferredoxin as an electron entry path to the complex.[89] The HDCR complex may be more widespread than initially thought, as similar gene clusters of as yet unknown function have been identified in sulfur-reducing *Desulfovibrio* sp.

5.5 Concluding Remarks

The wealth of new information concerning members of the Mo/W-*bis*PGD enzyme superfamily has been tremendous during the last decades with the availability of a number of high-resolution X-ray crystal structures and advances in the biochemistry and phylogenetic relationships between these enzymes. Most spectacular is elucidation of the exquisite structural and functional diversity of the superfamily, establishing the Mo/W-*bis*PGD structural motif as the workhorse of prokaryotic catalysis. With a very limited number of basic building blocks, Nature modulates the assembly of diverse Mo/W-*bis*PGD enzymes involved in bioenergetics in prokaryotes. What is fascinating is that with regards to a very limited combination of chemical reactivity at the active site, an uncovered diversity of reactivities has arisen from subtle modifications that have not yet been fully disclosed. The associated catalytic mechanisms remain to be fully elucidated, but it is clear that the diversity in reactivity exhibited by the Mo/W-*bis*PGD enzymes is not likely to be due to any single factor. Examination of the wealth of structural, spectroscopic, kinetic and computational data emphasizes the extraordinarily high catalytic plasticity of these enzymes that is responsible for their evolutionary success in prokaryotes. Remarkably, the presently available information allows us to now develop new approaches encompassing the diversity of the superfamily and establish common principles. Nature has apparently selected alternative solutions to solve enzymatic reactivity in the Mo/W-*bis*PGD enzymes superfamily, as reflected, for instance, in the phylogenetically distant Nap and Nar enzymes, which exhibit identical overall reaction despite a very different chemistry at their active site (Figures 5.2 and 5.4). In this chapter, we have proposed a new approach exploiting information coming from evolutionary diversity analysis to fuel research aimed at solving this structural conundrum.

Acknowledgements

We acknowledge all colleagues who have contributed to this work directly or through stimulating discussions. We thank the French National Research Agency (ANR, project MC2, grant number 11-BSV5-005-01), the A*Midex foundation of Aix-Marseille University (project MicrobioE, grant number ANR-11-IDEX-0001-02), The Région PACA (project MocoCAT APOG2013) and the CNRS for the project "Émergence CO_2" from the "Mission interdisciplinarité" for their financial support. We would also like to dedicate this chapter in memory to our colleague Prof. Dr Simon De Vries.

References

1. M. K. Johnson, D. C. Rees and M. W. Adams, *Chem. Rev.*, 1996, **96**, 2817–2840.
2. A. Kletzin and M. W. Adams, *FEMS Microbiol. Rev.*, 1996, **18**, 5–63.
3. Y. Zhang and V. N. Gladyshev, *J. Mol. Biol.*, 2008, **379**, 881–899.
4. A. Cvetkovic, A. L. Menon, M. P. Thorgersen, J. W. Scott, F. L. Poole 2nd, F. E. Jenney Jr., W. A. Lancaster, J. L. Praissman, S. Shanmukh, B. J. Vaccaro, S. A. Trauger, E. Kalisiak, J. V. Apon, G. Siuzdak, S. M. Yannone, J. A. Tainer and M. W. Adams, *Nature*, 2010, **466**, 779–782.
5. O. Einsle, F. A. Tezcan, S. L. Andrade, B. Schmid, M. Yoshida, J. B. Howard and D. C. Rees, *Science*, 2002, **297**, 1696–1700.

6. J. Wilcoxen, B. Zhang and R. Hille, *Biochemistry*, 2011, **50**, 1910–1916.
7. R. Hille, *Chem. Rev.*, 1996, **96**, 2757–2816.
8. G. Schwarz, R. R. Mendel and M. W. Ribbe, *Nature*, 2009, **460**, 839–847.
9. F. Schneider, J. Lowe, R. Huber, H. Schindelin, C. Kisker and J. Knablein, *J. Mol. Biol.*, 1996, **263**, 53–69.
10. S. Grimaldi, B. Schoepp-Cothenet, P. Ceccaldi, B. Guigliarelli and A. Magalon, *Biochim. Biophys. Acta*, 2013, **1827**, 1048–1085.
11. R. H. Holm, E. I. Solomon, A. Majumdar and A. Tenderhold, *Coord. Chem. Rev.*, 2011, **255**, 22.
12. S. Metz and W. Thiel, *Coord. Chem. Rev.*, 2011, **255**, 1085–1103.
13. M. J. Pushie and G. N. George, *Coord. Chem. Rev.*, 2011, **255**, 1055–1084.
14. Y. Zhang, S. Rump and V. N. Gladyshev, *Coord. Chem. Rev.*, 2011, **255**, 1206–1217.
15. H. Dobbek, *Coord. Chem. Rev.*, 2011, **255**, 1104–1116.
16. W. Hagen, *Coord. Chem. Rev.*, 2011, **255**, 1117–1128.
17. S. Leimkuhler, M. Wuebbens and K. Rajagopalan, *Coord. Chem. Rev.*, 2011, **255**, 1129–1144.
18. A. Magalon, J. G. Fedor, A. Walburger and J. H. Weiner, *Coord. Chem. Rev.*, 2011, **255**, 1159–1178.
19. A. Majumdar and S. Sarkar, *Coord. Chem. Rev.*, 2011, **255**, 1039–1054.
20. T. Nilsson, M. Rova and A. Smedja Backlund, *Biochim. Biophys. Acta*, 2013, **1827**, 189–197.
21. R. Hille, J. Hall and P. Basu, *Chem. Rev.*, 2014, **114**, 3963–4038.
22. R. A. Rothery, G. J. Workun and J. H. Weiner, *Biochim. Biophys. Acta*, 2008, **1778**, 1897–1929.
23. R. A. Rothery and J. H. Weiner, *JBIC, J. Biol. Inorg. Chem.*, 2015, **20**, 349–372.
24. M. J. Pushie, J. J. Cotelesage and G. N. George, *Metallomics*, 2014, **6**, 15–24.
25. L. B. Maia, J. J. Moura and I. Moura, *JBIC, J. Biol. Inorg. Chem.*, 2015, **20**, 287–309.
26. F. Baymann, E. Lebrun, M. Brugna, B. Schoepp-Cothenet, M. T. Giudici-Orticoni and W. Nitschke, *Philos. Trans. R. Soc., B*, 2003, **358**, 267–274.
27. B. Schoepp-Cothenet, R. van Lis, A. Atteia, F. Baymann, L. Capowiez, A. L. Ducluzeau, S. Duval, F. ten Brink, M. J. Russell and W. Nitschke, *Biochim. Biophys. Acta*, 2013, **1827**, 79–93.
28. B. Schoepp-Cothenet, R. van Lis, P. Philippot, A. Magalon, M. J. Russell and W. Nitschke, *Sci. Rep.*, 2012, **2**, 263.
29. M. J. Romao, *Dalton Trans.*, 2009, 4053–4068.
30. F. Ten Brink, *Met. Ions Life Sci.*, 2014, **14**, 15–35.
31. A. Hochheimer, D. Linder, R. K. Thauer and R. Hedderich, *Eur. J. Biochem.*, 1996, **242**, 156–162.
32. R. K. Thauer, K. Jungermann and K. Decker, *Bacteriol. Rev.*, 1977, **41**, 100–180.
33. A. Brune and B. Schink, *J. Bacteriol.*, 1990, **172**, 1070–1076.
34. B. Philipp and B. Schink, *J. Bacteriol.*, 1998, **180**, 3644–3649.
35. F. Kapralek, *J. Gen. Microbiol.*, 1972, **71**, 133–139.

36. O. Kniemeyer and J. Heider, *J. Biol. Chem.*, 2001, **276**, 21381–21386.
37. J. Dermer and G. Fuchs, *J. Biol. Chem.*, 2012, **287**, 36905–36916.
38. B. W. Vink, *Chem. Geol.*, 1996, **130**, 9.
39. J. Buc, C. L. Santini, R. Giordani, M. Czjzek, L. F. Wu and G. Giordano, *Mol. Microbiol.*, 1999, **32**, 159–168.
40. H. R. Beller, *Appl. Environ. Microbiol.*, 2005, **71**, 2170–2174.
41. F. Maixner, M. Wagner, S. Lucker, E. Pelletier, S. Schmitz-Esser, K. Hace, E. Spieck, R. Konrat, D. Le Paslier and H. Daims, *Environ. Microbiol.*, 2008, **10**, 3043–3056.
42. L. J. Bird, V. Bonnefoy and D. K. Newman, *Trends Microbiol.*, 2011, **19**, 330–340.
43. C. Coelho, P. J. Gonzalez, J. G. Moura, I. Moura, J. Trincao and M. Joao Romao, *J. Mol. Biol.*, 2011, **408**, 932–948.
44. R. Thome, A. Gust, R. Toci, R. Mendel, F. Bittner, A. Magalon and A. Walburger, *J. Biol. Chem.*, 2012, **287**, 4671–4678.
45. N. M. Cerqueira, P. J. Gonzalez, P. A. Fernandes, J. J. Moura and M. J. o. Ramos, *Acc. Chem. Res.*, 2015, **48**, 2875–2884.
46. T. Prisner, S. Lyubenova, Y. Atabay, F. MacMillan, A. Kroger and O. Klimmek, *JBIC, J. Biol. Inorg. Chem.*, 2003, **8**, 419–426.
47. F. Biaso, B. Burlat and B. Guigliarelli, *Inorg. Chem.*, 2012, **51**, 3409–3419.
48. M. Jormakka, S. Tornroth, B. Byrne and S. Iwata, *Science*, 2002, **295**, 1863–1868.
49. M. G. Bertero, R. A. Rothery, M. Palak, C. Hou, D. Lim, F. Blasco, J. H. Weiner and N. C. Strynadka, *Nat. Struct. Biol.*, 2003, **10**, 681–687.
50. M. Jormakka, D. Richardson, B. Byrne and S. Iwata, *Structure*, 2004, **12**, 95–104.
51. D. P. Kloer, C. Hagel, J. Heider and G. E. Schulz, *Structure*, 2006, **14**, 1377–1388.
52. M. J. Axley, A. Bock and T. C. Stadtman, *Proc. Natl. Acad. Sci. U. S. A.*, 1991, **88**, 8450–8454.
53. H. C. Raaijmakers and M. J. Romao, *JBIC, J. Biol. Inorg. Chem.*, 2006, **11**, 849–854.
54. M. Leopoldini, S. G. Chiodo, M. Toscano and N. Russo, *Chemistry*, 2008, **14**, 8674–8681.
55. C. S. Mota, M. G. Rivas, C. D. Brondino, I. Moura, J. J. Moura, P. J. Gonzalez and N. M. Cerqueira, *JBIC, J. Biol. Inorg. Chem.*, 2011, **16**, 1255–1268.
56. C. A. Trieber, R. A. Rothery and J. H. Weiner, *J. Biol. Chem.*, 1996, **271**, 27339–27345.
57. J. C. Hilton, C. A. Temple and K. V. Rajagopalan, *J. Biol. Chem.*, 1999, **274**, 8428–8436.
58. V. V. Pollock and M. J. Barber, *J. Biol. Chem.*, 2000, **275**, 35086–35090.
59. K. E. Johnson and K. V. Rajagopalan, *J. Biol. Chem.*, 2001, **276**, 13178–13185.
60. T. Satoh and F. N. Kurihara, *J. Biochem.*, 1987, **102**, 191–197.
61. P. Schrapers, T. Hartmann, R. Kositzki, H. Dau, S. Reschke, C. Schulzke, S. Leimkuhler and M. Haumann, *Inorg. Chem.*, 2015, **54**, 3260–3271.

62. H.-K. Li, C. Temple, K. V. Rajagopalan and H. Schindelin, *J. Am. Chem. Soc.*, 2000, **122**, 7673–7680.
63. R. C. Bray, B. Adams, A. T. Smith, B. Bennett and S. Bailey, *Biochemistry*, 2000, **39**, 11258–11269.
64. S. Dementin, P. Arnoux, B. Frangioni, S. Grosse, C. Leger, B. Burlat, B. Guigliarelli, M. Sabaty and D. Pignol, *Biochemistry*, 2007, **46**, 9713–9721.
65. V. Fourmond, B. Burlat, S. Dementin, M. Sabaty, P. Arnoux, E. Etienne, B. Guigliarelli, P. Bertrand, D. Pignol and C. Leger, *Biochemistry*, 2010, **49**, 2424–2432.
66. A. P. Srivastava, J. P. Allen, B. J. Vaccaro, M. Hirasawa, S. Alkul, M. K. Johnson and D. B. Knaff, *Biochemistry*, 2015, **54**, 5557–5568.
67. J. P. Ridge, K. F. Aguey-Zinsou, P. V. Bernhardt, G. R. Hanson and A. G. McEwan, *FEBS Lett.*, 2004, **563**, 197–202.
68. N. Cobb, C. Hemann, G. A. Polsinelli, J. P. Ridge, A. G. McEwan and R. Hille, *J. Biol. Chem.*, 2007, **282**, 35519–35529.
69. V. V. Pollock, R. C. Conover, M. K. Johnson and M. J. Barber, *Arch. Biochem. Biophys.*, 2003, **409**, 315–326.
70. N. Cobb, T. Conrads and R. Hille, *J. Biol. Chem.*, 2005, **280**, 11007–11017.
71. J. G. Jacques, V. Fourmond, P. Arnoux, M. Sabaty, E. Etienne, S. Grosse, F. Biaso, P. Bertrand, D. Pignol, C. Leger, B. Guigliarelli and B. Burlat, *Biochim. Biophys. Acta*, 2014, **1837**, 277–286.
72. P. Ceccaldi, J. Rendon, C. Leger, R. Toci, B. Guigliarelli, A. Magalon, S. Grimaldi and V. Fourmond, *Biochim. Biophys. Acta*, 2015, **1847**, 1055–1063.
73. R. A. Rothery, B. Stein, M. Solomonson, M. L. Kirk and J. H. Weiner, *Proc. Natl. Acad. Sci. U. S. A.*, 2012, **109**, 14773–14778.
74. S. Y. Wu, R. A. Rothery and J. H. Weiner, *J. Biol. Chem.*, 2015, **290**, 25164–25173.
75. B. R. Williams, Y. Fu, G. P. Yap and S. J. Burgmayer, *J. Am. Chem. Soc.*, 2012, **134**, 19584–19587.
76. M. J. Russell and A. J. Hall, *J. Geol. Soc. (London, U. K.)*, 1997, **154**, 377–402.
77. B. E. Erickson and G. R. Helz, *Geochim. Cosmochim. Acta*, 2000, **64**, 1149–1158.
78. S. Lucker, M. Wagner, F. Maixner, E. Pelletier, H. Koch, B. Vacherie, T. Rattei, J. S. Damste, E. Spieck, D. Le Paslier and H. Daims, *Proc. Natl. Acad. Sci. U. S. A.*, 2010, **107**, 13479–13484.
79. J. F. Stolz and P. Basu, *ChemBioChem*, 2002, **3**, 198–206.
80. S. Duval, A. L. Ducluzeau, W. Nitschke and B. Schoepp-Cothenet, *BMC Evol. Biol.*, 2008, **8**, 206.
81. T. Hettmann, R. A. Siddiqui, C. Frey, T. Santos-Silva, M. J. Romao and S. Diekmann, *Biochem. Biophys. Res. Commun.*, 2004, **320**, 1211–1219.
82. R. A. Rothery, M. G. Bertero, T. Spreter, N. Bouromand, N. C. Strynadka and J. H. Weiner, *J. Biol. Chem.*, 2010, **285**, 8801–8807.
83. A. E. Cozen, M. T. Weirauch, K. S. Pollard, D. L. Bernick, J. M. Stuart and T. M. Lowe, *J. Bacteriol.*, 2009, **191**, 782–794.

84. M. Guiral, P. Tron, C. Aubert, A. Gloter, C. Iobbi-Nivol and M. T. Giudici-Orticoni, *J. Biol. Chem.*, 2005, **280**, 42004–42015.
85. F. F. Morpeth and D. H. Boxer, *Biochemistry*, 1985, **24**, 40–46.
86. C. Avazeri, R. J. Turner, J. Pommier, J. H. Weiner, G. Giordano and A. Vermeglio, *Microbiology*, 1997, **143**(Pt 4), 1181–1189.
87. A. Nilavongse, T. H. Brondijk, T. W. Overton, D. J. Richardson, E. R. Leach and J. A. Cole, *Microbiology*, 2006, **152**, 3227–3237.
88. J. S. McDowall, B. J. Murphy, M. Haumann, T. Palmer, F. A. Armstrong and F. Sargent, *Proc. Natl. Acad. Sci. U. S. A.*, 2014, **111**, E3948–E3956.
89. K. Schuchmann and V. Muller, *Science*, 2013, **342**, 1382–1385.
90. V. Bonnefoy, J. Ratouchniak, F. Blasco and M. Chippaux, *FEMS Microbiol. Lett.*, 1997, **147**, 147–149.
91. G. B. Seiffert, G. M. Ullmann, A. Messerschmidt, B. Schink, P. M. Kroneck and O. Einsle, *Proc. Natl. Acad. Sci. U. S. A.*, 2007, **104**, 3073–3077.
92. M. Sabaty, C. Avazeri, D. Pignol and A. Vermeglio, *Appl. Environ. Microbiol.*, 2001, **67**, 5122–5126.
93. H. Schindelin, C. Kisker, J. Hilton, K. V. Rajagopalan and D. C. Rees, *Science*, 1996, **272**, 1615–1621.
94. M. Czjzek, J. P. Dos Santos, J. Pommier, G. Giordano, V. Mejean and R. Haser, *J. Mol. Biol.*, 1998, **284**, 435–447.
95. S. Gon, J. C. Patte, V. Mejean and C. Iobbi-Nivol, *J. Bacteriol.*, 2000, **182**, 5779–5786.
96. V. V. Pollock and M. J. Barber, *J. Biol. Chem.*, 1997, **272**, 3355–3362.
97. A. Messerschmidt, H. Niessen, D. Abt, O. Einsle, B. Schink and P. M. Kroneck, *Proc. Natl. Acad. Sci. U. S. A.*, 2004, **101**, 11571–11576.
98. J. T. Lin, B. S. Goldman and V. Stewart, *J. Bacteriol.*, 1993, **175**, 2370–2378.
99. A. J. Gates, V. M. Luque-Almagro, A. D. Goddard, S. J. Ferguson, M. D. Roldan and D. J. Richardson, *Biochem. J.*, 2011, **435**, 743–753.
100. T. Hartmann and S. Leimkuhler, *FEBS J.*, 2013, **280**, 6083–6096.
101. J. I. Oh and B. Bowien, *J. Biol. Chem.*, 1998, **273**, 26349–26360.
102. D. R. Jollie and J. D. Lipscomb, *J. Biol. Chem.*, 1991, **266**, 21853–21863.
103. M. Laukel, L. Chistoserdova, M. E. Lidstrom and J. A. Vorholt, *Eur. J. Biochem.*, 2003, **270**, 325–333.
104. A. P. Shuber, E. C. Orr, M. A. Recny, P. F. Schendel, H. D. May, N. L. Schauer and J. G. Ferry, *J. Biol. Chem.*, 1986, **261**, 12942–12947.
105. P. Arnoux, M. Sabaty, J. Alric, B. Frangioni, B. Guigliarelli, J. M. Adriano and D. Pignol, *Nat. Struct. Biol.*, 2003, **10**, 928–934.
106. J. M. Dias, M. E. Than, A. Humm, R. Huber, G. P. Bourenkov, H. D. Bartunik, S. Bursakov, J. Calvete, J. Caldeira, C. Carneiro, J. J. Moura, I. Moura and M. J. Romao, *Structure*, 1999, **7**, 65–79.
107. B. J. Jepson, S. Mohan, T. A. Clarke, A. J. Gates, J. A. Cole, C. S. Butler, J. N. Butt, A. M. Hemmings and D. J. Richardson, *J. Biol. Chem.*, 2007, **282**, 6425–6437.
108. J. C. Boyington, V. N. Gladyshev, S. V. Khangulov, T. C. Stadtman and P. D. Sun, *Science*, 1997, **275**, 1305–1308.

109. H. Raaijmakers, S. Macieira, J. M. Dias, S. Teixeira, S. Bursakov, R. Huber, J. J. Moura, I. Moura and M. J. Romao, *Structure*, 2002, **10**, 1261–1272.
110. M. Jormakka, K. Yokoyama, T. Yano, M. Tamakoshi, S. Akimoto, T. Shimamura, P. Curmi and S. Iwata, *Nat. Struct. Mol. Biol.*, 2008, **15**, 730–737.
111. M. Hensel, A. P. Hinsley, T. Nikolaus, G. Sawers and B. C. Berks, *Mol. Microbiol.*, 1999, **32**, 275–287.
112. C. Dahl, B. Franz, D. Hensen, A. Kesselheim and R. Zigann, *Microbiology*, 2013, **159**, 2626–2638.
113. N. K. Heinzinger, S. Y. Fujimoto, M. A. Clark, M. S. Moreno and E. L. Barrett, *J. Bacteriol.*, 1995, **177**, 2813–2820.
114. B. J. Jepson, L. J. Anderson, L. M. Rubio, C. J. Taylor, C. S. Butler, E. Flores, A. Herrero, J. N. Butt and D. J. Richardson, *J. Biol. Chem.*, 2004, **279**, 32212–32218.
115. K. Ogawa, E. Akagawa, K. Yamane, Z. W. Sun, M. LaCelle, P. Zuber and M. M. Nakano, *J. Bacteriol.*, 1995, **177**, 1409–1413.
116. A. J. Gates, C. S. Butler, D. J. Richardson and J. N. Butt, *Biochem. Soc. Trans.*, 2011, **39**, 236–242.
117. P. Lanciano, A. Savoyant, S. Grimaldi, A. Magalon, B. Guigliarelli and P. Bertrand, *J. Phys. Chem. B*, 2007, **111**, 13632–13637.
118. R. A. Rothery, M. G. Bertero, R. Cammack, M. Palak, F. Blasco, N. C. Strynadka and J. H. Weiner, *Biochemistry*, 2004, **43**, 5324–5333.
119. T. Krafft, A. Bowen, F. Theis and J. M. Macy, *DNA Sequence*, 2000, **10**, 365–377.
120. C. A. McDevitt, P. Hugenholtz, G. R. Hanson and A. G. McEwan, *Mol. Microbiol.*, 2002, **44**, 1575–1587.
121. H. D. Thorell, K. Stenklo, J. Karlsson and T. Nilsson, *Appl. Environ. Microbiol.*, 2003, **69**, 5585–5592.
122. H. Tang, R. A. Rothery, J. E. Voss and J. H. Weiner, *J. Biol. Chem.*, 2011, **286**, 15147–15154.
123. P. J. Ellis, T. Conrads, R. Hille and P. Kuhn, *Structure*, 2001, **9**, 125–132.
124. T. P. Warelow, M. Oke, B. Schoepp-Cothenet, J. U. Dahl, N. Bruselat, G. N. Sivalingam, S. Leimkuhler, K. Thalassinos, U. Kappler, J. H. Naismith and J. M. Santini, *PLoS One*, 2013, **8**, e72535.
125. C. A. Trieber, R. A. Rothery and J. H. Weiner, *J. Biol. Chem.*, 1996, **271**, 4620–4626.
126. D. Y. Sorokin, S. Lucker, D. Vejmelkova, N. A. Kostrikina, R. Kleerebezem, W. I. Rijpstra, J. S. Damste, D. Le Paslier, G. Muyzer, M. Wagner, M. C. van Loosdrecht and H. Daims, *ISME J.*, 2012, **6**, 2245–2256.
127. S. J. Coulthurst, A. Dawson, W. N. Hunter and F. Sargent, *Biochemistry*, 2012, **51**, 1678–1686.
128. A. P. Hinsley, N. R. Stanley, T. Palmer and B. C. Berks, *FEBS Lett.*, 2001, **497**, 45–49.
129. A. Lieutaud, R. van Lis, S. Duval, L. Capowiez, D. Muller, R. Lebrun, S. Lignon, M. L. Fardeau, M. C. Lett, W. Nitschke and B. Schoepp-Cothenet, *J. Biol. Chem.*, 2010, **285**, 20433–20441.
130. J. M. Santini and R. N. vanden Hoven, *J. Bacteriol.*, 2004, **186**, 1614–1619.

131. R. J. Turner, A. L. Papish and F. Sargent, *Can. J. Microbiol.*, 2004, **50**, 225–238.
132. B. Ize, S. J. Coulthurst, K. Hatzixanthis, I. Caldelari, G. Buchanan, E. C. Barclay, D. J. Richardson, T. Palmer and F. Sargent, *Microbiology*, 2009, **155**, 3992–4004.
133. R. M. Martinez-Espinosa, E. J. Dridge, M. J. Bonete, J. N. Butt, C. S. Butler, F. Sargent and D. J. Richardson, *FEMS Microbiol. Lett.*, 2007, **276**, 129–139.
134. A. Marietou, D. Richardson, J. Cole and S. Mohan, *FEMS Microbiol. Lett.*, 2005, **248**, 217–225.
135. J. T. Lin, B. S. Goldman and V. Stewart, *J. Bacteriol.*, 1994, **176**, 2551–2559.
136. E. Lebrun, J. M. Santini, M. Brugna, A. L. Ducluzeau, S. Ouchane, B. Schoepp-Cothenet, F. Baymann and W. Nitschke, *Mol. Biol. Evol.*, 2006, **23**, 1180–1191.
137. R. van Lis, W. Nitschke, S. Duval and B. Schoepp-Cothenet, *Biochim. Biophys. Acta*, 2013, **1827**, 176–188.
138. R. Blasco, F. Castillo and M. Martinez-Luque, *FEBS Lett.*, 1997, **414**, 45–49.
139. K. Schuchmann and V. Muller, *J. Biol. Chem.*, 2012, **287**, 31165–31171.
140. C. Costa, M. Teixeira, J. LeGall, J. J. G. Moura and I. Moura, *JBIC, J. Biol. Inorg. Chem.*, 1997, **2**, 198–208.
141. C. Sebban, L. Blanchard, M. Bruschi and F. Guerlesquin, *FEMS Microbiol. Lett.*, 1995, **133**, 143–149.
142. K. S. Bender, C. Shang, R. Chakraborty, S. M. Belchik, J. D. Coates and L. A. Achenbach, *J. Bacteriol.*, 2005, **187**, 5090–5096.
143. F. Cava, O. Zafra and J. Berenguer, *Mol. Microbiol.*, 2008, **70**, 507–518.
144. O. Zafra, F. Cava, F. Blasco, A. Magalon and J. Berenguer, *J. Bacteriol.*, 2005, **187**, 3990–3996.
145. O. Zafra, S. Ramirez, P. Castan, R. Moreno, F. Cava, C. Valles, E. Caro and J. Berenguer, *FEBS Lett.*, 2002, **523**, 99–102.
146. L. Stoffels, M. Krehenbrink, B. C. Berks and G. Unden, *J. Bacteriol.*, 2012, **194**, 475–485.
147. J. Simon and M. G. Klotz, *Biochim. Biophys. Acta*, 2012, **1827**, 114–135.
148. S. Afshar, E. Johnson, S. de Vries and I. Schroder, *J. Bacteriol.*, 2001, **183**, 5491–5495.
149. S. de Vries, M. Momcilovic, M. J. Strampraad, J. P. Whitelegge, A. Baghai and I. Schroder, *Biochemistry*, 2010, **49**, 9911–9921.
150. E. J. Dridge, C. A. Watts, B. J. Jepson, K. Line, J. M. Santini, D. J. Richardson and C. S. Butler, *Biochem. J.*, 2007, **408**, 19–28.
151. I. Schroder, S. Rech, T. Krafft and J. M. Macy, *J. Biol. Chem.*, 1997, **272**, 23765–23768.
152. C. A. McDevitt, G. R. Hanson, C. J. Noble, M. R. Cheesman and A. G. McEwan, *Biochemistry*, 2002, **41**, 15234–15244.
153. H. Hussain, J. Grove, L. Griffiths, S. Busby and J. Cole, *Mol. Microbiol.*, 1994, **12**, 153–163.
154. W. Dietrich and O. Klimmek, *Eur. J. Biochem.*, 2002, **269**, 1086–1095.

155. J. H. Weiner, G. Shaw, R. J. Turner and C. A. Trieber, *J. Biol. Chem.*, 1993, **268**, 3238–3244.
156. M. L. Rodrigues, T. F. Oliveira, I. A. Pereira and M. Archer, *EMBO J.*, 2006, **25**, 5951–5960.
157. V. Mejean, C. Iobbi-Nivol, M. Lepelletier, G. Giordano, M. Chippaux and M. C. Pascal, *Mol. Microbiol.*, 1994, **11**, 1169–1179.
158. T. Ujiiye, I. Yamamoto, H. Nakama, A. Okubo, S. Yamazaki and T. Satoh, *Biochim. Biophys. Acta*, 1996, **1277**, 1–5.
159. M. D. Roldan, H. J. Sears, M. R. Cheesman, S. J. Ferguson, A. J. Thomson, B. C. Berks and D. J. Richardson, *J. Biol. Chem.*, 1998, **273**, 28785–28790.
160. L. Pieulle, X. Morelli, P. Gallice, E. Lojou, P. Barbier, M. Czjzek, P. Bianco, F. Guerlesquin and E. C. Hatchikian, *J. Mol. Biol.*, 2005, **354**, 73–90.
161. R. O. Louro, *JBIC, J. Biol. Inorg. Chem.*, 2007, **12**, 1–10.
162. T. M. Iverson, D. M. Arciero, A. B. Hooper and D. C. Rees, *JBIC, J. Biol. Inorg. Chem.*, 2001, **6**, 390–397.
163. S. R. Starkenburg, P. S. Chain, L. A. Sayavedra-Soto, L. Hauser, M. L. Land, F. W. Larimer, S. A. Malfatti, M. G. Klotz, P. J. Bottomley, D. J. Arp and W. J. Hickey, *Appl. Environ. Microbiol.*, 2006, **72**, 2050–2063.
164. S. R. Starkenburg, F. W. Larimer, L. Y. Stein, M. G. Klotz, P. S. Chain, L. A. Sayavedra-Soto, A. T. Poret-Peterson, M. E. Gentry, D. J. Arp, B. Ward and P. J. Bottomley, *Appl. Environ. Microbiol.*, 2008, **74**, 2852–2863.
165. M. Kern and J. Simon, *Biochim. Biophys. Acta*, 2009, **1787**, 646–656.
166. T. H. Brondijk, A. Nilavongse, N. Filenko, D. J. Richardson and J. A. Cole, *Biochem. J.*, 2004, **379**, 47–55.
167. Y. V. Nancharaiah and P. N. Lens, *Microbiol. Mol. Biol. Rev.*, 2015, **79**, 61–80.
168. M. Boll, B. Schink, A. Messerschmidt and P. M. Kroneck, *Biol. Chem.*, 2005, **386**, 999–1006.
169. S. Schnell, F. Bak and N. Pfennig, *Arch. Microbiol.*, 1989, **152**, 556–563.
170. A. Tschech and B. Schink, *Arch. Microbiol.*, 1985, **143**, 8.
171. C. Kluge, A. Tschech and G. Fuchs, *Arch. Microbiol.*, 1990, **155**, 7.
172. P. I. Darley, J. A. Hellstern, J. I. Medina-Bellver, S. Marques, B. Schink and B. Philipp, *J. Bacteriol.*, 2007, **189**, 3824–3833.
173. H. A. Ball, H. A. Johnson, M. Reinhard and A. M. Spormann, *J. Bacteriol.*, 1996, **178**, 5755–5761.
174. O. Kniemeyer and J. Heider, *Arch. Microbiol.*, 2001, **176**, 129–135.
175. I. Luque, A. Herrero, E. Flores and F. Madueno, *Mol. Gen. Genet.*, 1992, **232**, 7–11.
176. J. E. Frias, E. Flores and A. Herrero, *J. Bacteriol.*, 1997, **179**, 477–486.
177. Q. Wang, H. Li and A. F. Post, *J. Bacteriol.*, 2000, **182**, 1764–1767.
178. A. C. Martiny, S. Kathuria and P. M. Berube, *Proc. Natl. Acad. Sci. U. S. A.*, 2009, **106**, 10787–10792.
179. J. E. Frias and E. Flores, *J. Bacteriol.*, 2010, **192**, 2769–2778.
180. M. Wyman and C. Bird, *Appl. Environ. Microbiol.*, 2007, **73**, 3028–3033.
181. P. M. Berube, S. J. Biller, A. G. Kent, J. W. Berta-Thompson, S. E. Roggensack, K. H. Roache-Johnson, M. Ackerman, L. R. Moore, J. D. Meisel, D. Sher, L. R. Thompson, L. Campbell, A. C. Martiny and S. W. Chisholm, *ISME J.*, 2015, **9**, 1195–1207.

182. Q. Wu and V. Stewart, *J. Bacteriol.*, 1998, **180**, 1311–1322.
183. J. C. Gutierrez, F. Ramos, L. Ortner and M. Tortolero, *Mol. Microbiol.*, 1995, **18**, 579–591.
184. C. Pino, F. Olmo-Mira, P. Cabello, M. Martinez-Luque, F. Castillo, M. D. Roldan and C. Moreno-Vivian, *Biochem. Soc. Trans.*, 2006, **34**, 127–129.
185. J. Wang and G. P. Zhao, *Biochem. Biophys. Res. Commun.*, 2009, **386**, 77–81.
186. Z. Shao, J. Gao, X. Ding, J. Wang, J. Chiao and G. Zhao, *Arch. Microbiol.*, 2011, **193**, 463–477.
187. V. M. Luque-Almagro, A. J. Gates, C. Moreno-Vivian, S. J. Ferguson, D. J. Richardson and M. D. Roldan, *Biochem. Soc. Trans.*, 2011, **39**, 1838–1843.
188. V. M. Luque-Almagro, V. J. Lyall, S. J. Ferguson, M. D. Roldan, D. J. Richardson and A. J. Gates, *J. Biol. Chem.*, 2013, **288**, 29692–29702.
189. P. Cabello, M. D. Roldan and C. Moreno-Vivian, *Microbiology*, 2004, **150**, 3527–3546.
190. R. M. Martinez-Espinosa, F. C. Marhuenda-Egea and M. J. Bonete, *FEMS Microbiol. Lett.*, 2001, **204**, 381–385.
191. R. M. Martinez-Espinosa, F. C. Marhuenda-Egea and M. J. Bonete, *FEMS Microbiol. Lett.*, 2001, **196**, 113–118.
192. B. Lledo, F. C. Marhuenda-Egea, R. M. Martinez-Espinosa and M. J. Bonete, *Gene*, 2005, **361**, 80–88.
193. M. J. Bonete, R. M. Martinez-Espinosa, C. Pire, B. Zafrilla and D. J. Richardson, *Saline Syst.*, 2008, **4**, 9.
194. R. M. Martinez-Espinosa, B. Lledo, F. C. Marhuenda-Egea, S. Diaz and M. J. Bonete, *Extremophiles*, 2009, **13**, 785–792.
195. B. L. Dalsing and C. Allen, *J. Bacteriol.*, 2014, **196**, 949–960.
196. E. Flores, J. E. Frias, L. M. Rubio and A. Herrero, *Photosynth. Res.*, 2005, **83**, 117–133.
197. S. H. Baek, G. Rajashekara, G. A. Splitter and J. P. Shapleigh, *J. Bacteriol.*, 2004, **186**, 6025–6031.
198. N. E. Van Alst, K. F. Picardo, B. H. Iglewski and C. G. Haidaris, *Infect. Immun.*, 2007, **75**, 3780–3790.
199. K. R. Barth, V. M. Isabella and V. L. Clark, *Microbiology*, 2009, **155**, 4093–4103.
200. M. J. Filiatrault, G. Tombline, V. E. Wagner, N. Van Alst, K. Rumbaugh, P. Sokol, J. Schwingel and B. H. Iglewski, *PLoS One*, 2013, **8**, e55594.
201. S. E. Winter, M. G. Winter, M. N. Xavier, P. Thiennimitr, V. Poon, A. M. Keestra, R. C. Laughlin, G. Gomez, J. Wu, S. D. Lawhon, I. E. Popova, S. J. Parikh, L. G. Adams, R. M. Tsolis, V. J. Stewart and A. J. Baumler, *Science*, 2013, **339**, 708–711.
202. S. Grimaldi, P. Lanciano, P. Bertrand, F. Blasco and B. Guigliarelli, *Biochemistry*, 2005, **44**, 1300–1308.
203. R. Arias-Cartin, S. Lyubenova, P. Ceccaldi, T. Prisner, A. Magalon, B. Guigliarelli and S. Grimaldi, *J. Am. Chem. Soc.*, 2010, **132**, 5942–5943.
204. J. Rendon, E. Pilet, Z. Fahs, F. Seduk, L. Sylvi, M. Hajj Chehade, F. Pierrel, B. Guigliarelli, A. Magalon and S. Grimaldi, *Biochim. Biophys. Acta*, 2015, **1847**, 739–747.

205. M. Aminuddin and D. J. Nicholas, *Biochim. Biophys. Acta*, 1973, **325**, 81–93.
206. H. R. Beller, P. S. Chain, T. E. Letain, A. Chakicherla, F. W. Larimer, P. M. Richardson, M. A. Coleman, A. P. Wood and D. P. Kelly, *J. Bacteriol.*, 2006, **188**, 1473–1488.
207. H. R. Beller, T. E. Letain, A. Chakicherla, S. R. Kane, T. C. Legler and M. A. Coleman, *J. Bacteriol.*, 2006, **188**, 7005–7015.
208. H. R. Beller, T. C. Legler, F. Bourguet, T. E. Letain, S. R. Kane and M. A. Coleman, *Biodegradation*, 2009, **20**, 45–53.
209. K. L. Straub, M. Benz, B. Schink and F. Widdel, *Appl. Environ. Microbiol.*, 1996, **62**, 1458–1460.
210. J. Bosch, K. Y. Lee, G. Jordan, K. W. Kim and R. U. Meckenstock, *Environ. Sci. Technol.*, 2012, **46**, 2095–2101.
211. S. E. Hoeft, J. S. Blum, J. F. Stolz, F. R. Tabita, B. Witte, G. M. King, J. M. Santini and R. S. Oremland, *Int. J. Syst. Evol. Microbiol.*, 2007, **57**, 504–512.
212. K. A. Weber, J. Pollock, K. A. Cole, S. M. O'Connor, L. A. Achenbach and J. D. Coates, *Appl. Environ. Microbiol.*, 2006, **72**, 686–694.
213. A. Chakraborty, E. E. Roden, J. Schieber and F. Picardal, *Appl. Environ. Microbiol.*, 2011, **77**, 8548–8556.
214. H. K. Carlson, I. C. Clark, R. A. Melnyk and J. D. Coates, *Front. Microbiol.*, 2012, **3**, 57.
215. S. E. Winter, P. Thiennimitr, M. G. Winter, B. P. Butler, D. L. Huseby, R. W. Crawford, J. M. Russell, C. L. Bevins, L. G. Adams, R. M. Tsolis, J. R. Roth and A. J. Baumler, *Nature*, 2010, **467**, 426–429.
216. J. B. Bliska and A. W. van der Velden, *mBio*, 2012, **3**, e00226-12.
217. C. A. Lopez, S. E. Winter, F. Rivera-Chavez, M. N. Xavier, V. Poon, S. P. Nuccio, R. M. Tsolis and A. J. Baumler, *mBio*, 2012, **3**, e00143-12.
218. S. Kuntumalla, Q. Zhang, J. C. Braisted, R. D. Fleischmann, S. N. Peterson, A. Donohue-Rolfe, S. Tzipori and R. Pieper, *BMC Microbiol.*, 2011, **11**, 147.
219. S. P. Nuccio and A. J. Baumler, *mBio*, 2014, **5**, e00929–e00914.
220. S. E. Winter and A. J. Baumler, *Gut Microbes*, 2014, **5**, 71–73.
221. A. M. Spees, T. Wangdi, C. A. Lopez, D. D. Kingsbury, M. N. Xavier, S. E. Winter, R. M. Tsolis and A. J. Baumler, *mBio*, 2013, **4**, e00430-13.
222. C. E. Vine, S. K. Purewal and J. A. Colc, *FEMS Microbiol. Lett.*, 2011, **325**, 108–114.
223. D. Seth, A. Hausladen, Y. J. Wang and J. S. Stamler, *Science*, 2012, **336**, 470–473.
224. N. J. Gilberthorpe and R. K. Poole, *J. Biol. Chem.*, 2008, **283**, 11146–11154.
225. L. B. Maia and J. J. Moura, *JBIC, J. Biol. Inorg. Chem.*, 2015, **20**, 403–433.
226. J. O. Lundberg, E. Weitzberg, J. A. Cole and N. Benjamin, *Nat. Rev. Microbiol.*, 2004, **2**, 593–602.
227. C. E. Vine and J. A. Cole, *FEMS Microbiol. Lett.*, 2011, **325**, 99–107.
228. G. Rowley, D. Hensen, H. Felgate, A. Arkenberg, C. Appia-Ayme, K. Prior, C. Harrington, S. J. Field, J. N. Butt, E. Baggs and D. J. Richardson, *Biochem. J.*, 2012, **441**, 755–762.

229. A. Arkenberg, S. Runkel, D. J. Richardson and G. Rowley, *Biochem. Soc. Trans.*, 2011, **39**, 1876–1879.
230. A. Trivedi, N. Singh, S. A. Bhat, P. Gupta and A. Kumar, *Adv. Microb. Physiol.*, 2012, **60**, 263–324.
231. T. Hoffmann, B. Troup, A. Szabo, C. Hungerer and D. Jahn, *FEMS Microbiol. Lett.*, 1995, **131**, 219–225.
232. M. M. Nakano and P. Zuber, *Annu. Rev. Microbiol.*, 1998, **52**, 165–190.
233. E. Härtig and D. Jahn, *Adv. Microb. Physiol.*, 2012, **61**, 195–216.
234. H. I. Boshoff and C. E. Barry 3rd, *Nat. Rev. Microbiol.*, 2005, **3**, 70–80.
235. M. P. Tan, P. Sequeira, W. W. Lin, W. Y. Phong, P. Cliff, S. H. Ng, B. H. Lee, L. Camacho, D. Schnappinger, S. Ehrt, T. Dick, K. Pethe and S. Alonso, *PLoS One*, 2010, **5**, e13356.
236. I. Weber, C. Fritz, S. Ruttkowski, A. Kreft and F. C. Bange, *Mol. Microbiol.*, 2000, **35**, 1017–1025.
237. C. D. Sohaskey, *J. Bacteriol.*, 2008, **190**, 2981–2986.
238. C. A. Andreae, R. W. Titball and C. S. Butler, *Res. Microbiol.*, 2014, **165**, 41–49.
239. S. Ramirez-Arcos, L. A. Fernandez-Herrero and J. Berenguer, *Biochim. Biophys. Acta*, 1998, **1396**, 215–227.
240. F. Cava, O. Zafra, A. Magalon, F. Blasco and J. Berenguer, *J. Biol. Chem.*, 2004, **279**, 45369–45378.
241. F. Cava, O. Zafra, M. S. da Costa and J. Berenguer, *Environ. Microbiol.*, 2008, **10**, 522–533.
242. D. Heintz, S. Gallien, S. Wischgoll, A. K. Ullmann, C. Schaeffer, A. K. Kretzschmar, A. van Dorsselaer and M. Boll, *Mol. Cell. Proteomics*, 2009, **8**, 2159–2169.
243. K. T. Finneran, M. E. Housewright and D. R. Lovley, *Environ. Microbiol.*, 2002, **4**, 510–516.
244. L. I. Hochstein and F. Lang, *Arch. Biochem. Biophys.*, 1991, **288**, 380–385.
245. S. Bickelsandkötter and M. Ufer, *Z. Naturforsch., C: Biosci.*, 1995, **50**, 365–372.
246. B. Lledo, R. M. Martinez-Espinosa, F. C. Marhuenda-Egea and M. J. Bonete, *Biochim. Biophys. Acta*, 2004, **1674**, 50–59.
247. K. Yoshimatsu, T. Sakurai and T. Fujiwara, *FEBS Lett.*, 2000, **470**, 216–220.
248. K. Yoshimatsu, O. Araya and T. Fujiwara, *Extremophiles*, 2007, **11**, 41–47.
249. P. Volkl, R. Huber, E. Drobner, R. Rachel, S. Burggraf, A. Trincone and K. O. Stetter, *Appl. Environ. Microbiol.*, 1993, **59**, 2918–2926.
250. W. G. Zumft, *Microbiol. Mol. Biol. Rev.*, 1997, **61**, 533–616.
251. D. J. Richardson, B. C. Berks, D. A. Russell, S. Spiro and C. J. Taylor, *Cell. Mol. Life Sci.*, 2001, **58**, 165–178.
252. L. Potter, H. Angove, D. Richardson and J. Cole, *Adv. Microb. Physiol.*, 2001, **45**, 51–112.
253. H. P. Liu, S. Takio, T. Satoh and I. Yamamoto, *Biosci., Biotechnol., Biochem.*, 1999, **63**, 530–536.
254. L. Bedzyk, T. Wang and R. W. Ye, *J. Bacteriol.*, 1999, **181**, 2802–2806.

255. L. C. Potter, P. Millington, L. Griffiths, G. H. Thomas and J. A. Cole, *Biochem. J.*, 1999, **344**(1), 77–84.
256. V. Stewart, Y. Lu and A. J. Darwin, *J. Bacteriol.*, 2002, **184**, 1314–1323.
257. M. S. Pittman and D. J. Kelly, *Biochem. Soc. Trans.*, 2005, **33**, 190–192.
258. C. A. Lopez, F. Rivera-Chavez, M. X. Byndloss and A. J. Baumler, *Infect. Immun.*, 2015, **83**, 3470–3478.
259. J. Simon, *FEMS Microbiol. Rev.*, 2002, **26**, 285–309.
260. C. Cruz-Garcia, A. E. Murray, J. A. Klappenbach, V. Stewart and J. M. Tiedje, *J. Bacteriol.*, 2007, **189**, 656–662.
261. H. Gao, Z. K. Yang, S. Barua, S. B. Reed, M. F. Romine, K. H. Nealson, J. K. Fredrickson, J. M. Tiedje and J. Zhou, *ISME J.*, 2009, **3**, 966–976.
262. D. J. Richardson, G. F. King, D. J. Kelly, A. G. McEwan, S. J. Ferguson and J. B. Jackson, *Arch. Microbiol.*, 1988, **150**, 131–137.
263. D. J. Richardson and S. J. Ferguson, *Arch. Microbiol.*, 1992, **157**, 535–537.
264. M. D. Roldan, F. Reyes, C. Morenovivian and F. Castillo, *Curr. Microbiol.*, 1994, **29**, 241–245.
265. H. J. Sears, S. Spiro and D. J. Richardson, *Microbiology*, 1997, **143**, 3767–3774.
266. H. J. Sears, G. Sawers, B. C. Berks, S. J. Ferguson and D. J. Richardson, *Microbiology*, 2000, **146**(Pt 11), 2977–2985.
267. M. Gavira, M. D. Roldan, F. Castillo and C. Moreno-Vivian, *J. Bacteriol.*, 2002, **184**, 1693–1702.
268. F. Reyes, M. D. Roldan, W. Klipp, F. Castillo and C. Moreno-Vivian, *Mol. Microbiol.*, 1996, **19**, 1307–1318.
269. F. Reyes, M. Gavira, F. Castillo and C. Moreno-Vivian, *Biochem. J.*, 1998, **331**(Pt 3), 897–904.
270. H. P. Koops and A. Pommerening-Roser, *FEMS Microbiol. Ecol.*, 2001, **37**, 1–9.
271. S. E. Vlaeminck, A. G. Hay, L. Maignien and W. Verstraete, *Environ. Microbiol.*, 2011, **13**, 283–295.
272. A. Teske, E. Alm, J. M. Regan, S. Toze, B. E. Rittmann and D. A. Stahl, *J. Bacteriol.*, 1994, **176**, 6623–6630.
273. K. Kirstein and E. Bock, *Arch. Microbiol.*, 1993, **160**, 447–453.
274. H. Sundermeyer-Klinger, W. Meyer, B. Warninghoff and E. Bock, *Arch. Microbiol.*, 1984, **140**, 153–158.
275. A. Freitag, M. Rudert and E. Bock, *FEMS Microbiol. Lett.*, 1987, **48**, 105–109.
276. E. Bock, P. A. Wilderer and A. Freitag, *Water Res.*, 1988, **22**, 245–250.
277. T. Yamanaka and Y. Fukumori, *FEMS Microbiol. Rev.*, 1988, **4**, 259–270.
278. L. van Niftrik and M. S. Jetten, *Microbiol. Mol. Biol. Rev.*, 2012, **76**, 585–596.
279. M. Strous, E. Pelletier, S. Mangenot, T. Rattei, A. Lehner, M. W. Taylor, M. Horn, H. Daims, D. Bartol-Mavel, P. Wincker, V. Barbe, N. Fonknechten, D. Vallenet, B. Segurens, C. Schenowitz-Truong, C. Medigue, A. Collingro, B. Snel, B. E. Dutilh, H. J. Op den Camp, C. van der Drift, I. Cirpus, K. T. van de Pas-Schoonen, H. R. Harhangi, L. van Niftrik,

M. Schmid, J. Keltjens, J. van de Vossenberg, B. Kartal, H. Meier, D. Frishman, M. A. Huynen, H. W. Mewes, J. Weissenbach, M. S. Jetten, M. Wagner and D. Le Paslier, *Nature*, 2006, **440**, 790–794.
280. B. Kartal, W. J. Maalcke, N. M. de Almeida, I. Cirpus, J. Gloerich, W. Geerts, H. J. Op den Camp, H. R. Harhangi, E. M. Janssen-Megens, K. J. Francoijs, H. G. Stunnenberg, J. T. Keltjens, M. S. Jetten and M. Strous, *Nature*, 2011, **479**, 127–130.
281. B. Kartal, M. M. Kuypers, G. Lavik, J. Schalk, H. J. Op den Camp, M. S. Jetten and M. Strous, *Environ. Microbiol.*, 2007, **9**, 635–642.
282. J. van de Vossenberg, J. E. Rattray, W. Geerts, B. Kartal, L. van Niftrik, E. G. van Donselaar, J. S. Sinninghe Damste, M. Strous and M. S. Jetten, *Environ. Microbiol.*, 2008, **10**, 3120–3129.
283. R. Hedderich, O. Klimmek, A. Kroger, R. Dirmeier, M. Keller and K. O. Stetter, *FEMS Microbiol. Rev.*, 1999, **22**, 353–381.
284. A. Kletzin, T. Urich, F. Muller, T. M. Bandeiras and C. M. Gomes, *J. Bioenerg. Biomembr.*, 2004, **36**, 77–91.
285. G. D. Fauque and L. L. Barton, *Adv. Microb. Physiol.*, 2012, **60**, 1–90.
286. M. Guiral, L. Prunetti, C. Aussignargues, A. Ciaccafava, P. Infossi, M. Ilbert, E. Lojou and M. T. Giudici-Orticoni, *Adv. Microb. Physiol.*, 2012, **61**, 125–194.
287. J. Simon and P. M. Kroneck, *Adv. Microb. Physiol.*, 2013, **62**, 45–117.
288. I. Schroder, A. Kroger and J. M. Macy, *Arch. Microbiol.*, 1988, **149**, 572–579.
289. T. Krafft, M. Bokranz, O. Klimmek, I. Schroder, F. Fahrenholz, E. Kojro and A. Kroger, *Eur. J. Biochem.*, 1992, **206**, 503–510.
290. T. Krafft, R. Gross and A. Kroger, *Eur. J. Biochem.*, 1995, **230**, 601–606.
291. J. M. Macy, I. Schroder, R. K. Thauer and A. Kroger, *Arch. Microbiol.*, 1986, **144**, 147–150.
292. O. Klimmek, A. Kroger, R. Steudel and G. Holdt, *Arch. Microbiol.*, 1991, **155**, 177–182.
293. M. Yamamoto, S. Nakagawa, S. Shimamura, K. Takai and K. Horikoshi, *Environ. Microbiol.*, 2010, **12**, 1144–1153.
294. S. Laska, F. Lottspeich and A. Kletzin, *Microbiology*, 2003, **149**, 2357–2371.
295. D. P. Moser and K. H. Nealson, *Appl. Environ. Microbiol.*, 1996, **62**, 2100–2105.
296. J. L. Burns and T. J. DiChristina, *Appl. Environ. Microbiol.*, 2009, **75**, 5209–5217.
297. E. L. Barrett and M. A. Clark, *Microbiol. Rev.*, 1987, **51**, 192–205.
298. A. P. Hinsley and B. C. Berks, *Microbiology*, 2002, **148**, 3631–3638.
299. F. Fischer, W. Zillig, K. O. Stetter and G. Schreiber, *Nature*, 1983, **301**, 511–513.
300. L. F. Feinberg, R. Srikanth, R. W. Vachet and J. F. Holden, *Appl. Environ. Microbiol.*, 2008, **74**, 396–402.
301. S. T. Fitz-Gibbon, H. Ladner, U. J. Kim, K. O. Stetter, M. I. Simon and J. H. Miller, *Proc. Natl. Acad. Sci. U. S. A.*, 2002, **99**, 984–989.

302. P. P. Chan, A. D. Holmes, A. M. Smith, D. Tran and T. M. Lowe, *Nucleic Acids Res.*, 2012, **40**, D646–D652.
303. M. Hinojosa-Leon, M. Dubourdieu, J. A. Sanchez-Crispin and M. Chippaux, *Biochem. Biophys. Res. Commun.*, 1986, **136**, 577–581.
304. S. E. Winter and A. J. Baumler, *Gut Microbes*, 2011, **2**, 58–60.
305. S. E. Winter, C. A. Lopez and A. J. Baumler, *EMBO Rep.*, 2013, **14**, 319–327.
306. S. P. Hanlon, R. A. Holt, G. R. Moore and A. G. McEwan, *Microbiology*, 1994, **140**, 6.
307. U. Kappler and H. Schafer, *Met. Ions Life Sci.*, 2014, **14**, 279–313.
308. J. H. Weiner, R. A. Rothery, D. Sambasivarao and C. A. Trieber, *Biochim. Biophys. Acta*, 1992, **1102**, 1–18.
309. U. Wissenbach, D. Ternes and G. Unden, *Arch. Microbiol.*, 1992, **158**, 68–73.
310. A. L. Shaw, A. Hochkoeppler, P. Bonora, D. Zannoni, G. R. Hanson and A. G. McEwan, *J. Biol. Chem.*, 1999, **274**, 9911–9914.
311. D. A. Saffarini, S. L. Blumerman and K. J. Mansoorabadi, *J. Bacteriol.*, 2002, **184**, 846–848.
312. C. Gruber, A. Legat, M. Pfaffenhuemer, C. Radax, G. Weidler, H. J. Busse and H. Stan-Lotter, *Extremophiles*, 2004, **8**, 431–439.
313. A. Oren and H. G. Trüper, *FEMS Microbiol. Lett.*, 1990, **70**, 33–36.
314. J. A. Muller and S. DasSarma, *J. Bacteriol.*, 2005, **187**, 1659–1667.
315. J. A. Gralnick, H. Vali, D. P. Lies and D. K. Newman, *Proc. Natl. Acad. Sci. U. S. A.*, 2006, **103**, 4669–4674.
316. M. N. Alves, S. E. Neto, A. S. Alves, B. M. Fonseca, A. Carrelo, I. Pacheco, C. M. Paquete, C. M. Soares and R. O. Louro, *Front. Microbiol.*, 2015, **6**, 665.
317. K. D. Bewley, M. A. Firer-Sherwood, J. Y. Mock, N. Ando, C. L. Drennan and S. J. Elliott, *Biochem. Soc. Trans.*, 2012, **40**, 1268–1273.
318. S. L. McCrindle, U. Kappler and A. G. McEwan, *Adv. Microb. Physiol.*, 2005, **50**, 147–198.
319. D. Richardson and G. Sawers, *Science*, 2002, **295**, 1842–1843.
320. G. Sawers, *Antonie Van Leeuwenhoek*, 1994, **66**, 57–88.
321. M. J. Almendra, C. D. Brondino, O. Gavel, A. S. Pereira, P. Tavares, S. Bursakov, R. Duarte, J. Caldeira, J. J. Moura and I. Moura, *Biochemistry*, 1999, **38**, 16366–16372.
322. D. J. Steenkamp and H. D. Peck Jr, *J. Biol. Chem.*, 1981, **256**, 5450–5458.
323. L. L. Barton, J. LeGall, J. M. Odom and H. D. Peck Jr, *J. Bacteriol.*, 1983, **153**, 867–871.
324. S. M. da Silva, I. Pacheco and I. A. Pereira, *JBIC, J. Biol. Inorg. Chem.*, 2012, **17**, 831–838.
325. I. A. Pereira, A. R. Ramos, F. Grein, M. C. Marques, S. M. da Silva and S. S. Venceslau, *Front. Microbiol.*, 2011, **2**, 69.
326. D. R. Lovley, D. E. Holmes and K. P. Nevin, *Adv. Microb. Physiol.*, 2004, **49**, 219–286.

327. D. R. Lovley, T. Ueki, T. Zhang, N. S. Malvankar, P. M. Shrestha, K. A. Flanagan, M. Aklujkar, J. E. Butler, L. Giloteaux, A. E. Rotaru, D. E. Holmes, A. E. Franks, R. Orellana, C. Risso and K. P. Nevin, *Adv. Microb. Physiol.*, 2011, **59**, 1–100.
328. C. R. Myers and J. M. Myers, *J. Bacteriol.*, 1997, **179**, 1143–1152.
329. L. Shi, T. C. Squier, J. M. Zachara and J. K. Fredrickson, *Mol. Microbiol.*, 2007, **65**, 12–20.
330. A. Niggemyer, S. Spring, E. Stackebrandt and R. F. Rosenzweig, *Appl. Environ. Microbiol.*, 2001, **67**, 5568–5580.
331. J. Friedebold and B. Bowien, *J. Bacteriol.*, 1993, **175**, 4719–4728.
332. R. Cramm, *J. Mol. Microbiol. Biotechnol.*, 2009, **16**, 38–52.
333. K. Hartwich, A. Poehlein and R. Daniel, *PLoS One*, 2012, **7**, e51662.
334. S. Wang, H. Huang, J. Kahnt and R. K. Thauer, *Appl. Environ. Microbiol.*, 2013, **79**, 6176–6179.
335. A. M. Appel, J. E. Bercaw, A. B. Bocarsly, H. Dobbek, D. L. DuBois, M. Dupuis, J. G. Ferry, E. Fujita, R. Hille, P. J. Kenis, C. A. Kerfeld, R. H. Morris, C. H. Peden, A. R. Portis, S. W. Ragsdale, T. B. Rauchfuss, J. N. Reek, L. C. Seefeldt, R. K. Thauer and G. L. Waldrop, *Chem. Rev.*, 2013, **113**, 6621–6658.
336. W. Buckel and R. K. Thauer, *Biochim. Biophys. Acta*, 2013, **1827**, 94–113.
337. K. C. Costa, P. M. Wong, T. Wang, T. J. Lie, J. A. Dodsworth, I. Swanson, J. A. Burn, M. Hackett and J. A. Leigh, *Proc. Natl. Acad. Sci. U. S. A.*, 2010, **107**, 11050–11055.
338. M. Karrasch, G. Borner, M. Enssle and R. K. Thauer, *FEBS Lett.*, 1989, **253**, 226–230.
339. M. Karrasch, G. Borner, M. Enssle and R. K. Thauer, *Eur. J. Biochem.*, 1990, **194**, 367–372.
340. M. Karrasch, G. Borner and R. K. Thauer, *FEBS Lett.*, 1990, **274**, 48–52.
341. P. A. Bertram and R. K. Thauer, *Eur. J. Biochem.*, 1994, **226**, 811–818.
342. A. Hochheimer, R. A. Schmitz, R. K. Thauer and R. Hedderich, *Eur. J. Biochem.*, 1995, **234**, 910–920.
343. S. Wang, H. Huang, J. Kahnt, A. P. Mueller, M. Kopke and R. K. Thauer, *J. Bacteriol.*, 2013, **195**, 4373–4386.
344. B. Schoepp-Cothenet, R. van Lis, A. Atteia, F. Baymann, L. Capowiez, A. L. Ducluzeau, S. Duval, F. Ten Brink, M. J. Russell and W. Nitschke, *Biochim. Biophys. Acta*, 2013, **1827**, 79–93.

CHAPTER 6

Enzymes of the Xanthine Oxidase Family

TAKESHI NISHINO[a,b], KEN OKAMOTO[a], AND SILKE LEIMKÜHLER*[c]

[a]Department of Biochemistry, Nippon Medical School, 1-1-5 Sendagi, Bunkyo-ku, Tokyo 113-8602, Japan; [b]Department of Applied Biological Chemistry, Graduate School of Agricultural and Life Sciences, University of Tokyo, 1-1-1 Yayoi, Bunkyo-Ku, Tokyo, 113-8657, Japan; [c]Department of Molecular Enzymology, Institute for Biochemistry and Biology, University of Potsdam, Karl-Liebknecht Str. 24-25, 14476 Potsdam, Germany
*E-mail: sleim@uni-potsdam.de

6.1 Introduction

The xanthine oxidase family of enzymes consists of the well-studied enzymes xanthine oxidoreductase (XOR) and aldehyde oxidase (AOX) from eukaryotes as well as several prokaryotic counterparts including xanthine dehydrogenase, CO dehydrogenase and nicotinate dehydrogenase (see below).[1] Historically, these enzymes were among the first to be investigated in a purified form.[2,3] Bovine xanthine oxidoreductase is the prototypical member of this enzyme family and was first described as aldehyde (formaldehyde) oxidizing enzyme (AOX) in the presence of methylene blue in cow's milk in 1902[4] as "Schardinger enzyme". On the other hand, the enzyme that converted hypoxanthine to xanthine and xanthine to uric acid under both aerobic and

RSC Metallobiology Series No. 5
Molybdenum and Tungsten Enzymes: Biochemistry
Edited by Russ Hille, Carola Schulzke, and Martin L. Kirk

Published by the Royal Society of Chemistry, www.rsc.org

anaerobic conditions (with methylene blue) was found in cow's milk and also various organs of the rat and ox.[5] It was, however, shown in 1926 that cow's milk XO was identical to the previously named AOX, which had been described by kinetics studies with the partially purified enzyme.[6] Thus, milk contains one enzyme later named XO. The purified XO enzyme was obtained in 1936 and was shown to be a flavin-containing iron–sulfur protein from the absorption spectrum.[7] The enzyme was subsequently found to contain molybdenum, which was proven to be essential for the activity.[8,9] Studies on cow's milk XOR have thus been carried out for over 110 years since its discovery in 1902[4] and the enzyme is one of the best-studied molybdenum-containing flavoproteins. Much pioneering work has been carried out with this enzyme, including the application of EPR to investigate the enzyme's molybdenum center[10,11] and the discovery of its ability to generate superoxide.[12] Various recent reviews about the enzyme itself are available which have been published in the latest half decade.[1,13–16]

Enzymes from the xanthine oxidase family in general catalyze hydroxylations at either aromatic or aliphatic carbon centers and are in general very complex enzymes that are inhibited by CN^-,[17] and invariably possessing additional redox-active sites for electron transfer from the molybdenum cofactor (Moco) center: minimally a pair of spinach-ferredoxin-like [2Fe2S] clusters, and usually FAD.[17,18] However, exceptions occur.[15,18–20] The most commonly used electron acceptors are O_2 or NAD^+, which react at the FAD site, while the substrate reacts at the molybdenum center. As a result, intramolecular electron transfer between the Moco and FAD *via* the two iron–sulfur centers occurs during turnover. The two iron–sulfur centers not only provide a path for electron transfer, but also act as an electron sink for making the reaction most efficient as shown by studies on xanthine oxidase:[21] two electrons received from the substrate at the molybdenum atom are transferred to the two FeS centers so that the oxidized Moco can react readily with another substrate molecule, while the two electrons in the iron–sulfur centers reduce the FAD center, which can then react with the electron acceptor substrates (NAD^+ or O_2).[21,22] Without an electron sink between two substrate active sites (Moco and FAD) the reaction would be inefficient. The electron transfer rates among the several redox-active centers are rapid[21,23,24] and the overall rate-limiting step is the release of the product at the Moco site, as demonstrated for chicken liver XDH[25] and bovine XOR.[26,27]

Like the other molybdoenzymes from the families of sulfite oxidase (see Chapter 7) and DMSO reductase (see Chapter 5) (Figure 6.1), Moco-containing hydroxylases from the xanthine oxidase family use water as the ultimate source of the oxygen atom incorporated into the product, and generate reducing equivalents in the course of substrate hydroxylation.[28] The physiological electron acceptors can be molecular oxygen, NAD^+, $NADP^+$ or quinones; however, in several enzymes the electron acceptor remains unknown.

The typical feature of enzymes of the xanthine oxidase family is the presence of an essential sulfur atom[29–31] present as sulfide in the equatorial position of the molybdenum ion (Figure 6.1).[32] The initial observation of

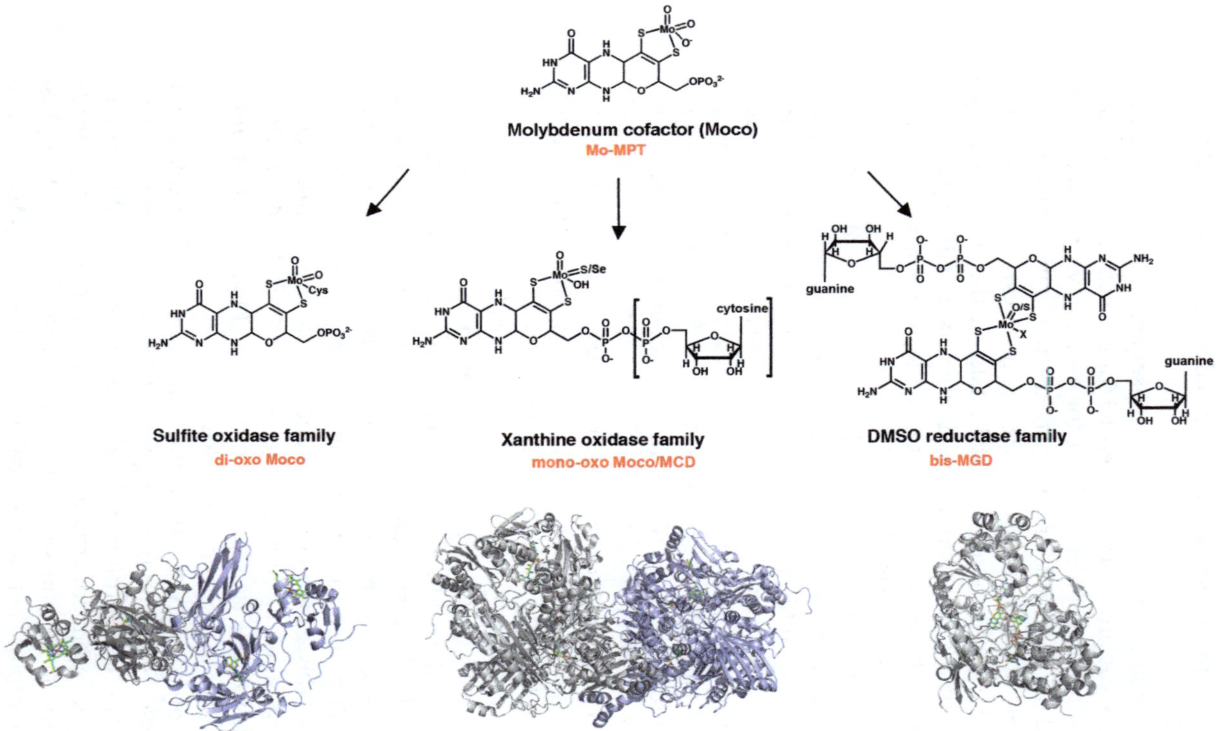

Figure 6.1 The families of molybdoenzymes. The basic form of Moco is a pyranopterin, named Mo-MPT, which coordinates the molybdenum atom by the characteristic dithiolene group at the C1′ and C2′ positions of the pyranopterin ring. Mo-MPT (shown in the tri-oxo structure can be further modified and three different molybdenum-containing enzyme families are classified according to their coordination at the molybdenum atom: the xanthine oxidase, sulfite oxidase and DMSO reductase families. The sulfite oxidase family is characterized by a MPT-$Mo^{VI}O_2$Cys ligand sphere. The xanthine oxidase family contains a MPT–Mo^{VI}OS(OH) core. In bacteria, the MPT core can be modified by an additional CMP nucleotide at the phosphate group, forming MCD. The DMSO reductase family contains a MGD_2–Mo^{VI}XY core, with X being either a sulfur or an oxygen ligand and Y being either a hydroxo or an amino acid ligand (Ser, Cys, Sec and Asp ligands have been identified so far). Shown are structures of representative enzymes from each family: chicken sulfite oxidase (pdb 1SOX), bovine xanthine dehydrogenase (pdb 1FIQ) and *Shewanella massilia* TMAO reductase (pdb 1TMO).

inhibition by cyanide[33] was subsequently followed up with the observation that the sulfide atom essential for catalytic activity of bovine milk XO[29] was released by CN^- treatment as SCN^-, resulting in formation of an inactive desulfo form of the enzyme that occurs naturally in significant amounts, at least in eukaryotes.[29,30] The cyanolyzable sulfur was subsequently shown to be a ligand to Mo, identified as being present as a Mo=S group by EPR studies.[34] Fully active bovine milk XOR can be reproducibly isolated using a folate affinity chromatography procedure coupled with allopurinol,[35,36] and the as-isolated enzymes from other organisms (notably chicken under particular conditions[37] and the bacterium *Rhodobacter capsulatus*[38]) typically have full activity. The remaining ligands to the molybdenum ion are the two dithiolene sulfurs from the molybdopterin (MPT)-moiety, an oxo ligand and a hydroxo ligand.[32] Mo-MPT as well as the molybdopterin cytosine dinucleotide (MCD) form of the molybdenum cofactor have been identified in bacterial enzymes of this family (Figure 6.1, Table 6.1, and also recently reviewed by Hille *et al.*, 2014).[1]

6.2 An Overview of Enzymes from the Xanthine Oxidase Family

Enzymes of the xanthine oxidase family are present in both eukaryotes and prokaryotes. The eukaryotic enzymes of the xanthine oxidase family include xanthine oxidase (XO) (EC 1.17.3.2), xanthine dehydrogenase (XDH) (EC 1.17.1.4) and aldehyde oxidase (E.C. 1.2.3.1), which in general are α_2 homodimers (Table 6.1).[15,39] Mammalian XO and XDH are products of the same gene, with post-translational interconversion occurring as described in detail below. These proteins contain three cofactor-binding domains on a single polypeptide: an N-terminal domain, which binds two [2Fe2S] clusters FeSI and FeSII, a central domain, which binds FAD, and a C-terminal domain, which binds Moco in the form of Mo-MPT[40,41] as predicted from the primary amino acid sequence analyses (Figure 6.2).[42–44] Each domain is connected by a flexible linker region. In bacteria and archaea, the enzymes of the xanthine oxidase family are more diverse and include the CO dehydrogenases[45] as well as enzymes metabolizing quinoline-related compounds[46] and derivatives of nicotinic acid.[47] *Rhodobacter capsulatus* xanthine dehydrogenase,[48] *Desulfovibrio gigas* aldehyde oxidoreductase,[49] *Pseudomonas putida* quinoline-2-oxidoreductase,[46] *Eubacterium barkeri* nicotinate dehydrogenase,[50] *Oligotropha carboxidovorans* CO dehydrogenase[51] and *Thaurea aromatica* 4-hydroxybenzoyl-CoA reductase[52] are among the best-characterized bacterial enzymes from this family and crystal structures exist for each (Table 6.1). In general, the redox-active centers of bacterial and archaeal members of the xanthine oxidase family are found in separate subunits, and exceptions from the general cofactor composition have been identified (Figure 6.2). Among the bacterial xanthine oxidoreductases, the enzymes characterized from *R. capsulatus*,[53] *Comamonas acidovorans*[54] and *Veillonella atypica*[55] are representative. These

Table 6.1 Characterized members of the xanthine oxidase family of molybdoenzymes.

Enzymes	Subunit composition	Gene locus/organism	Moco form present in the catalytic subunit (+ unusual features)	Redox couple	Physiological electron acceptor	PDB	References
Xanthine oxidase family							
Eukaryotic xanthine oxidoreductases							
Xanthine oxidoreductase	α_2	Bovine	Mo-MPT	Uric acid/xanthine	NAD^+/O_2	1FIQ	Enroth *et al.*, 2000[40]
Xanthine oxidoreductase	α_2	Human	Mo-MPT	Uric acid/xanthine	NAD^+/O_2	2EIQ	Yamaguchi *et al.*, 2007[128]
Xanthine oxidoreductase	α_2	Rat	Mo-MPT	Uric acid/xanthine	NAD^+/O_2	1WYG	Nishino *et al.*, 2005[78]
Xanthine dehydrogenase	α_2	Chicken	Mo-MPT	Uric acid/xanthine	NAD^+	—	Sato *et al.*, 1995[44]
Procaryotic xanthine oxidoreductases							
Xanthine dehydrogenase	$(\alpha\beta)_2$	XdhAB/*xdhABC*/*R. capsulatus*	Mo-MPT	Uric acid/xanthine	NAD^+	1JRO	Truglio *et al.*, 2002[48]
Xanthine dehydrogenase	$(\alpha\beta)_2$	XdhABC/*xdhABC*/*Comamonas acidovorans*	Mo-MPT	Uric acid/xanthine	NAD^+	—	Xiang and Edmondson, 1996[54]
Xanthine dehydrogenase	$\alpha\beta\gamma$	XdhABC/*xdhABC*/*Veillonella atypica*	MCD	Uric acid/xanthine	NAD^+	—	Gremer and Meyer, 1996[55]
Eukaryotic aldehyde oxidases							
Aldehyde oxidase	α_2	Human AOX1	Mo-MPT	Aldehydes/carboxylic acid	O_2	—	Hartmann *et al.*, 2012[185]
Aldehyde oxidase	α_2	Mouse AOX1	Mo-MPT	Aldehydes/carboxylic acid	O_2	—	Schumann *et al.*, 2009[190]
Aldehyde oxidase	α_2	Mouse AOX2	Mo-MPT	Aldehydes/carboxylic acid	O_2	—	Terao *et al.*, 2001[168]

Aldehyde oxidase	α_2	Mouse AOX3	Mo-MPT	Aldehydes/carboxylic acid	O_2	3ZYV	Coelho *et al.*, 2013[41]
Aldehyde oxidase	α_2	Mouse AOX4	Mo-MPT	Aldehydes/carboxylic acid	O_2	—	Kurosaki *et al.*, 2004[171]
Prokaryotic aldehyde/*N*-heterocyclic ring oxidoreductases							
Aldehyde oxidoreductase	$\alpha\beta\gamma$	$(PaoABC)_2$/*paoABCD*/*E. coli*	MCD	Benzoic acid/benzaldehyde	O_2	—	Neumann *et al.*, 2009[122]
Aldehyde oxidoreductase	α_2	MOP/*Desulvovibrio gigas*	MCD No FAD subunit	Carboxylate/aldehyde	?	1VLB	Romão *et al.*, 1995[49]
Quinaldine-4-oxidase	$(\alpha\beta\gamma)_2$	QoxLMS/*qoxMLS*/*Arthrobacter* sp. *Rü61a*	MCD	1*H*-4-Oxoquinaldine/quinaldine	O_2	—	Stephan *et al.*, 1996[57]
Quinoline-2-oxidoreductase	$(\alpha\beta\gamma)_2$	QorLMS/*qorMLS*/*Pseudomonas putida*	MCD	Quinolin-2(1*H*)-one/quinoline	?	1T3Q	Blase *et al.*, 1996[196]; Frerichs-Deeken *et al.*, 2003[197]; Bonin *et al.*, 2004[46]
Isoquinoline 1-oxidoreductase	$\alpha\beta$	IorAB/*iorAB*/*Brevundimonas diminuta*	MCD	Isoquinolin-1(2*H*)-one/isoquinoline	?	—	Lehmann *et al.*, 1995[60]
Quinoline-4-carboxylate-2-oxidoreductase	$(\alpha\beta\gamma)_2$	*Agrobacterium* sp. *Pseudomonas* sp.	MCD	2-Oxo-1,2-dihydroquinoline-4-carboxylate/quinoline-4-carboxylate	?	—	Bauer and Lingens, 1992[58]
Quinaldic acid 4-oxidoreductase	$\alpha\beta$ $(\alpha\beta\gamma)_2$	*Serratia marcecens* *Pseudomonas* sp.	MCD	Kynurenate/quinaldate	?	—	Fetzner and Lingens, 1993[59]

(*continued*)

Table 6.1 (*continued*)

Enzymes	Subunit composition	Gene locus/organism	Moco form present in the catalytic subunit (+ unusual features)	Redox couple	Physiological electron acceptor	PDB	References
Prokaryotic proteins with unique properties							
Nicotinate dehydrogenase	$(\alpha\beta\gamma\delta)_2$	NdhFSLM/*ndhFSLM*/*Eubacterium barkeri*	MCD Mo=Se ligand	6-Hydroxynicotinate/nicotinate	$NADP^+$	3HRD	Wagener *et al.*, 2009[50]
CO dehydrogenase	$(\alpha\beta\gamma)_2$	CoxMSL/*coxMSL-DEFG*/*Oligotropha carboxidovorans*/*Hydrogenophaga pseudoflava*	MCD [$CuSMoO_2$] center	CO_2/CO	Quninones	1N5W 1SU6	Dobbek *et al.*, 2002[51]
4-Hydroxybenzoyl-CoA reductase	$(\alpha\beta\gamma)_2$	HcrABC/*h-crCAB*/*Thaurea aromatica*	MCD Additional [4Fe4S] cluster	4-Hydroxybenzoyl-CoA/benzoyl-CoA	Ferredoxin (e^- donor)	1SB3	Unciuleac *et al.*, 2004[52]

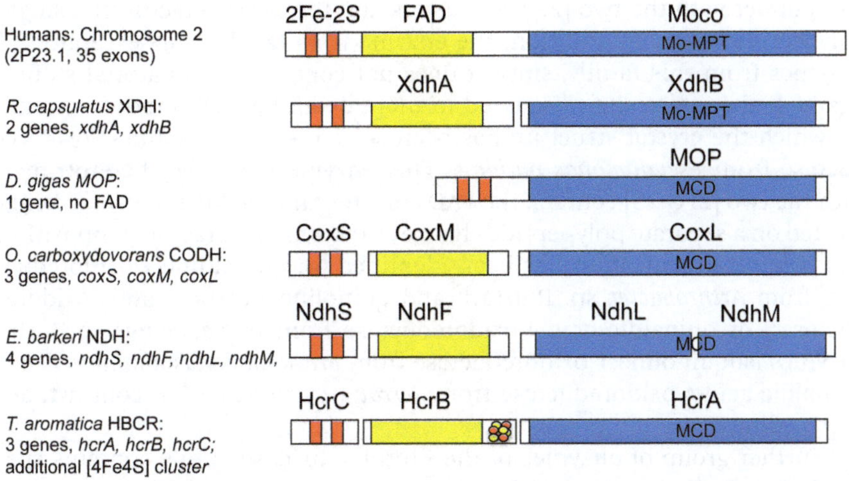

Figure 6.2 Overview of subunit composition and cofactor organization of selected enzymes of the XO family. The binding sites for the [2Fe2S] clusters are shown in red, the binding sites for the FAD cofactor are colored in yellow and the binding sites for Moco are colored schematically in blue (the subunits are not drawn to scale). The FAD domain is absent in the *D. gigas* aldehyde oxidoreductase. In the *R. capsulatus* xanthine dehydrogenase the iron–sulfur and flavin-binding portions of the protein constitute one subunit (XdhA), and the Mo-MPT binding portion a second (XdhB) subunit. In the *O. carboxidovorans* CO dehydrogenase, the iron–sulfur centers are together in one subunit (CoxS), the flavin in a second (CoxM) and the Moco as MCD in a third (CoxL) subunit. In the *E. barkeri* nicotinate dehydrogenase, the iron–sulfur centers are together in one subunit (NdhS), the flavin in a second (NdhF) and the Moco as MCD in a third (NdhL) and fourth subunit (NdhM). In the *T. aromatica* HBCR, the iron–sulfur centers are together in one subunit (HcrC), the flavin and a [4Fe4S] cluster in a second (HcrB) and the Moco as MCD in a third (HcrA) subunit.

enzymes are organized either as $(\alpha\beta)_2$ heterodimers and contain Mo-MPT as cofactor (*R. capsulatus*, *C. acidovorans*) or as $\alpha\beta\gamma$ heterotrimers containing the MCD cofactor (*V. atypica*) (Table 6.1). In the enzymes from *R. capsulatus* and *C. acidovorans*, the cofactor-containing domains for the two [2Fe–2S] clusters and FAD are found in one subunit, while the Mo-MPT cofactor is found in a second. In the enzyme from *V. atypica*, the two [2Fe–2S] cofactors, the FAD cofactor and the MCD cofactor are each found in separate subunits. Surprisingly, a dimerization *via* the Moco-containing domain/subunit has not been reported for the enzyme from *V. atypica*.

Further enzymes from the xanthine oxidase family of bacterial origin include aldehyde oxidoreductases[49] and quinoline/quinaldine oxidases.[46,56] The first X-ray structure of an enzyme from the xanthine oxidase family was obtained from aldehyde oxidoreductase (MOP) from *Desulfovibrio gigas*.[49] This enzyme lacks the FAD cofactor domain and is synthesized as an α_2 dimer with the two [2Fe2S] clusters and the MCD cofactor in a single polypeptide chain. In addition, the enzyme is so far the only exception of enzymes from this family, since it does not contain the equatorial sulfido ligand at the active site (discussed in more detail below). Another enzyme for which the crystal structure has been solved is the quinoline-2-oxidoreductase from *Pseudomonas putida.*[46] This enzyme is a $(\alpha\beta\gamma)_2$ heterotrimer, with the two [2Fe–2S] centers, the FAD cofactor and the MCD cofactor being located on a separate polypeptide chain each. A similar cofactor composition and subunit organization has been identified for quinaldine-4-oxidoreductase from *Arthrobacter* sp. Rü61a,[57] and quinoline-4-carboxylate-2-oxidoreductase[58] or quinaldic acid-4-oxidoreductase from *Pseudomonas* sp.[59] The enzymes isoquinoline-1-oxidoreductase from *Brevundimonas diminuta*[60] and quinaldic acid-4-oxidoreductase from *Serratia marcescens*,[59] in contrast, are abeterodimers.

A further group of enzymes of the xanthine oxidase family contains special features either at the molybdenum active site or in the cofactor/subunit composition. These enzymes include CO dehydrogenase from *O. carboxidovorans*,[51] nicotinate dehydrogenase from *E. barkeri*[50] and the 4-hydroxybenzoyl-CoA reductase from *Thaurea aromatica.*[52] The CO dehydrogenase is unique in that it contains a binuclear active site with a copper atom coordinated to the molybdenum atom *via* a μ-sulfido bridge, forming a binuclear Mo–S–Cu-(S-Cys388) center.[61] The Moco is present as a MCD cofactor and the enzyme forms a $(\alpha\beta\gamma)_2$ heterotrimer.[62] In contrast, at the MCD-containing active site of nicotinate dehydrogenase, the sulfido ligand is replaced by a selenido ligand.[50] This enzyme is also unique in being a $(\alpha\beta\gamma\delta)_2$ heterotetramer, with the molybdenum-binding portion of the protein being split into two separate polypeptides. The 4-hydroxybenzoyl-CoA-reductase from *T. aromatica* is an αβγ heterotrimer, and also does not dimerize *via* the MCD-containing Moco subunit.[52,63] Here, an additional [4Fe–4S] cluster has been identified in the FAD-containing subunit that is presumably involved in transferring electrons from ferredoxin for the reduction of 4-hydroxybenzoyl-CoA to the MCD active site.

6.3 Xanthine Oxidoreductases from Eukaryotes and Bacteria

Xanthine oxidoreductases (XOR) exist in a great variety of organisms from bacteria to higher plants and humans.[1,15,18] The enzymes catalyze the metabolic conversion of hypoxanthine to xanthine and xanthine to urate, which are the last steps in purine metabolism in humans. Mechanistically, each reaction is an oxidative hydroxylation, and two electrons are passed from the substrate to the enzyme in each step (reviewed in a recent article by Hille *et al.*).[1] Under normal physiological circumstances, NAD^+ is the physiological electron acceptor and NADH is formed. In mammals, however, under certain conditions a conformational change of the enzyme can be induced and the dehydrogenase transformed to an oxidase (XO) that uses molecular oxygen as electron acceptor. This so-called D/O (dehydrogenase to oxidase) conversion occurs reversibly by oxidation of thiols to disulfides, or irreversibly converted by mild proteolysis.[16,64] Because the oxidase and dehydrogenase forms are products of the same gene, the enzyme is referred to as xanthine oxidoreductase (XOR) here. By contrast to the mammalian enzyme, enzymes from other sources (chicken,[65] *Drosophila*[66] and bacteria[48]) exist as a stable dehydrogenase and are not converted to the oxidase.

Mammalian XOR exists as a homodimer of 150 kDa subunits, with the crystal structure of the bovine enzyme shown in Figure 6.3.[40] Each of the subunits is composed of three domains. The largest domain contains the Moco, the intermediate domain contains the FAD cofactor and the small domain contains the two [2Fe–2S] clusters as predicted by the amino acid sequence comparisons from various sources.[17,18] The iron–sulfur centers are designated FeSI and FeSII, based on their reduction potentials and EPR signals, with FeSII having the higher potential.[67–69] The order of electron transfer in the enzyme is Moco → FeSI → FeSII → FAD as described further below.[22,40] The arrangement of redox-active centers is consistent with the role of the two iron–sulfur centers as an electron sink as described above.

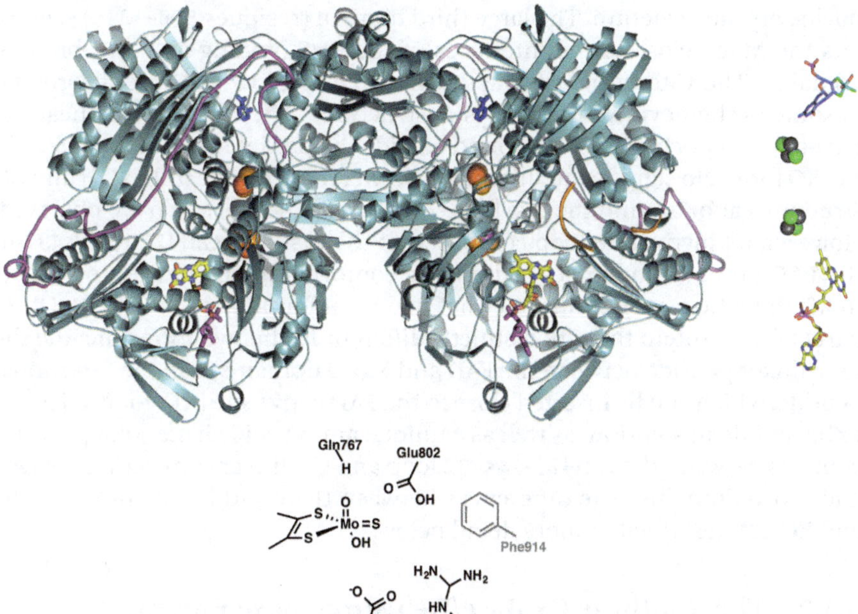

Figure 6.3 Overall structure, cofactor arrangement of bovine xanthine oxidase. The structure of the *bovine XOR* enzyme (1FIQ). The cofactor arrangement is highlighted on the right side. Bovine XOR oxidizes hypoxanthine to xanthine and xanthine to uric acid at the Moco and transfers electron to NAD^+, forming NADH + H^+(XDH) or to O_2, forming H_2O_2 and O_2^- (XO).

The electron transfer rates among them are rapid[21] and the overall rate-limiting step has again been shown to be the release of the product at the Moco site with chicken liver XDH,[25] bovine XOR[27] and the enzyme isolated from *R. capsulatus*.[70,71]

6.3.1 The Crystal Structure of Bovine XOR

The first crystal structure was reported for bovine XOR in 2000 by Enroth *et al.*[40] The structure exhibits a butterfly shape, with the dimer interface formed between the large Moco-binding domains (Figure 6.3). The dimensions of the entire enzyme molecule are 155 × 90 × 70 Å. The cofactor arrangement in the two α_2 subunits is approximately linear, with a closest approach of the two independently functioning chains of ~50 Å (between the two Mo atoms).[72] As discussed above, the structure showed that each monomer is arranged in three domains.[40] The small N-terminal domain (residues 3–164) contains the two [2Fe–2S] cofactors and is connected to the second, FAD-containing domain (residues 192–527) by a long, partially disordered segment consisting of residues 165–192. The FAD domain is connected to the third domain by another extended segment (residues 528–589), which is also partially disordered; residues 532–536 are not represented by electron density in the bovine crystal structure. The large third domain (residues 590–1315) sequesters the Moco close to the interfaces of the Fe/S-binding and FAD-binding domains. The C-terminal residues 1315–1320 to 1332–1331 are disordered in some of the bovine enzyme crystal structures. They are inserted near the FAD-binding pocket,[73] but are found outside the pocket in a mutant locked in the XO form. No significant differences between the bovine milk XDH and XO structures at or around the two [2Fe–2S] centers and the Moco are observed. However, a polypeptide loop composed of the residues Gln422 to Lys432 in the FAD domain undergoes a dramatic conformational change upon XDH to XO transition, very similar to what had been found with milk XOR. Other parts of the protein that are distinctly different in the two forms include the long linker peptide between the FAD and Moco domains, and the C-terminal peptide, which can be inserted close to the FAD active site (XDH-locked form) or be mobile in solution, as well as a unique amino acid cluster that provides contacts between the Gln422–Lys432 loop and the linker connecting the FAD and Moco domains. The differences between the crystal structures of XDH and XO are described in more detail below.

6.3.2 The Xanthine Oxidase/Dehydrogenase Enigma

The transformation of the dehydrogenase form of XOR to the oxidase form, either by cysteine oxidation or by proteolytic cleavage, has been shown to play an important role in physiological and pathological processes.[74,75] The main differences in the catalytic properties of XDH and XO are the changed reactivities of the reduced $FADH_2$ towards the final electron acceptors NAD^+ and molecular oxygen. Whereas XDH reacts rapidly with NAD^+, but only slowly with oxygen, XO reacts in the opposite manner.[76] During the D/O conversion, the FAD cofactor is the only redox-active center that shows significant changes in reactivity.[40,77,78] It has been shown that in XDH, the reduction potential of the neutral FAD semiquinone (FADH*)/$FADH_2$ couple (410 mV in bovine XDH) is much lower than that of the FAD/FADH* couple (270 mV in bovine XDH).[79,80] Additionally, the neutral semiquinone FADH• is more thermodynamically stable in XDH than in XO.[81] It has been suggested that $FADH_2$ reacts with molecular oxygen to produce H_2O_2 and oxidized FAD, while the semiquinone FADH• reacts with molecular oxygen to produce O_2^- plus FAD as shown by stopped flow experiments with stable chicken XDH.[82] Since the semiquinone is stabilized in XDH and not in XO, XDH produces more O_2^- per electron introduced to the enzyme than XO does.[26,81]

Crystal structures at 1.65 and 2.3 Å resolution for both XDH and XO, respectively, have revealed the difference in the protein environment and the hydrogen bonding network surrounding the FAD binding site for XDH and XO. The two structures show that the two forms differ in particular in the orientation of a loop region (Gln423–Lys433) near the flavin binding site.[40] It has been suggested that the highly charged loop near the flavin-binding site is the main factor that causes the change in reduction potential of the FAD in XDH *versus* XO (Figure 6.4). Asp429 (Asp428 rat) is very close to the C_6 position of the flavin isoalloxazine ring, whereas it was replaced by Arg426 (Arg425 rat) in XO. Such a difference in electrostatic environments is consistent with the previous active-site probe studies with artificial FADs.[83,84] It has further been shown that NAD^+ binding to XDH triggers a change in orientation of the loop that increases the reduction potential of the FAD by ~180 mV to further facilitate the forward electron transfer from FeSII to the flavin.[85,86] The influence of the positive NAD^+-ring charge is partly compensated by (i) deprotonation of the Asp429 and (ii) the van der Waals volume of the NAD^+ molecule.[86] In the dehydrogenase form, the protonation state of Asp429 changes significantly in response to flavin reduction, but this does not occur in the oxidase form as it is unable to bind NAD^+ due to steric interference by the Gln423-Lys433 loop.[40]

The D/O transition has been studied for a long time, including site-directed mutagenesis work as well as the X-ray crystal structures that demonstrated the structural differences between the two forms, and has recently been reviewed.[16] It has been shown that the D/O transformation occurs rather readily by the interaction of the C-terminal peptide with the membrane or with other proteins, without proteolysis or disulfide formation (Figure 6.4). It has also been shown that the C-terminal peptide plays a role not only in NAD^+ binding, but also in stabilizing the XDH conformation through its insertion into the cavity, since a protein variant in which the C-terminal peptide was deleted converts the enzyme to a form intermediate between the dehydrogenase and oxidase.[73] As described in the next section, XOR is mostly found in the cytosol of liver cells,[87] but its interaction with the membrane of capillary endothelial cells has also been reported.[88–90] It is tempting to speculate that the C-terminal peptide may be involved in such interactions.

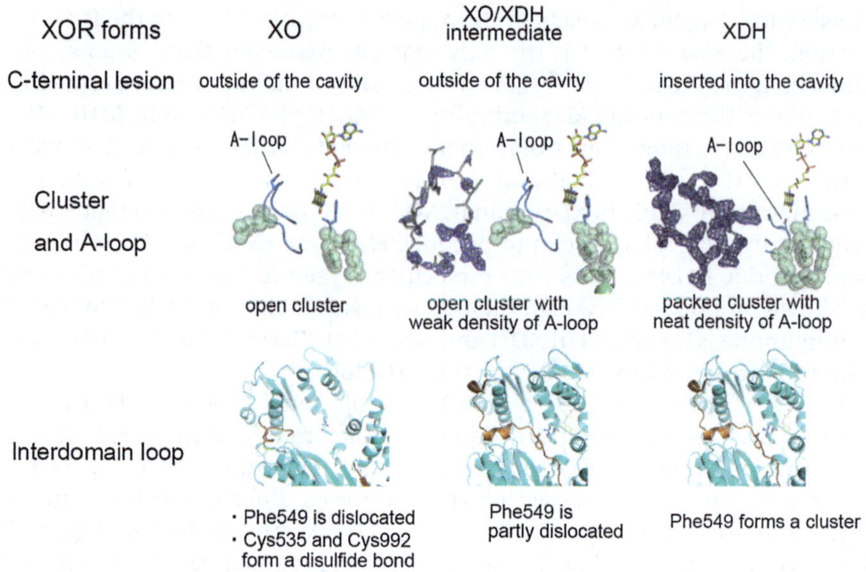

Figure 6.4 The proposed structural changes of XOR from XDH to XO forms. Top: positioning of the C-terminal peptide, it is outside of the FAD active site cavity in the XO form but is inside in the XDH forms. Middle: the arrangement of A-loop around FAD. The electron density of A-loop (dark blue) is clear and Asp428 is close to C6 of isoalloxazine ring of FAD in XDH but is not clear in XO due to being disordered in XO. The amino acids cluster, shown as green space filled model, is packed in XDH but is disrupted in XO. Bottom: the inter-domain long linker between FAD and Moco domains is clearly observed in XDH form and helix structures, brown ribbon model, are observed near the origin of the linker. The linker is mostly disordered in XO. Phe549 is away from the cluster, and the disulfide bond between Cys 535 and Cys 992 is formed after complete conversion from XDH to XO. The intermediate form of XDH/XO can exist when the C-terminal peptide is outside the peptide. XDH, XO and XDH/XO are illustrated from PDP codes 4YTY, 4YTZ and 4YTZ, respectively.

A variant of rat XOR has been generated containing the amino acid exchanges W335A/F336L, which resulted in a form of the enzyme that is stably locked in the oxidase form. This enzyme variant produces O_2^- and H_2O_2 in a ratio of 6 : 1.[77] This enzyme variant had a higher FAD midpoint potential that allowed this mutant to become such an efficient producer of superoxide ions. In native XOR, the reduction potential of the FeSII cluster and that of FAD are close together. Based on the thermodynamic equilibrium of the electrons introduced into the wild-type enzyme, the FAD cofactor thus exists in both the $FADH_2$ and FADH• states, which in turn leads to H_2O_2 and O_2^- production.[26,76] Due to the increase in the FAD midpoint potential in the W335A/F336L variant, the electrons are more easily transferred from FeSII to the FAD in a one-electron transfer step thereby forming more FADH• which results in a higher O_2^- production.[77]

6.3.3 The Physiological Role of Xanthine Oxidoreductase in Mammals

XOR is widely distributed in various species and the physiological roles of the enzyme are species-specific, *e.g.* in uricotelic animals such as mammals, which excrete nitrogen as urea. In mammals, the main role of XOR is in the catabolism of adenine- and guanine-containing nucleotides while in uricotelic animals such as birds, its main role is the excretion of nitrogen from amino acid degradation after formation of uric acid *via* the *de novo* synthesis of IMP. In insects, XOR is additionally involved in the metabolism of pigments.[91] Consequently, gene expression and regulation differs from one species to the next. While expression of the XOR gene is regulated in various animals, however, it seems to be constitutively expressed in mammals. We focus here mainly on mammals, since the enzymological characterization, including determination of the crystal structure, has been extensively studied and the physiological and pathological roles are hot topics at the moment because XOR is the target for drugs against hyperuricemia, gout and a growing number of other diseases in humans.[92,93] In mammals, XOR is distributed in various organs including the liver and endothelial cells of vascular vessels, and is present in high concentration in the mammary gland.[88,94] The enzyme is thought to be localized in cytosol[87] rather than in peroxisome,[95] but seems to be membrane-associated in endothelial cells[96] and complexed with other proteins in milk.[97] It should be noted, however, that XOR-deficient patients in humans do not show any serious symptoms other than xanthinuria, which is described in more detail in Section 2.5. This suggests that the XOR enzyme might not play major roles in maintaining asymptomatic health.

XOR has also been shown to be involved in the secretion of milk fat globules,[98] and is known to be a major protein component of the milk fat globule membrane (MFGM) that envelops fat droplets in freshly expressed milk.[99] The MFGM is derived from the apical membrane of the secretory cell and surrounds the fat droplet as in the secretion process.[100,101] Two studies have provided experimental evidence for the long-suspected involvement of XOR in milk secretion. McManaman *et al.* (2002)[102] have shown that in the mouse mammary gland, transcription of XOR increases with lactation, and that at the onset of secretion the enzyme moves from the cytoplasm of the secretory cell to the apical cell membrane, where it colocalizes with adipophilin (ADPH) and butyrophilin (Btn). ADPH is associated with the lipid droplet,[103] and Btn is an integral membrane protein of the plasma membrane.[104] XOR forms a complex with these proteins, suggesting that the enzyme mediates the interaction of the lipid droplet and the apical plasma membrane in the secretion process. Additionally, Vorbach *et al.* (2002)[105] have examined mice with a targeted disruption in the XOR gene. While homozygous (−/−) mice died at 6 weeks or less after birth, heterozygous (+/−) mice survived and matured with normal fertility and litter size. However, heterozygous (+/−) females were defective in lactation, and all their pups died of starvation soon after birth. Electron microscopy showed that XOR was necessary to envelop the milk fat droplet during secretion. These studies have suggested that XOR plays a structural role in milk fat droplet formation that is independent of its enzymatic activity, consistent with observation that the enzyme is present mostly in the inactive demolybdo form in the milk fat droplets.[106]

Still, an antimicrobial function of the enzyme in milk has been suggested[107] on the basis of the observation that the oxidase activity of XOR peaks dramatically in the first few weeks postpartum,[108] at a time when the antimicrobial activity of milk is most needed by the infant. A protective role for XOR is also underlined by the fact that upon specific inhibition of XOR, the microbial activity in milk is increased. Samples of fresh human milk and bovine milk have each been shown to inhibit the metabolic activity of *E. coli*, demonstrating the specific role of XOR in this reaction.[75,109] Obviously, these arguments run counter reports that the XO is mostly demolybdo form in human milk.[106] Thus two roles are suggested, one role of the demolybdo form in milk fat droplet formation and one role of the active molybdo form as antimicrobial agent, likely by the formation of reactive oxygen species (ROS).

XOR catalyzes the oxidation of a wide range of substrates and, in the oxidase form, can pass electrons to molecular oxygen, generating ROS.[110] XOR has been implicated in ROS-mediated ischemia-reperfusion injury, and it has been suggested that XOR-derived ROS plays a role in innate immunity, specifically in the inflammatory response and in the antimicrobial defense of the gastrointestinal tract.[111] Additionally, several groups have reported that XOR can reduce nitrite to yield nitric oxide (NO),[112–117] a potent vasodilator. XO reduces nitrite at the molybdenum center and produces NO under hypoxic conditions with NADH as the electron donor. It has been suggested that the XOR-produced NO might play a protective role in ischemia,[118,119] *e.g.* as part of a defense mechanism or a regulatory response under some pathogenic conditions, such as tissue ischemia.[117] However, it should be noted that the reported NO production by XO is very low even under anaerobic conditions. It is not surprising from a chemical point of view that the water-exchangeable hydroxyl group at OH–Mo(IV) can be replaced by NO_2 to produce NO.[13] Although formation of NO from nitrite or formation of xanthine from urate by XOR is chemically feasible,[13] it is not yet clear whether these extremely slow reactions have any physiological significance.[120]

6.3.4 The Bacterial XDH from *Rhodobacter capsulatus*

A prokaryotic XDH with similar activity to the eukaryotic form can be isolated from *R. capsulatus*; it is a cytoplasmic enzyme with an $(\alpha\beta)_2$ heterotetrameric structure and a molecular mass of 275 kDa. The cofactors have been shown to be located on two different subunits, with the iron–sulfur clusters and the FAD in the XdhA subunit, and the Moco in XdhB (Figure 6.2).[121] The amino acid sequence of XDH has a high degree of similarity to eukaryotic XORs. Furthermore, analysis of the Moco in XDH has demonstrated the presence of the MPT form as cofactor, as found in all eukaryotic Moco-containing enzymes. This is in contrast to most bacterial molybdoenzymes from the XO family in which the terminal phosphate of MPT is covalently attached to an additional CMP nucleotide to generate the molybdopterin cytosine dinucleotide (MCD) form of the cofactor.[122] Like chicken XDH, the *R. capsulatus* enzyme does not undergo conversion to the oxidase form.[123] *R. capsulatus* XDH has at least six times the specific activity of the bovine enzyme.[70]

The crystal structure of *R. capsulatus* XDH at a resolution of 2.7 Å has been reported,[124] being solved by molecular replacement using bovine milk XDH (pdb code 1FO4 [40]) as a search model. The butterfly-shaped heterotetramer is composed of two functionally independent αβ heterodimers related by a two-fold axis of symmetry. The N-terminal Fe/S-containing domain of XdhA (residues a1 to a153) contains two [2Fe2S] clusters and is attached to the C-terminal FAD-binding domain (residues a185 to a462) *via* a 31 amino acid extended peptide linker. XdhB (residues b1 to b777) is the Moco-binding subunit and the site of dimerization between heterodimers. It also contains the binding site for the substrate from which two electrons are transferred to the oxidized Mo(VI). The two [2Fe–2S] clusters constitute a pathway for electron transfer to the FAD, the site of NAD^+ reduction. The two active sites of the heterotetramer are 52 Å (Mo–Mo distance) apart with no other cofactors in between, suggesting two independent catalytic units. No electron density is observed for residues a167 to a179 of the linker between the two domains of XdhA and for residues b382 to b397 of the Moco-binding subunit XdhB.

The FAD domain present in the *R. capsulatus* XdhA subunit shares only 22% sequence identity to the FAD domain of bovine XDH, but is structurally very similar.[40] The FAD molecule is bound deeply within the enzyme in an extended conformation. It has been shown that proteolytic cleavage or cysteine modification induces changes in the FAD active site of the bovine enzyme that prevents the reaction with NAD^+, but not with dioxygen[125] and results in conversion of the dehydrogenase form (XDH) into the oxidase form (XO). In contrast, *R. capsulatus* XDH has not been observed to lose its ability to react with NAD^+.[70] Inspection of the sequence and structure of *R. capsulatus* XDH showed that the elements that trigger the D/O conversion in the bovine enzyme are not present in the bacterial enzyme. In addition, the cysteines that are oxidized to convert XDH to XO are not present in *R. capsulatus* XDH (C535 is replaced by P_A462, the last residue of XdhA, while C992 is replaced by L_B421). The proteolytic sites in bovine XDH are located after L219 and K569 (pancreatin cleavage) or after L551 (tryptic cleavage).[40] Both L569 and L551 are located in the linker connecting the FAD and Moco domains, which in *R. capsulatus* XDH are not connected but rather are separate subunits.

The protein modifications that cause the D/O conversion in the bovine enzyme are distant from the FAD active site. Upon proteolytic cleavage, a 13-residue (a422 to a434) "variable loop" swings in front of the isoalloxazine ring of FAD, changing the electrostatic environment and sterically interfering with NAD^+ binding.[40] It has been shown that the loop conformation observed in *R. capsulatus* XDH superimposes surprisingly well with that of the XDH-form of the bovine enzyme.[124]

In accordance with the crystallographic data, *R. capsulatus* XDH (like avian XDH) is isolated with high reactivity toward NAD^+ and low reactivity toward oxygen as electron acceptor, and does not convert to the oxidase form.[70] The absorption spectrum observed when the enzyme reacts with xanthine under aerobic conditions indicates that it is largely reduced in the steady state, consistent with a lower overall reactivity toward O_2 as compared with NAD^+ as oxidizing substrate.[70] The predominant flavin species is the neutral semiquinone form, as evidenced by the increase in absorption at 620 nm relative to oxidized enzyme, which is typical for the dehydrogenase form of the XORs at pH 7.8. It is known, however, that the dehydrogenase forms of the vertebrate enzymes produce larger amounts of O_2^- than do the oxidase forms.[126,123] During turnover with xanthine and O_2, approximately 38% of the electron flux through the *R. capsulatus* XDH generates O_2^- radicals, which is comparable to the chicken and rat enzymes (producing 40–44% and 34% O_2^- radicals, respectively).[126] Access to the molybdenum center of the *R. capsulatus* enzyme is only possible through a narrow, rectangular opening in XdhB with dimensions of approximately 5 Å (length) × 3 Å (width) and 5 Å (depth), a perfect geometry for small planar molecules like xanthine. The bottom of the pocket is lined by the molybdenum with its oxo ligand and by the side chain carboxylate of E_B730.[48] The rectangular sides of the pocket are lined by charged residues E_B232 and R_B310 on the short edges, and hydrophobic residues F_B344 and F_B459 on the long edges. Further away from the Moco, the pocket opens up to a funnel shaped entrance. The crystal structure of the enzyme in complex with alloxanthine (the oxidized form of the XDH inhibitor allopurinol used for gout treatment) shows that there are numerous interactions between the protein and the inhibitor. Functional groups of the 6-membered ring of alloxanthine interacted with E_B232, R_B310 and, *via* a water molecule, with E_B730. The two charged residues are well conserved among the XDH sequences from different bacterial species. N8 of the 5-membered ring of alloxanthine formed a direct coordination to the molybdenum atom of Moco.

6.3.5 The Mechanism of Substrate Conversion at the XOR Active Site

XOR catalyzes the conversion of hypoxanthine to xanthine, and further the oxidation of xanthine to uric acid, the final two steps in the purine degradation pathway. The enzyme also has limited activity towards purine analogs and aldehydes. The oxidized form of the molybdenum ion is in the VI oxidation state, coordinated by an oxo (=O) ligand, one hydroxo (–OH) ligand and one sulfido (=S) ligand, in addition to the two sulfur ligands originating from the pterin dithiolene group. Crystallographic analysis of fully active bovine milk XDH has shown that the double-bonded oxygen ligand occupies the apical position in the square-pyramidal metal coordination sphere, whereas the other four ligands lie in the equatorial plane.[40,127]

The active site of bovine XOR contains amino acid residues which are highly conserved in all eukaryotic and prokaryotic XORs: Phe 914, Phe 1009,

Glu 802, Glu 1261, Gln 767 and Arg 880,[40] and the *R. capsulatus* enzyme provides a convenient system in which to conduct mutational studies. Due to the differing composition of subunits, the numbering in the bacterial enzyme is as follows: Phe 344, Phe 459, Glu 232, Glu 730, Gln 197 and Arg 310.[48] Based upon crystallographic data, substrate intercalates between the phenylalanines, which serve to orient the substrate for catalysis. Glutamate 1261 (730 in the bacterial enzyme) lies beneath the molybdenum center.

Reaction mechanisms have been suggested for both the bovine XOR and *R. capsulatus* XDH, the two XOR enzymes of this enzyme family.[15] For bovine XOR it has been shown that substrate hydroxylation is initiated by a base-catalyzed mechanism in which the side chain of a glutamic acid Glu1261 in the bovine sequence deprotonates the Mo–OH group.[19] The oxyanion then attacks the electrophilic C8 carbon to yield the hydroxylated product. The deprotonated Glu1261 is stabilized by a hydrogen bond to the N-1 atom of the substrate.[32] The Mo=S ligand subsequently accepts a hydride ion from the reactive carbon atom to produce Mo–SH and a reduced Mo^{IV} center. An important contribution of this Glu residue to catalysis has been confirmed by mutating Glu1261 in the human enzyme with the resulting proteins being devoid of any hydroxylation activity.[128] This general mechanism is also consistent with the studies of the *R. capsulatus* enzyme by site-directed mutagenesis study (see below).[128,129] It has been suggested that proton transfer from Mo–OH to the Glu is induced upon binding of the substrate as shown by theoretical studies (Figure 6.5).[130]

In contrast to the well-accepted role of Glu1261, that for Glu802 (232 in the bacterial enzyme) and Arg880 (residues 310 in the bacterial enzyme) remains in dispute. Two contradictory models have been proposed for the binding mode of the substrate xanthine and its activation mechanism by XOR as illustrated in Figure 6.5C and D. The difference of two models based variously on experimental results,[128,131] theoretical studies[132,133] and several review papers are published,[13,20] but the problem seems to be more complex than originally assumed and remains to be clarified. The points of discussion are complex and can be summarized as follows:

(a) How to resolve the contradictory crystal structures of various substrate binding modes with various compounds under various conditions: the resolution of these X-ray structures, clean and reproducible electron density models of the X-ray structures in addition to appropriate forms of enzyme. *e.g.* desulfo-form,[13,134,135] demolybdo-form[136] and possible intermediate forms of reactions.[32,136] (b) How to explain the activation energy of substrates by interaction with the amino acid residues in the conserved amino acid residues in the active site cavity, hydrogen-bond network[128,132,133] or ionic interaction.[131,137] (c) How to explain the metabolic sequence, *i.e.* hydroxylation at the 2-position of hypoxanthine precedes that at the 8-position.[13,16,135] (d) Ascertaining whether Glu802 is protonated or deprotonated under the condition of substrates in addition to the inhibitors that is important for the binding mode substrate and the role of it.[32,71,127,138] (e) Establishing the factors involved in product release (the

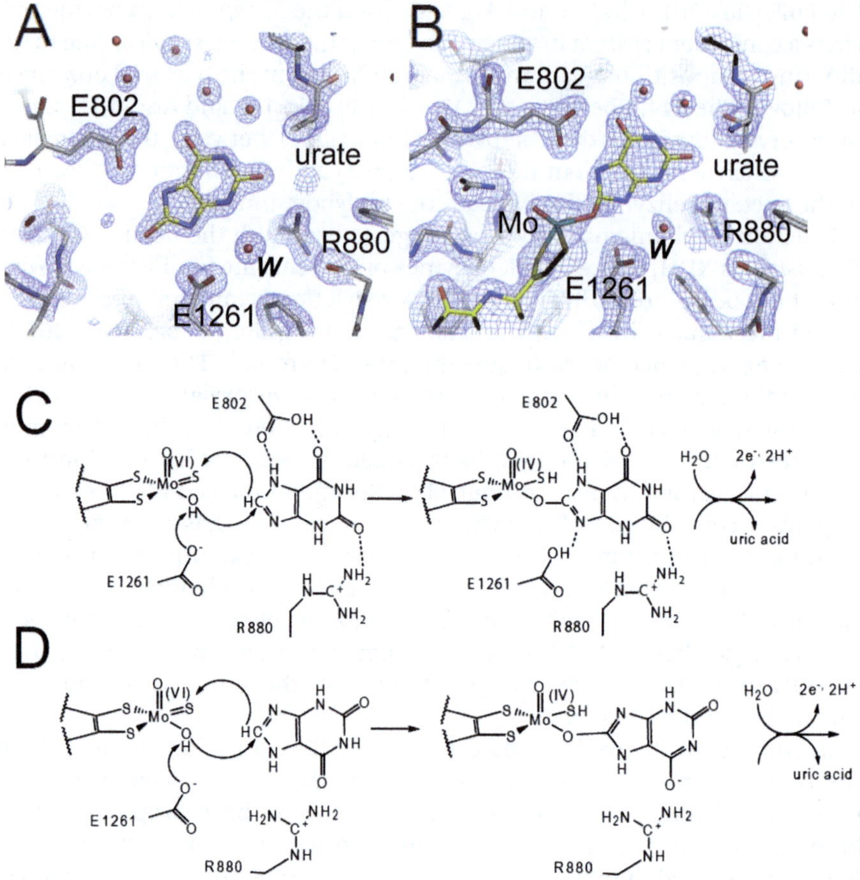

Figure 6.5 Binding mode of urate to the XOR. (A) Structure of urate bound form to demolybdo rat XOR (3AN1). (B) Structure of reduced (MoIV) of bovine XOR with urate, forming Mo–O–C bond (3AM2). (C and D) Reaction mechanism of XOR and binding modes of the substrate xanthine proposed by Yamaguchi *et al.* (C) and Pauff *et al.* (D). Amino acid numbers are shown as those of bovine XOR. *W* indicates water molecule near the urate 6,7 positions.

rate-limiting step for the reaction with most substrates).[71,135] These points have been discussed based on the mutation studies,[128,137,139] kinetic and structural studies of various forms in addition to the theoretical studies such as density functional theory[132] or QM/MM studies.[133,140] Instead of the firstly argued problem of the preference of the activation energy in the two binding modes,[137] a recent paper has argued that in order for glutamate 802 to facilitate tautomerization, the residue must be deprotonated regardless of other compound bound forms of XOR.[71] In order to further investigate the role of Glu 802 and the effect of its protonation, kinetic studies on the reductive half-reaction of E232Q variant (corresponding to E802 in bovine

XOR) of the recombinant *R. capsulatus* enzyme have been conducted.[71] The studies showed that the E232Q mutant exhibited a 12-fold decrease in k_{red}, comparable to the effect that was observed in previous published kinetic studies of the E232A variant. On the other hand, the decrease of k_{red} in this variant was predicted alternatively by QM/MM studies[140] and it has been argued that the role of such proton exchange can be replaced by the water molecule that exists near C_3 and C_9 positions at hydrogen distances in urate binding structure.[136] The question can be raised as to why the rate of product release from the E232Q variant of the *R. capsulatus* enzyme[71] is similar to that of native bovine XOR under similar conditions in case the tautomerization of the product urate is only due to Glu232. The failure to form xanthine from urate of the *R. capsulatus* enzyme under anaerobic conditions in the presence of NADH is not surprising, because the extent of formation of the Mo(IV)–urate complex is expected to be extremely low due to much higher rate of release of urate in *R. capsulatus* enzyme[71] and clear but limited occupancy of Mo(IV)-*O*-xanthine is low due to its instability even in bovine XDH (~60%) as described.[136] Since the rate of product release (and also the behavior for the structure based inhibitors) is quite different between the *R. capsulatus* and bovine enzymes, it is not surprising that no formation of xanthine is observed in the former case. This difference of urate binding seems to come from the molecular dynamics of the proteins.[32] Further studies are necessary in the near future including whether the crystal structure of the urate bound form to the reduced Mo(IV), as illustrated in Figure 6.5B, is indeed the intermediate structure of the hydroxylation of xanthine.[136]

6.3.6 Medical Relevance of XOR

Xanthinuria, elevated blood levels of xanthine, is a disease in humans involving abnormalities of XDH.[142] It presents itself as three subtypes: classical xanthinuria type I, which is caused by XDH deficiency alone; classical xanthinuria type II, which represents both XDH and aldehyde oxidase deficiencies;[143] and Moco deficiency, leading to the concomitant loss of function of all molybdoenzymes. It is known that in xanthinuria II the enzyme responsible for attaching the terminal sulfur ligand at the molybdenum-center of the two enzymes has been mutated. In contrast, in type I classical xanthinuria the defect is in the structural gene for XDH and thus only XDH activity is absent.[142] Xanthinuria I and II both result in the failure to convert hypoxanthine and xanthine to urate. Classical xanthinuria is often benign and may not be diagnosed, although some patients develop kidney malfunction, with fatality in extreme cases. The two types of classical xanthinuria can be distinguished by the ability of type I but not type II subjects to convert allopurinol to oxipurinol and 1-methylnicotinamide to the 2- and 4-pyridones, the latter of which is a substrate exclusively to AO.[142] Since allopurinol inhibits XDH activity,[144] but not AO activity,[145] allopurinol (4-OH-pyrazolo-pyrimidine) has been widely used for the clinical control of uric acid production in conditions of hyperuricaemia (Figure 6.6). Because XOR is the major source of urate, and elevated levels of urate in the blood can lead to gout and compromised kidney function, the enzyme is the primary target for therapeutic intervention.[146] Inhibitors of XOR are effective and generally well-tolerated drugs for the treatment of gout. Some inhibitors inhibit the enzyme by forming a stable reaction intermediate or analog of this intermediate with reduced molybdenum. Allopurinol, which has been used as an antigout drug for over 40 years, is an isomer of hypoxanthine in which the 8-position carbon atom is replaced with the adjacent 7-position nitrogen atom. It was initially thought to work as a simple competitive inhibitor that binds to the molybdenum center competitively with respect to xanthine, but subsequently found to inhibit in a more complicated fashion; the inhibition progresses time-dependently eventually leading to formation of a tightly bound complex of inhibitor with the reduced enzyme (Figure 6.6). Allopurinol itself is in fact a substrate of XO and is transformed to oxipurinol (alloxanthine; 4,6-dihydroxypyrazolopyrimidine), which forms a tight complex with the reduced molybdenum center.[144] In the crystal structures of the allopurinol-inhibited form with *R. capsulatus* XDH at 3 Å resolution and the oxipurinol-bound form of reduced bovine XO, the equatorial oxygen of Mo(IV) is replaced by the nitrogen atom at the 2-position (corresponding to the 8-position of the purine base) of oxipurinol. The formation of this tight oxipurinol–molybdenum complex is the basis for the potent inhibitory effect of allopurinol, though the oxipurinol eventually dissociates upon reoxidation of the enzyme in air ($t_{1/2}$ = 300 min at 25 °C).[147] Until recently, allopurinol was the most effective inhibitor to treat hyperuricaemia clinically.[148] However, a novel structure-based and highly effective inhibitor of XOR with the generic name febuxostat has been approved for the

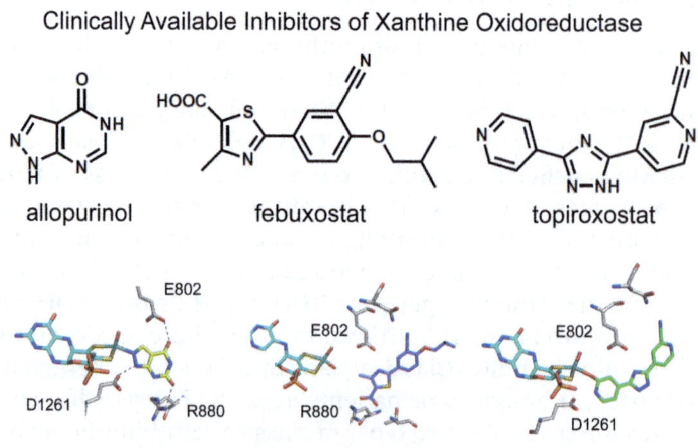

Figure 6.6 The chemical structures of inhibitors of XOR and their bound forms in the Moco active site in bovine XDH. Numbers of inhibitors are reported, but only clinically available ones for gout and hyperuricaemia are shown here. Structures of allopurinol (oxipurinol), febuxostat and topiroxostat are obtained from PDB codes 3BDJ, 1N5X and 1V97, respectively.

treatment of gout in EU countries (in 2008), the USA (2009) and subsequently in various other countries, including Japan (2011) (Figure 6.6).[149] Febuxostat has a more potent and longer lasting effect on lowering blood urate levels than allopurinol in mammalian species.[141] Clinical efficacy and tolerance to febuxostat have been confirmed, and the drug is available as Adenuric (EU), Uloric (US) or Feburic (Japan) for the chronic management of hyperuricaemia in patients with gout.[149] Febuxostat is a larger molecule than allopurinol, and the binding mechanism to XOR is quite different. Febuxostat fills most of the solvent access channel cavity and substrate binding pocket of XOR, acting as a structure-based inhibitor *via* multiple interactions, including ionic bonding of its carboxyl group with Arg880, hydrogen bonding of the nitrogen atom of the thiazole with Glu802, and sandwiching of the thiazole ring between Phe914 and Phe1009 in bXOR (Figure 6.6).[150]

6.4 Aldehyde Oxidases

Aldehyde oxidases (AOX) and XOR are very closely related enzymes sharing a similar overall protein architecture and a high amino acid sequence and structural identity and similar kinetic properties.[151] It has been shown that the two enzymes are evolutionary conserved and originated through at least two asynchronous gene duplication events.[152] Despite their structural similarities and ability to oxidize many of the same substrates, AOX and XOR vary in their substrate specificities and their response to inhibitors. XOR and AOX generally catalyze oxidation reactions that involve nucleophilic attack of electrophilic carbon centers, including aldehyde groups and carbons of N-heterocycles that are para to or adjacent to the ring nitrogen (Figure 6.7),[153–155] these normally being the most electropositive (*e.g.* in pyrazine, purine, pyrimidine, quinoline, pyridine and pteridine). While the physiologic role of XOR is the conversion of purines xanthine and hypoxanthine to uric acid, AOX is incapable of catalyzing this reaction.[156,157] AOX on the other hand converts *N*1-methylnicotinamide to its pyridone derivative, which is not a good substrate of XOR.[158]

AOX in general has a broader substrate spectrum than XO and oxidizes different types of aldehydes from their corresponding acids or is involved in the hydroxylation of N-heterocyclic ring systems possessing –CH=N– moiety (*e.g.* phthalazine) and purine or aromatic and non-aromatic charged heterocycles with a $-CH=N^+$ group (*e.g.* *N*1-methylnicotinamide and *N*-methylphthalazinium) (Figure 6.7). Thus, the term "aldehyde oxidase" might not always be appropriate, since these enzymes do not necessarily act on substrates containing an aldehyde functionality.[159,160]

AOX enzymes are present in both prokaryotes and eukaryotes. The mammalian AOX enzymes that have been examined to date are exclusively oxidases; NAD^+ is not used as a substrate and no D/O conversion occurs in these enzymes.[41] Thus, AOX enzymes use molecular oxygen exclusively as oxidizing substrate, producing hydrogen peroxide and superoxide. At present, little is known about the physiological relevance of AOX in humans and other

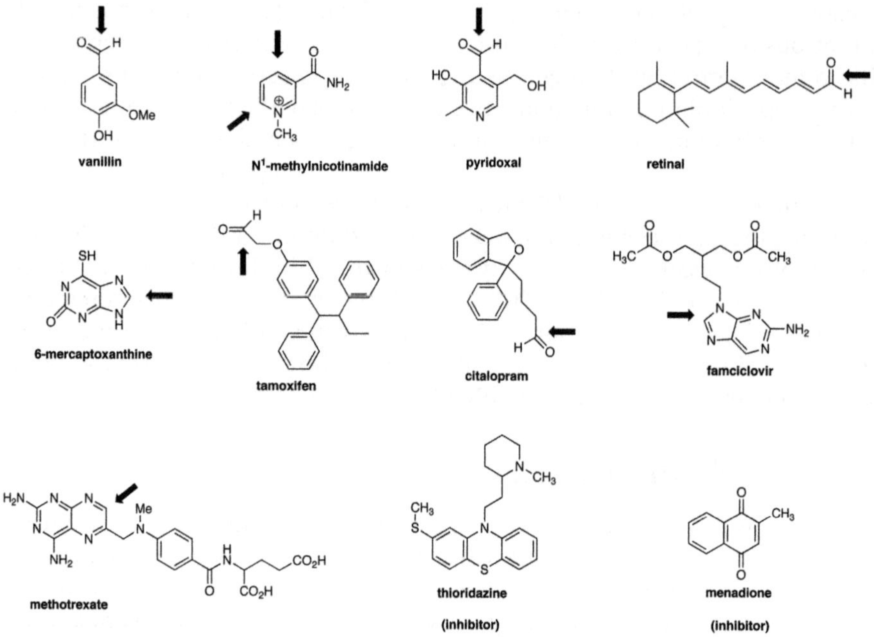

Figure 6.7 Aldehyde-containing substrates, heterocycle-containing substrates and inhibitors of AOX enzymes. The position of substrate oxidation/hydroxylation is shown by the black arrow.

mammals, although these enzymes are known to play a role in the metabolism of drugs and compounds of toxicological relevance in the liver. AOX was described to be important in the metabolism of xenobiotics and endogenous compounds.[161–163] Two notable endogenous substrates for AOX are retinaldehyde and pyridoxal. Retinaldehyde is the principal component of visual pigments and AOX has been suggested to be an important player in the overall visual process since it catalyzes the biotransformation of this aldehyde to its corresponding carboxylic acid, retinoic acid, which is the active form of vitamin A (Figure 6.7).[164] AOX enzymes were also described to be present in plants, where they catalyze the final steps in the biosynthesis of the important plant hormones abcissic acid and indole-3-acetic acid.[165] Since none of the plant enzymes were studied in detail with respect to their catalytic mechanism, this review focuses on the better studied mammalian aldehyde oxidases.

6.4.1 The Mammalian Aldehyde Oxidases

6.4.1.1 Gene Structure and Expression

Different animal species are characterized by a different complement of *AOX* genes.[152,166,167] While humans contain only a single active gene, marsupials and rodents are characterized by four genes encoding AOX with different substrate specificities, and the genes are clustered in a short distance on the same chromosome (Figure 6.8).

The first hint of the existence of multiple AOX enzymes in certain animal species came from an early analysis of the limited number of mouse expressed sequence tags (ESTs) available in the NCBI database,[152] as first identified in the group of Enrico Garattini (Milano). This has allowed them to identify and isolate two other mouse cDNAs encoding catalytically active proteins highly related to bovine, mouse and rat AOX1, which had been characterized before.[168–172] In total, rodents contain four *AOX* genes, mapping with the original AOX1 orthologous genes to a small region of 350 kilobases on mouse chromosome 1 and rat chromosome 9.[152,166,167] The genes have been designated *Aox1*, *Aox3*, *Aox4* and *Aox2*, with *Aox1* encoding an enzyme most closely homologous to AOXs in other mammalian species, like humans (Figure 6.8). The evolutionary history of the *AOX* gene cluster in mammals is characterized by a sudden and species-specific shift from multiplication to suppression/deletion. *Bos taurus* seems to have maintained three active *AOX* genes (*AOX1*, *AOX4* and *AOX2*) on chromosome 2. In humans, only one functional *AOX* gene is present on chromosome 2q, namely *AOX1*. *AOX3* has been deleted altogether and *AOX4* and *AOX2* transformed into inactive pseudogenes (Figure 6.8). The AOX1 gene contains 35 conserved coding exons. Human AOX1 and mouse AOX1 and AOX3 have an overlapping tissue distribution and both enzymes are found predominantly in the liver and the lung. In contrast, the expression of mouse AOX4 and AOX2 is much more restricted. The richest source of AOX4 is the Harderian gland in the orbit of the eye, a gland not present in humans or primates.[166,167,173] The Harderian gland is involved in diverse homeostatic functions, including thermoregulation, lubrication of the eye surface and regulation of the circadian rhythm. A

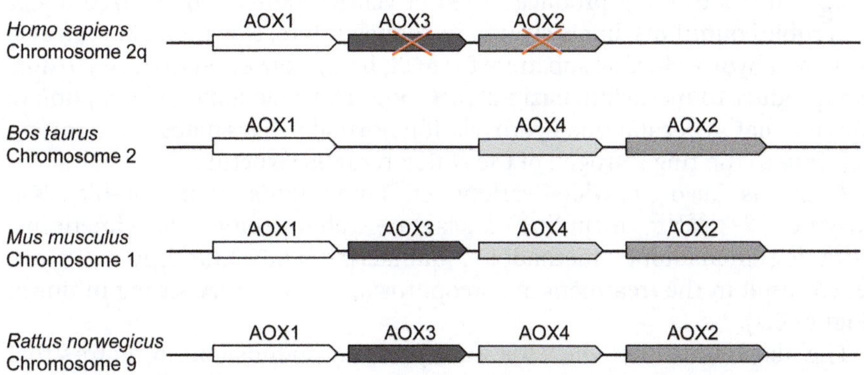

Figure 6.8 The organization of aldehyde oxidase genes in vertebrates. A schematic representation of gene organization of aldehyde oxidase genes and pseudogenes present in the genomes of *Homo sapiens*, *Bos taurus*, *Mus musculus* and *Rattus norwegicus*. Orthologous genes are indicated by boxes showing the same shading in gray. Pseudogenes are crossed through. The boxes are not drawn to scale.

second rich source of AOX4 is the epidermal layer of the skin. AOX2, on the other hand, is restricted to the secretory cells of the Bowman's gland, the most important exocrine gland of the olfactory mucosa.[164]

6.4.1.2 Substrates of Mammalian Aldehyde Oxidases

Until now, little has been known about the physiological function of AOX in mammals. The better known function of mammalian AOX enzymes involves their role as drug-metabolizing enzymes. AOX is involved in the phase I metabolism of numerous compounds of both medical and toxicological relevance, potentially acting in concert with the microsomal P450 system,[154] and AOX enzymes have an important role in the metabolism of the antitumor and immunosuppressive agents methotrextate, 6-mercaptopurine and azathioprine (Figure 6.7).[174–176] Further, AOX enzymes metabolize the antimalaria agent quinine and the antiviral drug famciclovir.[177,178] The prominent drug tamoxifen (Figure 6.7), used in the treatment of breast cancer, as well as the antidepressant citalopram are transformed into aldehydic intermediates by AOX.[162] Drugs metabolized by AOX1 are predominantly hydroxylated on their N-heterocyclic ring system.

The physiological role of AOX has been suggested to be involved in the metabolism of vitamin B_3 (nicotinamide), B_6 (pyridoxal phosphate) and A (retinol),[167] and retinaldehyde has been suggested to be one of the best substrates for mouse mAOX1, mAOX3, mAOX4 and mAOX2. Other enzymes that perform the same reaction are the three cofactorless aldehyde dehydrogenases ALDH1A1, ALDH1A2 and ALDH1A3, which have comparable K_M and V_{max} values. Vanillin is an example of a primary xenobiotic containing an aldehyde group that is rapidly transformed into vanillic acid by the AOX enzymes. Phenazine methosulfate (PMS) serves as an electron acceptor of AOX.[179] It is a dye that produces a red or yellow color when oxidized under anaerobic conditions, but is clear in its reduced state. However, PMS can also serve as a hydroxylatable substrate for AOX, being converted to the red oxidation product 10-methylphenazin-2(10*H*)-one. This reaction is an exception to the rule that AOX-catalyzed hydroxylation normally takes place at the carbon adjacent to the ring nitrogen of the N-heterocyclic structure.

There is also a wide variety of compounds that inhibit AOX enzymes,[161,163,180,181] including amsacrine, chlorpromazine, cimetidine, estradiol, menadione, methadone, quinacrine and thioridazine. RaloxifeneA, used in the treatment of osteoporosis, is also a very strong inhibitor (Figure 6.7).

The above illustrates the wide variety of compounds and inhibitors that act on AOX enzymes. The physiological substrate(s) remain to be definitively identified, and it has been suggested that AOX1 does not have an essential role for humans since individuals with a genetic lesion in the molybdenum cofactor sulfurase that is required by both XOR and AOX have relatively minor clinical symptoms and no evident developmental abnormalities.[182] This also might indicate that AOX does not play an essential role in the biosynthesis of retinoic acid. Interestingly, it has recently been shown that human AOX1 expression in the adipose tissue may be involved in fat deposition.[183] It has been shown that human AOX interacts with the ABCA-1 transporter involved in lipid uptake in the liver, suggesting a potential role of AOX1 in lipid metabolism.[184]

6.4.1.3 *Single Nucleotide Polymorphisms*

A major source of inter-individual variability in the human population is allelic variants of the human AOX1 gene containing single nucleotide exchanges, resulting in single nucleotide polymorphisms (SNPs) encoding proteins with increased or decreased catalytic activity.[185,186] Missense single nucleotide polymorphisms in the coding region of the hAOX1 gene may affect the catalytic function of the enzyme in either a positive or a negative fashion. Numerous SNPs of the human AOX1 gene are available in the NCBI dbSNP database (http://www.ncbi.nlm.nih.gov/snp). Allelic frequency data are available from a cohort of 180 volunteers representative of the Italian population. In this population, one nonsense mutation, one synonymous and five non-synonymous mutations were identified.[185] The most frequent missense mutation identified is a G/A substitution resulting in the R1297K amino acid exchange. Another relatively frequent SNP is a T/C substitution resulting in the amino acid exchange L1271S and a G/A substitution, which results in the amino acid exchange N1135S. Two other substitutions are rather rare and result in the amino acid exchanges R802C and R921H. These AOX1 variants resulting from SNPs have been purified from a heterologous expression system in *E. coli* and characterized biochemically. This work has shown that the amino acid exchanges R802C and R921H result in proteins with reduced catalytic activities, and the individuals bearing these alleles may be compromised in their ability to oxidize aldehydes. The N1135S and H1297R amino acid exchanges result in protein variants with increased catalytic activities, and these individuals might be classified as fast metabolizers. Further studies are necessary to analyze the influence on the other SNPs available in the database on enzyme activity.

6.4.1.4 *The Crystal Structure of mAOX3*

The first crystal structure of a mammalian AOX has been reported for mAOX3, some 12 years after the first crystal structure for bovine XOR (Figure 6.9). As expected from the high amino acid sequence identity of 50% shared by the two enzymes, the overall topologies of mouse AOX3 and bovine XORs are very similar.[41] The homodimeric mAOX3 can be divided into three major domains, which are involved in the binding of each cofactor type. The small N-terminal domain (domain I: 20 kDa, 169 amino acids) harbors the two FeS clusters, designated FeSI and FeSII, according to their distinct EPR signals.[187,188] The domain II (40 kDa, 302 amino acids) binds the FAD and the isoalloxazine ring of FAD is stacked between two hydrophobic side chains:

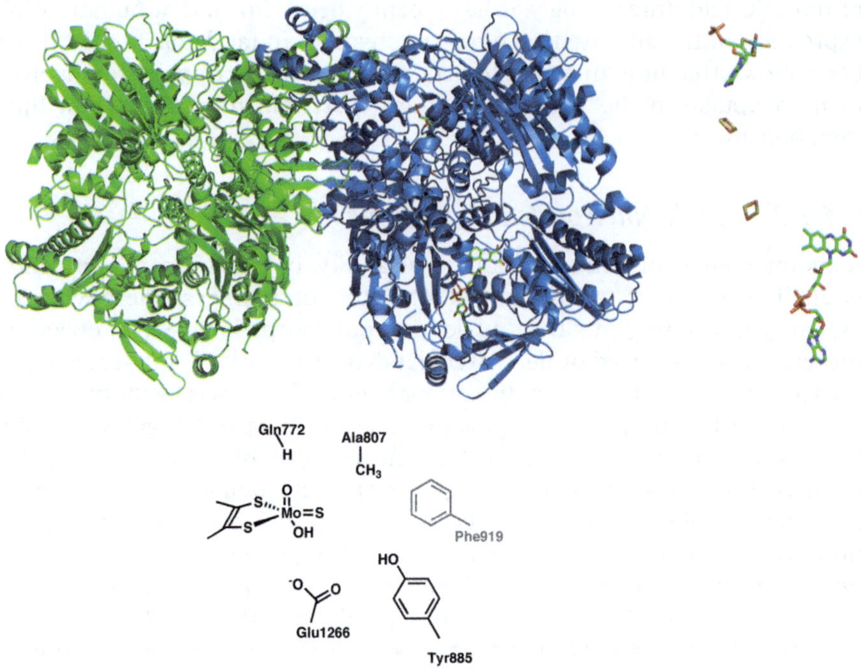

Figure 6.9 (AOX3): overall structure, cofactor arrangement and active site composition of mouse mAOX3 (3ZYV). Shown is the active-site structure of Moco and the relevant residues (mAOX3 numbering) in the active site proposed to be involved in catalysis and substrate binding. The Gln772 and Glu1266 are highly conserved in the enzymes of the xanthine oxidase family.

Leu344 (Phe in XOR) and Phe438 (Ile in XOR). Domain I and domain II are connected by a poorly conserved linker region (61 amino acids).[41] The C-terminal and largest domain III (90 kDa, 559 amino acids) contains the Moco binding site, with the Mo catalytic center located at the bottom of a ~15 Å deep and wide pocket, directly leading to the catalytic center. Domain III also contains the dimerization binding sites. The protein cofactor disposition is consistent with the expected electron transfer pathway: electrons released from the hydroxylation reaction at the Mo catalytic center are transferred to the two FeS clusters and the FAD in sequence. Subsequently, electrons are transferred to the final acceptor, O_2.

As in all members of the molybdo-flavoenzyme XOR family, the active site molybdenum of mAOX3 adopts a distorted square pyramidal coordination geometry, with the two sulfur atoms from the dithiolene in the equatorial plane. The two other equatorial positions include the catalytically essential sulfur atom (=S) and the equatorial –OH group that is the source of the transferred oxygen in substrate oxidation. An oxygen atom (=O) occupies the apical position. Like in the structure of bXOR, the highly conserved residues are also present in mAOX3: Gln772, Glu1266, Phe919 and Phe1014

(Figure 6.9). It has been suggested for XOR that the two conserved Phe residues (Phe919 and Phe1014) are responsible for the orientation of substrate and inhibitor molecules through stacking interactions.[189] As in bXOR, Gln772 of AOX3 makes a hydrogen bond with the apical oxygen ligand and Glu1266, suggested to act as active site base in abstracting a hydrogen from the Mo–OH group, is in contact with this ligand.[188] Other key residues located in the catalytic core are not conserved between AOX3 and XOR, and are expected to be responsible for most of the differences in substrate binding, specificity and catalysis.

Homology modeling and molecular dynamics simulations of the four mouse AOX isoforms have shown that the major differences of the four proteins are in the substrate-binding site region.[173] These studies predict that mAOX1 can accept a broader range of substrates of different shape, size and chemical nature than the other isoforms,[168,190] with mAOX4 seeming to have the most restricted range of substrates. Consistent with the tissue distribution and evolutionary origin, overlapping substrate specificities are predicted for the pairs mAOX2/mAOX4 and mAOX3/mAOX1.[164,171] The mAOX1 isoform possesses a variety of polar and charged amino acids in its active site, consistent with its predicted broader substrate specificity. In contrast, mAOX4 has the narrowest specificity region and its active site is mainly composed of hydrophobic residues, suggesting that it is limited to action on small, linear and hydrophobic substrates. The mAOX3 and mAOX2 isoforms are very similar but with the presence of more hydrophobic amino acid residues in the mAOX2 form, suggesting that the substrate specificity of mAOX2 should be closer to the one from mAOX4 and the mAOX3 more similar to mAOX1.[173,191]

6.4.1.5 Catalytic Mechanism

Heterologous expression systems for mammalian AOX enzymes in *E. coli* have been developed, which have allowed site directed mutagenesis approaches to study the role of distinct amino acid residues in the active site. So far, mAOX1, mAOX3 and hAOX1 have been expressed successfully in *E. coli* in an active form.[185,187,190] In both mAOX1 and mAOX3, the active site glutamate has proved to be essential for catalysis, since exchange to glutamine results in inactive enzyme (E1265Q in mAOX1 and E1266Q in mAOX3).[188,190] However, for the mAOX3-E1266Q variant a residual activity with benzaldehyde is observed (60% reduction of activity in comparison to the wild-type enzyme), while the enzyme is inactive with N-heterocyclic compounds.[188] The residual activity might be explained by the higher electrophilicity of the carbonyl carbon atom in aldehydes, as compared to N-heterocyclic compounds, which are substrates of XOR.

The residues suggested to be involved in substrate binding and orientation in bXOR, E802 and R880 are replaced by a valine and a methionine, respectively, in all AOX1 enzymes.[190] In mAOX3, these two residues are A807 and Y885. For human XOR and *R. capsulatus* XDH it was shown before that exchanges of the glutamate and the arginine to valine and methionine, respectively, as found in AOX1, resulted in loss of activity towards purines as substrates and gain of activity towards aldehydes, showing that indeed these residues contribute to substrate binding and determine the substrate specificity in the active site.[128] However, the reverse exchange of Val 806 to Glu and Met 884 to Arg in mouse AOX1 also abolished the activity with both purines and aldehydes as substrate.[190] However, since the activity of mAOX1 was generally very low, the complete loss of activity might have been the result of the restricted detection limit of the steady-state kinetic determination.

To study the influence of the residues involved in substrate binding at the active site of mAOX3, Ala807 has been exchanged to valine and Tyr885 to methionine, the two conserved residues in all AOX1 enzymes.[187] The A807V mutation does not affect the kinetic constants with smaller substrates like benzaldehyde or phthalazine, but the affinity for bulkier substrates like phenanthridine is decreased and the catalytic efficiency slightly raised. Upon conversion of Tyr885 to methionine the kinetic constants also remain largely the same, with small hydrophobic substrates like benzaldehyde and phthalazine. These substrates are converted by the variant with unchanged rate constants from wild-type enzyme and only a minor increase in K_M (possibly due to the smaller hydrophobic pocket in the Y885M variant). On the other hand, bulkier substrates such as phenanthridine and more charged substrates like *N*1-methylnicotinamide are converted with higher efficiency by the variant. This may be due to the greater flexibility of the methionine side chain in comparison to tyrosine, which may facilitate binding of these substrates. The effects are much more pronounced in the double variants.

In a further study the mode of substrate selectivity and the basis of catalytic activity in mAOX3, a variant containing numerous amino acid substitutions in and near the active site to residues found in XOR and other AOX isoforms, has been prepared.[188] The variant mAOX3-F776K/A807E/D878L/L881S/Y885R/P1015T/Y1019L shows drastically decreased activity towards all substrates tested. Additional introduction of K889H substitution rescues AOX activity and results in XOR-like activity with hypoxanthine as substrate. This illustrates the importance of Lys889 in the active site by retaining the interaction between E1266 and the hydroxyl group of the molybdenum atom. Lys889 is conserved in all AOX enzymes but replaced by His in XORs (His884). Lys889 lies ~10 Å away from the Mo center and ~6 Å away from Glu1266. This lysine adopts a position similar to the conserved Arg880 of XOR. Molecular dynamics simulations demonstrated that substrate docking into the active site causes Lys889 to move from its original position, establishing new interactions with Glu1266 and/or the substrate itself. Consequently, the nucleophilic attack of the Mo–OH moiety or the stabilization of the transition state is affected. Additionally, it is concluded that bulkier substrates or substrates unable to interact with Lys889 *via* an oxygen or nitrogen (such as phenanthridine or *N*1-methylnicotinamide) are unable to bind in the histidine variant and consequently not in XORs.

Additionally, structural features along the substrate access funnel may also account for differences in substrate specificities between mAOX3 and

XOR. In AOX3 the funnel is ~20 Å wide at the surface, and becomes tighter towards the active site, where it is approximately 8 Å wide.[41] Several residues along the funnel are divergent in the two proteins, particularly: Arg717, Asp878, Glu880, Leu881 and Thr1081 of AOX3 are replaced by Leu712, Leu873, His875, Ser876 and Pro1076 in XOR. Comparison of the molecular surfaces of mAOX3 and bXOR indicates that the shape and width of the substrate access funnel is more anionic and wider in the case of mAOX3. Taken together this implies greater access of larger and bulkier substrates for AOXs enzymes in comparison to XORs.

The presence of some highly conserved residues at the catalytic center (Glu1266, Phe919, Gln 772, in mAOX3 numbering), as well as the type of reaction catalyzed, suggests that the reaction mechanism for both AOXs and XORs is similar, though not identical, since AOXs exhibit much broader substrate specificity.[41,173,188] The reaction mechanism that is proposed for AOXs starts by the nucleophilic attack of the activated Mo–OH ligand (activated by Glu1266) on the carbon atom of the substrate (carbon atom adjacent to an aromatic nitrogen atom) (i). Concomitantly, hydride transfer occurs to the sulfido ligand and an intermediate species is formed. This intermediate is stabilized by hydrogen bonding interactions with residues of the active site, namely Glu1266 and Lys889. This concerted mechanism is favored by experimental evidence with substituted N-heterocycles as well as by recent DFT calculations.[192] In step (ii), the product is released from the reduced Mo site and a water molecule replenishes the vacant coordination position. The reaction cycle is completed (iii) once Mo is re-oxidized and the two reducing equivalents are transferred to molecular oxygen *via* the two [2Fe2S] and the FAD cofactor. It is expected that the proton transfer from the hydroxyl ligand of Mo to Glu1266 disrupts the interaction of Glu1266 with Lys889. This forces Lys889 to move away from Glu1266 and the Mo center where it can interact with the product before the latter is released from the active site.

6.4.1.6 The FAD Site

Surprisingly, while mAOX3 (like all eukaryotic AOX enzymes) is an obligatory oxidase that cannot use NAD^+ as oxidizing substrate, its FAD-binding configuration resembles that of the dehydrogenase rather than oxidase form of the bovine enzyme.[41] The reversible D/O conversion involves the formation of a new disulfide bond, between two cysteine residues (Cys535 and Cys992), which are not conserved in mAOX3 and are substituted by Tyr542 and Phe997, respectively. As discussed above, both reversible and irreversible inter-conversion of XDH into XO requires the structural rearrangement of an 11-residue loop (Gln423-Lys433 in bovine XOR[40]), which is in close proximity to the FAD cofactor. The loop movement leads to a change in the electrostatic environment around the FAD cofactor[78] and, in the oxidase configuration, blocks access of NAD^+ to its binding site. As a consequence, the XO form does not bind NAD^+ and uses oxygen as the final electron acceptor instead. In mAOX3 the final electron acceptor is always oxygen, and it was anticipated that the structure of the corresponding "variable loop" would be similar to the one observed in the XO form of XOR.[40] In contrast, the mouse AOX3 loop turned out to be totally superimposable with the one present in the XDH form. The 11 residues that comprise the loop are conserved in bovine and human XOR but not in AOXs. Of the 11 residues in the highly charged XOR loop (*Q*^{423}*A*S*R*REDDI*A*K^{433}) only four are conserved in mouse AOX3 (*Q*^{430}*A*P*R*QQNAF*A*T^{440}) or human AOX1. Interestingly, while 6 out of the 11 residues are charged in XOR, only the conserved Arg433 is charged in AOXs. In addition, XOR Arg427, which is substituted by Gln434 in mAOX3, belongs to the cluster of conserved amino acid residues (Arg335, Trp336, Arg427, Phe549, bXO numbering) responsible for the D/O conversion.[16,193] These residues are also not conserved in *Rc*XDH, which cannot be converted to an oxidase form either. Arg427 corresponds to a Gln in AOXs and to an Asp in *Rc*XDH. Trp336 (bXOR numbering) plays a major role in the inter-conversion mechanism and is substituted by a Thr in AOXs and by an Arg in *Rc*XDH. All these structural elements involved in the D/O conversion are absent in the mouse and human AOXs, which might be related to the inability of an inter-conversion mechanism by AOX.

6.4.2 Bacterial Aldehyde Oxidoreductases

6.4.2.1 *The Aldehyde Oxidoreductase MOP from* Desulfovibrio gigas

The aldehyde : ferredoxin oxidoreductase from the sulfate-reducing bacterium *Desulfovibrio gigas* (frequently abbreviated MOP, for molybdenum protein) was the first member of the molybdenum hydroxylase family to be characterized crystallographically (Figure 6.10).[49] MOP is a dimer composed of two identical 907-residue subunits that each contain a molybdenum coordinated by a single molybdopterin cytosine dinucleotide cofactor (MCD), and two different [2Fe–2S] clusters. The observed structural organization in MOP is conserved in all enzymes of the XO family, although MOP lacks the FAD domain found in most other family members. Structural comparisons showed that the FAD domain of XOR is positioned in the region of the MOP sequence corresponding to the connecting segment between the iron–sulfur- and molybdenum-binding portions of the polypeptide.

The active site structure of MOP reveals that the Mo is coordinated by the dithiolene sulfurs of the MCD cofactor, and three non-protein oxygen ligands. In the absence of a sulfido group to the Mo, this state of MOP was first suggested to correspond to an inactive desulfo form. Somewhat surprisingly, based on inhibition studies (*e.g.* cyanide and ethylene glycol) and high-resolution crystal structures, it has been suggested that the *D. gigas* enzyme is indeed active in the absence of the Mo=S group (Figure 6.10).[194] This proposition is under large debate, especially since the activity of XOR enzymes to oxidize aldehydes is strictly dependent on the Mo=S being present, since the accompanied hydride transfer from the substrate to the sulfido-ligand

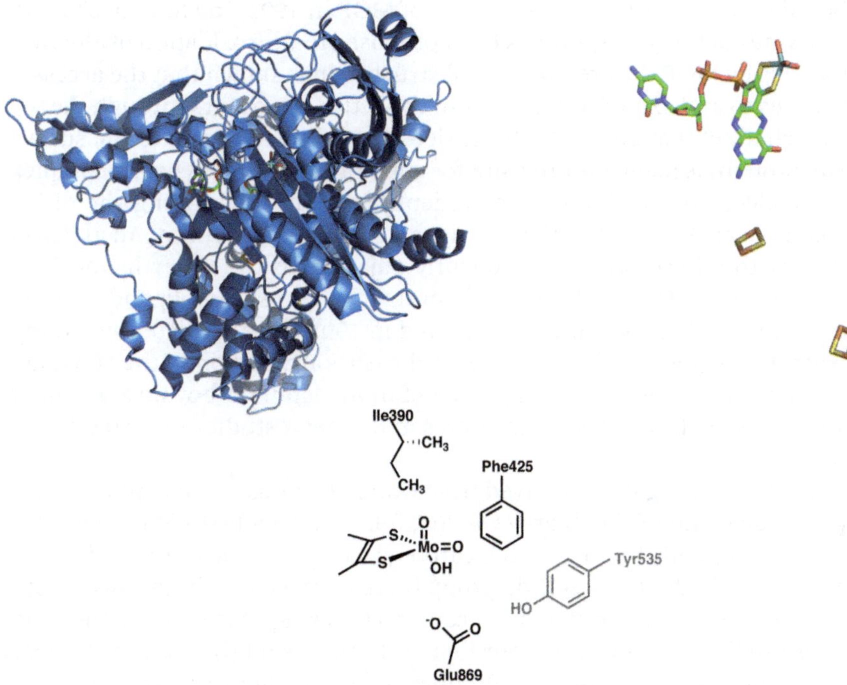

Figure 6.10 (MOP): overall structure, cofactor arrangement and active site composition of *D. gigas* MOP (1SIJ). Shown is the active-site structure of Moco and the relevant residues (MOP numbering) in the active site proposed to be involved in catalysis and substrate binding. MOP was crystallized as a monomer; however, the active protein is expected to be a dimer. The FAD subunit is missing in MOP. The conserved glutamine is exchanged by Ile390 but the Glu869 is present.

might not be possible or is unstable when it is exchanged by an oxygen. It is thus highly controversial whether the desulfo *D. gigas* enzyme is indeed functional, but should this prove to be the case then the active site of the enzymes of the XO family would require a reexamination and conclusive generalizations about the reaction mechanism are not possible.

The two [2Fe–2S] clusters in MOP have been investigated in detail by EPR. These studies have shown that in MOP, the redox-active iron in FeSI is coordinated by Cys100 and Cys139, while that in FeSII is coordinated by Cys40 and Cys45.[195] Similar work, in conjunction with site-directed mutagenesis studies involving replacement of the cysteines ligating the iron ions to serines, have been performed with quinaldine 4-oxidase from *Arthrobacter nitroguajacolicus Rü61a*, demonstrating the same positions of the redox-active irons in the two [2Fe–2S] clusters of this protein.[56] Given the extremely high homology of the iron–sulfur centers in all enzymes of the XO family, the reducible iron sites of FeSI and FeSII are likely to be the same in all enzymes of the XO family.[22]

Based on the early structural studies of MOP in 1995, the first mechanism for enzymes of the XO family has been proposed for the oxidation of aldehyde substrates by MOP.[19] Here, the crystal structure has shown that the access of substrates to and exit of products from the active site occurs through the ~15 Å long channel that connects the buried molybdenum center to the surface of the protein. A likely binding site for substrates in the Michaelis complex, prior to electron transfer, has been identified from the binding site of iso-propanol near the equatorial water ligand of the molybdenum. An aldehyde placed at this location is consequently adjacent to the water ligand (now identified to be –OH) and further from the oxo and sulfido ligands. On the basis of this orientation, it was suggested in 1995 that this water ligand represents the oxygen species that attacks the carbonyl carbon of the substrate. This mechanism has now been proven in more detail for bovine XOR and *R. capsulatus* XDH from crystal structures and kinetic studies (as explained in detail above).

While the –OH ligand is derived from water, it acts as a nucleophile attacking the side chain of the nearby Glu869, followed by a hydride transfer. The hydrogen originally bonded to the carbonyl carbon of the substrate is transferred as a hydride to the sulfido group (here suggested to be an oxo-group), resulting in reduction of the metal center. Following these events, the product carboxylic acid would be coordinated to the metal (in the Mo(IV) state) through the transferred oxygen. Product is then further released. After completion of the reductive half-cycle, the Mo(VI) state would be regenerated through a series of one-electron transfers to the [2Fe2S] clusters of MOP, with the electrons ultimately passed to an external acceptor (the physiological electron acceptor still remains unknown in the case of MOP).

6.4.2.2 *Quinoline 2-Oxidoreductase from* Pseudomonas putida *86*

A number of bacterial strains that can utilize quinoline or quinoline derivatives as a sole source of carbon and energy have been isolated.[57] Quinoline 2-oxidoreductases catalyze the first step in the catabolism of quinoline, introducing a hydroxyl group originating from water into quinoline to yield 2-oxo-1,2-dihydroquinoline, a reaction that involves cleavage of a CH bond. Quinoline 2-oxidoreductase (Qor) from *Pseudomonas putida* 86 is encoded by the *qorMSL* genes and comprises an $(\alpha\beta\gamma)_2$ hexamer with small, medium and large subunits possessing a pair of [2Fe–2S] centers, FAD and a molybdenum center as MCD cofactor, respectively (Figure 6.11).[196,197] The molybdenum ion is coordinated by the two enedithiolate sulfur atoms of MCD, two oxo-ligands and a catalytically crucial equatorial sulfido-ligand. The substrate quinoline can gain access to the active site through a channel that is about 17 Å deep and 9 Å wide.[46] The entrance is located at the N-terminal domain of QorL near the dimer interface. Mostly aromatic residues (*e.g.* F542, F228, W331 and Y370) create a hydrophobic environment for the substrate's two-ring system. Threonine residues (T90, T231, T469, T481 and T482) situated

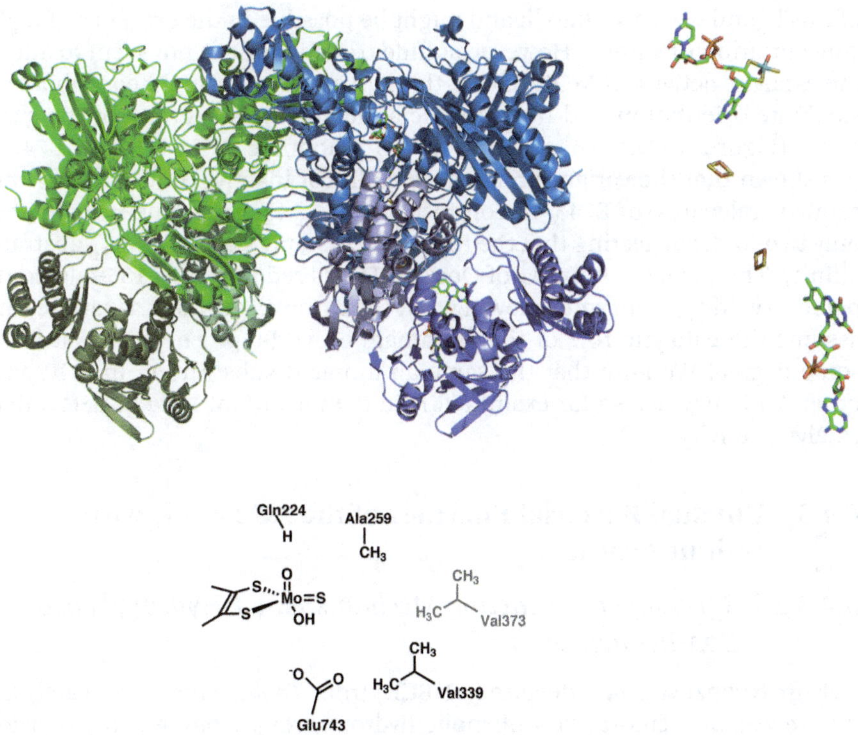

Figure 6.11 (Qor): overall structure, cofactor arrangement and active site composition of *P. putida* quinoline oxidoreductase (Qor) (1T3Q). Qor is composed of QorMLS as an $(\alpha\beta\gamma)_2$ heterotrimer, with QorL shown in dark-blue/green, QorM shown in medium-blue/green and QorS shown in light-blue/green. Shown is the active-site structure of Moco (MCD) and the relevant residues (Qor numbering) in the active site proposed to be involved in catalysis and substrate binding. The Gln224 and Glu743 are highly conserved in the enzymes of the xanthine oxidase family.

near the entrance of the channel may contribute to the substrate entrance/product release. In the case of MOP it was suggested that residues lining the channel, specifically F425, F494, L497 and L626, might change their conformation to permit access of different aldehyde substrates to the substrate binding pocket.[49] The corresponding residues in Qor are small amino acid residues like A259, T332, L334 and G470, making the channel amenable to larger molecules like quinoline. The three highly conserved active site amino acid residues Q224, R371 and E743 maintain the same conformation as in all the structures of XO family members, with the exception of MOP where no equivalent of the glutamine residue that hydrogen-bonds to the apical Mo=O group is present. The specific role of this glutamine appears to be stabilization of the oxo-ligand during catalysis and biogenesis to prevent its exchange against water or a sulfido-ligand by the formation of a hydrogen bond. As MOP lacks this stabilizing interaction for the apical oxo ligand, replacement of this ligand with a sulfido ligand might be possible in the presence of high concentration of sulfide. However, sulfide treatment does not seem to affect the catalytic activity of MOP. In Qor, the active site E743 might contribute to substrate orientation and nucleophilic attack of the oxo-group of the substrate (Figure 6.11).[46] Substitution of the glutamate to E743V and E743D has shown that the variant proteins were almost inactive, underscoring the catalytic relevance of E743. In contrast, the K_{Mapp} for quinoline is increased only two-fold, suggesting that the replacements only slightly affect substrate affinity. The turnover number of QorE743D was reduced only 155-fold compared to wild-type enzyme, showing that the carboxyl group of aspartate can assume the catalytic role of the glutamate of wild-type enzyme to at least some degree. We note that the same amino acid substitution in XOR and other AOX enzymes so far examined results in a variant lacking detectable catalytic activity.

6.4.3 Unusual Bacterial Enzymes of the XO Family with Unique Features

6.4.3.1 Reversing the Reaction Mechanism in 4-Hydroxybenzoyl-CoA-Reductase

4-Hydroxybenzoyl-CoA-reductase (4-HBCR) from *Thauera aromatica* catalyzes the irreversible removal of a phenolic hydroxy group from 4-hydroxybenzoyl-CoA, a reductive process yielding benzoyl-CoA and water.[63] The reaction catalyzed by 4-HBCR plays an important role in the anaerobic degradation of phenolic compounds such as phenol, *p*-cresol or low-molecular-mass aromatic products derived from lignin degradation.[198] The enzyme is so far unique to the XO family of molybdenum enzymes, since it is a reductase rather than a dehydrogenase or oxidase.[1] Thus, the reaction is principally the reverse reaction of that catalyzed by other XO enzymes, with the reducing equivalents required for the reaction provided by a 2*x*[4Fe–4S] bacterial ferredoxin. The pyranopterin cofactor of the molybdenum center is present as the MCD form of Moco. The 270 kDa enzyme is an $(\alpha\beta\gamma)_2$ heterotrimer with Moco in the large subunit, two [2Fe–2S] clusters in the small subunit and a FAD as well as an additional [4Fe–4S] cluster in the medium subunit.[52] This [4Fe–4S] cluster is unique to 4-HBCR and is found in a 41-amino acid insert in the FAD subunit (Figure 6.12). The [4Fe–4S] cluster has been suggested to be the point of entry of reducing equivalents into 4-HBCR from the ferredoxin. The [4Fe–4S] cluster of 4-HBCR is at a distance of 16.5 Å from the isoalloxazine ring of the FAD, which is rather far away. In contrast to other enzymes of the XO family, it has been suggested that the enzyme operates *via* a radical-based Birch-like mechanism, in which a first reducing equivalent is added to molybdenum-coordinated substrate, followed by protonation at C4 of substrate followed by the addition of a second reducing equivalent, which leads to dehydroxylation and rearomatization of the substrate.[63] However, details of the exact reaction mechanism of the enzyme still need to be proven.

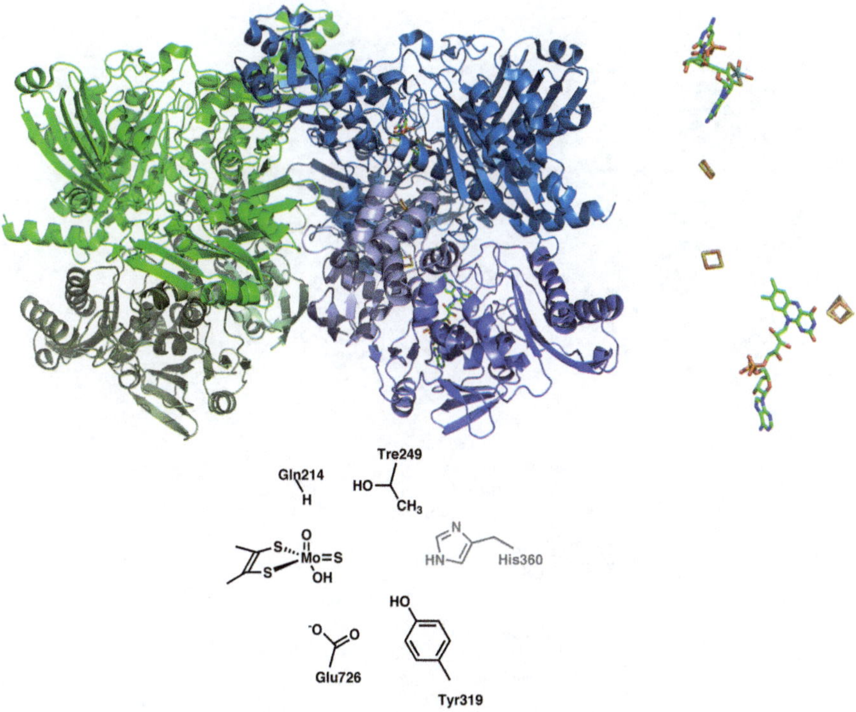

Figure 6.12 (HBCR): overall structure, cofactor arrangement and active site composition of *T. aromatica* 4-hydroxybenzoyl-CoA-reductase (HBCR) (1SB3). HBCR is composed of HcrABC as an $(\alpha\beta\gamma)_2$ heterotrimer, with HcrA shown in dark-blue/green, HcrB shown in medium-blue/green and HcrC shown in light-blue/green. HrrB binds an additional [4Fe4S] cluster not present in the other enzymes of the xanthine oxidase family. Shown is the active-site structure of Moco (MCD) and the relevant residues (HBCR numbering) in the active site proposed to be involved in catalysis and substrate binding. The Gln214 and Glu726 are highly conserved in the enzymes of the xanthine oxidase family.

6.4.3.2 *A Selenide Ligand at the Active Site of Nicotinate Dehydrogenase*

Eubacterium barkeri is an anaerobic soil bacterium able to ferment nicotinate to propionate, acetate, carbon dioxide and ammonia utilizing a pathway that is initiated by the hydroxylation of nicotinate to 6-hydroxynicotinate.[199] Nicotinate dehydrogenase (NDH) from *E. barkeri* catalyzes the hydroxylation of nicotinate to 6-hydroxynicotinate.[200] The enzyme is unique among the crystallographically characterized enzymes of the XO family in that it possesses a catalytically essential Mo=Se ligand rather than a Mo=S (Figure 6.13).[50] So far, selenium has been identified as selenocysteine at the active site of several molybdoenzymes, *e.g.* in the formate dehydrogenase H from *E. coli*, a member of the DMSO reductase family of enzymes.[1,201] Other enzymes that have

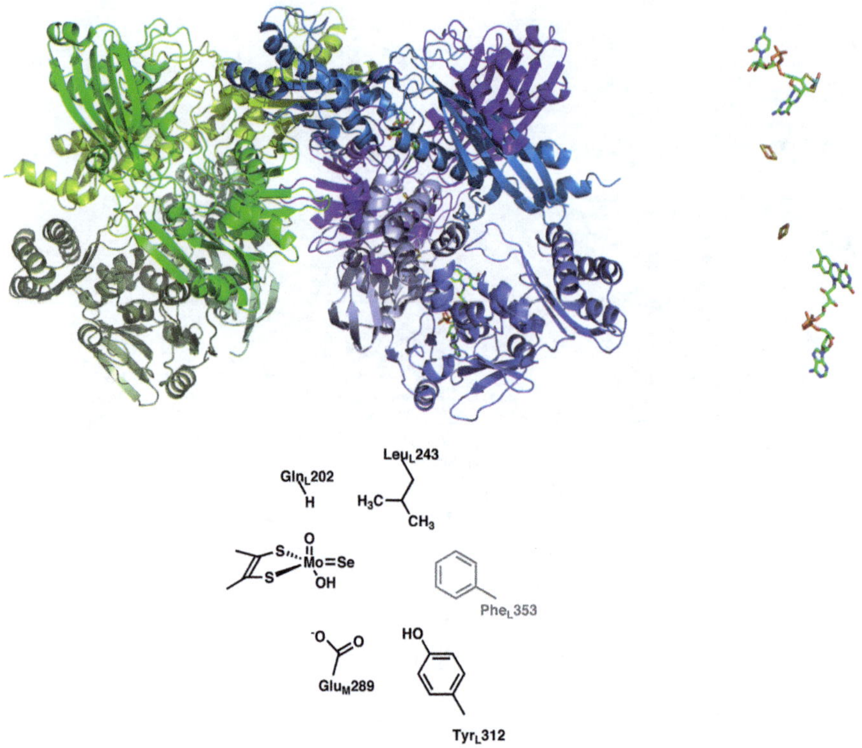

Figure 6.13 (NDH): overall structure, cofactor arrangement and active site composition of *E. barkeri* nicotinate dehydrogenase (NDH) (3HRD). NDH is composed of NdhFSLM as an $(\alpha\beta\gamma\delta)_2$ heterotetramer, with NdhL and NdhM shown in two blue/green colors, NdhF shown in medium-blue/green and NdhS shown in light-blue/green. Shown is the active-site structure of Moco (MCD) with the equatorial selenide ligand and the relevant residues (NDH numbering) in the active site proposed to be involved in catalysis and substrate binding. The NdhL and NdhM subunit contributes to the binding of Moco with the Gln$_L$202 and Glu$_M$289 being highly conserved in the enzymes of the xanthine oxidase family.

been identified as possibly possessing a Mo=Se ligand include the XOR from *Clostridium purinolyticum*, *C. acidiurici* and *E. barkeri*, and the purine hydroxylase from *C. purinolyticum*.[202–204] Nicotinate dehydrogenase from *E. barkeri* is encoded by the *ndhFSLM* gene cluster and the enzyme has an $(\alpha\beta\gamma\delta)_2$ composition.[50] Here, the NdhS subunit (23 kDa) contains the two [2Fe–2S] clusters, the NdhF subunit (33 kDa) and the FAD, and the molybdenum-binding domain is split into two separate subunits, NdhL (50 kDa) and NdhM (37 kDa), which together bind the MCD form of the molybdenum cofactor (Figure 6.13). The role of selenium in catalysis has been assessed by density functional calculations, which indicate that the transition state for the critical hydride transfer from substrate to the molybdenum center is stabilized by an additional 3 kcal mol^{-1} with selenium rather than sulfur in the molybdenum coordination sphere. Conclusively, incorporation of selenium rather than sulfur in the active site increases the rate of the chemical step of the reaction to about two orders of magnitude. However, the sulfide substituted enzyme has not been produced so far to prove the role of the selenide ligand and its role in rate acceleration of the reaction. In addition, there is at present very little information as to how the selenium atom is inserted into the active site of these enzymes.

6.4.3.3 *The Only Binuclear Molybdenum Enzyme: Copper and Molybdenum at the Active Site of CODH*

The molybdenum-containing CO dehydrogenases harbor a unique binuclear metallo-cluster, which uniquely joins the elements Mo, S and Cu in the form of a bimetallic [$CuSMoO_2$] cluster (Figure 6.14).[51] So far, the enzymes from *Oligotropha carboxidovorans* and *Hydrogenophaga pseudoflava* have been characterized in detail.[205] These organisms are aerobes that are able to grow with CO as sole source of both carbon and energy.[61] CO dehydrogenase catalyzes the

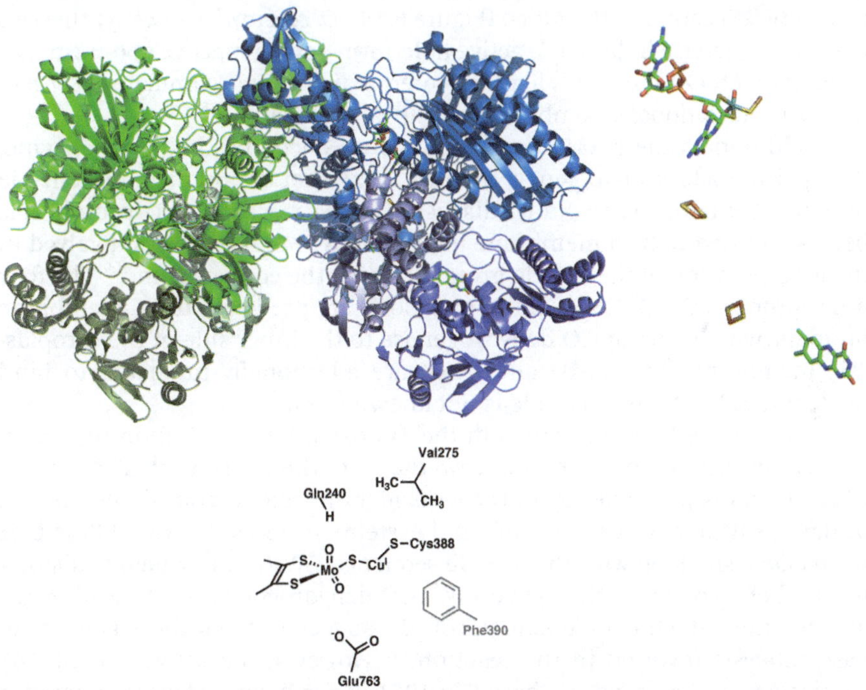

Figure 6.14 (CODH): overall structure, cofactor arrangement and active site composition of *O. carboxidovorans* CO dehydrogenase (CODH) (1N5W). CODH is composed of CoxMSL as an $(\alpha\beta\gamma)_2$ heterotrimer, with CoxL shown in dark-blue/green, CoxM shown in medium-blue/green and CoxS shown in light-blue/green. Shown is the active-site structure of Moco (MCD) containing the sulfido-bridged Cu, additionally coordinated by Cys388. The relevant residues (Cox numbering) in the active site proposed to be involved in catalysis and substrate binding are shown with the Gln240 and Glu763 being highly conserved in the enzymes of the xanthine oxidase family.

reaction $CO + H_2O = CO_2 + 2e^- + 2H^+$. The CO_2 generated in this pathway can subsequently be fixed non-photosynthetically *via* the reductive pentose phosphate pathway.[206] This enables bacteria like *O. carboxidovorans* to be involved in the clearance of ~2 × 10^8^ metric tons of CO from the environment.[1] The *O. carboxidovorans* CO dehydrogenase is encoded by the *coxBCMSLDEFGHIK* gene cluster, which is localized on the 128 kb megaplasmid pHCG3 that additionally encodes hydrogenase and enzymes of autotrophic CO_2 fixation.[207] The 277 kDa CO dehydrogenase from *O. carboxidovorans* is a dimer of CoxLSM heterotrimers. The L-subunit accommodates the unique [$Cu^{I}SMo^{VI}O_2$] cluster in the catalytic site, which is buried ~17 Å below the solvent-accessible surface of the enzyme.[51] The molybdenum ion is coordinated by the MCD-form of Moco.[62] The molybdenum has the same square-pyramidal coordination geometry seen in other members of this enzyme family, with an apical Mo=O and an equatorial plane consisting of two sulfurs from the pyranopterin cofactor, an equatorial Mo=O and the sulfur that bridges to the copper. The M-subunit contains the FAD-cofactor and the S-subunit binds one proximal and another distal [2Fe2S] center to the Moco (Figure 6.14). *O. carboxidovorans* synthesizes CO dehydrogenase in both the active holo-form and an inactive apo-form that contains a [MoO_3] site and is likely an intermediate that accumulates *in vivo* in the post-translational assembly of the [$Cu^{I}SMo^{VI}O_2$] cluster.[51]

In addition to the *coxMSL* structural genes encoding the $(\alpha\beta\gamma)_2$ enzyme, the operon includes two genes, *coxF* and *coxI*, that are predicted to encode proteins homologous to XdhC discussed above.[208] The CoxD protein has been suggested to be a membrane integral AAA + ATPase that is involved in the incorporation of the sulfido group bridging the copper atom.[209] The four other proteins CoxB, CoxC, CoxH and CoxK are predicted to be membrane proteins, which anchor CO dehydrogenase to the inner side of the cytoplasmic membrane.[210,211] CoxD and CoxK are additionally predicted to bind nucleotides, but their exact role is not known so far.[207]

In the oxidized enzyme Mo is in the +VI oxidation state; upon reduction with CO or dithionite it is reduced *via* the +V to the +IV state.[212] The Cu-ion, which remains permanently in the +1 oxidation state, is coordinated by the bridging sulfur, a water molecule and cysteine residue 388. Cys 388 is part of an active site loop with the unique sequence VAYRCSER, which is distinct from other enzymes of the XO family.[51] CO dehydrogenase is reduced by CO in a reaction which is independent of pH, suggesting that there is no acid-base catalysis involved in the reaction.[212] However, the active site Glu763 (equivalent to the active site base Glu 1261 in the bovine XOR) is conserved in CO dehydrogenase (Figure 6.14). A very high-resolution (1.1 Å) structure of CO dehydrogenase in complex with the inhibitor *n*-butylisonitrile has been reported. In this structure, the bridging sulfur bond to the copper is revealed to be cleaved, showing that the inhibitor has been inserted between the Mo and Cu atoms.[51] From this structure, a mechanism has been proposed in which CO is inserted between the bridging sulfur and copper in a similar manner, which results in the formation of a thiocarbamate at the active site that is subsequently hydrolyzed by solvent, regenerating the sulfur bridge.

Another study has suggested an alternative mechanism, which involves the nucleophilic attack by the equatorial Mo=O on an initial formed Cu–CO complex, followed by release of CO_2.[61]

It has been shown that when the bridging sulfur of CO dehydrogenase is removed by cyanide the copper is also lost. A reconstitution protocol has been developed that utilizes Cu(I)-thiourea as the source of copper. Additionally, it was shown that copper can be replaced by other metals in the active site, like silver, which partially restores activity.[213] The silver reconstituted enzyme reduces CO with a 6-fold reduced reaction rate in comparison to the copper enzyme. Further studies showed that CO dehydrogenase is also able to react with H_2 as substrate.[214] It has been suggested that H_2 is coordinated and polarized by the copper atom, followed by deprotonation and formation of a copper hydride, which subsequently deprotonates. As electron acceptors several quinone species have been identified that effectively reoxidize CO dehydrogenase. Thus, consistent with localization of the enzyme at the membrane, the physiological oxidant for CO dehydrogenase at the FAD site is most likely ubiquinone. The FAD-site of CO dehydrogenase resembles the dehydrogenase rather than the oxidase form of the bovine XOR. Since the enzyme is membrane-bound the CoxM subunit might interact with a membrane protein encoded by the *cox* gene cluster.

6.5 Conclusions

The enzymes of the xanthine oxidase family are the longest studied molybdoenzymes and their investigations started more than 100 years ago. Much has been learned from spectroscopic, kinetic and X-ray crystallographic studies and especially the pioneering early studies on XOR have paved the way for further studies on the group of molybdoenzymes from the other families. Work over the past several years has yielded a clear picture of the reactions catalyzed by these enzymes leading to substrate hydroxylation at the Moco active site. The reaction is mainly dictated by the reactivity of the molybdenum atom at the active site, and leads to transfer of a Mo–OH from the metal coordination sphere to substrate, with concomitant hydride transfer in the reverse direction. There is clear evidence for general base catalysis in the course of the reaction, and crystallographic studies have provided important contributions to how specific amino acid residues in the active site participate in catalysis. Still, open questions remain, even after 100 years of research, which should be encouraging for young scientists to enter the field by standing on the shoulders of giants.

Acknowledgements

The authors thank all current and former members of the research group in addition to collaboration partners who were involved in the work over the past years. SL especially thanks K.V. Rajagopalan for his great support over the years and the fruitful discussions. The work was mainly supported by continuous grants of the Deutsche Forschungsgemeinschaft (to SL) and from the Japanese Ministry of Education, Science and Sports (TN & KO).

References

1. R. Hille, J. Hall and P. Basu, *Chem. Rev.*, 2014, **114**, 3963–4038.
2. J. Horbaczewski, *Monatsh. Chem.*, 1891, **12**, 221–275.
3. W. Spitzer, *Arch. Ges Physiol.*, 1899, **76**, 192–203.
4. F. Schardinger, *Journal*, 1902, **5**, 1113–1121.
5. E. J. Morgan, C. P. Stewart and F. G. Hopkins, *Proc. R. Soc. London, Ser. B*, 1922, **94**, 109–131.
6. M. Dixon and S. Thurlow, *Biochem. J.*, 1924, **18**, 989–992.
7. E. G. Ball, *J. Biol. Chem.*, 1939, **128**, 51–67.
8. E. C. De Renzo, E. Kaleita, P. G. Heytler, J. J. Olson, B. L. Huchings and J. H. Williams, *Arch. Biochem. Biophys.*, 1953, **45**, 247–253.
9. D. A. Richert and W. W. Westerfeld, *J. Biol. Chem.*, 1953, **203**, 915–923.
10. R. C. Bray, *Biochem. J.*, 1959, **73**, 690–694.
11. G. B. Palmer and R. C. Beinert, *J. Biol. Chem.*, 1964, **239**, 2657–2666.
12. J. M. McCord and I. Fridovich, *J. Biol. Chem.*, 1968, **243**, 5753–5760.
13. K. Okamoto, T. Kusano and T. Nishino, *Curr. Pharm. Des.*, 2013, **19**, 2606–2614.
14. K. Ichida, Y. Amaya, K. Okamoto and T. Nishino, *Int. J. Mol. Sci.*, 2012, **13**, 15475–15495.
15. R. Hille, T. Nishino and F. Bittner, *Coord. Chem. Rev.*, 2011, **255**, 1179–1205.
16. T. Nishino, K. Okamoto, B. T. Eger, E. F. Pai and T. Nishino, *FEBS J.*, 2008, **275**, 3278–3289.
17. T. Nishino, *J. Biochem.*, 1994, **116**, 1–6.
18. R. Hille and T. Nishino, *FASEB J.*, 1995, **9**, 995–1003.
19. R. Huber, P. Hof, R. O. Duarte, J. J. Moura, I. Moura, M. Y. Liu, J. LeGall, R. Hille, M. Archer and M. J. Romão, *Proc. Natl. Acad. Sci. U. S. A.*, 1996, **93**, 8846–8851.
20. R. Hille, *Dalton Trans.*, 2013, **42**, 3029–3042.
21. J. S. Olson, D. P. Ballou, G. Palmer and V. Massey, *J. Biol. Chem.*, 1974, **249**, 4363–4382.
22. T. Nishino and K. Okamoto, *J. Inorg. Biochem.*, 2000, **82**, 43–49.
23. R. Hille and R. F. Anderson, *J. Biol. Chem.*, 1991, **266**, 5608–5615.
24. K. Kobayashi, M. Miki, K. Okamoto and T. Nishino, *J. Biol. Chem.*, 1993, **268**, 24642–24646.
25. L. M. Schopfer, V. Massey and T. Nishino, *J. Biol. Chem.*, 1988, **263**, 13528–13538.
26. J. Hunt and V. Massey, *J. Biol. Chem.*, 1992, **267**, 21479–21485.
27. C. M. Harris and V. Massey, *J. Biol. Chem.*, 1997, **272**, 28335–28341.
28. R. Hille, *Chem. Rev.*, 1996, **96**, 2757–2816.
29. V. Massey and D. Edmondson, *J. Biol. Chem.*, 1970, **245**, 6595–6598.
30. T. Nishino, R. Ito and K. Tsushima, *Biochim. Biophys. Acta*, 1975, **403**, 17–22.
31. R. C. Wahl and K. V. Rajagopalan, *J. Biol. Chem.*, 1982, **257**, 1354–1359.
32. K. Okamoto, K. Matsumoto, R. Hille, B. T. Eger, E. F. Pai and T. Nishino, *Proc. Natl. Acad. Sci. U. S. A.*, 2004, **101**, 7931–7936.

33. M. Dixon and D. Keilin, *Proc. R. Soc. B*, 1936, **119**, 159–190.
34. S. Gutteridge, S. J. Tanner and R. C. Bray, *Biochem. J.*, 1978, **175**, 887–897.
35. T. Nishino, T. Nishino and K. Tsushima, *FEBS Lett.*, 1981, **131**, 369–372.
36. T. Nishino and K. Tsushima, *J. Biol. Chem.*, 1986, **261**, 11242–11246.
37. T. Nishino, *Biochim. Biophys. Acta*, 1974, **341**, 93–98.
38. S. Leimkühler, M. Kern, P. S. Solomon, A. G. McEwan, G. Schwarz, R. R. Mendel and W. Klipp, *Mol. Microbiol.*, 1998, **27**, 853–869.
39. R. Hille, *Arch. Biochem. Biophys.*, 2005, **433**, 107–116.
40. C. Enroth, B. T. Eger, K. Okamoto, T. Nishino, T. Nishino and E. F. Pai, *Proc. Natl. Acad. Sci. U. S. A.*, 2000, **97**, 10723–10728.
41. C. Coelho, M. Mahro, J. Trincao, A. T. Carvalho, M. J. Ramos, M. Terao, E. Garattini, S. Leimkuhler and M. J. Romao, *J. Biol. Chem.*, 2012, **287**, 40690–40702.
42. Y. Amaya, K. Yamazaki, M. Sato, K. Noda, T. Nishino and T. Nishino, *J. Biol. Chem.*, 1990, **265**, 14170–14175.
43. K. Ichida, Y. Amaya, K. Noda, S. Minoshima, T. Hosoya, O. Sakai, N. Shimizu and T. Nishino, *Gene.*, 1993, **133**, 279–284.
44. A. Sato, T. Nishino, K. Noda, Y. Amaya and T. Nishino, *J. Biol. Chem.*, 1995, **270**, 2818–2826.
45. H. Dobbek, L. Gremer, O. Meyer and R. Huber, *Proc. Natl. Acad. Sci. U. S. A.*, 1999, **96**, 8884–8889.
46. I. Bonin, B. M. Martins, V. Purvanov, S. Fetzner, R. Huber and H. Dobbek, *Structure*, 2004, **12**, 1425–1435.
47. J. S. Holcenberg and E. R. Stadtman, *J. Biol. Chem.*, 1969, **244**, 1194–1203.
48. J. J. Truglio, K. Theis, S. Leimkühler, R. Rappa, K. V. Rajagopalan and C. Kisker, *Structure*, 2002, **10**, 115–125.
49. M. J. Romão, M. Archer, I. Moura, J. J. G. Moura, J. LeGall, R. Engh, M. Schneider, P. Hof and R. Huber, *Science*, 1995, **270**, 1170–1176.
50. N. Wagener, A. J. Pierik, A. Ibdah, R. Hille and H. Dobbek, *Proc. Natl. Acad. Sci. U. S. A.*, 2009, **106**, 11055–11060.
51. H. Dobbek, L. Gremer, R. Kiefersauer, R. Huber and O. Meyer, *Proc. Natl. Acad. Sci. U. S. A.*, 2002, **99**, 15971–15976.
52. M. Unciuleac, E. Warkentin, C. C. Page, M. Boll and U. Ermler, *Structure*, 2004, **12**, 2249–2256.
53. S. Leimkühler, M. Kern, P. S. Solomon, A. G. McEwan, G. Schwarz, R. R. Mendel and W. Klipp, *Mol. Microbiol.*, 1998, **27**, 853–869.
54. Q. Xiang and D. Edmondson, *Biochemistry*, 1996, **35**, 5441–5450.
55. L. Gremer and O. Meyer, *Eur. J. Biochem.*, 1996, **238**, 862–866.
56. R. Kappl, S. Sielker, K. Ranguelova, J. Wegner, K. Parschat, J. Huttermann and S. Fetzner, *Biochemistry*, 2006, **45**, 14853–14868.
57. I. Stephan, B. Tshisuaka, S. Fetzner and F. Lingens, *Eur. J. Biochem.*, 1996, **236**, 155–162.
58. G. Bauer and F. Lingens, *Biol. Chem. Hoppe-Seyler*, 1992, **373**, 699–705.
59. S. Fetzner and F. Lingens, *Biol. Chem. Hoppe-Seyler*, 1993, **374**, 363–376.
60. M. Lehmann, B. Tshisuaka, S. Fetzner and F. Lingens, *J. Biol. Chem.*, 1995, **270**, 14420–14429.

61. R. Hille, S. Dingwall and J. Wilcoxen, *JBIC, J. Biol. Inorg. Chem.*, 2015, **20**, 243–251.
62. O. Meyer and K. V. Rajagopalan, *J. Bacteriol.*, 1984, **157**, 643–648.
63. M. Boll, G. Fuchs, C. Meier, A. Trautwein, A. El Kasmi, S. W. Ragsdale, G. Buchanan and D. J. Lowe, *J. Biol. Chem.*, 2001, **276**, 47853–47862.
64. E. D. Corte and F. Stirpe, *Biochem. J.*, 1967, **102**, 520–524.
65. K. V. Rajagopalan and P. Handler, *J. Biol. Chem.*, 1967, **242**, 4097–4107.
66. W. A. Doyle, J. F. Burke, A. Chovnick, F. L. Dutton, J. R. Whittle and R. C. Bray, *Eur. J. Biochem.*, 1996, **239**, 782–795.
67. T. Iwasaki, K. Okamoto, T. Nishino, J. Mizushima and H. Hori, *J. Biochem.*, 2000, **127**, 771–778.
68. R. Hille, W. R. Hagen and W. R. Dunham, *J. Biol. Chem.*, 1985, **260**, 10569–10575.
69. G. Palmer and V. Massey, *J. Biol. Chem.*, 1969, **244**, 2614–2620.
70. S. Leimkühler, A. L. Stockert, K. Igarashi, T. Nishino and R. Hille, *J. Biol. Chem.*, 2004, **279**, 40437–40444.
71. J. Hall, S. Reschke, H. Cao, S. Leimkuhler and R. Hille, *J. Biol. Chem.*, 2014, **289**, 32121–32130.
72. L. A. Tai and K. C. Hwang, *Biochemistry*, 2004, **43**, 4869–4876.
73. T. Nishino, K. Okamoto, Y. Kawaguchi, T. Matsumura, B. T. Eger and E. F. Pai, *FEBS J.*, 2015, **282**, 3075–3090.
74. E. D. Corte and F. Stirpe, *Biochem. J.*, 1972, **126**, 739–745.
75. R. Harrison, *Drug Metab. Rev.*, 2004, **36**, 363–375.
76. T. Saito and T. Nishino, *J. Biol. Chem.*, 1989, **264**, 10015–10022.
77. R. Asai, T. Nishino, T. Matsumura, K. Okamoto, K. Igarashi, E. F. Pai and T. Nishino, *J. Biochem.*, 2007, **141**, 525–534.
78. T. Nishino, K. Okamoto, Y. Kawaguchi, H. Hori, T. Matsumura, B. T. Eger, E. F. Pai and T. Nishino, *J. Biol. Chem.*, 2005, **280**, 24888–24894.
79. J. Hunt, V. Massey, W. R. Dunham and R. H. Sands, *J. Biol. Chem.*, 1993, **268**, 18685–18691.
80. J. Hunt and V. Massey, *J. Biol. Chem.*, 1992, **267**, 21479–21485.
81. T. Saito and T. Nishino, *J. Biol. Chem.*, 1989, **264**, 10015–10022.
82. T. Nishino, T. Nishino, L. M. Schopfer and V. Massey, *J. Biol. Chem.*, 1989, **264**, 2518–2527.
83. V. Massey, L. M. Schopfer, T. Nishino and T. Nishino, *J. Biol. Chem.*, 1989, **264**, 10567–10573.
84. T. Saito, T. Nishino and V. Massey, *J. Biol. Chem.*, 1989, **264**, 15930–15935.
85. T. Nishino and T. Nishino, *J. Biol. Chem.*, 1989, **264**, 5468–5473.
86. H. Ishikita, B. T. Eger, K. Okamoto, T. Nishino and E. F. Pai, *J. Am. Chem. Soc.*, 2012, **134**, 999–1009.
87. M. Ichikawa, T. Nishino and A. Ichikawa, *J. Histochem. Cytochem.*, 1992, **40**, 1097–1103.
88. A. Kooij, K. S. Bosch, W. M. Frederiks and C. J. Van Noorden, *Virchows Arch. B*, 1992, **62**, 143–150.
89. A. Kooij, *Histochem. J.*, 1994, **26**, 889–915.
90. G. Bruder, H. Heid, E. D. Jarasch, T. W. Keenan and I. H. Mather, *Biochim. Biophys. Acta*, 1982, **701**, 357–369.

91. R. K. Hughes, W. A. Doyle, A. Chovnick, J. R. Whittle, J. F. Burke and R. C. Bray, *Biochem. J.*, 1992, **285**(Pt 2), 507–513.
92. C. E. Berry and J. M. Hare, *J. Physiol.*, 2004, **555**, 589–606.
93. P. Pacher, A. Nivorozhkin and C. Szabó, *Pharmacol. Rev.*, 2006, **58**, 87–114.
94. E. D. Jarasch, C. Grund, G. Bruder, H. W. Heid, T. W. Keenan and W. W. Franke, *Cell*, 1981, **25**, 67–82.
95. S. Angermüller, G. Bruder, A. Völkl, H. Wesch and H. D. Fahimi, *Eur. J. Cell Biol.*, 1987, **45**, 137–144.
96. M. G. Battelli, A. Bolognesi and L. Polito, *Biochim. Biophys. Acta*, 2014, **1842**, 1502–1517.
97. I. H. Mather, *J. Dairy Sci.*, 2000, **83**, 203–247.
98. M. P. Gandhi and S. P. Ahuja, *Zentralbl. Veterinarmed. A.*, 1979, **26**, 765–769.
99. I. H. Mather, *J. Dairy Sci.*, 2000, **83**, 203–247.
100. T. W. Keenan, *Adv. Exp. Med. Biol.*, 2001, **501**, 125–136.
101. I. H. Mather and T. W. Keenan, *J. Mammary. Gland. Biol. Neoplasia.*, 1998, **3**, 227–232.
102. J. L. McManaman, C. A. Palmer, R. M. Wright and M. C. Neville, *J. Physiol.*, 2002, **545**, 567–579.
103. H. W. Heid, M. Schnolzer and T. W. Keenan, *Biochem. J.*, 1996, **320**(Pt 3), 1025–1030.
104. L. R. Banghart, C. W. Chamberlain, J. Velarde, I. V. Korobko, S. L. Ogg, L. J. Jack, V. N. Vakharia and I. H. Mather, *J. Biol. Chem.*, 1998, **273**, 4171–4179.
105. C. Vorbach, A. Scriven and M. R. Capecchi, *Genes Dev.*, 2002, **16**, 3223–3235.
106. B. Godber, S. Sanders, R. Harrison, R. Eisenthal and R. C. Bray, *Biochem. Soc. Trans.*, 1997, **25**, 519S.
107. H. M. Martin, J. T. Hancock, V. Salisbury and R. Harrison, *Infect. Immun.*, 2004, **72**, 4933–4939.
108. A. M. Brown, M. Benboubetra, M. Ellison, D. Powell, J. D. Reckless and R. Harrison, *Biochim. Biophys. Acta*, 1995, **1245**, 248–254.
109. J. T. Hancock, V. Salisbury, M. C. Ovejero-Boglione, R. Cherry, C. Hoare, R. Eisenthal and R. Harrison, *Antimicrob. Agents Chemother.*, 2002, **46**, 3308–3310.
110. M. C. Lee, M. Velayutham, T. Komatsu, R. Hille and J. L. Zweier, *Biochemistry*, 2014, **53**, 6615–6623.
111. S. L. Linas, D. Whittenburg and J. E. Repine, *Am. J. Physiol.*, 1990, **258**, F711–F716.
112. T. M. Millar, C. R. Stevens, N. Benjamin, R. Eisenthal, R. Harrison and D. R. Blake, *FEBS Lett.*, 1998, **427**, 225–228.
113. Z. Zhang, D. Naughton, P. G. Winyard, N. Benjamin, D. R. Blake and M. C. Symons, *Biochem. Biophys. Res. Commun.*, 1998, **249**, 767–772.
114. B. L. Godber, J. J. Doel, G. P. Sapkota, D. R. Blake, C. R. Stevens, R. Eisenthal and R. Harrison, *J. Biol. Chem.*, 2000, **275**, 7757–7763.
115. H. Li, A. Samouilov, X. Liu and J. L. Zweier, *J. Biol. Chem.*, 2001, **276**, 24482–24489.

116. L. B. Maia and J. J. Moura, *JBIC, J. Biol. Inorg. Chem.*, 2011, **16**, 443–460.
117. L. B. Maia, V. Pereira, L. Mira and J. J. Moura, *Biochemistry*, 2015, **54**, 685–710.
118. A. Webb, R. Bond, P. McLean, R. Uppal, N. Benjamin and A. Ahluwalia, *Proc. Natl. Acad. Sci. U. S. A.*, 2004, **101**, 13683–13688.
119. J. E. Baker, J. Su, X. Fu, A. Hsu, G. J. Gross, J. S. Tweddell and N. Hogg, *J. Mol. Cell. Cardiol.*, 2007, **43**, 437–444.
120. T. Nishino and K. Okamoto, *JBIC, J. Biol. Inorg. Chem.*, 2015, **20**, 195–207.
121. S. Leimkühler, M. Kern, P. S. Solomon, A. G. McEwan, G. Schwarz, R. R. Mendel and W. Klipp, *Mol. Microbiol.*, 1998, **27**, 853–869.
122. M. Neumann, G. Mittelstädt, F. Seduk, C. Iobbi-Nivol and S. Leimkühler, *J. Biol. Chem.*, 2009, **284**, 21891–21898.
123. T. Nishino, L. M. Schopfer and V. Massey, *J. Biol. Chem.*, 1989, **264**, 2518–2527.
124. J. J. Truglio, K. Theis, S. Leimkühler, R. Rappa, K. V. Rajagopalan and C. Kisker, *Structure*, 2002, **10**, 115–125.
125. R. Hille and T. Nishino, *FASEB J.*, 1995, **9**, 995–1003.
126. C. M. Harris and V. Massey, *J. Biol. Chem.*, 1997, **272**, 8370–8379.
127. K. Okamoto, B. T. Eger, T. Nishino, S. Kondo, E. F. Pai and T. Nishino, *J. Biol. Chem.*, 2003, **278**, 1848–1855.
128. Y. Yamaguchi, T. Matsumura, K. Ichida, K. Okamoto and T. Nishino, *J. Biochem.*, 2007, **141**, 513–524.
129. S. Leimkühler, A. L. Stockert, K. Igarashi, T. Nishino and R. Hille, *J. Biol. Chem.*, 2004, **279**, 40437–40444.
130. T. Amano, N. Ochi, H. Sato and S. Sakaki, *J. Am. Chem. Soc.*, 2007, **129**, 8131–8138.
131. J. M. Pauff, J. Zhang, C. E. Bell and R. Hille, *J. Biol. Chem.*, 2008, **283**, 4818–4824.
132. C. A. Bayse, *Dalton Trans.*, 2009, 2306–2314.
133. S. Metz and W. Thiel, *J. Am. Chem. Soc.*, 2009, **131**, 14885–14902.
134. J. M. Pauff, H. Cao and R. Hille, *J. Biol. Chem.*, 2009, **284**, 8760–8767.
135. H. Cao, J. Hall and R. Hille, *Biochemistry*, 2014, **53**, 533–541.
136. K. Okamoto, Y. Kawaguchi, B. T. Eger, E. F. Pai and T. Nishino, *J. Am. Chem. Soc.*, 2010, **132**, 17080–17083.
137. J. M. Pauff, C. F. Hemann, N. Jünemann, S. Leimkühler and R. Hille, *J. Biol. Chem.*, 2007, **282**, 12785–12790.
138. A. Fukunari, K. Okamoto, T. Nishino, B. T. Eger, E. F. Pai, M. Kamezawa, I. Yamada and N. Kato, *J. Pharmacol. Exp. Ther.*, 2004, **311**, 519–528.
139. A. Glatigny, P. Hof, M. J. Romão, R. Huber and C. Scazzocchio, *J. Mol. Biol.*, 1998, **278**, 431–438.
140. S. Metz and W. Thiel, *J. Phys. Chem. B*, 2010, **114**, 1506–1517.
141. H. Kikuchi, H. Fujisaki, T. Furuta, K. Okamoto, S. Leimkühler and T. Nishino, *Sci. Rep.*, 2012, **2**, 331.
142. H. A. Simmonds, S. Reiter and T. Nishino, in *The Metabolic and Molecular Bases of Inherited Disease*, 7th edn, ed. C. R.Scriver, A. L.Beaudet, W. S.Sly and D.Valle, McGraw-Hill, New York, 1995, pp. 1781–1797.

143. S. Reiter, H. A. Simmonds, N. Zollner, S. L. Braun and M. Knedel, *Clin. Chim. Acta*, 1990, **187**, 221–234.
144. V. Massey, H. Komai, G. Palmer and G. B. Elion, *J. Biol. Chem.*, 1970, **245**, 2837–2844.
145. M. K. Wolpert, J. R. Althaus and D. G. Johns, *J. Pharmacol. Exp. Ther.*, 1973, **185**, 202–213.
146. G. B. Elion, *Science*, 1989, **244**, 41–47.
147. V. Massey, H. Komai, G. Palmer and G. B. Elion, *Vitam. Horm.*, 1970, **28**, 505–531.
148. G. B. Elion, S. Callahan, H. Nathan, S. Bieber, R. W. Rundles and G. H. Hitchings, *Biochem. Pharmacol.*, 1963, **12**, 85–93.
149. T. Nishino and K. Okamoto, *JBIC, J. Biol. Inorg. Chem.*, 2015, **20**, 195–207.
150. K. Okamoto and T. Nishino, *J. Nippon Med. Sch.*, 2008, **75**, 2–3.
151. E. Garattini, R. Mendel, M. J. Romao, R. Wright and M. Terao, *Biochem. J.*, 2003, **372**, 15–32.
152. E. Garattini, M. Fratelli and M. Terao, *Cell. Mol. Life Sci.*, 2008, **65**, 1019–1048.
153. C. Beedham, *Drug Metab. Rev.*, 1985, **16**, 119–156.
154. C. Beedham, *Pharm. World Sci.*, 1997, **19**, 255–263.
155. C. Beedham, *Journal*, 1998, **25**, 39–52.
156. T. A. Krenitsky, *Adv. Exp. Med. Biol.*, 1973, **41**, 57–64.
157. T. A. Krenitsky, *Biochem. Pharmacol.*, 1978, **27**, 2763–2764.
158. T. A. Krenitsky, S. M. Neil, G. B. Elion and G. H. Hitchings, *Arch. Biochem. Biophys.*, 1972, **150**, 585–599.
159. E. Garattini and M. Terao, *Drug Metab. Rev.*, 2011, **43**, 374–386.
160. E. Garattini and M. Terao, *Expert Opin. Drug Metab. Toxicol.*, 2012, **8**, 487–503.
161. M. Zientek, Y. Jiang, K. Youdim and R. S. Obach, *Drug Metab. Dispos.*, 2010, **38**, 1322–1327.
162. D. C. Pryde, D. Dalvie, Q. Hu, P. Jones, R. S. Obach and T. D. Tran, *J. Med. Chem.*, 2010, **53**, 8441–8460.
163. R. S. Obach, P. Huynh, M. C. Allen and C. Beedham, *J. Clin. Pharmacol.*, 2004, **44**, 7–19.
164. M. Terao, M. Kurosaki, M. M. Barzago, M. Fratelli, R. Bagnati, A. Bastone, C. Giudice, E. Scanziani, A. Mancuso, C. Tiveron and E. Garattini, *Mol. Cell. Biol.*, 2009, **29**, 357–377.
165. R. Hille, T. Nishino and F. Bittner, *Coord. Chem. Rev.*, 2011, **255**, 1179–1205.
166. M. Kurosaki, M. Bolis, M. Fratelli, M. M. Barzago, L. Pattini, G. Perretta, M. Terao and E. Garattini, *Cell. Mol. Life Sci.*, 2013, **70**, 1807–1830.
167. E. Garattini, M. Fratelli and M. Terao, *Hum. Genomics*, 2009, **4**, 119–130.
168. M. Terao, M. Kurosaki, M. Marini, M. A. Vanoni, G. Saltini, V. Bonetto, A. Bastone, C. Federico, S. Saccone, R. Fanelli, M. Salmona and E. Garattini, *J. Biol. Chem.*, 2001, **276**, 46347–46363.
169. M. Terao, M. Kurosaki, G. Saltini, S. Demontis, M. Marini, M. Salmona and E. Garattini, *J. Biol. Chem.*, 2000, **275**, 30690–30700.

170. M. Kurosaki, S. Demontis, M. M. Barzago, E. Garattini and M. Terao, *Biochim. Biophys. Acta*, 1999, **341**, 71–80.
171. M. Kurosaki, M. Terao, M. M. Barzago, A. Bastone, D. Bernardinello, M. Salmona and E. Garattini, *J. Biol. Chem.*, 2004, **279**, 50482–50498.
172. S. Demontis, M. Kurosaki, S. Saccone, S. Motta, E. Garattini and M. Terao, *Biochim. Biophys. Acta*, 1999, **1489**, 207–222.
173. N. M. Cerqueira, C. Coelho, N. F. Bras, P. A. Fernandes, E. Garattini, M. Terao, M. J. Romao and M. J. Ramos, *JBIC, J. Biol. Inorg. Chem.*, 2015, **20**, 209–217.
174. C. G. Jordan, M. R. Rashidi, H. Laljee, S. E. Clarke, J. E. Brown and C. Beedham, *J. Pharm. Pharmacol.*, 1999, **51**, 411–418.
175. M. R. Rashidi, C. Beedham, J. S. Smith and S. Davaran, *Drug Metab. Pharmacokinet.*, 2007, **22**, 299–306.
176. A. H. Chalmers, P. R. Knight and M. R. Atkinson, *Aust. J. Exp. Biol. Med. Sci.*, 1969, **47**, 263–273.
177. S. E. Clarke, A. W. Harrell and R. J. Chenery, *Drug Metab. Dispos.*, 1995, **23**, 251–254.
178. K. V. Rajagopalan, I. Fridovich and P. Handler, *J. Biol. Chem.*, 1962, **237**, 922–928.
179. K. V. Rajagopalan and P. Handler, *Biochem. Biophys. Res. Commun.*, 1962, **8**, 43–47.
180. R. S. Obach and R. L. Walsky, *J. Clin. Psychopharmacol.*, 2005, **25**, 605–608.
181. J. M. Hutzler, R. S. Obach, D. Dalvie and M. A. Zientek, *Expert Opin. Drug Metab. Toxicol.*, 2013, **9**, 153–168.
182. H. Peretz, M. S. Naamati, D. Levartovsky, A. Lagziel, E. Shani, I. Horn, H. Shalev and D. Landau, *Mol. Genet. Metab.*, 2007, **91**, 23–29.
183. J. Weigert, M. Neumeier, S. Bauer, W. Mages, A. A. Schnitzbauer, A. Obed, B. Groschl, A. Hartmann, A. Schaffler, C. Aslanidis, J. Scholmerich and C. Buechler, *FEBS Lett.*, 2008, **582**, 2965–2972.
184. A. Sigruener, C. Buechler, E. Orso, A. Hartmann, P. J. Wild, L. Terracciano, M. Roncalli, S. R. Bornstein and G. Schmitz, *Horm. Metab. Res.*, 2007, **39**, 781–789.
185. T. Hartmann, M. Terao, E. Garattini, C. Teutloff, J. F. Alfaro, J. P. Jones and S. Leimkuhler, *Drug Metab. Dispos.*, 2012, **40**, 856–864.
186. M. Adachi, K. Itoh, A. Masubuchi, N. Watanabe and Y. Tanaka, *J. Biochem. Mol. Biol.*, 2007, **40**, 1021–1027.
187. M. Mahro, C. Coelho, J. Trincao, D. Rodrigues, M. Terao, E. Garattini, M. Saggu, F. Lendzian, P. Hildebrandt, M. J. Romao and S. Leimkuhler, *Drug Metab. Dispos.*, 2011, **39**, 1939–1945.
188. M. Mahro, N. F. Bras, N. M. Cerqueira, C. Teutloff, C. Coelho, M. J. Romao and S. Leimkühler, *PLoS One*, 2013, **8**, e82285.
189. K. Okamoto, Y. Kawaguchi, B. T. Eger, E. F. Pai and T. Nishino, *J. Am. Chem. Soc.*, 2010, **132**, 17080–17083.
190. S. Schumann, M. Terao, E. Garattini, M. Saggu, F. Lendzian, P. Hildebrandt and S. Leimkuhler, *PLoS One*, 2009, **4**, e5348.

191. R. Vila, M. Kurosaki, M. M. Barzago, M. Kolek, A. Bastone, L. Colombo, M. Salmona, M. Terao and E. Garattini, *J. Biol. Chem.*, 2004, **279**, 8668–8683.
192. J. F. Alfaro and J. P. Jones, *J. Org. Chem.*, 2008, **73**, 9469–9472.
193. Y. Kuwabara, T. Nishino, K. Okamoto, T. Matsumura, B. T. Eger and E. F. Pai, *Proc. Natl. Acad. Sci. U. S. A.*, 2003, **100**, 8170–8175.
194. T. Santos-Silva, F. Ferroni, A. Thapper, J. Marangon, P. J. Gonzalez, A. C. Rizzi, I. Moura, J. J. Moura, M. J. Romao and C. D. Brondino, *J. Am. Chem. Soc.*, 2009, **131**, 7990–7998.
195. C. More, M. Asso, G. Roger, B. Guigliarelli, J. Caldeira, J. Moura and P. Bertrand, *Biochemistry*, 2005, **44**, 11628–11635.
196. M. Blase, C. Bruntner, B. Tshisuaka, S. Fetzner and F. Lingens, *J. Biol. Chem.*, 1996, **271**, 23068–23079.
197. U. Frerichs-Deeken, B. Goldenstedt, R. Gahl-Janssen, R. Kappl, J. Huttermann and S. Fetzner, *Eur. J. Biochem.*, 2003, **270**, 1567–1577.
198. J. Johannes, M. C. Unciuleac, T. Friedrich, E. Warkentin, U. Ermler and M. Boll, *Biochemistry*, 2008, **47**, 4964–4972.
199. J. S. Holcenberg and E. R. Stadtman, *J. Biol. Chem.*, 1969, **244**, 1194–1203.
200. T. Schrader, B. Thiemer and J. R. Andreesen, *Appl. Microbiol. Biotechnol.*, 2002, **58**, 612–617.
201. J. C. Boyington, V. N. Gladyshev, S. V. Khangulov, T. C. Stadtman and P. D. Sun, *Science*, 1997, **275**, 1305–1308.
202. R. Wagner and J. R. Andreesen, *Arch. Microbiol.*, 1979, **121**, 255–260.
203. T. Schrader, A. Rienhofer and J. R. Andreesen, *Eur. J. Biochem.*, 1999, **264**, 862–871.
204. W. T. Self and T. C. Stadtman, *Proc. Natl. Acad. Sci. U. S. A.*, 2000, **97**, 7208–7213.
205. O. Meyer, S. Jacobitz and B. Krüger, FEMS Microbiol, *Rev.*, 1986, **39**, 161–179.
206. G. M. King and C. F. Weber, *Nat. Rev. Microbiol.*, 2007, **5**, 107–118.
207. B. Santiago, U. Schubel, C. Egelseer and O. Meyer, *Gene*, 1999, **236**, 115–124.
208. M. Neumann and S. Leimkühler, *Biochem. Res. Int.*, 2011, **2011**, 850924.
209. A. Pelzmann, M. Ferner, M. Gnida, W. Meyer-Klaucke, T. Maisel and O. Meyer, *J. Biol. Chem.*, 2009, **284**, 9578–9586.
210. F. Spreitler, C. Brock, A. Pelzmann, O. Meyer and J. Kohler, *ChemBioChem*, 2010, **11**, 2419–2423.
211. A. M. Pelzmann, F. Mickoleit and O. Meyer, *JBIC, J. Biol. Inorg. Chem.*, 2014, **19**, 1399–1414.
212. B. Zhang, C. F. Hemann and R. Hille, *J. Biol. Chem.*, 2010, **285**, 12571–12578.
213. J. Wilcoxen, S. Snider and R. Hille, *J. Am. Chem. Soc.*, 2011, **133**, 12934–12936.
214. J. Wilcoxen and R. Hille, *J. Biol. Chem.*, 2013, **288**, 36052–36060.

CHAPTER 7

The Sulfite Oxidase Family of Molybdenum Enzymes

ULRIKE KAPPLER*[a] AND GUENTER SCHWARZ[b]

[a]School of Chemistry and Molecular Biosciences, The University of Queensland, St. Lucia, Qld 4072, Australia; [b]Institute of Biochemistry, Department of Chemistry and Center for Molecular Medicine Cologne and Cologne Excellence Cluster on Cellular Stress Responses in Aging-Associated Diseases, University of Cologne, Zuelpicher Str. 47, 50674 Koeln, Germany

*E-mail: u.kappler@uq.edu.au

7.1 Introduction

The sulfite oxidase (SO) family of enzymes is one of the three originally proposed families of mononuclear molybdenum enzymes and takes its name from the best-studied representatives of this enzyme family, the vertebrate SOs.[1] While initially enzymes of this family were believed to be exclusively found in eukaryotic organisms,[2] it has been recognized over the past 15 years that there is also an abundant group of enzymes originating from bacteria that belong to this enzyme family, and many of these bacterial enzymes are currently uncharacterized.[3–6] Both reductases and oxidases/dehydrogenases have been identified in the SO family, and so far all of these enzymes convert sulfur- or nitrogen-containing substrates. The key feature of these enzymes is the binding of a molybdopterin-type molybdenum cofactor (Moco), which is covalently bound *via* a highly conserved cysteine residue to the enzyme[7] (Figure 7.1).

RSC Metallobiology Series No. 5
Molybdenum and Tungsten Enzymes: Biochemistry
Edited by Russ Hille, Carola Schulzke, and Martin L. Kirk

Published by the Royal Society of Chemistry, www.rsc.org

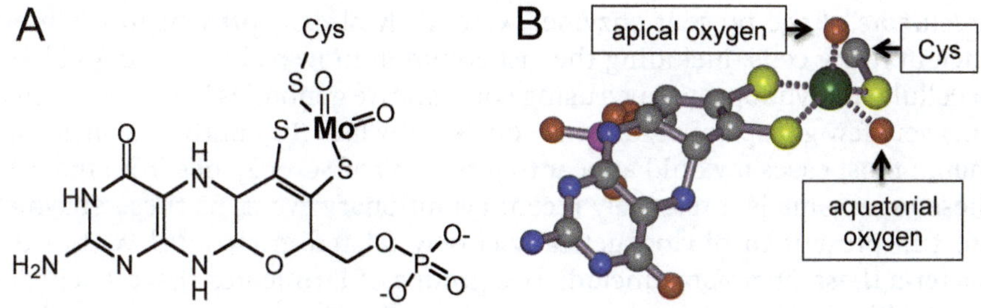

Figure 7.1 Structure of the molybdenum cofactor (Moco) in enzymes of the SO family. (A) Chemical structure and (B) three-dimensional structure of Moco from chicken SO.

The basic reaction catalyzed by SO-type Mo-enzymes is a two-electron oxidation or reduction of the substrate (eqn (7.1)) that leads to incorporation or abstraction of oxygen into or from the substrate, with the oxygen being transferred to or taken from water.

$$\text{substrate}_{red} + H_2O \leftarrow\rightarrow \text{substrate}_{ox} + 2e^- + 2H^+ \tag{7.1}$$

At present, the best characterized representatives of the SO family are the vertebrate and plant sulfite oxidases, the plant nitrate reductases and several types of bacterial sulfite dehydrogenases, along with a large number of less well-characterized bacterial and archaeal enzymes, some of which are thought to be reductases. The recently discovered mammalian aldoximine reductases (mARC)[8,9] have also been proposed to be members of the sulfite oxidase enzyme family[10] (see below), but are likely representatives of a new family of Mo enzymes.[11]

Despite the seeming disparity of the reactions catalyzed by the SO family enzymes, there are several common features shared by all characterized representatives of this enzyme family. These include a central, molybdenum-cofactor-binding protein domain exhibiting the so-called "SUOX" fold (SUOX for sulfite oxidase, the first structurally characterized enzyme of the family),[12] the presence of a single equivalent of the pyranopterin cofactor (present as the mononucleotide form) coordinated to the metal, and a cysteine as a direct amino acid ligand to the catalytic Mo centre (Figure 7.1). The following sections will review the occurrence and function of sulfite oxidase family enzymes in pro- and eukaryotic organisms, and will also provide an overview of key catalytic and spectroscopic properties of these enzymes.

7.2 Phylogenetic Structure of the SO Enzyme Family

The fact that genes encoding proteins with significant homologies to characterized sulfite oxidase family proteins are found in all three domains of life[13,14] indicates that the SO enzyme family is evolutionarily ancient, and that precursors of the present enzymes were likely already present in the most ancient living cells, including the last common universal ancestor. Analyses of cellular molybdoproteomes using comparative genomics has revealed that only very few groups of organisms consistently lack SO family enzymes, and that in most cases it would appear that the loss of SO enzymes from many of these organisms is a relatively recent evolutionary event, as these enzymes are still present in phylogenetically closely related groups.[13,14] Within the Bacteria these organisms include two groups of Firmicutes (Lactobacillales and Mollicutes), the Spirochaetes, Fusobacteria, Thermotoga group and the Chlamydiae, with the latter having a highly derived lifestyle, accompanied by a strong reduction of genome contents.[13,14] In the Archaea, SO family enzymes occur in all Crenarchaeota and Euryarchaeota except the methanogenic lineages, and are also found to be absent in Nanoarchaeota.[13,14] Data on the Thaumarchaeota are not presently available in the literature,[13,14] but our analyses indicate that SO family enzymes are indeed present in Thaumarchaeota. In the Eukarya, SO family enzymes are absent in those few groups that have lost the ability to synthesize Moco, including several types of Fungi, Alveolata and others.[13,14]

The central feature that defines the members of this large enzyme family is the Moco-binding domain, which retains the same three-dimensional fold (Figure 7.2) as well as key features of the active site structure common to all members. The SUOX fold consists of a characteristic combination of 9 to 12 α-helices and 2 to 3 β-sheets and so far has been found in all structurally characterized representatives of the SO family,[12] which includes four sulfite-converting enzymes originating from chicken liver, the plant *Arabidopsis thaliana* and two different species of Proteobacteria, as well as the catalytic domain of a plant nitrate reductase and a bacterial reductase (YedY) with a possible function in methionine sulfoxide repair.[15–19]

The detailed structure of the Mo-domain including its exact size and the amino acid signatures around the active site can be diagnostic for specific subgroups of enzymes.[12,20] These small but significant variations in the Mo domain structure are clearly reflected in the conserved domains database,[21] where the SO family enzymes appear as a superfamily (cd00321) with four subfamilies (cd2107-2110), the YedY-like Moco; the Bacterial SO family Moco group, the Archaeal/Bacterial SO Moco group and the SO family Moco Dimer group. The SO-Moco Dimer group is the only group that contains four further subgroups, the Eukaryotic SO and NR enzymes, and the bacterial SoxC and SorA Moco.

One of the distinguishing features of the four major subfamilies within the SO enzyme superfamily is the size of their Mo-domain (Figure 7.3). The YedY-like Moco and the Bacterial SO family Moco groups have domain sizes of ~35 and ~30 kDa, respectively, while the Archaeal/Bacterial SO family Moco has the smallest Moco domain of about 26 kDa. Enzymes of the SO family Moco dimer group can be distinguished from all other subgroups of the SO family in that their Moco domain always occurs in tandem with a highly conserved dimerization domain (~40 kDa for these two domains, additional domains can be present).

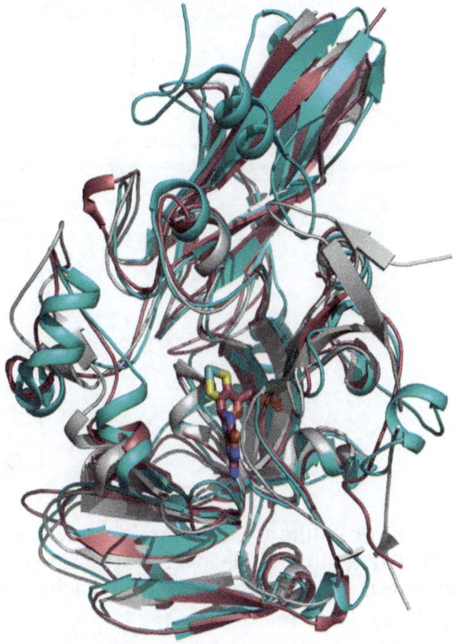

Figure 7.2 The SUOX fold. Overlay of the Mo domains of chicken SO, Pichia NR and SorA. Chicken sulfite oxidase (PDB 1SOX, red); *S. novella* sulfite dehydrogenase (PDB 2BPB, gray); *P. angusta* assimilatory nitrate reductase (PDB 2BIH, teal).

Of the four major protein groups found in the SO family (Figure 7.3), the SO family Moco dimer group contains most of the structurally and biochemically characterized enzymes, with crystal structures of at least one representative of each of the four subgroups having been reported.[16,17,22,23]

The only other major group containing a structurally characterized enzyme is the YedY-like Moco group.[15] Neither the Bacterial SO family Moco group nor the Archaeal/Bacterial SO Moco group currently contain any biochemically or structurally studied representatives.

Phylogenetic analyses based on the conserved SUOX domain indicate that bacterial enzymes dominate the SO enzyme family and are found in all four subfamilies, but at least two other systems for the classification of SO family enzymes have also been reported. The first large-scale analysis classifies the SO family enzymes into three major groups based on the structure of the molybdenum domain, with one of the groups having two subgroups.[20] Conserved amino acid motifs around the active site have also been reported. The four groups identified using this method correspond to the four major conserved domain groups discussed above. Additional diversity of the sequences is apparent within each of the major groups, but no assignments of function have been attempted given the lack of biochemical information. Several previously unrecognized forms of SO enzymes have also been described, including cytoplasmic enzymes and

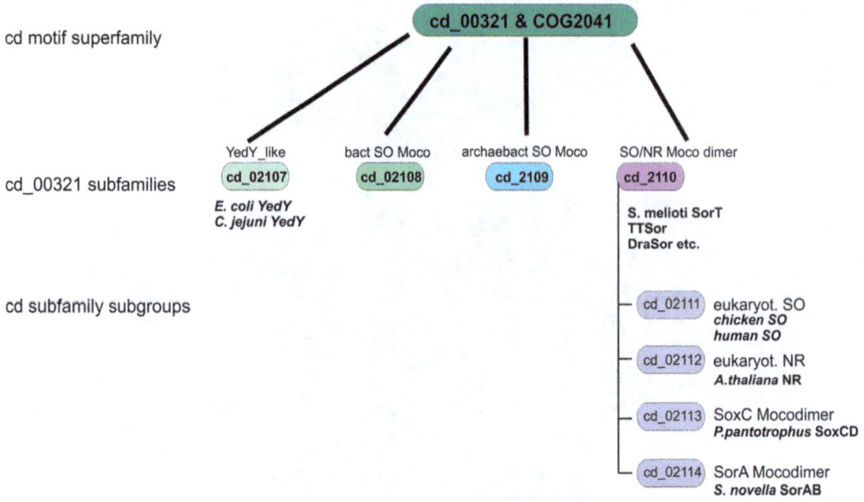

Figure 7.3 Classification of Sulfite Oxidase family enzymes based on conserved domains (CDD database). The overall structure of the family is shown, with characterized enzymes contained in each group shown in bold.

proteins containing integral membrane domains.[20] Another classification system has been proposed based on the fold of the central molybdenum domain using comparisons of SUOX fold secondary structure elements in all sequences available at the time.[12] This study also compared the active site residues as well as the presence of export signals. Based on these characteristics, the enzymes of the SO family are proposed to form eight clades, two of which are classified as "True YedY" and "YedY-related" groups, while all of the groups are classified as "SUOX" or SUOX and NIA (assimilatory nitrate reductase) groups.

Again, the above analyses have clearly highlighted the diversity of the proteins in this enzyme family, as well as the fact that there are several groups of proteins for which no function is known, but which are likely to catalyze new reactions, as indicated *e.g.* by a cytoplasmic localization (Clade 7). The structure-based clade system and the Mo domain-based classifications are related, and a scheme of equivalencies has been proposed.[4]

An interesting case are the recently discovered mARC/MOSC molybdenum enzymes, whose members catalyze Moco sulfuration or convert amidoxime substrates (mARC = mitochondrial amidoxime reducing component). A clear classification for these enzymes has so far not been established, but some reports have indicated that they belong to the sulfite oxidase enzyme family based on the fact that mARC enzymes appear to have an active site that is similar to those found in SO family enzymes. For a mARC-like enzyme from *Chlamydomonas*, mutation in a particular cysteine residue abolished activity, which has been taken as evidence that, as in the SO family enzymes, this cysteine is a ligand to the Mo cofactor in these

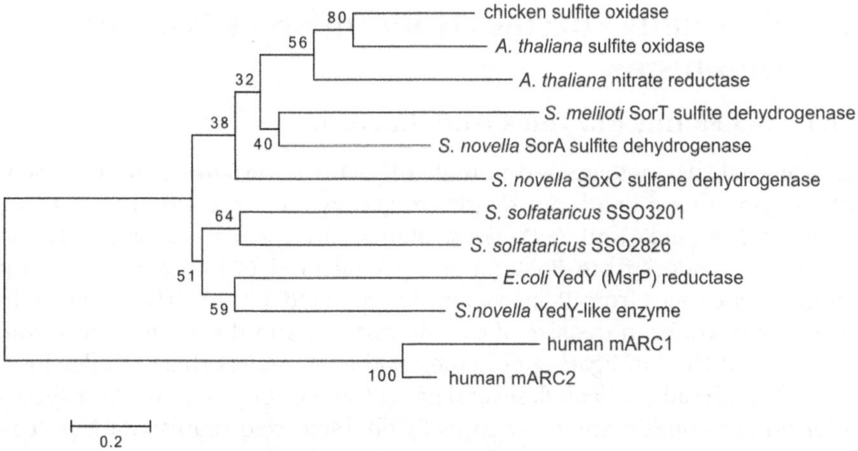

Figure 7.4 Phylogenetic analysis of established members of the SO fold in comparison to mARC proteins shows a clear separation of the two groups. The phylogenetic analysis was carried out using Mega 6.0 and the Neighbour-joining algorithm. Bootstrap values are indicated at the branchpoints and used 500 replicates. All sequences were retrieved from publically available databases. Two sequences from uncharacterized archaeal enzymes of the SO family (SSO2826 and 3201 from *Sulfolobus solfatariucs*) were included to represent this group of uncharacterized enzymes.

enzymes. This has recently been confirmed in studies of the active site of two MOSC enzymes (pmARC, HMCS) using EXAFS,[24] as well as mutational analyses involving a Cys273 in the human mARC enzymes that abolished enzyme function.[25]

We have carried out a small-scale phylogenetic analysis of characterized SO family enzymes, representatives of known SO family enzyme groups and the mARC1 and 2 enzymes from humans (Figure 7.4). The two mARC enzymes clearly group in a separate cluster from all the other enzymes, and sequence homologies were just above 20%, with amino acid identities as low as 6–8%. When inspecting the alignment, the cysteine ligands of all SO family enzymes are found in the same position, but in the mARC enzymes no cysteine is observed in that particular position. There is a cysteine motif in the mARC sequences but it is located at a distance of 14 residues from the SO family Mo-ligating cysteine and could be the natural amino acid ligand to the Mo centre: recent spectroscopic analyses have confirmed that mARC-type enzymes have a Mo active site that appears to be identical to that of the SO family enzymes, while other representatives of the MOSC proteins appear to have a cysteine-persulfide amino acid ligand to the Mo centre.[25] Overall this appears to indicate that the mARC enzymes, while probably originating from the same ancestral enzyme as the SO/NR family enzymes, form a new family of molybdenum enzymes, rather than being part of the sulfite oxidase family, as has also been proposed by ref. 11.

7.3 SO Family Enzymes from Different Types of Organisms

7.3.1 SO Family Enzymes from Bacteria

The bacterial enzymes clearly dominate the SO enzyme family, with representatives being found in all four major groups: two of these groups, cd_02107 and cd_02108 (YedY and YedY-like proteins), are exclusive to Bacteria, one contains sequences from Bacteria and Archaea (cd_02109) and one group contains sequences from Bacteria and Eukarya (cd_02110). The enzymes in these groups differ in the size of the Mo domain and the amino acid signatures around the Mo ligating cysteine residue as well as their cellular location.[3,4,20] As already indicated, several of the known subgroups of SO enzymes so far do not contain any biochemically characterized representatives. The different groups and what is known about their properties will be discussed below, additional details are available in ref. 4 and 20.

7.3.1.1 *The cd_02107 Group: YedY and Related Enzymes*

This group contains one characterized representative, the eponymous YedY gene product from *E. coli*. The YedY-like enzymes generally have a Mo domain size of 300–330 amino acids (~35 kDa), and the encoding genes occur together with a gene encoding a membrane bound *b*-type cytochrome, YedZ. YedZ functions as the external electron transfer partner to YedY and contains six transmembrane helices.[20] In the database, YedZ proteins are classified as "PRK05419 putative sulfite oxidase subunit YedZ" and COG2717, YedZ. The subunits are distantly related to ferric reductase subunits. The annotation of the YedZ proteins in existing genome sequences varies and includes "sulfoxide reductase subunit YedZ" as well as "sulfite oxidase subunit YedZ" and "ferric reductase transmembrane component protein".

Bioinformatics analyses indicate that the YedY enzymes are located in the periplasmic space, and all protein sequences contain a Tat-leader peptide.[4,12] The *E. coli* YedY protein was first identified during a search of the *E. coli* genome sequence for new molybdenum enzymes.[15,26] The crystal structure of the enzyme showed the typical SUOX fold of SO enzymes, but there were significant differences in the amino acids surrounding the active site, indicating that the substrate of this enzyme is not sulfite.[15] Following determination of the reduction potentials of the YedY Mo centre (E_{m7} = 132 mV) and the YedZ heme group (E_{m7} = −8 mV), it was concluded that YedY most likely functioned as a reductase,[15,26] and a weak dimethylsulfoxide (DMSO) reductase activity was detected for purified YedY.

Spectroscopic studies of YedY have been undertaken by several groups,[27–29] revealing some unusual features: EPR-based redox titrations indicate that the YedY Mo centre does not normally assume the fully oxidized Mo(VI) state, and as isolated is in the partially reduced Mo(V) state. This is unusual, as it would limit the number of electrons that can be stored in/transferred to the Mo centre during catalysis to just one rather than two as is the case for most enzymes in the SO family. The coordination sphere of the Mo centre is similar to that seen for sulfite oxidases, with three equatorial sulfur ligands (two dithiolene sulfurs and one cysteine sulfur), one apical oxo and one equatorial hydroxyl/hydro ligand, with EXAFS studies indicating that a nearby Asn residue might be weakly coordinated to the Mo centre (bond length 2.64 Å) in the Mo(v) state.[27] This sixth ligand has not been observed in other SO family members. It has also been noted that the contribution of the dithiolene sulfur to the coordination sphere of the YedY Mo centre suggests a stronger involvement of these ligands in fine-tuning the reduction potential of the YedY Mo centre than in other SO family enzymes.[27,28]

Another more recent study using a similar set of spectroscopic methods also concluded that there is an additional ligand with a bonding distance of ~2.6 Å, which has been attributed to the same Asn residue or a Glu residue.[29] The authors of this work have also shown that exposure of YedY to DMSO did not lead to any changes in the coordination sphere of the Mo centre, while the alternative substrate TMAO, and the substrate analogue acetone, did bind to the Mo centre, presumably through the equatorial oxygen ligand.[29]

The amino acid residues around the YedY Mo active site resemble those found in *S*- and *N*-oxide reductases,[20] with tryptophan residues being particularly prominent, while the arginine residues found typically in SOEs are absent, which is in keeping with the weak DMSO reductase and *N*-oxide reductase activity reported for YedY.[15,26] However, DMSO reduction is unlikely to be the physiological function of YedY as YedY was unable to support anaerobic growth of *E. coli* with DMSO as respiratory substrate.[26] A recent paper found that in *E. coli*, YedY/YedZ form a periplasmic methionine sulfoxide detoxification system capable of converting both the *R*- and *S*-stereoisomer of methionine sulfoxide.[19] It was shown that the enzyme was induced specifically in response to hypochlorous acid, which is released by neutrophils during bacteria–host interactions. The authors of this work accordingly renamed the enzymes MsrP/MsrQ to better reflect the function of the enzyme.

Similarly, in *Campylobacter jejuni*, a strain carrying a gene knockout in a YedY-related enzyme had defects in colonizing chicken, which also suggests that YedY is important for host cell interactions.[30] However, the YedY knockout strain of *C. jejuni* was shown to have a nitrosative stress phenotype rather than a sulfoxide stress phenotype,[30] which matches the results of the spectroscopic studies above. Genes encoding enzymes related to YedY (MsrP) are found in many Enterobacteria and other bacterial species, many of which are known pathogens.[20]

7.3.1.2 The cd_02108 Group of Enzymes

The enzymes of this group have catalytic subunits of approximately 30 kDa, or 240–270 amino acids, and to the best of our knowledge there are at present no well-characterized representatives of this group. The smaller physical size of the Mo domain is due to a loss of parts of the C-terminal end of the domain as compared to YedY. In addition, there are clear differences in the cd_02108 enzymes and YedY in several key residues, including the amino acid signature surrounding the cysteine ligand to the Mo centre.[20] Enzymes of this type group in Clade 2 according to the classification of ref. 12.

Genomic analyses indicate that, similar to the situation seen with the YedY enzymes, the genes encoding the cd_02108 group enzymes generally occur together with genes encoding a membrane-bound cytochrome. These membrane-bound cytochromes differ from the YedZ proteins in that they have six transmembrane regions and belong to COG4117 (YdhU), annotated as "thiosulfate reductase cytochrome *b* subunit". It is unclear at present whether this group of proteins act as dehydrogenases or reductases. Based on sequence alignments and attempts to identify important amino acids based on their relative positions in the sequence, the active site environment appears to be less hydrophobic than seen in YedY enzymes, with fewer tryptophan residues being present. The general localization of these proteins appears to be periplasmic.[12] It is noteworthy that for some of the enzymes in this group, originating from *e.g. Mycobacterium avium* and *Rubrobacter xylanophilus*, it appears that a gene fusion between the Mo subunit and the membrane-bound cytochrome has occurred, essentially leading to a single, membrane-bound protein.[20] Another important issue to note is that in standard genome annotations, the proteins belonging to this group are generally annotated as "YedY" and "YedZ", respectively, for the catalytic and the heme-containing subunits (this is even the case for sequences used in the reference alignment for cd_02108 available through NCBI). This is a problem as the proteins in this group are clearly distinct from the "true" YedY proteins, in terms of their domain structure and the residues surrounding the active site, which indicates that these proteins likely convert specific, but as yet unknown, substrates, but are not likely to be *S*- and *N*-oxide reductases like the YedY proteins.

7.3.1.3 SO Family Proteins from Archaea and Proteobacteria (cd_02109)

The proteins in this group are the smallest known members of the SO family and consist of a single Moco domain of about 22–25 kDa through a further reduction of the N-terminal part of the domain. Again, to the best of our knowledge no representative of these enzymes has been studied and no proposal as to a specific physiological function has been made, although they occur in a variety of bacterial and archaeal phyla including both Cren- and Euryarchaeota and representatives of the high GC Gram positives, Firmicutes, the Thermus/Deinococcus group, cyanobacteria and various α-Proteobacteria.[20] Within cd_02109, the archaeal and bacterial enzymes form distinct groups that can be differentiated based on their sequence similarities, which drops to ~25–33% for comparisons across the two groups, with higher values being obtained for comparisons within the two groups. The enzymes of this type generally lack a signal peptide and are devoid of transmembrane helices, indicating a cytoplasmic location.

For one of the bacterial enzymes, SMb20584 from *Sinorhizobium meliloti* 1021, it has been shown that gene expression is increased in the presence of glucose or thiosulfate in the growth medium, while expression levels in the presence of taurine and methanesulfonic acid are lower.[31] This does not indicate a direct association of gene expression with the presence of sulfur compounds in the growth medium, suggesting that the enzyme may catalyze an unrelated reaction. The conserved residues around the active site resemble those found in the YedY and YedY-like enzymes.[20] It will be interesting in future to unravel the function of these enzymes and especially to establish whether they are involved in the metabolism of sulfur compounds.

7.3.1.4 The Bacterial Enzymes of the "Moco-dimer" Group – Bacterial Sulfite and Sulfur Dehydrogenases

Within the sulfite oxidase family, the cd_02110 group of Moco-dimer enzymes has several readily identifiable subgroups, two of which consist exclusively of bacterial enzymes (cd_02113/2114), while the other two are exclusive to eukaryotic organisms (cd_02111/2112) and contain sulfite oxidases and assimilatory nitrate reductases, respectively. The bacterial Moco-dimer enzymes also can be distinguished based on the reactions they catalyze: the cd02113 "bact_SoxC_Moco" group consists exclusively of the SoxCD "sulfur dehydrogenases" that are part of the bacterial Sox pathway for the oxidation of reduced sulfur compounds,[23] while the cd_02114 "bact_SorA_Moco" group contains bacterial sulfite-oxidizing enzymes such as SorAB.[32] It should be noted, however, that the existing subgroups are very specific and only contain very closely related groups of enzymes. In fact, most of the characterized bacterial sulfite-oxidizing enzymes show great homology to the cd_02110 domain pattern, but not to any of the subgroups.[4,31] This includes the SOEs from *Delftia acidovorans*, *Thermus thermophilus*, *Deinococcus radiodurans*, *Campylobacter jejuni* and *Sinorhizobium meliloti.*[3,4,31,33–37] Only the SorAB sulfite dehydrogenase from *Starkeya novella* and a related enzyme from *Cupriavidus necator* that has only partially purified clearly fall into the cd_02114 group of enzymes.[16,32,36] Thus, when naming bacterial sulfite oxidizing enzymes, care should be taken as clearly not every bacterial SOE is a "SorA". This topic has also been reviewed in ref. 3, 4 and 20.

The unifying characteristic of all these enzymes, regardless of origin, is that their Mo-binding domain always occurs fused to a dimerization domain, bringing the overall size of the central catalytic subunits for the bacterial enzymes to ~40 kDa. The dimerization domain has been named to indicate that in the crystal structures of the plant and vertebrate enzymes that are available it mediates most of the subunit interactions.[3,4,17] Another interesting observation is that almost all characterized bacterial enzymes in this group appear to be encoded in operons that also contain genes encoding *c*-type cytochromes.[4,20] However, despite this, these cytochromes fulfil various functions as either permanent subunits of the enzymes (SoxCD, SorAB) or as final electron acceptors (SorT)[16,32,33] (see detailed discussion below).

7.3.1.5 The Bacterial SoxCD Proteins (cd_02113)

The SoxCD sulfur or sulfane dehydrogenases are proteins that are composed of a catalytic Moco-dimer subunit (SoxC), and a cytochrome *c* subunit, SoxD, that can contain either one or two heme groups.[23,38] The SoxCD protein catalyzes the six-electron oxidation of a sulfane sulfur atom, bound to Cys110 of the SoxY protein, to a sulfone group as part of the reaction cycle of the thiosulfate-oxidizing Sox-multienzyme system.[23,39] In bacteria that completely oxidize thiosulfate and other reduced sulfur compounds to sulfate, this enzyme system has four essential components, SoxAX, SoxYZ, SoxB and SoxCD, and these tend to be encoded in a single large operon, with accessory and regulatory proteins encoded by nearby genes.[40] SoxAX catalyzes the attachment of the sulfur substrate to a conserved cysteine on the SoxYZ protein, SoxB hydrolyzes the outer sulfone groups of the cysteine-bound thiosulfate, leading to the formation of sulfate. SoxCD subsequently converts the remaining sulfane group to another sulfone group, followed by another reaction with SoxB, regenerating the SoxYZ protein and liberating another sulfate molecule.[40–43]

The SoxCD proteins were initially described as sulfur dehydrogenases, but recently this has been changed to sulfane dehydrogenase to better reflect the proposed reaction mechanism, where the main substrate of the enzyme is a SoxY-Cys110-persulfide, which is oxidized through sulfenic and sulfinic acid intermediates to a SoxY-Cys110-sulfonic acid.[23] While the presence of SoxCD is essential for the function of the overall Sox complex reaction, it is interesting to note that different forms of the SoxD protein exist that bind either one or two heme groups, and it was shown that for two heme forms of SoxD, only one of the two heme groups is essential for activity.[20,44,45]

The structure of the SoxCD enzyme from *Paracoccus pantotrophus* with the heme-2 of SoxD deleted has been recently reported, confirming that the protein is a heterotetramer in which the interactions between two individual SoxCD complexes are entirely mediated by the SoxC subunits.[23,39] The SoxC subunit has the typical SUOX fold found in all SO family enzymes, although interesting differences exist in the structure of the SoxCD active site and in the position of the SoxD protein relative to the active site/main body of the enzyme. While in the sulfite-oxidizing enzymes of the Cd_02110 group a conserved arginine hydrogen bonds to both redox-active cofactors, in SoxCD histidine 116 has been proposed to carry out this function. The conserved arginine 114 has been proposed to hydrogen bond to water molecules to position them in the correct orientation for oxygen atom transfer to the substrate. No reference has been made to a residue equivalent to the Tyr 236 in *e.g.* the SorAB sulfite dehydrogenase, in the SoxCD active site. This tyrosine is conserved in SorAB, CSO and SoxCD and in the former two enzymes hydrogen bonds to the axial oxo/hydroxo group of the Mo centre.

The SoxD subunit of the sulfane dehydrogenase has an exposed heme group that allows a close approach of the redox-active cofactors of the two subunits to an edge-to-edge distance of 10.4 Å and a Mo–Fe distance of

18.6 Å. Interestingly, the SoxD subunit is positioned differently from the SorB subunit in the SorAB sulfite dehydrogenase, the only other SO family enzyme having a permanent complex between a Mo and a heme subunit to have been crystallographically characterized (see below).[16] This is interesting, as the SorB binding site has been proposed to be identical to the site at which the mobile heme domain of the mammalian sulfite oxidases interacts with the Mo active site,[46] and to the interactions site between another bacterial sulfite oxidizing enzyme, SorT, and its electron acceptor, SorU[47] (see below).

The altered interaction site of SoxD with the SoxC Mo subunit leads to a change in the position of the bulk of the heme cofactor relative to the Mo redox centre, although the positions of the heme propionate groups that are involved in the closest approach between the two cofactors are nearly identical for the SoxCD and the SorAB complexes.[23] The unusual binding site for the SoxD subunit may be due to the nature of the SoxCD substrate, which requires the SoxYZ protein to approach the Mo active site very closely in order for the cysteine persulfide substrate, which is located on a mobile C-terminal motif of the SoxY protein, to be able to interact with the Mo active site. Unfortunately, no information is available regarding the relative position that the deleted second heme domain of the SoxD protein might occupy in the wild-type SoxCD protein. The available structure suggests a gating function for SoxD–Arginine 98 that would prevent substrates other than the SoxY C-terminus from entering the active site.[23]

Spectroscopic investigations of the SoxCD active site have shown that the Mo centre is very similar to that of the well-characterized SOEs in the Moco-dimer enzyme group, exhibiting the typical high- and low-pH EPR signatures found in the vertebrate SOs.[39] A main difference that has been noted, however, is the absence of hyperfine splitting in the low-pH EPR spectrum of SoxCD, indicating the absence of a strongly coupled proton that is typical of the vertebrate SOEs.[3,48]

7.3.1.6 The Bacterial "SorA"-type Sulfite Dehydrogenases (cd_02114)

The only well-characterized enzyme in this group is the heterodimeric SorAB sulfite dehydrogenase from *Starkeya novella.*[16,32] This enzyme was the first bacterial SOE to be purified to homogeneity and structurally characterized, and as a result of its unique homodimeric structure was the first heme containing SOE that allowed insights into the intramolecular electron transfer events that are essential for catalysis.[16] The SorAB SOE has been studied using a variety of spectroscopic techniques, including EPR and EXAFS, protein film voltammetry, laser flash photolysis as well as kinetic analyses of both steady-state and pre-steady-state parameters.[4,16,20,32,49–61] This work has revealed the roles of several active site residues, including a conserved arginine (Arg55) that hydrogen bonds to both the heme and Mo cofactors, in facilitating electron transfer between the redox-active centres. Substitutions of Arg 55 have a profound effect on substrate affinity, decreasing it by several orders of magnitude. His 57, located near the cofactors in SorAB, has also been found to modulate the substrate affinity of the enzyme. Arg 55 and His 57 are equivalent to Arg 114 and His 116 in SoxCD (discussed above) where they apparently have very different functions. Tyr 236 of SorAB, which hydrogen bonds to the catalytic oxo/hydroxo group of the Mo centre, is found to strongly affect turnover, with a moderate effect on substrate affinity. Interestingly, no equivalent to the third arginine found in the substrate access channel of vertebrate SOs (Arg450, CSO numbering) is present in SorAB, and yet high affinities for the substrate sulfite are observed. The available data on this enzyme have been reviewed several times.[3,4,20]

Because it was the first bacterial SOE to become available for detailed studies, SorAB is often (erroneously) regarded as a "prototype" for bacterial SOEs. In fact, SorAB is currently the only example of a bacterial SOE that exists as a permanent heterodimeric complex of a heme and a Mo subunit, as none of the SOEs purified from other bacteria to date contained a heme subunit. Instead, all of these additional bacterial enzymes contained only Mo redox centres (see below). It should also be noted that while genes encoding cytochrome components are often found downstream of SOE catalytic subunits, there is significant variation in the structure and type of these cytochrome components that well exceeds the variability of the Mo subunits.[20] The cytochrome subunits are only conserved in very closely related enzymes, while more distant relatives generally can be recognized through their differing associated cytochromes. In addition, the presence of a cytochrome-encoding gene in an SOE-encoding operon cannot be taken as an indication that the cytochrome is a part of the actual SOE (again, a misconception that often leads to annotations of bacterial SOEs as "SorAB") – most bacterial SOEs that have been purified to homogeneity to date are Mo-only enzymes, despite the fact that the encoding operons all include a gene encoding a cytochrome. In two of these cases it has been shown that these encoded cytochromes are external electron acceptors for these enzymes.[33,47,62] Thus, even within the cd_02114 group it cannot be taken for granted that a heme subunit is part of all enzyme as *e.g.* the SOE from *C. necator* has been reported to be a heme free enzyme (Table 7.1).

7.3.1.7 Other Bacterial Sulfite Oxidizing Enzymes (cd_02110)

Since the purification of SorAB, SOEs from various other bacterial species have been reported, and as already indicated none of these enzymes contains a heme (Table 7.1).[3,4,20] Except for the *C. necator* SOE, which is part of the cd_02114 group, the SOEs from *T. thermophilus*, *D. radiodurans*, *S. meliloti*, *C. jejuni* and *D. acidovorans* all have Mo domains similar to the cd_02110 Moco-dimer group of enzymes, but do not fall into any of the subgroups (cd_02113; cd_02114) already established within this category. It is thus likely

Table 7.1 Properties of bacterial Moco-dimer sulfite-oxidizing enzymes.[a]

	S. novella SorAB	*C. necator* SorA	*T. thermophilus* TTSor	*D. radiodurans* DraSOR	*S. meliloti* SorT
Subunit composition	Aβ	α	α2	α/α2	α2
Cofactors	Mo, heme *c*	Mo	Mo	Mo	Mo
Cdd group	cd_02114	cd_02114	cd_02110	cd_02110	cd_02110
$K_{M\ sulfite}$	27 μM	50–100 μM	10.7 μM	94.5 μM	15.5 μM
$K_{M\ cyt.c}$	2 μM	n.r.	n.r.	n.r.	31.8 μM[d]
$K_{M\ ferricyanide}$	n.d.	0.9 mM	6.3 μM	n.r.	3.4 μM
k_{cat}	345 s^{-1}	n.r.	53,318 s^{-1} [c]	n.r.	343 (139[d]) s^{-1}
V_{max}	422 U mg^{-1}	412.6 U mg^{-1} [b]	n.r.	n.r.	522 U mg^{-1}
Reference	32,53	36	35	34	31,47

[a]n.r. – not reported, n.d. – not determined.
[b]computed from the activity of 6850 mkat kg^{-1}
[c]value as in ref. 34, a specific activity of 56.67 U mg^{-1} was reported for the purified enzyme.
[d]value with SorU cytochrome as acceptor.

that additional subgroups will eventually be defined as part of the Moco-dimer type enzymes that make up cd_02110.

Like their counterparts from vertebrates, plants and the other bacterial SOE groups, all of these bacterial enzymes have good affinities for sulfite, and in two cases the natural electron acceptor has been shown to be *c*-type cytochromes.[47,62,63] This is of interest, as enzymes in this category were initially referred to as "atypical" SOEs in the literature because they did not show reactivity with horse heart cytochrome as an electron acceptor but rather showed high activities with the artificial electron acceptor ferricyanide.[64] It is now known that the acceptor cytochromes for these enzymes have an acidic pI rather than the highly basic pI of horse heart cytochrome *c*, explaining the lack of reactivity with the latter as an electron acceptor. Another property of these enzymes reported early on was their unusual molecular mass of ~1.5 monomers as determined by gel filtration. This has been observed for the enzymes from both *D. acidovorans* and *S. meliloti*,[31,33,64] and for the latter MALLS was used to show that the enzyme is a true homodimer. The enzymes have been reviewed in ref. 4, and therefore only an update will be presented here.

At present the two best-studied enzymes in this group are those from *Thermus thermophilus* and *Sinorhizobium meliloti*. The catalytic parameters for both of these enzymes are shown in Table 7.1, and for both enzymes it has been shown that a cytochrome *c* is the natural electron acceptor, confirming that the enzymes are linked to metabolic functions such as respiration.[33,62,63] The *S. meliloti* SorT SOE is the only enzyme in this group for which a crystal structure is available.[47] It confirms the homodimeric structure of the enzyme but also reveals an unprecedented arrangement of the two monomers in what is essentially an "upside down" dimer, *i.e.* the dimerization is not mediated by contacts between the dimerization domains, but by contacts between the dimerization and Moco domains.[47] The active site is very similar to those of other SOEs, showing the typical pentacoordinate Mo centre with a cysteine ligand attached and two arginine residues in the substrate access channel.[47] The unusual domain packing of the enzyme might explain the aberrant gelfiltration behaviour observed in earlier work. The SorU cytochrome that is encoded by a gene directly downstream of the *sorT* gene has been shown to be the natural electron acceptor for SorT.[33,47] A second redox-active protein, a pseudoazurin, is encoded directly downstream of the *sorU* gene.[33] Both proteins have been investigated for their ability to accept electrons from SorT, and only SorU is found to be an efficient electron acceptor for SorT.[33] The physiological role of the pseudoazurin remains unclear, although both SorU and the pseudoazurin are able to transfer electrons to respiratory chain components as indicated by increased oxygen consumption of reconstituted respiratory systems in the presence of the proteins.[33]

The interactions between SorT and SorU have been probed in detail, and a crystal structure of the electron transfer complex has recently become available.[47] This paper also reported a K_{MSorU} of 32 ± 5 μM and a turnover number of 140 ± 11 s^{-1} for the SorU-based SorT assay. The SorT–SorU complex has an unusual protein–protein interface, including a large number of hydrogen bonds that are normally associated with more permanent complexes. SorU also shows some conformational changes on complex formation with SorT, while the SorT structure is largely unchanged.[47]

For the *T. thermophilus* SOR (TTSOR), an electron transfer chain from sulfite to oxygen *via* the action of TTSOR in conjunction with a diheme cytochrome c_{550} and a monoheme cytochrome c_{552} directly to cytochrome oxidases of the aa_3 or ba_3 type has been established.[62] In addition, it has been shown that only the N-terminal domain of the cytochrome c_{550} interacts with TTSor, and that intramolecular electron transfer within the cytochrome c_{550} is rapid, indicating that both electrons liberated by sulfite oxidation can be transferred directly from TTSor to one molecule of cytochrome c_{550}.[63] This would represent a simplification of the enzyme mechanism (see below), as in all other SOEs that use *c*-type cytochromes as electron acceptors two separate interaction events involving single-heme cytochromes are required to fully reoxidize the SOEs.[3,4,32,33,65] The gene encoding cytochrome c_{550} is found directly downstream of the gene encoding TTSor.[62]

It is interesting that SOEs of this type appear to require an electron transfer partner with an acidic pI, *i.e.* require a surface charge structure that is different from those of typical "mitochondrial type" cytochromes such as horse heart cytochrome c_{550}, which have pIs close to ~10. The pI for SorU is 4.29, and we also computed the pI for the *T. thermophilus* cytochrome c_{550}, which is 6.15. In contrast, the secondary electron acceptor in *T. thermophilus*, cyt. c_{552}, has a pI of 9.52, in keeping with it being an electron shuttle that connects to the terminal reductases. A final observation for this class of enzyme is that in some cases the cytochromes may be membrane-bound.[36] We have not been able to confirm this analysis, however, and according to our data both of the cytochromes in question are soluble, periplasmic proteins.

7.3.2 SO Family Enzymes from Vertebrates

In vertebrates, SO catalyzes the two-electron oxidation of sulfite to sulfate coupled to the reduction of two molecules of cytochrome *c*.[66] The structure and reaction mechanism of SO has been extensively reviewed by Hille and colleagues.[1] Here we will focus more on the biological function and deficiency of SO. Sulfite oxidation represents the terminal step in the oxidative degradation of cysteine. The intracellular pool of cysteine is small as most of it is bound in the much larger pool of glutathione (GSH).[67] In the extracellular compartment, cysteine is oxidized to cystine (50–150 μM), which represents the major transport form of cysteine.[68] Cysteine and cystine are transported by different membrane carriers.[35] In the central nervous system, glial cells mainly import cysteine *via* the cystine-glutamate antiporter (xCT) providing the major route for GSH synthesis in the brain.

Vertebrate SOs are homodimers with each monomer harbouring an N-terminal cytochrome b_5-type heme domain, a catalytic Moco domain and a C-terminal dimerization domain (Figure 7.5).[17] In vertebrates, SO is localized in the mitochondrial intermembrane space where electrons derived from sulfite oxidation are directly passed to the physiological electron acceptor cytochrome *c*.[69] Until now, the contribution of sulfite to mitochondrial respiration has not been considered as a source of any significance. However, we have found that sulfite can increase oxygen consumption in mitochondria of wild-type fibroblasts, indicating that electrons derived from sulfite oxidation significantly contribute to the activity of cytochrome *c* oxidase (complex IV) as well as ATP synthesis (Schwarz *et al.*, unpublished results). In contrast, respiration in cells derived from patients with molybdenum cofactor (Moco) deficiency decreases with as little as 10 μM sulfite, demonstrating a strong inhibitory effect of sulfite on mitochondrial respiration in the absence of active SO. These findings are consistent with reports that mitochondrial glutamate dehydrogenase is a target of sulfite toxicity.[70] Future studies are needed to establish the interconnection between sulfite-driven

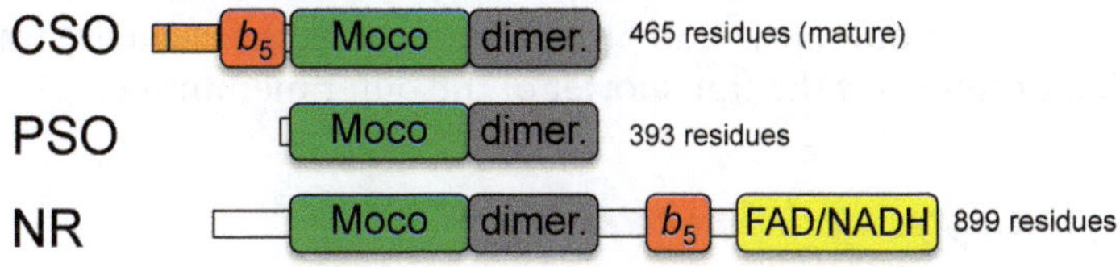

Figure 7.5 Domain structure of chicken SO (CSO), Arabidopsis SO (PSO) and nitrate reductase (NR). Bound cofactors are indicated in the respective domain. The dimerization domain is shown in grey. The orange extension in CSO indicates the bipartite mitochondrial targeting sequence of 80 residues.

mitochondrial respiration and sulfite-induced inhibition of ATP synthesis. One hypothesis might be a homeostatic link between the capacity of cytochrome *c* to accept reducing equivalents and the "feeding" of the respiratory chain by complex I, which needs to be reduced under conditions of limiting concentrations of oxidized cytochrome *c*.

7.3.2.1 Maturation of Mammalian SO

As a mitochondrial enzyme localized to the intermembrane space, SO requires a specific translocation and maturation pathway to function physiologically. This process is highly coordinated and involves targeting and mitochondrial processing of SO, integration of Moco and heme cofactors and homodimerization.[71] We have identified Moco as a key factor in SO assembly, which not only ensures mitochondrial localization but also directs heme integration and dimerization. In the absence of Moco, the majority of SO is not present within the mitochondrion but is instead mis-localized to the cytosol, although processing of the N-terminal mitochondrial-targeting peptide has shown that the protein has been imported into mitochondria to the site of proteolytic cleavage (inner membrane). Cytochrome *c* also requires the heme cofactor as well as the heme-integrating cytochrome *c* heme lyase for correct localization to the mitochondrial intermembrane space.[72] Such cofactor-dependent translocation defines the so-called folding-trap mechanisms for mitochondrial import.[73] In addition, a bipartite mitochondrial targeting signal in SO is required, which involves the pre-sequence pathway and further requires an intact membrane potential across the inner mitochondrial membrane and ATP.[74] While this signal peptide is essential, it is not sufficient for mitochondrial localization of SO, given that Moco must be present to ensure localization of SO within the mitochondrial intermembrane space. SO is thus unique among mitochondrial intermembrane proteins in combining conventional leader sequence-based translocation with the folding trap mechanism for which the presence of Moco is strictly required.[71] These data further demonstrate that Moco is integrated before the heme, suggesting that the apo-form of the heme domain might function as a co-chaperone in the folding of the Moco domain followed by subsequent dimerization. How Moco enters the mitochondrial intermembrane space remains unknown; however, the involvement of a carrier protein, which has been first reported in the alga Chlamydomonas,[75] can be excluded due to the exclusive transport of unfolded proteins *via* the transporter of the outer membrane.

7.3.2.2 Catalytic Cycle in Vertebrate SO

The catalytic mechanism of SO involves transfer of two electrons from sulfite to Moco ($Mo^{VI} \rightarrow Mo^{IV}$) followed by two single-electron transfer steps *via* the heme domain to cytochrome *c*. Within dimeric SO, both active sites are well separated with a Mo–Mo distance of 38 Å,[17] and are believed to operate independently. The crystal structure of chicken SO further shows a large distance of approx. 30 Å between the heme iron and the molybdenum centre of a given monomer,[17] with each heme domain in the dimer in a slightly different orientation relative to its respective Moco domain, suggesting a high spatial mobility of the heme domain.[17] The large heme distance to the active site molybdenum centre, which has also been confirmed in solution,[76] is in contrast to the high electron transfer rates that have been reported by laser flash photolysis.[77] Based on the inhibition of internal electron transfer by solution viscosity[77] and the use of deletions of the solvent-exposed tether connecting the heme and Moco domain, it has been suggested that during catalysis both domains undergo conformational changes in order to enable efficient electron transfer.[52] Enemark and coworkers have studied the electron transfer rate within chicken and human SO using the one-electron reduced state of the SO enzyme and found rates of up to 2000 s^{-1}, which decreases somewhat in the presence of anions or at low pH.[78] Therefore, a conformation with a much closer Mo-to-heme distance is required for efficient electron transfer. We have modelled such a conformation by overlaying the chicken SO heme domain with the one from SorB in a similar orientation to that seen in the SorAB complex (Figure 7.6). This approach is further supported by cross-linking studies in human SO, showing that the modelled interfaces are indeed in special proximity (Schwarz *et al.*, unpublished results). In contrast, Brody and Hille[79] investigated the reductive and oxidative half-reaction of chicken SO using steady-state and rapid kinetics and found a pH-independent rate constant for k_{red} of 200 s^{-1}, which involves both the reductive half-reaction (conversion of sulfite to sulfate) and the first electron transfer step from Mo^{IV} to heme. For human SO, somewhat lower values in the range of 70–100 s^{-1} have been determined.[80] Interestingly, rapid kinetics studies of the isolated human molybdenum domain have demonstrated a very fast formation of the enzyme–sulfite complex (k_{red} ~2000 s^{-1}) demonstrating product (sulfate) release and/or the first electron transfer from Mo to heme as rate-limiting step(s) (Figure 7.7).

Plant SO differs from vertebrate SO in lacking an N-terminal heme domain (see below) (Figure 7.5).[81] In the absence of the heme domain, catalytic electron transfer to cytochrome *c* is not possible and, as a result, plant SO is able to use molecular oxygen as electron acceptor. In a recent study, we asked the question whether vertebrate SO is also able to reduce oxygen and what the determinants are to enable or inhibit this type of reaction.[82] We have found that reducing or inhibiting electron transfer between the Moco and heme domain enables both mouse and human SO to transfer electrons to molecular oxygen in a sulfite-dependent manner. The highest activity in the production of H_2O_2 is seen with the isolated molybdenum domain, which is comparable to plant SO,[82] demonstrating that oxygen reactivity is an intrinsic property of eukaryotic SO molybdenum domains and is only suppressed in vertebrate SO by the efficient electron transfer from molybdenum to heme.

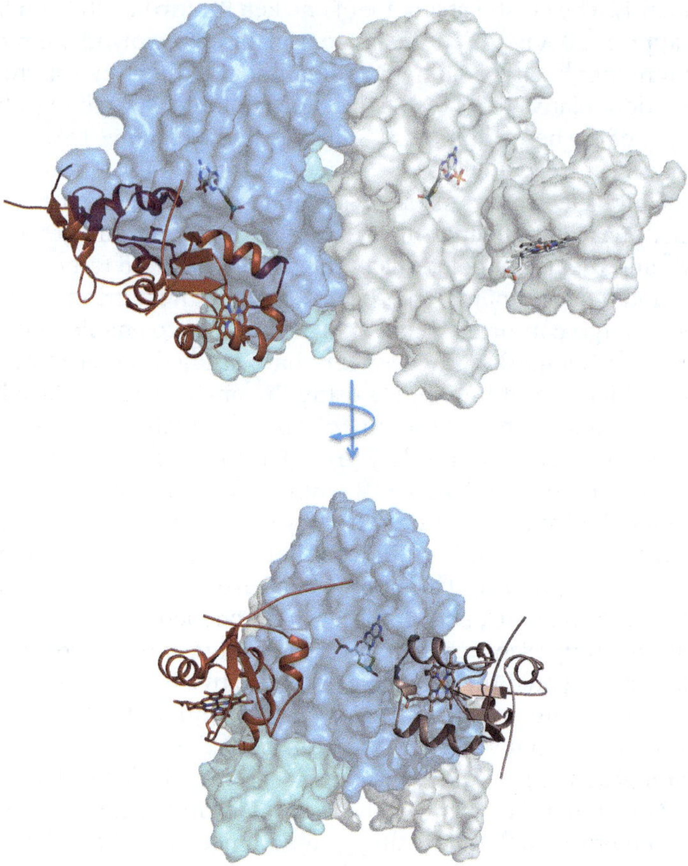

Figure 7.6 Structure of chicken SO with the experimentally determined confirmation of the heme domain and a modelled electron transfer-competent orientation. Shown is the dimer of chicken SO with one monomer in transparent surface representation and the other monomer in transparent surface representation for the Moco (blue) and dimerization domain (cyan). The heme domain in the coloured monomer is shown in the conformation as found in the CSO structure or modelled on the basis of the structure of SorB relative to SorA.

7.3.2.3 SO Deficiency

The first patient lacking SO activity was identified in 1967 at the National Institute of Health (Bethesda, MA, USA) by Mudd and coworkers, who reported on an infant with a severe metabolic disorder of unknown origin.[83] The accumulation of sulfite, thiosulfate and *S*-sulfocysteine in the urine and the lack of inorganic sulfate excretion pointed to a defect in SO. In the following decades, approx. 50 cases of SO deficiency have been reported in the literature,[84,85] some of which have later been recognized as patients with

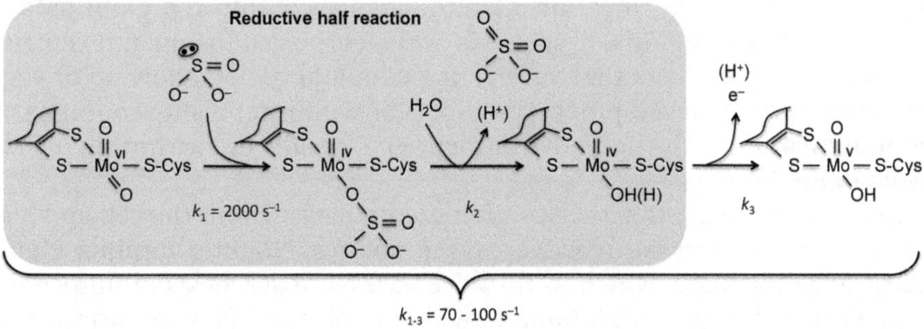

Figure 7.7 Reaction mechanism of sulfite oxidation at the Mo centre followed by one electron transfer step to heme. The reaction can be followed by stopped flow kinetics monitoring heme reduction ($k_{1\text{-}3}$). The rate for the reduction of the Mo was obtained for the isolated Mo domain of human SO (k_1).

Moco deficiency[86] who are lacking all molybdenum enzyme activities. Disease-causing mutations have been reported in many of the published cases. Biochemical studies in particular by the laboratories of John Enemark[17,66,87,88] and K.V. Rajagopalan[17,80,89] have disclosed the underlying disease-causing mechanism of various mutations causing SO deficiency. Mutations impairing Moco binding, catalysis including substrate binding and release as well as intramolecular electron transfer have been identified. Recently, we have participated in the diagnosis of an SO patient with a mutation that impaired Moco binding during the process of mitochondrial SO maturation, resulting in the accumulation of SO in the cytosol (Schwarz *et al.*, unpublished results). Very few cases with mild forms of SO deficiency have been reported, one of which showed very positive response to dietary restriction.[90] Besides the reduction of methionine in the diet, anticonvulsive treatment is the only therapy option, resulting in a very poor prognosis and early childhood death.[91] One possible treatment might be enzyme substitution therapy, but targeting SO to mitochondria is not possible. Therefore, the recently identified oxygen reactivity of mouse and human SO offers an opportunity to treat patients in the periphery by administration of an immune-shaded SO molybdenum domain, using PEGylation to reduce the antigenicity of the administered protein.[82]

Sulfite accumulation is a hallmark of both SO and Moco deficiency and is believed to be the major cause of the accompanying symptoms. In general, the earliest symptoms are observed within the first days of life, and include feeding difficulties accompanied with intractable seizures, a predominant opisthotonus and an exaggerated startle reaction.[92] Disease progression is accompanied by psychomotor retardation due to progressive cerebral atrophy and ventricular dilatation, which are typical in brain MRI scans of patients. In addition, major radiological features of the disease include global cerebral edema, cystic encephalomalacia, cortical and white matter atrophy, focal or bilateral changes within the globi pallidi and subthalamic regions, dysgenesis of the corpus callosum and ventriculomegaly.[93–95] Patients that survive the neonatal period show no or very delayed neuronal development, require tube feeding and show no signs of communication with their environment and usually die within their first years of life.[91]

In the absence of SO activity, sulfite accumulates within the cell and has been found to increase reactive oxygen species.[70] Sulfite inhibits glutamate dehydrogenase, which in turn decreases ATP synthesis in mitochondria when respiring on glutamate (which is of particular importance in the brain). Accumulated sulfite passes the plasma-membrane and spreads throughout the body *via* the circulation. Within the extracellular space, it can reduce disulfide bridges, primarily in membrane proteins, thus affecting their folding, stability and activity. The first scavenging mechanism is the reaction with cystine leading to the formation of *S*-sulfocysteine (SSC),[96] which is highly abundant in patients. SSC is structurally similar to glutamate, is able to bind to *N*-methyl-D-aspartate receptors and is proposed to be the cause of seizures and subsequent brain damage in the patients.[97] In fact, early studies in rats demonstrated that SSC induces the same type of brain damage as glutamate suggesting an excitotoxicity-based disease mechanism.[97] In this context we have studied the cellular mechanism of SSC-mediated neuronal cell death and found a receptor-mediated calcium influx followed by calpain activation and proteolytic cleavage of synaptic proteins. The biological relevance of this finding is underscored by the suppression of cell death in a cell culture model following the pharmacological blockage of any of the involved steps of the pathway and further extended to an animal model (Kumar *et al.*, unpublished results). Sulfite-dependent SSC formation results in cystine depletion,[91] which directly impacts glutathione synthesis (Figure 7.8), the major antioxidant in neuronal tissue and most abundant low-molecular-weight thiol in animal cells (0.5–10 mM).[68] Furthermore, the observed accumulation of thiosulfate suggests an increased reaction of sulfite with protein-bound persulfides, the latter being dependent on H_2S formation, derived from the non-oxidative way of cysteine elimination (Figure 7.8).[98] However, the important function of H_2S as a signalling molecule in the brain has only recently been recognized[99] and factors leading to lowering of H_2S levels are believed to contribute to the pathogenesis of age-associated neurodegenerations such as Huntington's disease.[100]

In summary, vertebrate SO represents an important enzyme in cysteine catabolism, which produces inorganic sulfate as a harmless end product and clears toxic sulfite that harms the cell. In addition, the contribution of SO in providing electrons for mitochondrial respiration has been underestimated. Therefore, understanding the molecular interplay of the different catabolic branches of cysteine will help to clarify the complex pathological picture seen in patients suffering from SO deficiency.

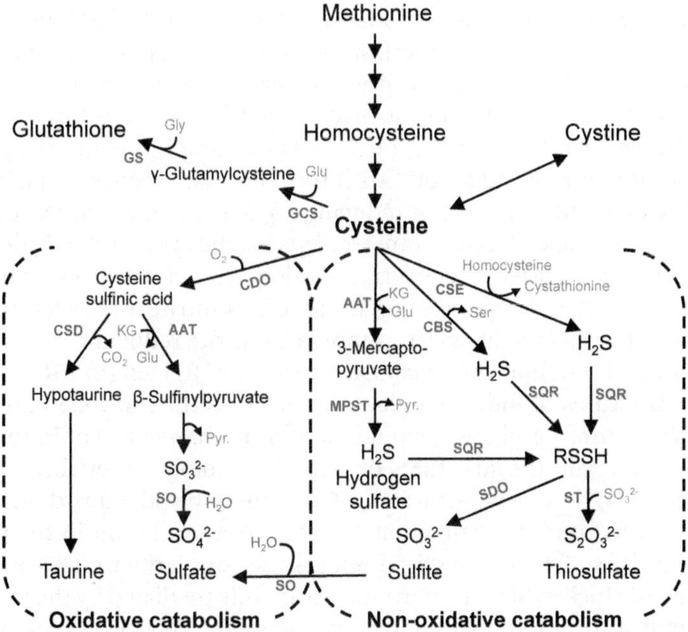

Figure 7.8 Cysteine catabolism in humans. Cysteine can be absorbed from diet or synthesized from methionine through several intermediates of the transmethylation pathway. Cysteine is the precursor of glutathione (GSH) synthesis in two enzymatic reactions catalyzed successively by the enzymes, γ-glutamylcysteine synthetase (GCS) and glutathione synthetase (GS). The catabolism of cysteine follows two major routes: an oxidative pathway leading to the formation of sulfate and taurine as end sulfur-containing products or a non-oxidative pathway leading to hydrogen sulfide formation, which can be further converted to thiosulfate. The enzyme abbreviations used are: CBS, cystathionine β-synthase; CSE, cystathionine γ-lyase (cystathionase); CDO, cysteine dioxygenase; CSD, cysteinesulfinate decarboxylase; AAT, aspartate aminotransferase; SO, sulfite oxidase; MPST, 3-mercaptopyruvate sulfurtransferase; SQR, quinone oxidoreductase; SDO, sulfur dioxygenase; ST, sulfurtransferase.

7.3.3 SO Family Enzymes from Plants

7.3.3.1 Nitrate Reductase

The best-studied plant enzyme of the SO family of molybdenum enzymes is nitrate reductase (NR), which is also found in fungi and algae.[101] In general, eukaryotic NR is a cytoplasmic, water-soluble, molybdo-iron-flavo-enzyme involved in the reduction of nitrate to nitrite as the first step in nitrate assimilation. It is a homodimeric protein of about 200–220 kDa with each monomer folded into five distinguishable domains (Figure 7.5) of which three are involved in the binding of prosthetic groups: the N-terminal Moco domain, the central cytochrome b_5-type heme domain and the C-terminal FAD domain. The additional domains are the dimerization and NAD(P)H-binding domains.[102] The dimerization, heme and FAD domains are connected by protease-sensitive and solvent-exposed loop sequences called hinge-I and hinge-II. NADH (in most higher plants and algae) or NADPH (in fungi) provides the reducing equivalents for nitrate reduction. An intramolecular electron transfer chain ensures the efficient transport of reducing equivalents from NAD(P)H to FAD followed by two single-electron transfer steps *via* the cytochrome b_5 domain to the molybdenum centre, where nitrate undergoes two-electron reduction to nitrite.[103] In plants and algae, the reaction product nitrite is translocated to the chloroplast where it is reduced to ammonia by nitrite reductase.[104]

In contrast to vertebrate SO, a crystal structure of full-length NR is still lacking. Only structures of individual domains, such as the molybdenum domain from *Pichia angusta*[22] and the flavin domain from *Zea mays*,[105] or in the case of the heme b_5 domain the structures of several homologous vertebrate proteins are available.[17] The crystal structure of the NR molybdenum domain (residues 1–484) has been determined at 1.7 and 2.6 Å resolution in the presence and absence of sulfate and shows a remarkable similarity to the corresponding domain of chicken SO.[22] Dimerization is mainly mediated by the C-terminal domain with some additional interactions between the Moco domains. There are differences to SO in the extended, mostly unstructured N- and C-terminal regions and in an additional loop, as well as alterations at the active site. In the 1.7 Å structure, in addition to the conformational change induced in the active site by the bound sulfate, in comparison to the apo-structure, four water molecules with very low B-factors can be used to guide a structure-based model for the NR reaction mechanism, depicting nitrate binding, formation of a penta-coordinated reaction intermediate and release of products (nitrite and water).

NR activity is tightly regulated by both transcriptional and post-transcriptional mechanisms: nitrate, light, CO_2, phytohormones and other metabolites of nitrogen and carbon metabolism are the main signals and effectors.[106] A conserved phosphoserine residue in hinge-I is involved in post-translational regulation of plant NR.[107] Upon the phosphorylation-dependent binding of the so-called 14-3-3 proteins to NR[108] enzyme activity is greatly decreased due to reduction of the intramolecular electron transfer rate between the heme and Moco domain.[109–111] This regulation represents a paradigm for 14-3-3-dependent control of enzyme activity. Given the high degree of conformational flexibility expected of holo-NR, studies of the 14-3-3 inhibited NR are likely to demonstrate a decreased domain motion, which may help to obtain a crystal structure of full-length NR. Alternatively, cryoEM might also provide further structural information on NR and the orientation of the different domains during the catalytic cycle.

7.3.3.2 SO in Plants

In contrast to chicken and human SO, the homologous enzyme in plants lacks the heme domain (Figure 7.5), while in *Chlamydomonas reinhardtii* SO shows the same primary structure as in vertebrate enzymes SO.[112] Also, an approximately 80–100 residue N-terminal extension suggests a mitochondrial targeting of the *Chlamydomonas* enzyme. The crystal structure of plant SO shows very high similarity to vertebrate SO, although one particular residue (Arg 374) is seen pointing away from the active site while the same conserved residue in chicken SO is oriented towards the active site; this has been interpreted as a substrate-/product-induced conformational change.[18] Due to the absence of other prosthetic groups, plant SO represents an ideal system in which to study the spectroscopic properties of the molybdenum centre. In the oxidized form the enzyme shows two maxima at 360 and 480 nm, while both bands convert to one signal at 400 nm following reduction.[81]

The mechanism of sulfite oxidation in plant SO has been shown to proceed in the same way as in vertebrate SO. The oxidative half-reaction, however, is fundamentally different and has been found to produce superoxide anions as direct products of the reaction following a rapid disproportionation to H_2O_2.[113–115] The rate constant for sulfite oxidation is larger than that seen with the molybdenum domain of human SO[116] and therefore too fast to be examined by conventional stopped-flow techniques. In light of the overall k_{cat} for the plant enzyme,[81] the oxidative half-reaction represents the rate-limiting step in plant SO. Due to the production of H_2O_2, plant SO is localized in peroxisomes.[117] Recent studies have proposed that plant SO plays a role in controlling the sulfate–sulfite cycle and contributes to the overall sulfur homeostasis in plants. Specifically, SO protects the plant from sulfite overload, particularly when exposed to atmospheric SO_2.[118,119]

7.4 General Aspects of Catalysis in SO Family Enzymes

For the majority of enzyme types in the SO family the reaction catalyzed is unclear, and for this reason catalytic details will only be discussed for SOEs and NRs. These enzymes are also the only ones for which extensive spectroscopic as well as crystallographic data are available that permit insights into the molecular details of catalysis. In light of the recent and extensive review of the literature by Hille and colleagues,[1] we will briefly summarize general aspects of SO catalysis and touch on a "moonlighting" activity of SO.

7.4.1 Sulfite Oxidizing Enzymes

The oxidation of sulfite to sulfate requires the transfer of two electrons from the substrate to the Mo centre of the enzyme in conjunction with the incorporation of an oxygen atom from water into the substrate to form sulfate. Following sulfite oxidation, the enzyme must be re-oxidized through by an external electron acceptor in order to be able to perform another reaction cycle (Figure 7.9). These two processes (sulfite oxidation to sulfate and reoxidation of the now-reduced enzyme) constitute the reductive and the oxidative half-reactions of the enzyme's reaction cycle.

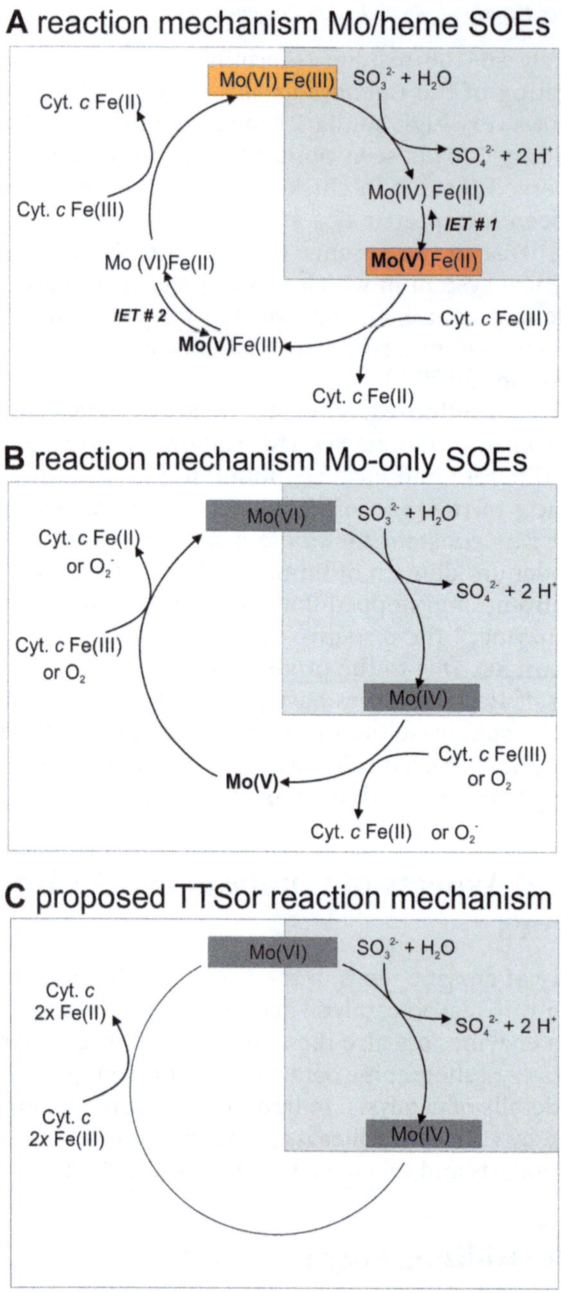

Figure 7.9 Reaction mechanism of sulfite converting SO family enzymes. Panel A: reaction mechanism for enzymes containing both a heme and a molybdenum centre (vertebrate SOs and bacterial SorAB enzyme). Panel B: reaction mechanism for enzymes containing only a Mo redox centre (plant SO, many bacterial SOEs *e.g.* SorT). Panel C: proposed reaction mechanism for the bacterial *Thermus thermophilus* SO (TTSor), which, unlike the enzymes shown in Panels A and B, interacts with a two-electron accepting external electron acceptor, essentially bypassing the formation of the EPR active Mo(V) form that has been used to probe the ligand environment of the Mo centre in other SO family enzymes.

During the reductive half-reaction, the incoming sulfite molecule attacks the electrophilic equatorial oxo group of the Mo centre, leading to the formation of a Mo-bound sulfate molecule, and the reduction of the Mo centre. Bound product is then displaced by solvent water, leading to the formation of sulfate and a hydroxyl/hydro-ligand on the Mo centre. The oxidative half-reaction then consists of two electron-transfer steps that can proceed either by transfer of reducing equivalents from the molybdenum centre to the b-type cytochrome (in the case of the vertebrate enzyme) or directly to O_2, to generate superoxide.

During turnover the Mo centre cycles between the Mo(VI) and Mo(IV) oxidation states, while heme centres in the SOEs (when present) cycle between the Fe(III) and Fe(II) states, requiring sequential oxidation of the fully reduced molybdenum centre. Regardless of the presence or absence of a heme in these enzymes, the reductive half-reaction at the molybdenum centre is considered to be fundamentally the same (Figure 7.9).

A hallmark of sulfite-oxidizing enzymes is in the absence of an oxidizing substrate, sulfite will react with enzyme to generate the two-electron reduced form, which is stable and can be isolated, a fact that has been exploited in the spectroscopy of SOEs.[3] This is typical of a ping-pong mechanism of catalysis, which is the generally accepted mechanism of catalysis in SOEs. In such a mechanism the first stage of the reaction (sulfite oxidation) should be complete, and the product (sulfate) already liberated before the transfer of the electrons stored in the Mo centre to an external acceptor, and for vertebrate enzymes diagnostic plots of kinetic data indicate the existence of such a mechanism.[65]

The catalytic parameters for different SOEs and their correlation with changes in the environment/ligand field of the Mo centre have recently been reviewed,[3] but an interesting and generally little discussed aspect of SOE catalysis is that while the enzyme clearly operates *via* a ping-pong mechanism, electron transfer out of the reduced Mo centre can occur while the sulfate molecule is still bound to the Mo centre, *i.e.* electron transfer out of the molybdenum centre is independent of the liberation of the sulfate product molecule so that the reductive and oxidative half-reactions of the catalytic cycle do not have to occur consecutively, as in standard ping-pong mechanisms, but can occur to some extent simultaneously. Particularly in enzymes with a second redox-active centre (CSO, SorAB), electron transfer from the fully reduced Mo(IV) centre to the heme can occur before the sulfate molecular has been liberated.[54] Evidence for this has come from studies of active site mutations in the *S. novella* SorAB SDH using flash photolysis,[54] as well as from replots of kinetic data analyses that failed to indicate a ping-pong mechanism (Kappler, unpublished results) in this enzyme, despite the existence of the characteristic, stable, two-electron reduced intermediate form of the enzyme.[32]

There are certain catalytic advantages in the alternative electron transfer pathways proposed by ref. 54, which include an overall faster catalytic cycle as the oxidative and reductive half-reactions can occur simultaneously rather than in sequence, and that might explain the very high turnover numbers seen especially for bacterial SOEs (Table 7.1). In the vertebrate SOs the catalytic cycle includes a domain rearrangement to bring the heme close to the molybdenum centre, which has the effect of slowing down the overall rate of catalysis and allowing the typical features of a ping-pong mechanism (*i.e.* sequential catalysis) to be observed. Another consideration is that in SOEs that do not contain a heme cofactor, both electrons stored in the enzyme as a result of sulfite oxidation can be transferred to the external acceptor simultaneously, if, as shown for the TTSor, this acceptor can take two electrons. This simplifies and potentially speeds up the reaction catalyzed, by eliminating the need for "docking" of a second molecule of the external acceptor. Thus significant diversity in the catalytic mechanism of SOEs exists, which depends on the exact type and number of redox centres present in the SOEs and in their external electron acceptors.

7.4.2 Nitrite Reduction by SO and Other Mo Enzymes

While plants, fungi and algae are all able to reduce nitrate to nitrite, vertebrates are unable to do so. However, nitrate and nitrite have been found to play important roles in animals, given that both nitrate and nitrite can be used to generate the bioactive nitric oxide (NO). Nitrate is either derived from NO oxidation or found in the diet, and nitrate reduction can be attributed to bacterial NRs in the gut. In contrast to nitrate, nitrite itself is bioactive, and is reduced to NO to regulate blood pressure,[120] hypoxic vasodilation,[121] cellular cytoprotection[122] and mitochondrial respiration and function under hypoxic or exercise stress.[123] Enzymes and proteins that can contribute to nitrite reduction are heme proteins, such as hemoglobin, myoglobin, neuroglobin, cytoglobin and plant hemoglobins.[124] In addition, molybdenum enzymes such as xanthine oxidase, aldehyde oxidase and, most recently, SO and mARC have been reported to function in nitrite reduction in vertebrates.[125]

SO, which is the closest homologue to plant nitrate reductase, has been investigated for its ability to reduce nitrite in great detail.[126] It has been shown that, similar to other molybdenum enzymes, nitrite reduction is dependent on Moco and, surprisingly, the presence of the heme domain inhibits NO synthesis. Together with EPR studies it has been demonstrated that SO-dependent nitrite reduction requires a fully reduced Mo^{IV} centre, while Mo^{V}, the one-electron oxidized Moco intermediate (generated by the first electron transfer during the oxidative half-reaction) is inert to nitrite.[126] As a result, there is no catalytic turnover due to the absence of a second electron acceptor to regenerate Mo^{VI} species for accepting the next sulfite molecule. Most importantly, studies with fibroblasts lacking either SO or all Moco enzyme activities have shown that nitrite-dependent cGMP synthesis (which requires NO production) is mainly dependent on SO, demonstrating that, at least in fibroblast, the contribution of other Mo enzymes or heme proteins to nitrite reduction is negligible.[126]

As discussed in Section 7.2, mARC proteins very likely form a related, but separate, family of Mo enzymes, as they lack significant sequence and folding similarities to the SO family of enzymes. Recent functional and spectroscopic studies have shown that vertebrate mARC enzymes are also able to reduce nitrite *in vitro* and in cells.[127] Furthermore, the underlying mechanism is found to involve either an oxyl or hydroxyl radical transfer based on EPR signals of the nitrite oxidase enzyme, which is very similar to the signal of the low pH form of SO.[128]

Mechanistically, it remains interesting and challenging to demonstrate under which conditions sulfite-dependent nitrite reduction becomes physiologically relevant. It is well established that nitrite-dependent NO synthesis is favoured under hypoxic conditions, which in turn would slow down electron transfer from SO heme domain to cytochrome *c*, thereby increasing the possibility of nitrite to accept the first electron from sulfite-reduced SO. We have preliminary evidence, however, that in the presence of sulfite and low concentrations of cytochrome *c*, catalytic NO synthesis is feasible. Future studies using animal models with organ-specific and inducible SO gene knockouts will be required to demonstrate the extent to which and under what conditions SO participates in nitrite-dependent NO synthesis. In line with the proposed role of SO and mARC in nitrite-dependent NO synthesis are findings in plants showing that nitrate reductase is an important contributor to plant NO synthesis.[129]

7.5 Concluding Remarks

The SO family of Mo enzymes was initially defined based on the active site structure of vertebrate SO almost 20 years ago.[7] While this family of enzymes was assumed to be only present in eukaryotes, with the identification of SorAB it was also found in prokaryotes and ever since the number of bacterial representatives is continuously growing. In this review we aimed to summarize the current knowledge on the different forms of SO enzymes that catalyze either oxidative or reductive reactions on sulfur and nitrogen atoms. While for many bacterial enzymes, their physiological function remains elusive, novel functions for "old" enzymes have been assigned, which could reflect the close relation of the different chemistries of various members of these enzymes (SO *versus* NR). While the reaction mechanism in the active site of SO enzymes is well established, the way in which the electrons are transferred from or to the substrate is still poorly understood. Enzymes with heme and other cofactors build intra-molecular electron transfer chains that seem to rely on a dynamic motion of the respective domains and therefore hamper crystallization of those enzymes. Interferences with these electron transfer chains either are used to regulate their catalytic function (NR) or might provide switch points for alternative functions (vertebrate SO in nitrite reduction). The complex role of SO in the vertebrate organism is highlighted by the dramatic metabolic stroke under conditions of SO deficiency causing severe brain damage. Understanding these processes will help to evaluate the cellular function of SO enzymes and might uncover their unrecognized contribution in many cellular processes being related to sulfur and nitrogen metabolism.

Acknowledgements

We thank our coworkers for their dedicated, passionate and hard work in our laboratories and the fruitful and inspiring discussion. We thank Nastassia Havarushka for preparing Figures 7.2 and 7.6. The authors also appreciate the funding they received from the various sources, such as the Australian Research Council (ARC), German Science Foundation (DFG), the EU FP7, the Federal Ministry for Education and Research (BMBF), the Center for Molecular Medicine Cologne (CMMC) and the Fonds der Chemischen Industrie (FCI).

References

1. R. Hille, J. Hall and P. Basu, *Chem. Rev.*, 2014, **114**, 3963–4038.
2. R. Hille, *Chem. Rev.*, 1996, **96**, 2757–2816.
3. U. Kappler and J. Enemark, *JBIC, J. Biol. Inorg. Chem.*, 2014, 1–12, DOI: 10.1007/s00775-014-1197-3.
4. U. Kappler, *Biochim. Biophys. Acta*, 2011, **1807**, 1–10.
5. U. Kappler and C. Dahl, *FEMS Microbiol. Lett.*, 2001, **203**, 1–9.
6. J. Simon and P. M. H. Kroneck, in *Adv. Microb. Phys.*, ed. K. P. Robert, Academic Press, 2013, vol. 62, pp. 45–117.
7. R. Hille, *Chem. Rev.*, 1996, **96**, 2757–2816.
8. G. Ott, D. Reichmann, C. Boerger, I. Cascorbi, F. Bittner, R. R. Mendel, T. Kunze, B. Clement and A. Havemeyer, *Drug Metab. Dispos.*, 2014, **42**, 718–725.
9. A. Havemeyer, J. Lang and B. Clement, *Drug Metab. Rev.*, 2011, **43**, 524–539.
10. A. Chamizo-Ampudia, A. Galvan, E. Fernandez and A. Llamas, *Eukaryotic Cell*, 2011, **10**, 1270–1282.
11. B. Wahl, D. Reichmann, D. Niks, N. Krompholz, A. Havemeyer, B. Clement, T. Messerschmidt, M. Rothkegel, H. Biester, R. Hille, R. R. Mendel and F. Bittner, *J. Biol. Chem.*, 2010, **285**, 37847–37859.
12. G. J. Workun, K. Moquin, R. A. Rothery and J. H. Weiner, *Microbiol. Mol. Biol. Rev.*, 2008, **72**, 228–248.
13. Y. Zhang, S. Rump and V. N. Gladyshev, *Coord. Chem. Rev.*, 2011, **255**, 1206–1217.
14. Y. Zhang and V. N. Gladyshev, *J. Mol. Biol.*, 2008, **379**, 881–899.
15. L. Loschi, S. J. Brokx, T. L. Hills, G. Zhang, M. G. Bertero, A. L. Lovering, J. H. Weiner and N. C. J. Strynadka, *J. Biol. Chem.*, 2004, **279**, 50391–50400.

16. U. Kappler and S. Bailey, *J. Biol. Chem.*, 2005, **280**, 24999–25007.
17. C. Kisker, H. Schindelin, A. Pacheco, W. A. Wehbi, R. M. Garrett, K. V. Rajagopalan, J. H. Enemark and D. C. Rees, *Cell*, 1997, **91**, 973–983.
18. N. Schrader, K. Fischer, K. Theis, R. R. Mendel, G. Schwarz and C. Kisker, *Structure*, 2003, **11**, 1251–1263.
19. A. Gennaris, B. Ezraty, C. Henry, R. Agrebi, A. Vergnes, E. Oheix, J. Bos, P. Leverrier, L. Espinosa, J. Szewczyk, D. Vertommen, O. Iranzo, J.-F. Collet and F. Barras, *Nature*, 2015, **528**, 409–412.
20. U. Kappler, in *Microbial Sulfur Metabolism*, Springer, Berlin, 2008, pp. 151–169.
21. A. Marchler-Bauer, J. B. Anderson, P. F. Cherukuri, C. DeWeese-Scott, L. Y. Geer, M. Gwadz, S. He, D. I. Hurwitz, J. D. Jackson, Z. Ke, C. Lanczycki, C. A. Liebert, C. Liu, F. Lu, G. H. Marchler, M. Mullokandov, B. A. Shoemaker, V. Simonyan, J. S. Song, P. A. Thiessen, R. A. Yamashita, J. J. Yin, D. Zhang and S. H. Bryant, *Nucleic Acids Res.*, 2005, **33**, D192–D196.
22. K. Fischer, G. G. Barbier, H. J. Hecht, R. R. Mendel, W. H. Campbell and G. Schwarz, *Plant Cell*, 2005, **17**, 1167–1179.
23. U. Zander, A. Faust, B. U. Klink, D. de Sanctis, S. Panjikar, A. Quentmeier, F. Bardischewsky, C. G. Friedrich and A. J. Scheidig, *J. Biol. Chem.*, 2011, **286**, 8349–8360.
24. L. J. Giles, C. Ruppelt, J. Yang, R. R. Mendel, F. Bittner and M. L. Kirk, *Inorg. Chem.*, 2014, **53**, 9460–9462.
25. C. E. Sparacino-Watkins, J. Tejero, B. Sun, M. C. Gauthier, J. Thomas, V. Ragireddy, B. A. Merchant, J. Wang, I. Azarov, P. Basu and M. T. Gladwin, *J. Biol. Chem.*, 2014, **289**, 10345–10358.
26. S. J. Brokx, R. A. Rothery, G. J. Zhang, D. P. Ng and J. H. Weiner, *Biochemistry*, 2005, **44**, 10339–10348.
27. M. J. Pushie, C. J. Doonan, K. Moquin, J. H. Weiner, R. Rothery and G. N. George, *Inorg. Chem.*, 2010, **50**, 732–740.
28. J. Yang, R. Rothery, J. Sempombe, J. H. Weiner and M. L. Kirk, *J. Am. Chem. Soc.*, 2009, **131**, 15612–15614.
29. K. G. V. Havelius, S. Reschke, S. Horn, A. Döring, D. Niks, R. Hille, C. Schulzke, S. Leimkühler and M. Haumann, *Inorg. Chem.*, 2011, **50**, 741–748.
30. A. Hitchcock, S. J. Hall, J. D. Myers, F. Mulholland, M. A. Jones and D. J. Kelly, *Microbiology*, 2010, **156**, 2994–3010.
31. J. J. Wilson and U. Kappler, *Biochim. Biophys. Acta*, 2009, **1787**, 1516–1525.
32. U. Kappler, B. Bennett, J. Rethmeier, G. Schwarz, R. Deutzmann, A. G. McEwan and C. Dahl, *J. Biol. Chem.*, 2000, **275**, 13202–13212.
33. L. Low, J. R. Kilmartin, P. V. Bernhardt and U. Kappler, *Front. Microbiol.*, 2011, **2**, 00058.
34. G. D'Errico, A. Di Salle, F. La Cara, M. Rossi and R. Cannio, *J. Bacteriol.*, 2006, **188**, 694–701.
35. A. Di Salle, G. D'Errico, F. La Cara, R. Cannio and M. Rossi, *Extremophiles*, 2006, **10**, 587–598.

36. K. Denger, S. Weinitschke, T. H. M. Smits, D. Schleheck and A. M. Cook, *Microbiology*, 2008, **154**, 256–263.
37. J. D. Myers and D. J. Kelly, *Microbiology*, 2005, **151**, 233–242.
38. A. Bagchi and P. Roy, *Biochem. Biophys. Res. Commun.*, 2005, **331**, 1107–1113.
39. S. C. Drew, E. Reijerse, A. Quentmeier, D. Rother, C. G. Friedrich and W. Lubitz, *Inorg. Chem.*, 2011, **50**, 409–411.
40. C. G. Friedrich, F. Bardischewsky, D. Rother, A. Quentmeier and J. Fischer, *Curr. Opin. Microbiol.*, 2005, **8**, 253–259.
41. V. Sauve, P. Roversi, K. J. Leath, E. F. Garman, R. Antrobus, S. M. Lea and B. C. Berks, *J. Biol. Chem.*, 2009, **284**, 21707–21718.
42. V. Sauve, S. Bruno, B. C. Berks and A. M. Hemmings, *J. Biol. Chem.*, 2007, **282**, 23194–23204.
43. U. Kappler and M. Maher, *Cell. Mol. Life Sci.*, 2013, **70**, 977–992.
44. F. Bardischewsky, A. Quentmeier, D. Rother, P. Hellwig, S. Kostka and C. G. Friedrich, *Biochemistry*, 2005, **44**, 7024–7034.
45. A. Quentmeier, R. Kraft, S. Kostka, R. Klockenkamper and C. G. Friedrich, *Arch. Microbiol.*, 2000, **173**, 117–125.
46. T. Utesch and M. A. Mroginski, *J. Phys. Chem. Lett.*, 2010, **1**, 2159–2164.
47. A. P. McGrath, E. L. Laming, G. P. Casas Garcia, M. Kvansakul, J. M. Guss, J. Trewhella, B. Calmes, P. V. Bernhardt, G. R. Hanson, U. Kappler and M. J. Maher, *eLife*, 2015, **4**, e09066.
48. H. Kimura, *Front. Physiol.*, 2012, **3**, 00101.
49. R. Lohmayer, A. Kappler, T. Lösekann-Behrens and B. Planer-Friedrich, *Appl. Environ. Microbiol.*, 2014, **80**, 3141–3149.
50. T. D. Rapson, U. Kappler, G. R. Hanson and P. V. Bernhardt, *Biochim. Biophys. Acta*, 2011, **1807**, 108–118.
51. P. Kalimuthu, J. Tkac, U. Kappler, J. J. Davis and P. V. Bernhardt, *Anal. Chem.*, 2010, **82**, 7374–7379.
52. T. D. Rapson, A. V. Astashkin, K. Johnson-Winters, P. V. Bernhardt, U. Kappler, A. M. Raitsimring and J. H. Enemark, *JBIC, J. Biol. Inorg. Chem.*, 2010, **15**, 505–514.
53. S. Bailey, T. Rapson, K. Winters-Johnson, A. V. Astashkin, J. H. Enemark and U. Kappler, *J. Biol. Chem.*, 2009, **284**, 2053–2063.
54. S. Emesh, T. D. Rapson, A. Rajapakshe, U. Kappler, P. V. Bernhardt, G. Tollin and J. H. Enemark, *Biochemistry*, 2009, **48**, 2156–2163.
55. A. V. Astashkin, E. L. Klein, D. Ganyushin, K. Johnson-Winters, F. Neese, U. Kappler and J. H. Enemark, *Phys. Chem. Chem. Phys.*, 2009, **11**, 6733–6742.
56. T. D. Rapson, U. Kappler and P. V. Bernhardt, *Biochim. Biophys. Acta, Bioenerg.*, 2008, **1777**, 1319–1325.
57. C. J. Doonan, U. Kappler and G. N. George, *Inorg. Chem.*, 2006, **45**, 7488–7492.
58. U. Kappler, S. Bailey, C. J. Feng, M. J. Honeychurch, G. R. Hanson, P. V. Bernhardt, G. Tollin and J. H. Enemark, *Biochemistry*, 2006, **45**, 9696–9705.

59. K. F. Aguey-Zinsou, P. V. Bernhardt, U. Kappler and A. G. McEwan, *J. Am. Chem. Soc.*, 2003, **125**, 530–535.
60. C. J. Feng, U. Kappler, G. Tollin and J. H. Enemark, *J. Am. Chem. Soc.*, 2003, **125**, 14696–14697.
61. U. Kappler and A. G. McEwan, *FEBS Lett.*, 2002, **529**, 208–214.
62. S. Robin, M. Arese, E. Forte, P. Sarti, A. Giuffre and T. Soulimane, *J. Bacteriol.*, 2011, **193**, 3988–3997.
63. S. Robin, M. Arese, E. Forte, P. Sarti, O. Kolaj-Robin, A. Giuffrè and T. Soulimane, *PLoS One*, 2013, **8**, e55129.
64. W. Reichenbecher, D. P. Kelly and J. C. Murrell, *Arch. Microbiol.*, 1999, **172**, 387–392.
65. K. V. Rajagopalan and M. P. Coughlan, in *Molybdenum and Molybdenum-Containing Enzymes*, Pergamon Press, Oxford, 1980, pp. 243–272.
66. C. Feng, G. Tollin and J. H. Enemark, *Biochim. Biophys. Acta*, 2007, **1774**, 527–539.
67. T. K. Chung, M. A. Funk and D. H. Baker, *J. Nutr.*, 1990, **120**, 158–165.
68. G. Wu, Y. Z. Fang, S. Yang, J. R. Lupton and N. D. Turner, *J. Nutr.*, 2004, **134**, 489–492.
69. H. J. Cohen, S. Betcher-Lange, D. L. Kessler and K. V. Rajagopalan, *J. Biol. Chem.*, 1972, **247**, 7759–7766.
70. X. Zhang, A. S. Vincent, B. Halliwell and K. P. Wong, *J. Biol. Chem.*, 2004, **279**, 43035–43045.
71. J. M. Klein and G. Schwarz, *J. Cell Sci.*, 2012, **125**, 4876–4885.
72. F. E. Nargang, M. E. Drygas, P. L. Kwong, D. W. Nicholson and W. Neupert, *J. Biol. Chem.*, 1988, **263**, 9388–9394.
73. W. Neupert and J. M. Herrmann, *Annu. Rev. Biochem.*, 2007, **76**, 723–749.
74. H. Ono and A. Ito, *J. Biochem.*, 1984, **95**, 345–352.
75. C. P. Witte, M. I. Igeno, R. Mendel, G. Schwarz and E. Fernandez, *FEBS Lett.*, 1998, **431**, 205–209.
76. A. V. Astashkin, A. Rajapakshe, M. J. Cornelison, K. Johnson-Winters and J. H. Enemark, *J. Phys. Chem. B*, 2012, **116**, 1942–1950.
77. C. Feng, R. V. Kedia, J. T. Hazzard, J. K. Hurley, G. Tollin and J. H. Enemark, *Biochemistry*, 2002, **41**, 5816–5821.
78. A. Pacheco, J. T. Hazzard, G. Tollin and J. H. Enemark, *J. Biol. Inorg. Chem.*, 1999, **4**, 390–401.
79. M. S. Brody and R. Hille, *Biochemistry*, 1999, **38**, 6668–6677.
80. H. L. Wilson and K. V. Rajagopalan, *J. Biol. Chem.*, 2004, **279**, 15105–15113.
81. T. Eilers, G. Schwarz, H. Brinkmann, C. Witt, T. Richter, J. Nieder, B. Koch, R. Hille, R. Hansch and R. R. Mendel, *J. Biol. Chem.*, 2001, **276**, 46989–46994.
82. A. A. Belaidi, J. Roper, S. Arjune, S. Krizowski, A. Trifunovic and G. Schwarz, *Biochem. J.*, 2015, **469**, 211–221.
83. S. H. Mudd, F. Irreverre and L. Laster, *Science*, 1967, **156**, 1599–1602.
84. W. H. Tan, F. S. Eichler, S. Hoda, M. S. Lee, H. Baris, C. A. Hanley, P. E. Grant, K. S. Krishnamoorthy and V. E. Shih, *Pediatrics*, 2005, **116**, 757–766.

85. G. Schwarz and A. A. Belaidi, *Metal ions in life sciences*, 2013, vol. 13, pp. 415–450.
86. M. Duran, F. A. Beemer, C. van de Heiden, J. Korteland, P. K. de Bree, M. Brink, S. K. Wadman and I. Lombeck, *J. Inherited Metab. Dis.*, 1978, **1**, 175–178.
87. C. Feng, H. L. Wilson, J. K. Hurley, J. T. Hazzard, G. Tollin, K. V. Rajagopalan and J. H. Enemark, *Biochemistry*, 2003, **42**, 12235–12242.
88. C. Feng, H. L. Wilson, J. K. Hurley, J. T. Hazzard, G. Tollin, K. V. Rajagopalan and J. H. Enemark, *J. Biol. Chem.*, 2003, **278**, 2913–2920.
89. H. L. Wilson, S. R. Wilkinson and K. V. Rajagopalan, *Biochemistry*, 2006, **45**, 2149–2160.
90. M. Del Rizzo, A. P. Burlina, J. O. Sass, F. Beermann, C. Zanco, C. Cazzorla, A. Bordugo, L. Giordano, R. Manara and A. B. Burlina, *Mol. Genet. Metab.*, 2013, **108**, 263–266.
91. J. L. Johnson and M. Duran, in *The metabolic and molecular bases of inherited disease*, ed. C. Scriver, A. Beaudet, W. Sly and D. Valle, McGraw-Hill, New York, 8th edn, 2001, pp. 3163–3177.
92. J. Reiss, U. Lenz, C. Aquaviva-Bourdain, S. Joriot-Chekaf, K. Mention-Mulliez and M. Holder-Espinasse, *Clin. Genet.*, 2011, **80**, 598–599.
93. E. Bayram, Y. Topcu, P. Karakaya, U. Yis, H. Cakmakci, K. Ichida and S. H. Kurul, *Eur. J. Paediatr. Neurol.*, 2013, **17**, 1–6.
94. K. Vijayakumar, R. Gunny, S. Grunewald, L. Carr, K. W. Chong, C. DeVile, R. Robinson, N. McSweeney and P. Prabhakar, *Pediatric Neurol.*, 2011, **45**, 246–252.
95. S. Arslanoglu, M. Yalaz, D. Goksen, M. Coker, S. Tutuncuoglu, M. Akisu, S. Darcan, N. Kultursay, M. Ciris and E. Demirtas, *Brain Dev.*, 2001, **23**, 815–818.
96. G. Schwarz, R. R. Mendel and M. W. Ribbe, *Nature*, 2009, **460**, 839–847.
97. J. W. Olney, C. H. Misra and T. de Gubareff, *J. Neuropathol. Exp. Neurol.*, 1975, **34**, 167–177.
98. B. D. Paul and S. H. Snyder, *Trends Biochem. Sci.*, 2015, **40**, 687–700.
99. B. D. Paul and S. H. Snyder, *Cell Cycle*, 2014, **13**, 2491–2493.
100. B. D. Paul, J. I. Sbodio, R. Xu, M. S. Vandiver, J. Y. Cha, A. M. Snowman and S. H. Snyder, *Nature*, 2014, **509**, 96–100.
101. L. P. Solomonson and M. J. Barber, *Annu. Rev. Plant Physiol. Plant Mol. Biol.*, 1990, **41**, 225–253.
102. W. H. Campbell, *Cell. Mol. Life Sci.*, 2001, **58**, 194–204.
103. L. Skipper, W. H. Campbell, J. A. Mertens and D. J. Lowe, *J. Biol. Chem.*, 2001, **276**, 26995–27002.
104. C. Masclaux-Daubresse, F. Daniel-Vedele, J. Dechorgnat, F. Chardon, L. Gaufichon and A. Suzuki, *Ann. Bot.*, 2010, **105**, 1141–1157.
105. G. Lu, W. H. Campbell, G. Schneider and Y. Lindqvist, *Structure*, 1994, **2**, 809–821.
106. W. H. Campbell, *Annu. Rev. Plant Physiol. Plant Mol. Biol.*, 1999, **50**, 277–303.
107. C. MacKintosh and S. E. Meek, *Cell. Mol. Life Sci.*, 2001, **58**, 205–214.
108. W. M. Kaiser and S. C. Huber, *J. Exp. Bot.*, 2001, **52**, 1981–1989.

109. I. Lambeck, J. C. Chi, S. Krizowski, S. Mueller, N. Mehlmer, M. Teige, K. Fischer and G. Schwarz, *Biochemistry*, 2010, **49**, 8177–8186.
110. I. C. Lambeck, K. Fischer-Schrader, D. Niks, J. Roeper, J. C. Chi, R. Hille and G. Schwarz, *J. Biol. Chem.*, 2012, **287**, 4562–4571.
111. J. C. Chi, J. Roeper, G. Schwarz and K. Fischer-Schrader, *JBIC, J. Biol. Inorg. Chem.*, 2015, **20**, 277–286.
112. S. Gerin, G. Mathy, A. Blomme, F. Franck and F. E. Sluse, *Biochim. Biophys. Acta*, 2010, **1797**, 994–1003.
113. K. Nowak, N. Luniak, C. Witt, Y. Wustefeld, A. Wachter, R. R. Mendel and R. Hansch, *Plant Cell Physiol.*, 2004, **45**, 1889–1894.
114. R. Hansch, C. Lang, E. Riebeseel, R. Lindigkeit, A. Gessler, H. Rennenberg and R. R. Mendel, *J. Biol. Chem.*, 2006, **281**, 6884–6888.
115. R. S. Byrne, R. Hansch, R. R. Mendel and R. Hille, *J. Biol. Chem.*, 2009, **284**, 35479–35484.
116. C. Hemann, B. L. Hood, M. Fulton, R. Hansch, G. Schwarz, R. R. Mendel, M. L. Kirk and R. Hille, *J. Am. Chem. Soc.*, 2005, **127**, 16567–16577.
117. K. Nowak, N. Luniak, C. Witt, Y. Wustefeld, A. Wachter, R. R. Mendel and R. Hansch, *Plant Cell Physiol.*, 2004, **45**, 1889–1894.
118. D. Randewig, D. Hamisch, C. Herschbach, M. Eiblmeier, C. Gehl, J. Jurgeleit, J. Skerra, R. R. Mendel, H. Rennenberg and R. Hansch, *Plant, Cell Environ.*, 2012, **35**, 100–115.
119. D. Yarmolinsky, G. Brychkova, R. Fluhr and M. Sagi, *Plant Physiol.*, 2013, **161**, 725–743.
120. A. Dejam, C. J. Hunter and M. T. Gladwin, *N. Engl. J. Med.*, 2007, **356**, 1590, author reply 1590.
121. K. Cosby, K. S. Partovi, J. H. Crawford, R. P. Patel, C. D. Reiter, S. Martyr, B. K. Yang, M. A. Waclawiw, G. Zalos, X. Xu, K. T. Huang, H. Shields, D. B. Kim-Shapiro, A. N. Schechter, R. O. Cannon, 3rd and M. T. Gladwin, *Nat. Med.*, 2003, **9**, 1498–1505.
122. M. T. Gladwin, A. N. Schechter, D. B. Kim-Shapiro, R. P. Patel, N. Hogg, S. Shiva, R. O. Cannon, 3rd, M. Kelm, D. A. Wink, M. G. Espey, E. H. Oldfield, R. M. Pluta, B. A. Freeman, J. R. Lancaster, Jr, M. Feelisch and J. O. Lundberg, *Nat. Chem. Biol.*, 2005, **1**, 308–314.
123. F. J. Larsen, T. A. Schiffer, S. Borniquel, K. Sahlin, B. Ekblom, J. O. Lundberg and E. Weitzberg, *Cell Metab.*, 2011, **13**, 149–159.
124. J. Tejero and M. T. Gladwin, *Biological Chem.*, 2014, **395**, 631–639.
125. L. B. Maia and J. J. Moura, *JBIC, J. Biol. Inorg. Chem.*, 2015, **20**, 403–433.
126. J. Wang, S. Krizowski, K. Fischer-Schrader, D. Niks, J. Tejero, C. Sparacino-Watkins, L. Wang, V. Ragireddy, S. Frizzell, E. E. Kelley, Y. Zhang, P. Basu, R. Hille, G. Schwarz and M. T. Gladwin, *Antioxid. Redox Signaling*, 2015, **23**, 283–294.
127. C. E. Sparacino-Watkins, J. Tejero, B. Sun, M. C. Gauthier, J. Thomas, V. Ragireddy, B. A. Merchant, J. Wang, I. Azarov, P. Basu and M. T. Gladwin, *J. Biol. Chem.*, 2014, **289**, 10345–10358.
128. J. Yang, L. J. Giles, C. Ruppelt, R. R. Mendel, F. Bittner and M. L. Kirk, *J. Am. Chem. Soc.*, 2015, **137**, 5276–5279.
129. J. Lozano-Juste and J. Leon, *Plant Physiol.*, 2010, **152**, 891–903.

CHAPTER 8

Nitrogenase Mechanism: Electron and Proton Accumulation and N_2 Reduction

LANCE C. SEEFELDT*[a], DENNIS R. DEAN[b], AND BRIAN M. HOFFMAN[c]

[a]Department of Chemistry and Biochemistry, Utah State University, Logan, UT 84322, USA; [b]Department of Biochemistry, Virginia Tech, Blacksburg, VA 24061, USA; [c]Department of Chemistry, Northwestern University, Evanston, IL 60208, USA

*E-mail: lance.seefeldt@usu.edu

8.1 Introduction

Nitrogenase is the enzyme found in diazotrophic bacteria that catalyzes the reduction of dinitrogen (N_2) to ammonia (NH_3). This biological activity represents the largest input of fixed N into the global biogeochemical N cycle.[1] Three different types of nitrogenase have been identified and these are designated as the Mo-dependent, the V-dependent or the Fe-only enzymes.[2] The most widely distributed and studied of these is the Mo-dependent nitrogenase,[3] which is the focus of this review. The Mo-dependent nitrogenase is comprised of two component proteins, called the MoFe protein and the Fe protein (Figure 8.1). The MoFe protein is an $\alpha_2\beta_2$ tetramer, with each $\alpha\beta$ unit

RSC Metallobiology Series No. 5
Molybdenum and Tungsten Enzymes: Biochemistry
Edited by Russ Hille, Carola Schulzke, and Martin L. Kirk

Published by the Royal Society of Chemistry, www.rsc.org

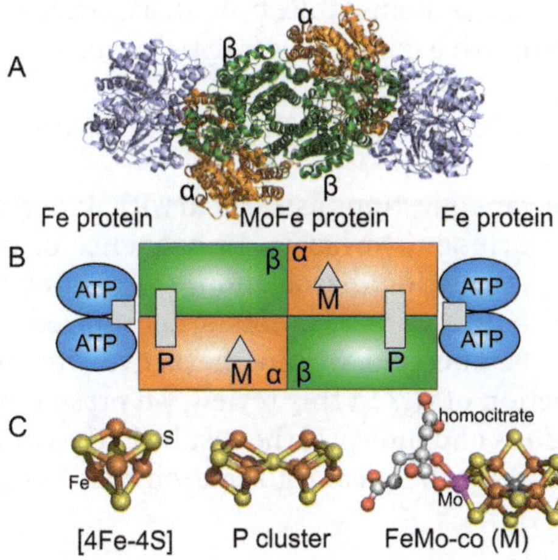

Figure 8.1 Mo-nitrogenase. (Panel A). Shown is the ribbon diagram for the nitrogenase MoFe protein–2(Fe protein (βγ-methylene-ATP)$_2$) complex (PDB ID 2AFK). The MoFe protein (center) alpha subunits are colored orange, while the beta subunits are colored green. The two Fe proteins (on each end) are shown in blue. (Panel B). A cartoon diagram of the nitrogenase complex with the [4Fe–4S] cluster shown as a gray box, the P cluster as a gray rectangle and the FeMo-co (M) as a gray triangle. (Panel C). Shown are the structures of the [4Fe–4S] cluster, the P cluster and FeMo-co with Fe in rust, S in yellow, C in gray, Mo in magenta and O in red.

constituting a catalytic unit. Each αβ unit of the MoFe protein contains a [8Fe–7S] cluster called the P cluster and an [7Fe–8S–Mo–C-homocitrate] cluster called FeMo-co (or M).[4] The FeMo-co is the site where N_2 binds and is reduced.

The other component protein of Mo-nitrogenase is called the Fe protein. The Fe protein is an α_2 homodimer, which has a single [4Fe–4S] cluster bridged between its identical subunits.[5] The Fe protein also contains two ATP binding sites, one site located within each subunit.[6,7] The Fe protein functions to donate electrons to the MoFe protein in a process coupled to the hydrolysis of the two bound ATP to ADP + P_i. To accomplish these events, one reduced-nucleotide bound Fe protein transiently associates with one αβ unit of the MoFe protein (Figure. 8.1). During this transient association, which lasts for approximately 200 ms, a cascade of events occurs resulting in the transfer of one electron from the Fe protein to the MoFe protein and the hydrolysis of two ATP molecules. At the end of the cycle, the oxidized Fe protein, now having two bound ADP molecules, dissociates from the MoFe protein. The free oxidized Fe protein is then ready for re-reduction and replacement of ADP by ATP, thereby making it available for another cycle of single electron delivery to the MoFe protein. In order to satisfy the stoichiometry of the overall biological reduction of N_2 to yield $2NH_3$, shown below (eqn (8.1)), 8 cycles of Fe protein–MoFe protein association and dissociation is necessary to accomplish a complete catalytic cycle.

$$N_2 + 8H^+ + 8e^- + 16ATP \rightarrow 2NH_3 + H_2 + 16ADP + 16P_i \quad (8.1)$$

Some of the significant questions associated with understanding the nitrogenase mechanism include: (i) what is the sequence of events involved in component protein association/dissociation, electron delivery and nucleotide hydrolysis, and how are these events synchronized? (ii) how are electrons and protons accumulated within the active site to accomplish the activation and reduction of N_2? In this review, we present recent results that address these questions and interpret the results in the context of a wealth of structural, kinetic and biochemical information that has accumulated over many decades.

8.2 Electron Transfer and ATP Hydrolysis in Nitrogenase

As noted above, the transient association of the Fe protein with the MoFe protein initiates a series of reactions including electron transfer (ET) and ATP hydrolysis. Here, we present what is currently known about some of the key events that occur during the association of the Fe protein–MoFe protein, starting with the docking of the two proteins.

8.2.1 Docking of Fe Protein to the MoFe Protein

The docking of an Fe protein to one αβ unit of the MoFe protein has been reported to be fast, having a second-order binding rate constant of 5×10^7 M^{-1} s^{-1} and a first-order dissociation rate constant of 15 s^{-1}.[8,9] These values were established for the nitrogenase proteins isolated from the microbe *Klebsiella pneumonia*. More recent mechanistic studies have focused on the highly similar nitrogenase isolated from the microbe *Azotobacter vinelandii*. While some differences in specifics are expected between these two nitrogenases, affinities and rate constants reported for one are likely to approximate the values for the other. Thus, from the association/dissociation rate constants, the association of the *A. vinelandii* components should be fast. It would be reasonable to assume from this speed that the association is a simple, single-step event. From the analysis of X-ray structures of the Fe protein–MoFe protein complex in different nucleotide bound states, it appears that complex association is actually complex, with an initial encounter complex followed by the Fe protein "sliding" across part of the surface of the MoFe protein into a state optimal for ET transfer position, followed by a further slide to a state favoring dissociation of the two proteins.[10,11] From this it can be concluded that the association process is likely to be complex, involving a number of different bound states. We are just beginning to see the broad strokes of this process, with many of the details remaining to be filled in.

8.2.2 Electron Transfer Events

One of the important events that occurs in the Fe protein–MoFe protein complex is the transfer of one electron from the Fe protein [4Fe–4S] cluster to the MoFe protein, ultimately ending up on the FeMo-cofactor. In one of the Fe protein–MoFe protein trapped states, the [4Fe–4S] cluster of the Fe protein is positioned 18 Å away from the P cluster in the MoFe protein, with the P cluster located between the Fe protein [4Fe–4S] cluster and the FeMo-co.[12] This location provides strong circumstantial evidence for the involvement of the P cluster in the ET events. However, gaining direct evidence for the role of the P cluster in ET has been challenging. There are a few studies that observed changes in oxidation state of the P cluster during the ET events, suggesting a redox role of the P cluster.[13–16] In one study, it was observed that in a MoFe protein with an amino acid substitution near the P cluster, a portion of the P cluster resides in an oxidized state even when resting in dithionite. Electron transfer to this protein was observed to be very fast, reducing the P^{ox} state back to the reduced state (P^N), followed by normal ET rate constants. From these observations, a counter-intuitive order of ET was proposed. In what is called a "deficit-spending" ET model, the first ET event is from the P cluster to FeMo-co (Figure 8.2), resulting in P^{ox} and M^R states.[17] Such a model is supported by the observation that the P^N state of the P cluster has all 8 Fe atoms in the ferrous (2+) oxidation state.[18] Given that no protein-bound FeS clusters have been observed to be reduced below this oxidation state, it is reasonable to conclude that the resting state of the P cluster (P^N) is the most

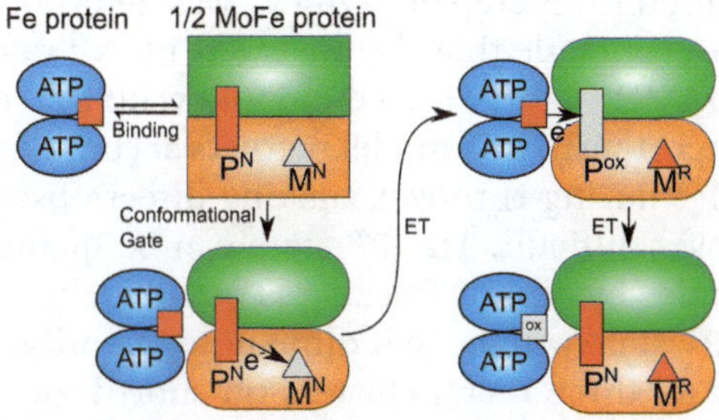

Figure 8.2 Proposed electron transfer events in nitrogenase. Shown is a cartoon of 1/2 of the MoFe protein with one bound Fe protein. In the first step (top left), an Fe protein $(ATP)_2$ associates with the MoFe protein. The [4Fe–4S] (box) is shown in its reduced (red) or oxidized (gray) state; the P cluster (rectangle) is shown in the P^N (red) or P^{ox} (gray) state; FeMo-co (triangle) is shown in the resting M^N (gray) or reduced M^R (red) state. The MoFe protein conformational changes are represented by changes in the shape of the αβ units.

reduced state accessed during turnover. Thus, it seems reasonable to consider that priming intermolecular electron transfer (ET from the Fe protein to the MoFe protein) could initially involve intramolecular electron transfer within the MoFe protein. In this model association of the reduced Fe protein and the resting state MoFe protein triggers the delivery of an electron from a fully reduced P cluster (P^N) to the resting form of FeMo-co (M^N) yielding an oxidized P cluster (P^{OX}) and reduced FeMo-co (M^R) (these proposed events are shown in Figure 8.2). The next step would involve back-fill ET from reduced Fe protein to the P^{ox} state, resulting in an oxidized Fe protein and the reduced P^N state (Figure 8.2). While such a "deficit-spending" ET model is attractive, studies to directly observe transitions of all the oxidation states of the three clusters during ET has not been accomplished.

Part of the challenge in gaining insights into roles of the three metal clusters during ET is the difficulty in observing oxidation state transitions of all three clusters on such a short time scale. The oxidation state of the Fe protein [4Fe–4S] cluster is relatively easy to monitor by following the increase in its visible absorption at 430 nm.[8] Using a stopped-flow spectrophotometer, it is possible to watch the time-dependent oxidation of the Fe protein when mixed with the MoFe protein. Fitting of the data to an exponential rise allows deduction of a rate constant for this ET event of ~140 s^{-1} at 30 °C. There is no equivalent direct observation of the rate constant for ET from the P cluster to the FeMo-co because there are no unique absorption maxima for these two clusters that can be followed in a stopped-flow spectrophotometer. Assuming a deficit-spending ET model, the observed ET from the Fe protein into the MoFe protein would represent both the ET from the P cluster to the FeMo-co and the ET from the Fe protein to the oxidized P cluster. Again, from the MoFe protein with the amino acid substitution near the P cluster, it was possible to directly observe the ET event from the Fe protein to the P^{ox} state. A rate constant for this step of >1700 s^{-1} was determined for this step.[19] Thus, it is logical to conclude that the observed 140 s^{-1} rate constant in fact represents the P to M ET event, with a rapid follow up ET from the Fe protein to the P cluster. Importantly, this model predicts accumulation of only a very small fraction of P^{ox} during turnover, making direct observation of the P^{ox} state during turnover difficult. The P^{ox} state can be quantified by EPR and thus trapping samples by rapidly freezing is possible, but the predicted low occupancy of the P^{ox} state will make it difficult to quantify. Developing some means to monitor all three metal clusters on short time scales during the pre-steady state is needed to clearly sort out the details of the ET events.

An important observation regarding the ET from the Fe protein into the MoFe protein is that this process is gated by large-scale protein conformational changes.[20] This was deduced by determining the effect of changing osmotic pressure on the rate constant for the ET event. From such a study, it was clear that large-scale (~800 Å^2) changes in the area of Fe protein/MoFe protein contact must proceed and gate the ET event. Assuming a deficit spending ET model, the conformational gate would apply to the P cluster to

FeMo-co ET event. This model then predicts that the binding of the Fe protein to the MoFe protein would induce conformational changes within the MoFe protein that facilitate ET from the P cluster to FeMo-co. It is hard to predict the nature or magnitude of such conformational changes. Unfortunately, none of the X-ray structures of Fe protein–MoFe protein complexes that have been solved[11] reveal any conformational changes deep within the MoFe protein that could account for the conformational control. More information about the nature of these protein conformational changes is needed to gain a better level of understanding of this part of the nitrogenase mechanism.

8.2.3 ATP and Nitrogenase

The ATP requirement for nitrogenase catalysis was originally discovered in some of the earliest studies on the enzyme by Burris and Mortenson[21,22] using cell free extracts. In those studies it was recognized that ATP is essential to achieve nitrogen fixation, and that catalysis required the hydrolysis of ATP to ADP and P_i. Following the successful purification of the enzyme, *in vitro* studies commenced revealing that a minimum stoichiometry of 2 ATP are consumed per electron transfer and there is also a requirement for a divalent metal ion, such as Mg^{++}. X-ray crystal structures of the Fe protein confirmed that the homodimeric Fe protein houses two ATP binding sites, one in each subunit, providing insight into the required stoichiometry.[5] The hydrolysis of 2 ATP per electron is a minimal stoichiometry under most assay conditions, but this ratio can be much higher under non-optimal conditions.[23] At the extreme, when the Fe protein is unable to transfer an electron (oxidized FeS cluster), ATP hydrolysis still occurs in a reaction that has been referred to as electron transfer independent ATP hydrolysis.

The binding of ATP to the Fe protein induces protein conformational changes that can be observed as changes in the properties of the protein and of its FeS cluster.[3] The latter changes include a lowering of the midpoint potential for the FeS cluster by over −100 mV and changes in the lineshape of its EPR spectrum. While the Fe protein binds two ATP molecules, with measured binding affinities,[24] it is not effective at catalyzing the hydrolysis of the bound ATP molecules. It is only upon binding of the Fe protein to the MoFe protein that ATP hydrolysis is activated. Later, the X-ray structure of an Fe protein–MoFe protein complex revealed that residues within the Fe protein are moved as a consequence of the association of the Fe protein with the MoFe protein that activate the hydrolysis reaction.[12]

While many of the details of ATP binding and hydrolysis by nitrogenase were sorted out through decades of studies, the exact role of ATP hydrolysis remains to be firmly established. One important question to address is the order of ATP hydrolysis and electron transfer. Several early studies were consistent with ATP hydrolysis occurring before electron transfer, leading to the logical proposal that the energy released from ATP hydrolysis ($\Delta G° = -30.5$ kJ mol^{-1}) was coupled to driving electron transfer and therefore substrate reduction. There were a few reports that suggested electron transfer might occur before ATP hydrolysis, pointing to the uncertainty in the order of these key events.[3] Part of the challenge is the difficulty in measuring ATP hydrolysis in the pre-steady state. Earlier work was able to quantify the early P_i release step. From careful kinetic studies, it became clear that P_i release significantly followed the electron transfer event, but this finding did not address the order of ATP hydrolysis and electron transfer.

Recently, an approach that allows quantitation of the extent of ATP hydrolysis in the pre-steady state was used to resolve this question.[25] The approach involved rapidly mixing nitrogenase proteins with radiolabeled ATP (α-^{32}P), quenching the reaction with acid at early times (3 ms), followed by separation of ATP from ADP by thin layer chromatography with quantitation of the radiolabeled ATP and ADP in each sample. By varying the time of reaction before quenching, it was possible to generate a pre-steady state time course for ATP hydrolysis (Figure 8.3). By placing the data for electron transfer, ATP hydrolysis, P_i release and protein–protein dissociation on the same graph, it is evident that ATP hydrolysis significantly follows the electron transfer, but is also significantly ahead of P_i release.

These measurements establish that ATP hydrolysis follows electron transfer, thereby revealing that the energy from hydrolysis is not used to drive electron transfer. This realization suggests a different model wherein ATP hydrolysis functions in nitrogenase, much like it does in several other

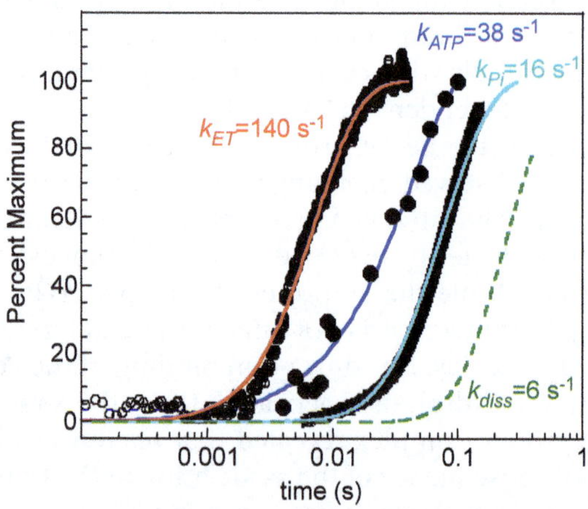

Figure 8.3 Order of events in nitrogenase. Pre-steady state time course analysis of electron transfer (o), ATP hydrolysis (•), and P_i release (□) are shown. The data are fit to a sequential model where ET, ATP hydrolysis, P_i release and dissociation occur stepwise with a single rate constant describing each step (k_{ET}, k_{ATP}, k_{Pi} and k_{diss}). The Fe protein dissociation (green trace) is simulated with a rate constant of 6 s^{-1}. Electron transfer is monitored by the change in absorbance at 430 nm. ATP hydrolysis is measured by the formation of the ^{32}P-ADP from ^{32}P-ATP. P_i release is measured using a real-time fluorescent phosphate binding protein sensor.

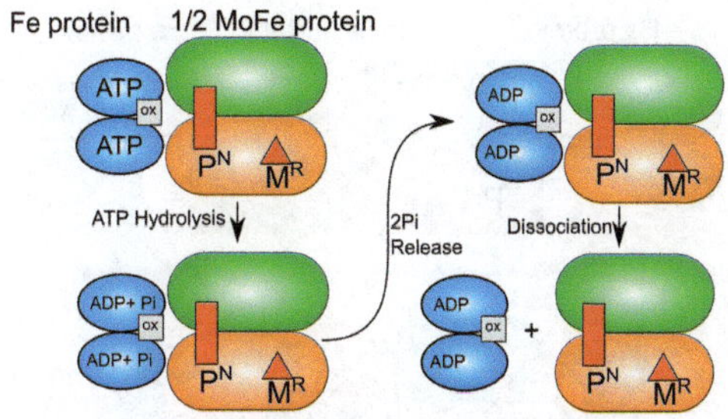

Figure 8.4 ATP hydrolysis in nitrogenase. Shown is a cartoon of 1/2 of the MoFe protein with one bound Fe protein. The first step (top left) picks up after all of the electron transfer reactions are complete. The next step is hydrolysis of the two bound ATP molecules resulting in two bound ADP + P_i (bottom left). The two P_i are released, creating an Fe protein with two bound ADP (top right). The final step is the dissociation of the Fe protein (ADP)2 complex from the MoFe protein (bottom right).

nucleotide utilizing enzymes, where the energy utilized from binding of ATP drives the association of the two proteins.[25] ATP hydrolysis is envisioned to activate release of the two proteins, the last step in the cycle (Figure 8.4). Clearly, while the order of key events seems fixed, there remains much work to fill in the details in this complex mechanism.

8.2.4 Negative Cooperativity in the Electron Transfer/ATP Hydrolysis Cycle

The native nitrogenase structure is composed of two identical catalytic halves (Figure 8.1), with each αβ unit of the MoFe protein binding an Fe protein. The traditional view has been that the two halves function as independent catalytic units, with all of the events in the catalytic cycle described in the above sections proceeding independently in each half.[3] A close inspection of the stoichiometry of electron transfer and ATP hydrolysis in the pre-steady state suggests that only approximately half of the electron transfer and ATP hydrolysis had occurred in the initial early times. A careful rapid quench study with EPR quantitation of the oxidation state of the metal clusters confirmed that only about half of the two Fe proteins are oxidized in the initial burst phase. Likewise, only two of the four bound ATP molecules are hydrolyzed in the initial burst. After the initial burst, the second electron from the Fe protein is seen to transfer, followed by the hydrolysis of the second set of ATP molecules. Thus, at least in the initial reaction, when the proteins are first mixed together, it appears that one side of the complex goes forward, while the other side is suppressed. In the limiting case of complete suppression, this phenomenon has been referred to as half-sites reactivity in other systems.[26] A global fit of the electron transfer, ATP hydrolysis and P_i release data reveals that the suppression of the second side is not complete, with a small leak of reactivity on the off side. Such an incomplete version of half sites reactivity

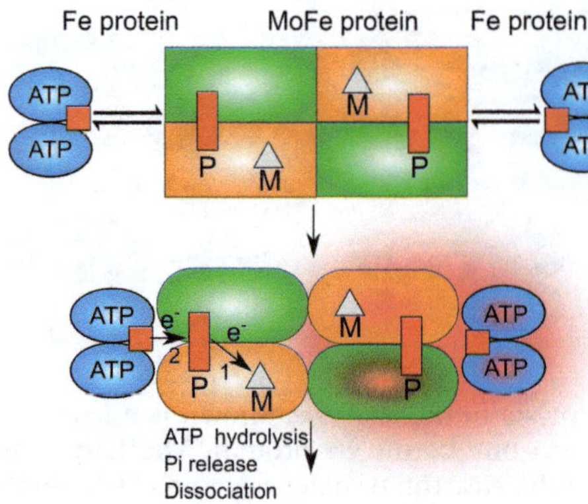

Figure 8.5 Negative cooperativity in nitrogenase. Shown is a cartoon of the MoFe protein with an Fe protein bound on each end. The left side is shown with deficit spending electron transfer between the clusters, which is followed by ATP hydrolysis, P_i release and dissociation on in the left half. The events associated with either binding or electron transfer in the left half are communicated to the right half to suppress electron transfer and ATP hydrolysis in a negative cooperativity mechanism. The right side is released to proceed with electron transfer and ATP hydrolysis in the next stage.

has been referred to as negative cooperativity. In fact, all of the observed data could be fit nicely to a negative cooperativity model (Figure 8.5).

While the details of how negative cooperativity is introduced into the system are poorly understood, computational studies of large-scale motions in nitrogenase show that the conformational changes in one half of nitrogenase are connected to the other half. This suggests that while one half is performing ET and ATP hydrolysis, the other half is in a conformational state that is not optimal for activity. This situation represents an example of long-range allosteric control where Fe protein binding, ET and ATP hydrolysis on one half of nitrogenase is able to impart negative cooperativity on the other half situated over 100 Å apart.

8.3 On the Mechanism of N_2 Reduction

In considering the mechanism for N_2 reduction catalyzed by nitrogenase, we are guided by the insights from our eminent colleague, Prof. Richard H. Holm, who has revealed the surprisingly difficult and complex task of defining a "mechanism" for chemical reactions with many steps. Inspired by his ideas, we summarize the key "*tiers*" in generating the nitrogenase mechanism, Chart 8.1. The "Activation Steps" involve the addition of electrons and protons to the active site to prepare it for substrate binding. The "Reaction Pathway" comprises the step-by-step sequence of N_2 *substrate* transformations. Such a pathway is largely descriptive, not explanatory, and independent

(i) **Activation Steps** – preparation of the *catalytic "machinery"*.
(ii) **Reaction Pathway** – the steps of N_2 reduction.
(iii) **Connection** – linking *(i)* and *(ii)*.
(iv) **Dynamics/Kinetics** – transformations between mechanistic steps.
(v) **Complete the Picture** – using experiment, models and theory.

Chart 8.1 "*Tiers*" in establishing the nitrogenase "mechanism"

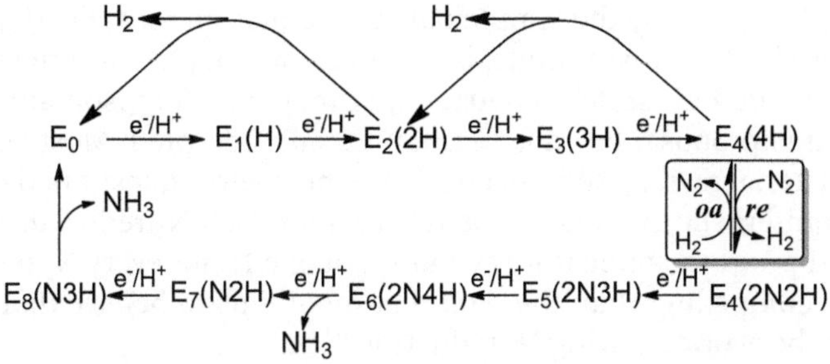

Figure 8.6 Simplified LT kinetic scheme highlighting correlated electron/proton delivery in eight steps, with inclusion of some of the possible pathways for decay by H_2 release. Parentheses added to the E_n notation to denote the stoichiometry of H/N bound to FeMo-co. When the molecular species corresponding to this stoichiometry is specified, it is noted; for example, E_4(2N2H) might have N_2 or N_2H_2 bound, and would be notated as $E_4(N_2$,2H) and $E_4(N_2H_2)$. *Box* surrounds the key mechanistic *tier (iv)* step, the *re*/*oa* equilibrium in which binding/release of N_2 is mechanistically coupled to the release/binding of H_2, as described in the text. The indicated steps of NH_3 release reflect the "Alternating" model we favor (see text).

of details of the catalytic machinery that performs these transformations. "Dynamics/Kinetics" includes, but is not limited to, a kinetic model for the transformations, and ideally would include the dynamics of each step. The "Complete the Picture" understanding that would be the outcome of *tier (v)* efforts couples the prior tiers through information about the active site derived from structure, experiment, theory and insights from model complexes, leading to an atomic-level understanding of the entire catalytic process.

Recent studies have yielded compelling proposals for the *tier (i, ii)* aspects of nitrogenase catalysis, and have revealed the mechanistic details of the step connecting them, *tier (iii)*.[27,28] This review begins with *tier (iv)*, the Lowe–Thorneley (LT) kinetic scheme for N_2 reduction, which is presented in simplified form in Figure 8.6. This "kinetic" foundation for a nitrogenase mechanism was developed by extensive studies in the 1970s and 1980s by many groups, especially Lowe and Thorneley and their coworkers.[3,29,30] The culmination of these extensive kinetic studies, which involved steady-state, stopped-flow and freeze-quench kinetics measurements, was the Lowe–Thorneley (LT) kinetic model for nitrogenase function, which describes the kinetics of transformations among catalytic intermediates (denoted E_n), where n is the number of steps of electrons/protons delivery to MoFe protein. Electron transfer from Fe protein to MoFe protein is driven by the binding and hydrolysis of two MgATP within the Fe protein as described above.

A central consequence of the kinetic measurements and a defining feature of this scheme, shown in a simplified version in Figure 8.6, is that the limiting enzymatic stoichiometry for enzyme-catalyzed nitrogen fixation is not what would be given by the simple balanced equation for reduction of N_2 to two NH_3 by six electrons/protons, but is given by eqn (8.1), a conclusion that is in agreement with stoichiometric experiments by Simpson and Burris.[31] However, this proposal has not been universally accepted. Most tellingly, in their magisterial review, Burgess and Lowe themselves questioned the mechanistic requirement that one H_2 be released for each N_2 reduced: "Thus the data that support the obligatory evolution of one H_2 for every N_2 reduced are much less compelling than the data that require us to believe that some H_2 will always be evolved during N_2 reduction."[3]

This section on nitrogenase mechanism touches lightly on the "reaction pathway", based on studies of catalytic intermediates that have been trapped and studied by EPR/ENDOR spectroscopies (*tiers (i, ii)*), but emphasizing a recent study that supports our proposed alternating pathway.[28] However, this review focuses primarily on a recent confirmation of our conclusion that the enzyme in fact exhibits the stoichiometry of eqn (8.1), with obligatory production of one H_2 per N_2 reduced.[27,28,32] We proposed that in the forward direction of the boxed equilibrium of Figure 8.6, the FeMo-co state that has accumulated four $[e^-/H^+]$, denoted $E_4(4H)$, is activated for the hydrogenation of N_2 to N_2H_2 by reductive elimination (*re*) of two $[e^-/H^+]$ in the form of H_2. In large part, this process was revealed through examining the consequences of the reverse of this equilibrium, the oxidative addition (*oa*) of H_2 with loss of N_2.[27] We summarize how this mechanism not only satisfies key constraints placed on mechanism by previous studies of nitrogenase function, but also successfully passes two stringent experimental tests suggested by the mechanism.

8.3.1 E_4: The "Janus Intermediate"

According to the simplified LT kinetic scheme of Figure 8.6, the first four of the eight $[e^-/H^+]$ of nitrogen fixation accumulate *prior* to N_2 binding/reduction, which occurs through reaction of $E_4(4H)$. The complete scheme[3,29,30] allows for N_2 binding at E_3 as well, but E_4 uniquely places the enzyme on the pathway to N_2 hydrogenation. The trapping and EPR/ENDOR characterization of the $E_4(4H)$ intermediate as having accumulated four $[e^-/H^+]$ in the form of hydrides bridging two Fe ions, namely as [Fe–H–Fe] moieties and two bound protons, Figure 8.7,[33–35] has been the key to recognizing the central role of hydrides in the mechanism for nitrogen fixation and to developing the mechanism of nitrogen fixation.[36] Indeed, we have depicted $E_4(4H)$ as the "Janus intermediate", Figure 8.7,

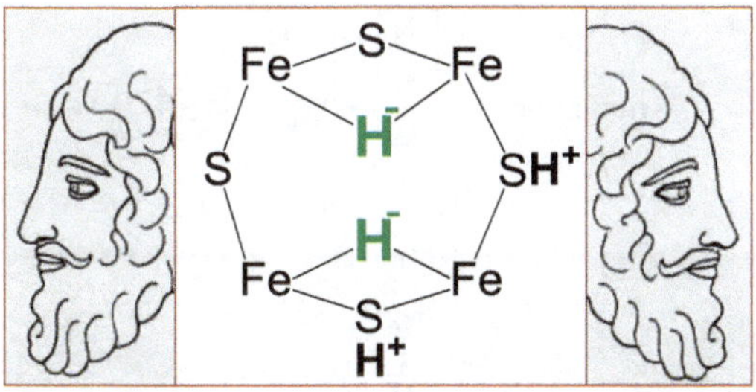

Figure 8.7 Depiction of E_4(4H) "Janus" intermediate, which contains two [Fe–H–Fe] moieties (their position is arbitrary), emphasizing its key role: it has accumulated four [e^-/H^+]; it looks forward to accumulation of four more and hydrogenation of N_2. Janus image adapted from: Janus12.jpg checkxstarinfinity.blogspot.com/. Figure is adapted with permission from ref. 28 (Hoffman *et al.*, 2013). Copyright 2013 American Chemical Society.

referring to the Roman God of transitions who is represented with two faces, one looking to the past and one looking to the future.[36] Looking "back" from E_4(4H) to the steps by which it is formed, E_4(4H) is the culmination of one-half of the electron/proton deliveries during N_2 fixation: four of the eight reducing equivalents are accumulated in E_4, before N_2 even becomes involved. Looking "forward", towards NH_3 formation, E_4(4H) is the state at which N_2 hydrogenation begins, and it is involved in one of the biggest puzzles in N_2 fixation: is H_2 lost upon N_2 binding, and if so, "why" and "how"?

Storage of the reducing equivalents accumulated in the E_4(4H) state as bridging hydrides has major consequences. A bridging hydride is less susceptible to protonation than a terminal hydride, and thus bridging hydride(s) diminish the tendency to lose reducing equivalents through the formation of H_2 (Figure 8.7), thereby facilitating the accumulation of reducing equivalents by FeMo-co. This mode also lowers the ability of the hydrides to undergo exchange with protons in the environment, a characteristic that is shown to be of central importance below. However, the bridging mode also lowers hydride reactivity towards substrate hydrogenation, relative to that of terminal hydrides.[37,38] How this "deactivated" intermediate becomes activated through the release of H_2 coupled to N_2 binding, the *tier (iii)* process which links the "electron-accumulation" steps of Figure 8.6 with the "substrate reduction" steps, is discussed below.

8.3.2 "Dueling" N_2 Reduction Pathways

Researchers have long considered two competing proposals for the second half of the LT kinetic scheme, the reaction pathway for N_2 reduction that begins with the Janus E_4 state.[3,39,40] These invoke distinctly different intermediates and different stages for release of the first NH_3, Figure 8.8, and computations suggest they likely involve different metal-ion sites on FeMo-co.[39] The "Distal" (*D*) pathway is associated with the Chatt[41,42] or Chatt–Schrock cycle[43] because

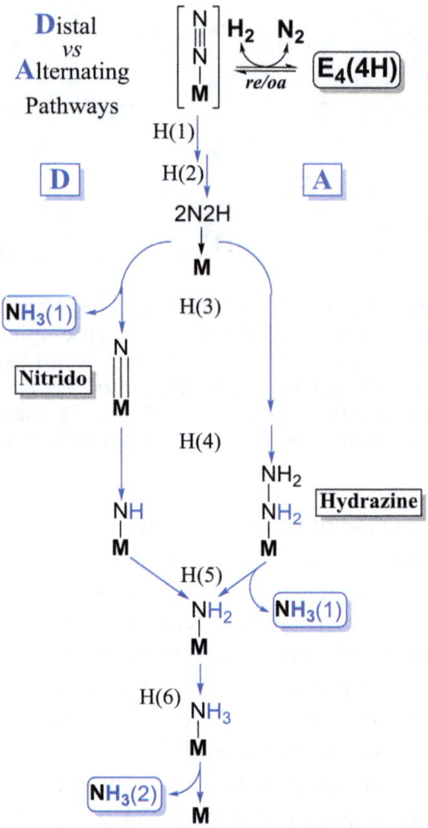

Figure 8.8 Comparison of traditionally called Distal and Alternating pathways for N_2 hydrogenation, most fundamentally distinguished by the different stages at which NH_3(1) is released, and by the presence of nitrido (D) *vs.* hydrazine (A) intermediates. The "2N2H" intermediate is at the reduction level of diazene, but may take the form of NH=NH or =N=NH_2. Adapted with permission from (Hoffman *et al.*, 2014). Copyright 2014 American Chemical Society.

it is utilized by inorganic Mo complexes discovered by these investigators to cleave N_2 (Chatt and coworkers[44,45]) and, most dramatically, to catalytically fix N_2 (Schrock and coworkers[46–49]). In this cycle, which has been suggested to apply to nitrogen fixation by nitrogenase with Mo as the active site,[39] a single N of N_2 is hydrogenated in *three* steps of N_2 hydrogenation until the first NH_3 is liberated, leaving a nitrido-N bound to FeMo-co; this is hydrogenated three more times to yield the second NH_3. In the "Alternating" (*A*) pathway, four hydrogenation steps form hydrazine; the first NH_3 is liberated only by the five hydrogenation steps (Figure 8.8).[27,28] As one can imagine alternative structures for the intermediates, the figure focuses on the defining difference between *D* and *A* pathways as being the release of the first NH_3 in the *D* as occurring after three hydrogenations of substrate, yielding a bound nitride, but only after the fifth hydrogenations, of bound hydrazine, in *A*.

Our study of trapped catalytic intermediates led us to assignments of E_4, E_7 and E_8, which made filling in the LT "boxes" for E_5, E_6 of the bottom, left-hand side of Figure 8.6 straightforward, thus unifying the reaction pathway for N_2 reduction with the LT scheme. We concluded that nitrogenase catalysis follows an "Alternating" pathway for the stages following N_2 binding and H_2 loss, most fundamentally distinguished by the late stage at which NH_3 is released, and by the presence of a hydrazine intermediate on the pathway,[27] with the final proposed pathway offering explanations for how the hydrogenated reaction intermediates, diazene and hydrazine, join the N_2 reduction pathway,[36] as shown in Figure 8.8 (below). This proposal was promptly supported by a study of the alternative substrate, NO_2^-.[50]

8.3.3 Nitrite and Hydroxylamine as Nitrogenase Substrates: Mechanistic Implications for the Pathway of N_2 Reduction

Nitrite (NO_2^-) is a six-electron substrate of nitrogenase that contains a single N atom and yields NH_3 as a reduction product,[51] and thus its reduction bears important similarities to the reduction of N_2 to two NH_3. We therefore surmised that it would be instructive to explore the catalytic pathway for NO_2^- reduction.[50] Fundamental to the comparison of this pathway with that of N_2 reduction is whether NO_2^- reduction, like that of N_2, produces H_2, and thus obeys the eight [e^-/H^+] stoichiometry of eqn (8.1). We showed[50] that NO_2^- reduction instead requires only six [e^-/H^+], together with an additional proton, to generate one NH_3 and two H_2O, eqn (8.2).

$$NO_2^- + 6e^- + 12ATP + 7H^+ \rightarrow NH_3 + 2H_2O + 12ADP + 12P_i \quad (8.2)$$

In addition, it was shown that NH_2OH is an excellent nitrogenase substrate.[50] The characterization of trapped intermediates during reduction of NO_2^- and NH_2OH catalyzed by remodeled nitrogenase MoFe protein, in conjunction with comparisons with the six-electron reduction of this substrate by the multi-heme enzyme cytochrome c nitrite reductase (ccNIR),[52–55] led us to the convergent alternating pathways for reduction of N_2, N_2H_2, N_2H_4, NO_2^- and NH_2OH presented in Figure 8.9.

8.3.4 Mechanistic Convergence and its Implications

The overall stoichiometries and early stages of N_2 and NO_2^- reduction by nitrogenase differ sharply, with reductive elimination of H_2 a central feature of N_2 fixation, but absent for NO_2^- reduction or reduction of any other nitrogenase substrate studied so far. However, ENDOR/ESEEM studies of trapped

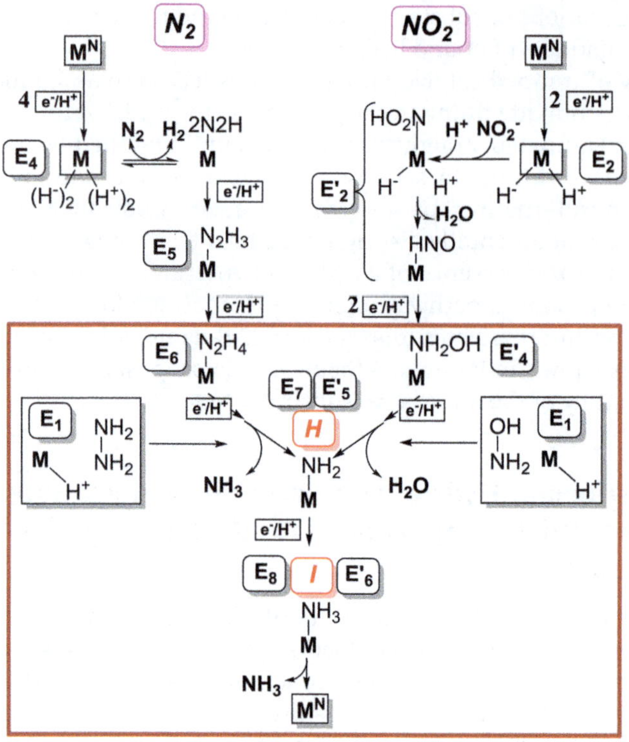

Figure 8.9 *Left*: Previously proposed pathway for N_2 reduction by nitrogenase (M represents FeMo-co).[4] An intermediate, labeled E_n, has accumulated n [e^-/H^+], $n = 0–8$. *Right*: Proposed dominant pathway for nitrite reduction by nitrogenase; intermediates of NO_2H reduction are labeled, E'_m, with accumulation of m [e^-/H^+]. *Boxed region*: convergence of pathways for nitrite and N_2 reduction by nitrogenase, as well as the most direct routes by which N_2H_4 and HN_2OH enter the pathway. Adapted with permission from Shaw *et al.*, 2014. Copyright 2014 American Chemical Society.

intermediates indicate that the reaction pathways converge to corresponding intermediates, which then react to form the identical states *H* and *I*, as captured in the boxed section of the two catalytic pathways, Figure 8.9. This figure also shows the most direct route by which both N_2H_4 and NH_2OH, independently acting as substrates, join their respective reaction pathways: they bind to E_1 and undergo bond cleavage, respectively liberating NH_3 and H_2O and generating the *H* intermediate.

Given the findings that not only N_2H_4 (and N_2H_2) but also NH_2OH are excellent nitrogenase substrates, and that common *H* and *I* intermediates form during nitrogenase turnover with both NO_2^- and NH_2OH on the one hand, and with dinitrogenous substrates N_2H_4 and N_2H_2 on the other, this study supported proposals that: (i) N_2H_4 and NH_2OH are indeed generated on the enzymatic pathways by which nitrogenase reduces the primary substrates, N_2 and NO_2^-; (ii) these results are consistent with an alternating pathway for nitrogen fixation in which the first NH_3 is generated by N–N bond cleavage upon delivery of the seventh [e^-/H^+] to the E_6 (N_2H_4-bound) intermediate to form $H = E_7$, and the second NH_3 by delivery of the eighth and last [e^-/H^+] to form $I = E_8$, Figure 8.9, (left); (iii) NO_2^- reduction proceeds by the analogous and convergent pathway, in which the H_2O is liberated by reductive cleavage of the NH_2–OH-bound intermediate to form $H = E'_5$ and NH_3 is produced by delivery of the sixth and last [e^-/H^+] $I = E'_6$, Figure 8.9 (right).

Overall, the parallels between N_2 fixation and NO_2^- and NH_2OH reduction by nitrogenase support the parallel nitrogenase pathways presented in Figure 8.9, with N–A bond cleavage after delivery of the fifth [e^-/H^+] to substrate, to yield the M-NH_2, *H* intermediate. This study supports a pathway in which N–N bond cleavage and liberation of the first NH_3 occur during formation of the E_7 intermediate (delivery of 5 [e^-/H^+] to substrate), *H* (Figures 8.8 and 8.9), and not earlier in the reduction process, through bond cleavage at E_5 (delivery of 3 [e^-/H^+] to substrate) with formation of a nitrido-bound intermediate.

8.3.5 Evolution of One H_2 per N_2 Reduced in Nitrogen Fixation is Obligatory

The pathway we propose is silent about the mechanism by which the E_4(4H) Janus intermediate might carry out the LT-proposed obligatory production of an H_2 molecule upon N_2 binding shown in Figure 8.9.[29,31,36] Consideration of the finding that E_4(4H) stores its four reducing equivalents as two bridging hydrides (Figure 8.7), viewed within the context of the well-known organometallic chemistries of hydrides[37,56] and dihydrogen,[57] led us to examine the two alternative mechanisms by which this state might bind and activate N_2 with release of H_2, then promptly proceeding to the formation of FeMo-co with a bound diazene-level species (N_2H_2) without additional accumulation of [e^-/H^+] as featured in the *A* reaction pathway, Figure 8.8. We then tested these mechanisms against the numerous constraints that are associated with the reaction of D_2 with the diazene-level $E_4(N_2/N_2H_2)$ state formed when N_2 binds to the cofactor and is reduced.[27]

The three principal constraints are listed in Chart 8.2. The first constraint is that when nitrogenase turns over in the presence of *both* N_2 and D_2, then two HD are formed through D_2 cleavage and solvent-proton reduction, with the stoichiometry summarized as constraint *(i)* of Chart 8.2.[3,58–60] Such HD formation *only* occurs in the presence of N_2, and not during enzymatic reduction of H^+ or *any* other substrate.[58,61,62] The second key constraint is the absence of exchange into solvent of D^+/T^+ derived from D_2/T_2 gas, Chart 8.2, *(ii)*.[58] The third constraint is that the enzyme be at the E_4 redox level, with a FeMo-co-bound N–N species at the reduction level of N_2H_2 or beyond, Chart 8.2, *(iii)*.[63]

(i) <u>*Stoichiometry:*</u>	**$D_2 + 2H^+ + 2e^- \Rightarrow 2HD$**
(ii) <u>*Scrambling:*</u>	**"*No*" T^+ released to solvent under T_2.**
(iii) <u>*Reduction Stage:*</u>	**D_2 reacts at 'N_2H_2' level.**

Chart 8.2 Key constraints on HD formation under N_2/D_2.

8.3.5.1 Hydride Protonation (hp) Mechanism

The most obvious *tier (iv)* mechanism for H_2 release mechanistically coupled to N_2 binding would involve the protonation of *one* bridging hydride by a sulfide-bound proton to form and release H_2; we called this the hydride protonation (*hp*) scheme. Such a mechanism for H_2 formation is invoked in discussions of hydrogenases,[57,64] However, we showed[27] that *hp* irretrievably violates the constraints of Chart 8.2, an argument that does not need to be repeated here.

8.3.5.2 Reductive Elimination (re) Mechanism

The second *tier (iv)* mechanism for H_2 loss upon N_2 binding connects E_4(4H) with E_4(2N2H) through the reversible, reductive elimination (*re*)/oxidative addition (*oa*), equilibrium binding of N_2 and release of H_2.[38] Of key importance, while the departing H_2 carries away only two of the four reducing equivalents stored in E_4, this process delivers two reducing equivalents to the FeMo-co core, which otherwise is reduced by at most one equivalent at a time during electron/proton activation. We now summarize the success in the *re*/*oa* mechanism in satisfying the constraints of Chart 8.2.[27] Figure 8.10 shows that the reverse of the *re* mechanism for N_2 activation, *oa* of H_2 and loss of N_2, satisfies all the constraints of Chart 8.2. D_2 binding and N_2 release would generate E_4(2D), the E_4 isotopomer in which *both* atoms of D_2 exist as [Fe–D–Fe] bridges. This state would relax with loss of HD to E_2(D), then to E_0 with loss of the second HD, thus satisfying the stoichiometry of eqn (8.1). If the reaction were carried out under T_2, essentially no T^+ would be lost to solvent because the bridging deuterides are deactivated for exchange with the protein environment and solvent, thus satisfying the "T^+ exchange" constraint, Chart 8.2. Consideration of alternative fates of the E_4(2D) formed by D_2 replacement of N_2 again shows the *re*/*oa* mechanism to be satisfactory.[27]

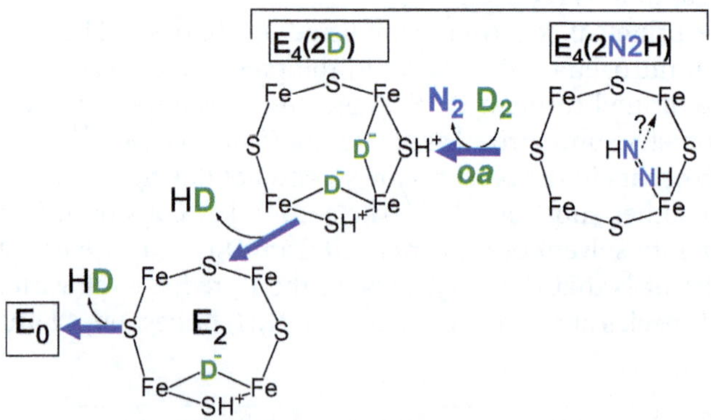

Figure 8.10 Heuristic visualization of reverse of *re* mechanism: *oa* of D_2 to E_4(2N2H) with loss of N_2. Shown, the Fe-2, 3, 6, 7 face of resting FeMo-co; the structure of FeMo-co must distort in different stages of catalysis.

8.3.6 First Experimental Test of the *re* Mechanism

Subsequent to formulation of the *re*/*oa* mechanism for the activation of FeMo-cofactor to reduce N_2,[27,36] we noted that addition of C_2H_2 to an N_2/D_2 reaction mixture should offer a rigorous test of the mechanism. We recognized that the *re* mechanism predicts (Figure 8.10) that turnover under $C_2H_2/D_2/N_2$ provides a test of the formation of the $E_4(2D)$ state with two bridging hydrides by *oa* of D_2 and the formation of $E_2(D)$ by subsequent HD release.[65] If $E_4(2D)$ is indeed formed, it would react with C_2H_2 to form $C_2H_2D_2$; the $E_4(2D)$ state also would relax through the loss of HD to form $E_2(D)$, an E_2 state whose unique isotopic composition can be generated in no other way. Interception of the $E_2(D)$ state by C_2H_2 would then generate C_2H_3D.[65]

We tested these predictions of an unprecedented involvement of gaseous D_2 in substrate reduction by using C_2H_2 reduction under $N_2/D_2/C_2H_2$ gas mixtures to intercept the $E_4(2D)$ and $E_2(D)$ states (Figure 8.11). As expected, the control reaction of turnover under D_2/C_2H_2 generates only C_2H_4, without formation of either C_2H_3D or $C_2H_2D_2$. In dramatic contrast, C_2H_2 reduction by nitrogenase under a $N_2/D_2/C_2H_2$ gas mixture in fact produces readily measured amounts of $C_2H_2D_2$ and even greater amounts of C_2H_3D.[65] This trapping of the *re*/*oa* intermediates $E_4(2D)$ and $E_4(2D)$ establishes the occurrence of the *oa* of D_2, the reverse of the *re* binding of N_2 with loss of H_2.

8.3.7 Second Experimental Test of *re*: Identification of the Key E_4(2N2H) Catalytic Intermediate

In developing the *re* mechanism, we used advanced paramagnetic techniques to characterize two nitrogenous E_n intermediates that are associated with states in the LT scheme subsequent to N_2 binding, $n \geq 4$ (Figure 8.9),

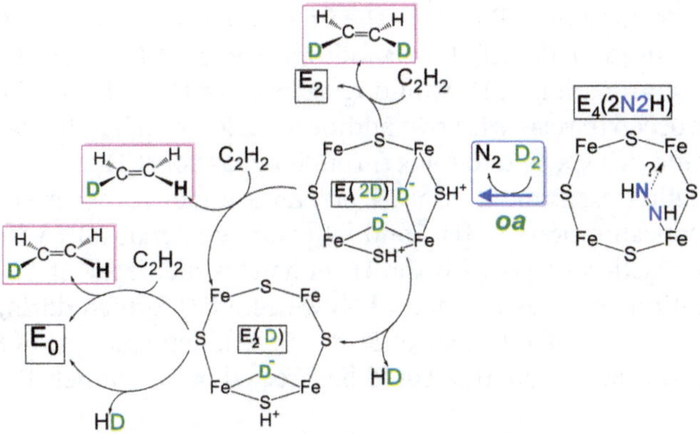

Figure 8.11 Cartoons showing that the "reverse" of *re* mechanism through displacement of N_2 coupled to *oa* of D_2 produces, successively, $E_4(2D)$ and $E_2(D)$, and how theses intermediates can be captured with C_2H_2. Adapted with permission from Hoffman *et al.*, 2014. Copyright 2014 American Chemical Society.

and that are common to turnover of remodeled nitrogenase with N_2H_2, Me–N_2H, N_2H_4, NO_2^- and H_2NOH. One is a non-Kramers ($S \geq 2$) state, denoted *H*, that is assigned as E_7, with bound [–NH_2]; the second is a Kramers (S = 1/2) state, denoted *I*, assigned as E_8, with bound NH_3.[2–4] During turnover of wild-type nitrogenase with N_2 an additional intermediate state with S = 1/2 was trapped, but not identified.[5–7] With a half-integer spin, like the E_0 resting state, this state must differ from E_0 by the accumulation of an n = *even* number of [e^-/H^+] to FeMo-co. This state, denoted as *EG*, must further correspond to an E_n state formed subsequent to binding N_2, n = 4, 6 or 8, because it gives ^{15}N ENDOR signals when trapped using $^{15}N_2$ substrate. The previous assignment of *I* as E_8[3,4] implies an assignment of *EG* to E_4 or E_6. The notation, *EG*, in fact was adopted because if the states of Figure 8.9 were to be labeled sequentially by letters beginning with $E_0 = A$, then E_4 would be E and E_6 would be *G*.

Because of the relatively low accumulation of *EG* in freeze-quenched samples, advanced paramagnetic resonance measurements had not yielded an assignment of *EG*. To identify *EG*, we therefore turned to the quench-cryoannealing relaxation protocol developed to determine the reduction level, *n*, of an EPR-active freeze-trapped E_n intermediate state.[8] The trapped state is allowed to relax to the resting E_0 state in frozen medium, at temperatures $T \leq -20$ °C, well below the melting temperature of the buffered samples, $T \sim 0$ °C. Keeping the sample frozen prevents any additional accumulation of reducing equivalents, because binding of reduced Fe protein to and release of oxidized protein from the MoFe protein both are abolished in a frozen solid; the frozen intermediate can only relax towards the resting state through steps that release a stable species from FeMo-co.[9] As the defining example of this approach, the freeze-trapped S = 1/2 intermediate shown by ENDOR spectroscopy to exhibit two [Fe–H–Fe] bridging hydrides[10,11] was identified as the E_4(4H) Janus intermediate by its relaxation to the resting state E_0 through the release of a total of four reducing equivalents in a two-step process, each step releasing H_2 (two equivalents), with formation of E_2(2H) in the first step (Figure 8.6).[8] This annealing protocol promised to identify *EG*, as follows. The n = 4 E_4(2N2H) state formed by reductive elimination of H_2 would relax by *oa* of H_2 and loss of N_2 (Scheme 8.1), then proceed to relax with two additional releases of H_2. In contrast, the n = 6 state would be expected to relax through release of N_2H_4.

Cryoannealing experiments on *EG* prepared under turnover conditions in which the concentrations of [H_2] and [N_2] were systematically varied demonstrated that reaction of *EG* with the H_2 enhances the decay of *EG* and limits its accumulation, whereas reaction of N_2 with E_4(4H) formed during the decay process slows the observed decay of *EG*.[32] In the kinetic scheme of Scheme 1.1 there is only one such state that could be affected by H_2, namely E_4(2N2H), the

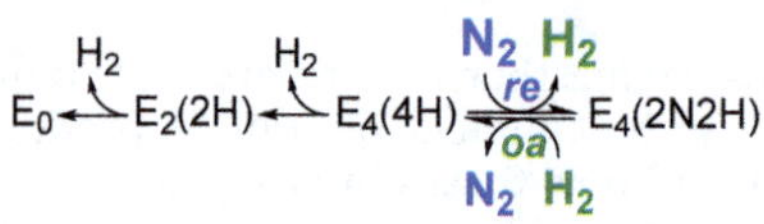

Scheme 8.1

key catalytic intermediate E_4(2N2H) formed upon reductive elimination of H_2 and binding of N_2, whose decay is to be understood as involving the reverse of the *re* activation equilibrium, Scheme 8.1: the oxidative-addition reaction of E_4(2N2H) with H_2 and loss of N_2 to form E_4(4H) (k_b). The competing regeneration of E_4(2N2H) from E_4(4H) through reaction with N_2 (k_r) and reductive elimination of H_2 in turn competes with the irreversible relaxation of E_4(4H) through loss of H_2 (k_d). This competition in the decay of *EG* was captured by a steady-state treatment for the concentration of E_4(4H) within the kinetic relaxation, Scheme 8.1.

The finding that decay of *EG* is accelerated by increasing $[H_2]$ and slowed by increasing $[N_2]$ in the frozen reaction mixture, and the resulting identification of *EG* as E_4(2N2H), directly demonstrates that activation of nitrogenase for N_2 binding and reduction involves the thermodynamically and kinetically reversible reductive-elimination/oxidative-addition (*re*/*oa*) activation equilibrium of Figure 8.6 as previously inferred,[27] and in so doing further confirms the limiting stoichiometry of eight electrons/protons for the reduction of N_2 to two NH_3 (eqn (8.1)). Said differently, this experiment joins the other tests of *re*/*oa* in confirming that formation of one H_2 per N_2 reduced is mechanistically required.

The test of the *re*/*oa* mechanism against historical constraints on nitrogenase function (Chart 8.2)[27] and the two experimental tests[32,65] summarized here establish the nitrogenase stoichiometry (eqn (8.1)) and the obligatory formation of H_2 upon N_2 binding. This mechanism provides answers to the long-standing question, "Does Nature 'waste' four ATP/two reducing equivalents through an obligatory loss of H_2 when N_2 binds, and, if so, why?" The answer is "yes", and the reason appears to be that only this process activates FeMo-co to cleave the N_2 bond.

8.4 Summary and Prospects

Electron Accumulation: It has been clear from the earliest studies that nitrogenase represents a complex molecular machine with many steps. One entire aspect of this complexity involves the association of the two nitrogenase proteins, the transfer of electrons, and the hydrolysis of ATP. Through a large number of studies conducted over more than four decades by many groups, it seems that the framework of the key events is set. It is equally clear that many of the details remain to be resolved over the coming years.

Mechanism: A second major aspect of nitrogenase is understanding how electrons accumulate to achieve the challenging reduction of N_2. Again, through studies by many groups, the picture for this process is becoming clearer. Recent studies have started to provide insights into the *tier (ii)* pathway for N_2 reduction, with a likely Alternating reduction pathway, which is distinguished by the appearance of N_2H_4 at the E_6 level and the "late" release of the first NH_3, upon hydrogenation of N_2H_4 and formation of E_7.[27,28,50] As described in this review, the breakthrough step has been in resolving the *tier (iii)* question of stoichiometry, 8[e^-/H+], not 6, and revealing the *re*/*oa* mechanism that activates FeMo-co for N_2 reduction at the cost of the release of H_2. Clearly, studies going forward will fill in the many gaps in our understanding of this complex set of events. For example, DeBeer, Neese and coworkers have shown through elegant Mo XAS measurements that the resting-state Mo must be formulated as Mo(III) in a non-Hund's rule configuration, rather than the previously proposed Mo(IV) valency.[66–68] Certainly the formulation of a self-consistent spin coupling model that incorporates this result will enhance our understanding of the electronic structure of FeMo-co. In parallel, the crystallographic discovery that FeMo-co loses an S atom when generating a kinetically non-competent CO-inhibited state adds weight to ideas formulated many times over the years, that FeMo-co might undergo structural distortions during catalysis.[69] And the understanding of how metal centers carry out N_2 binding and reduction is being immeasurably enriched by the explosion in biomimetic inorganic complexes studied by many groups.[70,71]

Acknowledgements

This work was supported by the Department of Energy, Office of Basic Energy Sciences (DRD and LCS) and the National Institutes of Health (GM111097 to BMH) and the National Science Foundation (MCB 1118613 to BMH).

References

1. V. Smil, *Enriching the Earth: Fritz Haber, Carl Bosch, and the Transformation of World Food Production*, The MIT Press, Cambridge, MA, 2004.
2. R. R. Eady, *Chem. Rev.*, 1996, **96**, 3013–3030.
3. B. K. Burgess and D. J. Lowe, *Chem. Rev.*, 1996, **96**, 2983–3012.
4. V. K. Shah and W. J. Brill, *Proc. Natl. Acad. Sci. U. S. A.*, 1977, **74**, 3249–3253.
5. M. M. Georgiadis, H. Komiya, P. Chakrabarti, D. Woo, J. J. Kornuc and D. C. Rees, *Science*, 1992, **257**, 1653–1659.
6. D. C. Rees and J. B. Howard, *Curr. Opin. Chem. Biol.*, 2000, **4**, 559–566.
7. L. C. Seefeldt and D. R. Dean, *Acc. Chem. Res.*, 1997, **30**, 260–266.
8. R. N. F. Thorneley and D. J. Lowe, in *Molybdenum Enzymes*, ed. T. G. Spiro, Wiley-Interscience Publications, New York, 1985, vol. 7, pp. 221–284.
9. P. E. Wilson, A. C. Nyborg and G. D. Watt, *Biophys. Chem.*, 2001, **91**, 281–304.
10. C. P. Owens, F. E. H. Katz, C. H. Carter, M. A. Luca and F. A. Tezcan, *J. Am. Chem. Soc.*, 2015, **137**, 12704–12712.
11. F. A. Tezcan, J. T. Kaiser, D. Mustafi, M. Y. Walton, J. B. Howard and D. C. Rees, *Science*, 2005, **309**, 1377–1380.
12. H. Schindelin, C. Kisker, J. L. Schlessman, J. B. Howard and D. C. Rees, *Nature*, 1997, **387**, 370–376.
13. R. W. Miller, B. E. Smith and R. R. Eady, *Biochem. J.*, 1993, **291**(3), 709–711.
14. J. M. Chan, J. Christiansen, D. R. Dean and L. C. Seefeldt, *Biochemistry*, 1999, **38**, 5779–5785.
15. D. J. Lowe, K. Fisher and R. N. Thorneley, *Biochem. J.*, 1993, **292**, 93–98.
16. D. Lowe, K. Fisher, R. Pau and R. Thorneley, *Curr. Plant Sci. Biotechnol. Agric.*, 1993, **17**, 95–100.

17. K. Danyal, D. R. Dean, B. M. Hoffman and L. C. Seefeldt, *Biochemistry*, 2011, **50**, 9255–9263.
18. W. R. Hagen, H. Wassink, R. R. Eady, B. E. Smith and H. Haaker, *Eur. J. Biochem.*, 1987, **169**, 457–465.
19. K. Danyal, D. R. Dean, B. M. Hoffman and L. C. Seefeldt, *Biochemistry*, 2011, **50**, 9255–9263.
20. K. Danyal, D. Mayweather, D. R. Dean, L. C. Seefeldt and B. M. Hoffman, *J. Am. Chem. Soc.*, 2010, **132**, 6894–6895.
21. L. E. Mortenson, *Proc. Natl. Acad. Sci. U. S. A.*, 1964, **52**, 272–279.
22. J. E. McNary and R. H. Burris, *J. Bacteriol.*, 1962, **84**, 598.
23. L. E. Mortenson and R. G. Upchurch, in *Current Perspectives in Nitrogen Fixation, Proceedings of the International Symposium*, Australia Academy of Science, Canberra, Australia, 1981, pp. 75–78.
24. W. N. Lanzilotta, V. D. Parker and L. C. Seefeldt, *Biochim. Biophys. Acta*, 1999, **1429**, 411–421.
25. S. Duval, K. Danyal, S. Shaw, A. K. Lytle, D. R. Dean, B. M. Hoffman, E. Antony and L. C. Seefeldt, *Proc. Natl. Acad. Sci. U. S. A.*, 2013, **110**, 16414–16419.
26. A. Levitzki, W. B. Stallcup and D. E. Koshland Jr, *Biochemistry*, 1971, **10**, 3371–3378.
27. B. M. Hoffman, D. Lukoyanov, Z.-Y. Yang, D. R. Dean and L. C. Seefeldt, *Chem. Rev.*, 2014, **114**, 4041–4062.
28. B. M. Hoffman, D. Lukoyanov, D. R. Dean and L. C. Seefeldt, *Acc. Chem. Res.*, 2013, **46**, 587–595.
29. R. N. F. Thorneley and D. J. Lowe, in *Molybdenum Enzymes*, ed. T. G. Spiro, Wiley, New York, 1985, pp. 221–284.
30. P. E. Wilson, A. C. Nyborg and G. D. Watt, *Biophys. Chem.*, 2001, **91**, 281–304.
31. F. B. Simpson and R. H. Burris, *Science*, 1984, **224**, 1095–1097.
32. D. Lukoyanov, Z.-Y. Yang, N. Khadka, D. R. Dean, L. C. Seefeldt and B. M. Hoffman, *J. Am. Chem. Soc.*, 2015, **137**, 3610–3615.
33. R. Y. Igarashi, M. Laryukhin, P. C. Dos Santos, H.-I. Lee, D. R. Dean, L. C. Seefeldt and B. M. Hoffman, *J. Am. Chem. Soc.*, 2005, **127**, 6231–6241.
34. D. Lukoyanov, Z.-Y. Yang, D. R. Dean, L. C. Seefeldt and B. M. Hoffman, *J. Am. Chem. Soc.*, 2010, **132**, 2526–2527.
35. D. Lukoyanov, B. M. Barney, D. R. Dean, L. C. Seefeldt and B. M. Hoffman, *Proc. Natl. Acad. Sci. U. S. A.*, 2007, **104**, 1451–1455.
36. B. M. Hoffman, D. Lukoyanov, D. R. Dean and L. C. Seefeldt, *Acc. Chem. Res.*, 2013, **46**, 587–595.
37. M. Peruzzini and R. Poli, *Recent Advances in Hydride Chemistry*, Elsevier Science B. V., Amsterdam, The Netherlands, 2001.
38. R. H. Crabtree, *The Organometallic Chemistry of the Transition Metals*, Wiley, Hoboken, NJ, 5th edn, 2009.
39. F. Neese, *Angew. Chem., Int. Ed.*, 2006, **45**, 196–199.
40. B. M. Hoffman, D. R. Dean and L. C. Seefeldt, *Acc. Chem. Res.*, 2009, **42**, 609–619.
41. C. J. Pickett, *J. Biol. Inorg. Chem.*, 1996, **1**, 601–606.

42. B. A. MacKay and M. D. Fryzuk, *Chem. Rev.*, 2004, **104**, 385–402.
43. H.-P. Jia and E. A. Quadrelli, *Chem. Soc. Rev.*, 2014, **43**, 547–564.
44. J. Chatt, J. R. Dilworth and R. L. Richards, *Chem. Rev.*, 1978, **78**, 589–625.
45. J. Chatt and R. L. Richards, *J. Organomet. Chem.*, 1982, **239**, 65–77.
46. R. R. Schrock, *Chem. Commun.*, 2003, 2389–2391.
47. D. V. Yandulov and R. R. Schrock, *Science*, 2003, **301**, 76–78.
48. R. R. Schrock, *Acc. Chem. Res.*, 2005, **38**, 955–962.
49. R. R. Schrock, *Angew. Chem., Int. Ed.*, 2008, **47**, 5512–5522.
50. S. Shaw, D. Lukoyanov, K. Danyal, D. R. Dean, B. M. Hoffman and L. C. Seefeldt, *J. Am. Chem. Soc.*, 2014, **136**, 12776–12783.
51. S. A. Vaughn and B. K. Burgess, *Biochemistry*, 1989, **28**, 419–424.
52. O. Einsle, A. Messerschmidt, R. Huber, P. M. H. Kroneck and F. Neese, *J. Am. Chem. Soc.*, 2002, **124**, 11737–11745.
53. D. Bykov and F. Neese, *J. Biol. Inorg. Chem.*, 2012, **17**, 741–760.
54. D. Bykov and F. Neese, *J. Biol. Inorg. Chem.*, 2011, **16**, 417–430.
55. D. Bykov, M. Plog and F. Neese, *J. Biol. Inorg. Chem.*, 2014, **19**, 97–112.
56. J. Ballmann, R. F. Munhá and M. D. Fryzuk, *Chem. Commun.*, 2010, **46**, 1013–1025.
57. G. J. Kubas, *Chem. Rev.*, 2007, **107**, 4152–4205.
58. B. K. Burgess, S. Wherland, W. E. Newton and E. I. Stiefel, *Biochemistry*, 1981, **20**, 5140–5146.
59. J.-L. Li and R. H. Burris, *Biochemistry*, 1983, **22**, 4472–4480.
60. B. B. Jensen and R. H. Burris, *Biochemistry*, 1985, **24**, 1141–1147.
61. G. E. Hoch, K. C. Schneider and R. H. Burris, *Biochim. Biophys. Acta*, 1960, **37**, 273–279.
62. E. K. Jackson, G. W. Parshall and R. W. F. Hardy, *J. Biol. Chem.*, 1968, **243**, 4952–4958.
63. K. Fisher, M. J. Dilworth and W. E. Newton, *Biochemistry*, 2000, **39**, 15570–15577.
64. C. Tard and C. J. Pickett, *Chem. Rev.*, 2009, **109**, 2245–2274.
65. Z.-Y. Yang, N. Khadka, D. Lukoyanov, B. M. Hoffman, D. R. Dean and L. C. Seefeldt, *Proc. Natl. Acad. Sci. U. S. A.*, 2013, **110**, 16327–16332.
66. R. Bjornsson, F. Neese, R. R. Schrock, O. Einsle and S. DeBeer, *J. Biol. Inorg. Chem.*, 2014, **20**, 447–460.
67. R. Bjornsson, F. A. Lima, T. Spatzal, T. Weyhermueller, P. Glatzel, E. Bill, O. Einsle, F. Neese and S. DeBeer, *Chem. Sci.*, 2014, **5**, 3096–3103.
68. K. M. Lancaster, M. Roemelt, P. Ettenhuber, Y. Hu, M. W. Ribbe, F. Neese, U. Bergmann and S. DeBeer, *Science*, 2011, **334**, 974–977.
69. T. Spatzal, K. A. Perez, O. Einsle, J. B. Howard and D. C. Rees, *Science*, 2014, **345**, 1620–1623.
70. I. Čorić, B. Q. Mercado, E. Bill, D. J. Vinyard and P. L. Holland, *Nature*, 2015, **526**, 96–99.
71. J. S. Anderson, G. E. Cutsail III, J. Rittle, B. A. Connor, W. A. Gunderson, L. Zhang, B. M. Hoffman and J. C. Peters, *J. Am. Chem. Soc.*, 2015, **137**, 7803–7809.

CHAPTER 9

Biosynthesis of the M-Cluster of Mo-Nitrogenase

J. A. WIIG[a], C. C. LEE[a], J. G. REBELEIN[a], N. S. SICKERMAN[a], K. TANIFUJI[a], M. T. STIEBRITZ[a], Y. HU*[a], AND M. W. RIBBE*[a,b]

[a]Department of Molecular Biology and Biochemistry, University of California, Irvine, CA 92697-3900, USA; [b]Department of Chemistry, University of California, Irvine, CA 92697-2025, USA
*E-mail: yilinh@uci.edu, mribbe@uci.edu

9.1 Introduction

Biological nitrogen fixation is the biochemical process that converts the inert, atmospheric dinitrogen (N_2) to a usable form of nitrogen, ammonia (NH_3).[1–3] This process is a key step in the global nitrogen cycle, where nitrogen becomes bio-accessible; therefore, it is essential for sustaining life on Earth. Nitrogen fixation also employs the powerful chemistry of breaking the N≡N triple bond, one of the strongest chemical bonds found in Nature. This challenging feat is accomplished by nitrogenases, a group of complex metalloenzymes that are unique in their ability to use protons and physiological electron donors for the reduction of N_2 to NH_3 at ambient temperature and pressure. As demonstrated by their essential function to all organic life forms, the evolutionary origins of nitrogenases are ancient, and these enzymes are now commonly found in free-living soil bacteria and symbiotic bacteria living at legume root nodules.[4,5]

RSC Metallobiology Series No. 5
Molybdenum and Tungsten Enzymes: Biochemistry
Edited by Russ Hille, Carola Schulzke, and Martin L. Kirk

Published by the Royal Society of Chemistry, www.rsc.org

Four classes of nitrogenases have been identified to date. The first three of them, namely the molybdenum (Mo)-dependent nitrogenase, the vanadium (V)-dependent nitrogenase and the iron (Fe)-only nitrogenase, are highly homologous to one another and mainly distinguished by the type of heterometal (or the absence thereof) at the active site. The fourth nitrogenase is dependent on superoxide and clearly different from the other three members of this enzyme family.[6–10] The Mo-nitrogenase is the most common form of the four known classes of nitrogenases. It consists of two components: one is referred to as the iron (Fe) protein; whereas the other is denoted as the molybdenum–iron (MoFe) protein. Encoded by *nifH*, the Fe protein (also termed NifH) is a homodimer containing one [Fe_4S_4] cluster at the interface of the identical subunits and one magnesium-adenosine triphosphate (MgATP) binding site within each subunit. Encoded by *nifD* and *nifK*, the MoFe protein (also termed NifDK) is an $\alpha_2\beta_2$-heterotetramer that houses a pair of unique metal centers per αβ-dimer: the P-cluster ([Fe_8S_7]), which is located at each αβ-subunit interface; and the M-cluster ([$MoFe_7S_9C$-homocitrate]), which is situated within each α-subunit. During catalysis, NifH repeatedly associates and dissociates with NifDK, transferring one electron in an ATP-dependent manner to the latter during each cycle of association/dissociation between NifH and NifDK.[11,12] Repetition of this cycle allows electrons to flow from the reduced [Fe_4S_4] cluster in NifH to the P-cluster and subsequently to the M-cluster of NifDK, where the substrate is reduced upon accumulation of a sufficient amount of electrons.

The M-cluster is one of the most unique and complex metal clusters in all known biochemistry.[13] This cluster is composed of a [$MoFe_7S_9C$] core, with the Mo center ligated by the organic molecule, *R*-homocitrate (HC).[13–16] The metal–sulfur core of the M-cluster can be viewed as [Fe_4S_3] and [$MoFe_3S_3$] subclusters joined together by three bridging sulfide atoms and one interstitial carbide atom. Within the protein matrix of NifDK, the M-cluster is anchored through only two amino acid residues: a histidine residue, which provides coordination through its nitrogen atom to the Mo center at one end of the cluster; and a cysteine residue, which provides coordination through its sulfur atom to the Fe atom at the opposite end of the cluster. The M-cluster can be extracted into organic solvents, such as *N*-methylformamide, as a complete entity and subsequently used to reconstitute the M-cluster-deficient NifDK (designated apo-NifDK) to achieve full activation of the protein.[17–19] Moreover, in the extracted state, the M-cluster is capable of performing catalytic substrate reduction in the presence of strong reductants.[20,21] The exact structure of the M-cluster has not yet been synthetically replicated, although topologs approximating its general architecture have been successfully prepared.[22–25] In the meantime, much progress has been made toward elucidating the biosynthetic mechanism of the M-cluster, particularly with regard to how the M-cluster is formed from the fusion of two [Fe_4S_4] clusters and how the insertion of the interstitial carbon atom takes place in this process.[16,26,27] This chapter will present an overview of the assembly pathway of nitrogenase M-cluster and highlight the recent advances toward a better understanding of this process.

9.2 M-Cluster Assembly

9.2.1 Overview

A number of *nif* gene products have been proposed to participate in the biosynthesis of the M-cluster.[28,29] However, based on gene knockout studies[30,31] and *in vitro* activity assays,[32–34] a minimum gene set has been identified, permitting definition of a core assembly pathway of the M-cluster (Figure 9.1). This pathway begins with the sequential synthesis of [Fe_2S_2] and [Fe_4S_4] clusters by the actions of NifS (*nifS* gene product), a pyridoxal phosphate-dependent cysteine desulfurase, and NifU (*nifU* gene product), a scaffold protein for Fe/S cluster assembly.[35] These small Fe/S clusters are transferred to NifB (*nifB* gene product), where a pair of [Fe_4S_4]-like clusters, termed the K-cluster, are formed.[36] Subsequently, NifB, a radical *S*-adenosyl-L-methionine (SAM) protein, transforms the K-cluster into a so-called L-cluster ([Fe_8S_9C]) in a radical-mediated process that involves the fusion of two [Fe_4S_4] clusters concomitant with the insertion of a ninth sulfur atom and a central carbide atom.[16,26] The L-cluster is then transferred to NifEN (*nifE* and *nifN* gene products), where it is matured into an M-cluster ([$MoFe_7S_9C$-homocitrate]) upon NifH-dependent replacement of a terminal Fe atom with Mo and homocitrate.[37–39] Finally, the newly formed M-cluster is delivered *via* direct protein–protein interactions from its low-affinity binding site in NifEN to its high-affinity binding site within the α-subunit of apo-NifDK to form the M-cluster-replete holo-NifDK.[40]

9.2.2 Formation of a Precursor to the M-Cluster

Assembly of the L-cluster is carried out by combined actions of NifU, NifS and NifB. Small Fe/S units are initially generated by the concerted actions of NifU and NifS, in which the pyridoxal-dependent cysteine desulfurase, NifS, provides sulfur units to the scaffold protein, NifU, to combine Fe and S into simple [Fe_2S_2] clusters. These [Fe_2S_2] clusters can then be arranged into [Fe_4S_4] clusters on NifU and transferred to NifB for subsequent processing.[35,41,42]

NifB plays a crucial role in forming the L-cluster, an 8Fe precursor to the M-cluster. When *nifB* is omitted from the genome of *Azotobacter vinelandii*, the resultant NifEN and NifDK proteins lack L-cluster and M-cluster, respectively.[43,44] Analysis of the primary sequence of NifB indicates that it contains a signature radical SAM motif (CxxxCxxC), as well as at least seven additional conserved cysteine residues that could be involved in the ligation of Fe/S cluster species. The SAM motif places NifB in a radical SAM superfamily that was first identified in 2001 by Sofia and coworkers.[36,45] Members of this enzyme family catalyze very diverse chemical transformations that rely on generation of a 5′-deoxyadenosyl radical (5′-dA•) and the subsequent radical cascade reactions.[46,47] Considering that the [Fe_8S_9C] L-cluster contains all the Fe and S atoms required for synthesis of the M-cluster, it has been hypothesized that NifB receives a pair of [Fe_4S_4] clusters from NifU and fuses

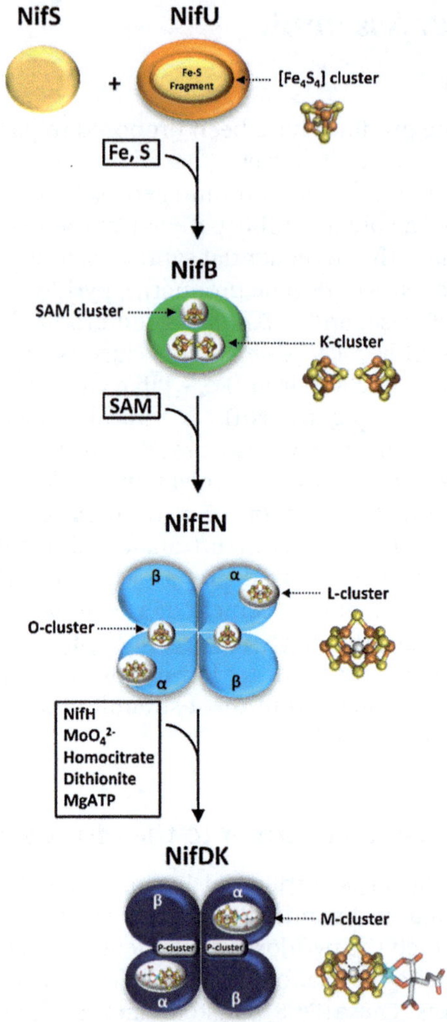

Figure 9.1 Biosynthesis of the M-cluster. This process begins with the combined action of NifS and NifU, which mobilize Fe and S for the formation of $[Fe_4S_4]$ clusters. These clusters are then transferred to NifB, forming a pair of adjacent $[Fe_4S_4]$ clusters (K-cluster). NifB couples and rearranges these two 4Fe clusters into an 8Fe cluster concomitant with the insertion of a carbide atom and an additional sulfur atom in a radical SAM-dependent process. These events result in the formation of an $[Fe_8S_9C]$ cluster (L-cluster), which is subsequently transferred to NifEN and matured into an M-cluster by NifH through ATP-dependent insertion of Mo and homocitrate (HC) in the presence of a reductant, dithionite. Finally, the M-cluster is delivered from NifEN to its target location in NifDK. Atoms of the clusters are colored as follows: Fe, orange; S, yellow; C, grey; O, red; Mo, cyan.

these clusters together with the insertion of a sulfur atom and a carbon atom *via* radical SAM chemistry. Owing to its apparent instability, researchers had faced significant challenges in the isolation and characterization of intact and functional NifB until a NifEN-B fusion protein was expressed in *A. vinelandii* a few years ago, which facilitated elucidation of the role of NifB in the assembly of the L-cluster (Figure 9.2(A)).[36]

Construction of the *A. vinelandii* NifEN-B fusion protein, wherein the N-terminus of NifB is fused with the C-terminus of NifN, was modeled after a naturally occurring NifEN-B fusion protein from *Clostridium pasteurianum*.[36] This engineered NifEN-B protein contains nearly twice the Fe and acid-labile sulfide content of NifEN, indicating the presence of additional Fe/S cluster species in NifB.[36] In addition, when reduced by dithionite, NifEN-B displays additional electron paramagnetic resonance (EPR) spectroscopic features at g = 2.02, 1.95 and 1.94, which are absent from the EPR spectrum of NifEN (Figure 9.2(B)). These features, which originate from the additional Fe/S cluster species in the NifB portion of NifEN-B, disappear after the addition of 40 mM SAM.[36] The difference spectrum between NifEN-B before and after addition of 40 mM SAM in the dithionite-reduced state reveals a distinct S = 1/2 signal with g values of 2.02, 1.95 and 1.90, suggesting that the additional, SAM-responsive Fe/S cluster species are of $[Fe_4S_4]$ nature (Figure 9.2(B) left). Spin quantitation data further reveal the presence of ~3 $[Fe_4S_4]$ clusters, which is consistent with the presence of a radical SAM cluster (*i.e.* an $[Fe_4S_4]$ cluster), as well as a K-cluster (*i.e.* an $[Fe_4S_4]$ cluster pair) in NifB, which collectively respond to SAM. More excitingly, concomitant with the disappearance of the SAM-responsive EPR signals in the dithionite-reduced NifEN-B, the magnitude of an L-cluster-specific, g = 1.94 signal increases in the IDS-oxidized NifEN-B (Figure 9.2(B) right), suggesting a conversion of the K-cluster to an L-cluster.[36] Activity analyses in which NifEN-B serves as an M-cluster donor to apo-NifDK reaffirm the presence of K-cluster on NifEN-B, which undergoes sequential transformation to L- and M-cluster before it is inserted into its binding site in apo-NifDK.[16]

Further studies of NifEN-B have identified the role played by NifB in the insertion of the central carbide during the process of K- to L-cluster conversion.[16] Incubation of SAM with NifEN-B results in two byproducts: *S*-adenosyl-homocysteine (SAH) and 5′-deoxyadenosine (5′-dAH) (Figure 9.3(A)). Although 5′-dAH is a common product of SAM cleavage by all radical SAM enzymes, only methylating or methylthiolating radical SAM enzymes, such as RlmN or RimO, have been shown to generate SAH as an additional product.[48–51] The observation of SAH as a byproduct of the reaction catalyzed by NifEN-B has led to the hypothesis that the *S*-methyl group of SAM is removed from SAM (leaving behind SAH) and processed initially by hydrogen abstraction by a 5′-deoxyadenosyl radical (5′-dA•) (generating 5′-dAH). Subsequently, two lines of evidence have provided strong support for these proposed events, which represent early steps along the pathway of central carbide insertion. First, incubation of NifEN-B with methyl-d_3-SAM yields a mixture of protonated 5′-dA (5′-dAH) and deuterated 5′-dA (5′-dAD), supplying proof

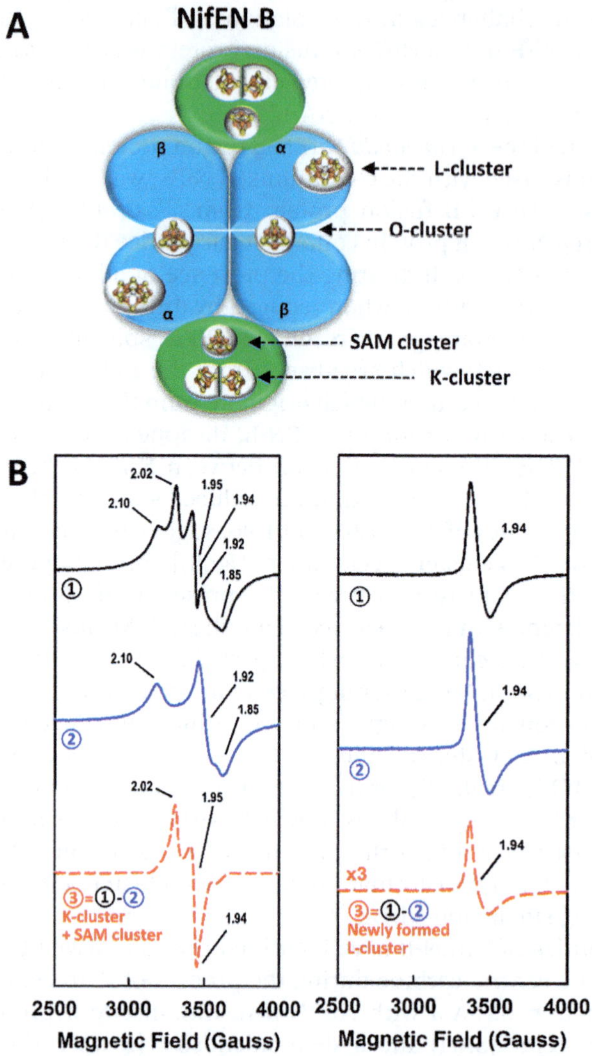

Figure 9.2 The NifEN-B fusion protein and the spectral features of its associated cluster species. (A) Schematic representation of the NifEN-B fusion protein. This fusion protein was constructed by fusing the 3′-end of *nifN* with the 5′-end of *nifB* and expressed in a *nifHDK*-deletion background of *A. vinelandii*, which resulted in accumulation of transient Fe/S cluster species on both the NifEN (*i.e.* the L-cluster) and the NifB (*i.e.* the K-cluster) portions of the fusion protein. (B) EPR spectra of NifEN-B (black), NifEN-B plus 40 mM SAM (blue) and the difference spectrum between NifEN-B and NifEN-B plus SAM (red) in dithionite-reduced (left) and IDS-oxidized (right) states. The *g* values are indicated.

for hydrogen abstraction of the *S*-methyl group by 5′-dA•; second, incubation of NifEN-B with methyl-^{14}C-SAM results in accumulation of the ^{14}C label on NifEN-B, pointing to incorporation of SAM-derived carbon during the process of K- to L-cluster conversion on this fusion protein (Figure 9.3(B) ① and ③).[16] The ^{14}C-labeled L-cluster on NifEN-B can be further matured into a ^{14}C-labeled M-cluster and transferred to apo-NifDK (Figure 9.3(B) ③). Moreover, the ^{14}C labeled L- and M-clusters can be extracted into an organic solvent, *N*-methylformamide (NMF), from their respective host proteins, and confirmed for identities based on activity assays (Figure 9.3(C)).[16] Together, these results firmly establish the *S*-methyl group of SAM as the source of central carbide that is incorporated into the L-cluster in a radical-chemistry-based reaction.

The transfer of the *S*-methyl group of SAM is not routed *via* a protein residue on NifEN-B, as the denatured peptides of NifEN-B do not accumulate ^{14}C label upon treatment with ^{14}C-SAM.[16] Additionally, post-translational

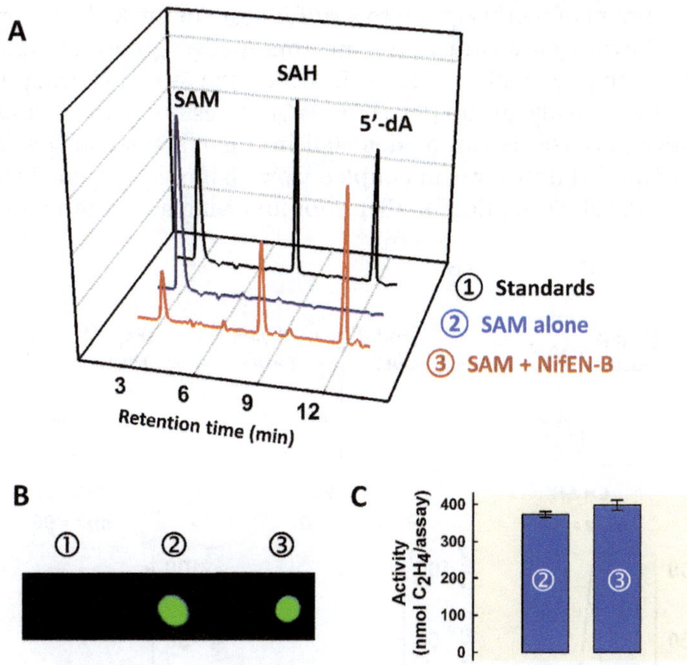

Figure 9.3 NifB-catalyzed carbide insertion into the NifEN-bound L-cluster. (A) The HPLC elution profiles of the SAM, SAH and 5′-dAH standards (①), SAM alone (②) and SAM plus NifEN-B (③). (B) Autoradiograph of NifEN-B in the absence (①) or presence (②) of methyl-^{14}C-SAM and NifDK, and NifDK re-isolated from an incubation mixture containing apo-NifDK, NifEN-B, methyl-^{14}C-SAM, NifH, MgATP, dithionite, molybdate and homocitrate (③). (C) The C_2H_2-reducing activity of NifDK reconstituted by the M-cluster derived from the L-cluster extracted from NifEN-B (in B, ②) upon incubation with NifH, MgATP, dithionite, molybdate and homocitrate; or by the M-cluster directly extracted from NifDK (in B, ③).

modification (PTM) analyses of NifEN-B reacted with SAM or methyl-d_3-SAM do not reveal methylation of any protein residue, implying a direct transfer of the *S*-methyl group of SAM to the K-cluster on NifEN-B.[16] Consistent with this suggestion, methanethiol (CH_3SH) and methane-d_3-thiol (CD_3SH) can be detected as the respective products by gas chromatography-mass spectrometry (GC-MS) when dithionite-reduced NifEN-B is incubated with SAM and methyl-d_3-SAM, respectively, and subsequently quenched by acid (Figure 9.4(A) and (B)), suggesting a direct transfer of the *S*-methyl group to an acid labile sulfur atom of the K-cluster.[26] Incubation of the Fe/Se-reconstituted NifEN-B (designated NifEN-B^{FeSe}), wherein the endogenous $[Fe_4S_4]$ clusters are removed by Fe chelator and replaced by $[Fe_4Se_4]$ clusters, with SAM, followed by acid quenching, results in the formation of methylselenol (CH_3SeH), the selenium-substituted form of methanethiol (Figure 9.4(C)), further establishing the sulfur atom of the K-cluster as the entry point of SAM-derived methyl group into the carbide insertion pathway.

The attachment of methyl group to a sulfur atom of the K-cluster precedes the event of hydrogen abstraction from this methyl group. Incubation of NifEN-B with an allyl analog of SAM, in which the *S*-methyl group (*S*-CH_3) is replaced by an *S*-allyl group (*S*-CH_2–CH=CH_2), results in the formation of SAH; however, no 5′-dAH can be detected in this reaction, suggesting that the event of methyl transfer is uncoupled from that of hydrogen abstraction by a 5′-dA• radical. Formation of allyl thiol following acid treatment of this

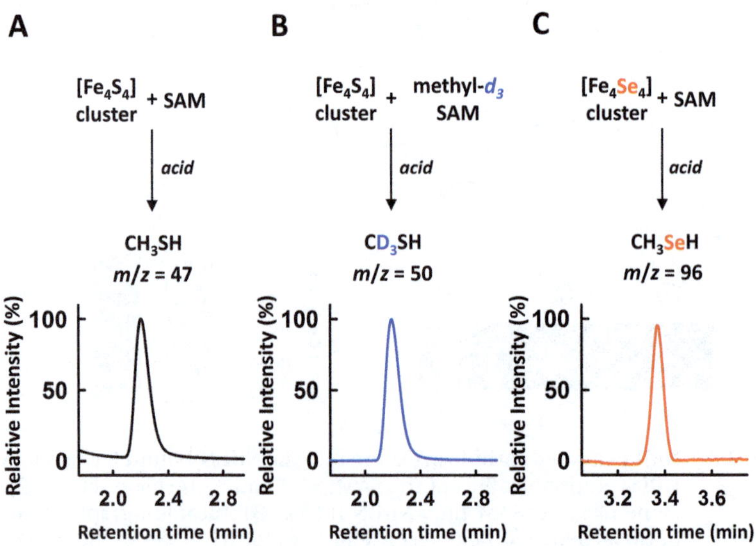

Figure 9.4 GC-MS analysis of acid-quenched products upon incubation of NifEN-B with SAM (A) or methyl-d_3-SAM (B), or upon incubation of NifEN-B^{FeSe} with SAM (C). The acid-quenched products were identified as methanethiol (A), methane-d_3-thiol (B) and methylselenol (C) based on their retention times, fragmentation patterns and *m*/*z* ratios.

incubation mixture further demonstrates the transfer of allyl group to an acid labile sulfur atom of the K-cluster, providing compelling evidence that methyl transfer occurs prior to hydrogen abstraction *via* an S_N2-type mechanism.[26] Interestingly, although a typical reaction of S_N2 methyl transfer does not require the presence of protein in a reduced state, transfer of methyl group to the K-cluster can only occur when NifEN-B is reduced by dithionite, suggesting that the K-cluster must be poised at a certain redox potential to render its associated sulfur atom more nucleophilic for methyl transfer *via* an S_N2-type nucleophilic substitution.[47]

Combined results from the characterization of NifEN-B fusion protein have led to the proposal of a carbide insertion pathway, which begins with transfer of a methyl group from SAM to the K-cluster, followed by hydrogen atom abstraction from this methyl group, which initiates the radical chemistry that is required for the coupling and rearrangement of the two 4Fe units of the K-cluster into an 8Fe L-cluster concomitant with further deprotonation of the SAM-derived carbon intermediate into a carbide that appears in the center of the L-cluster, as well as incorporation of an additional, ninth sulfur into this cluster (Figure 9.5). While the initial steps of this pathway have been elucidated, further work is needed to decipher the radical-based mechanism that produces the fully deprotonated μ_6-carbide at the heart of the L-cluster, as well as the source of the ninth sulfur atom and how it is inserted into the L-cluster, in order to gain a full understanding of the mechanism of the formation of this 8Fe precursor to the M-cluster.

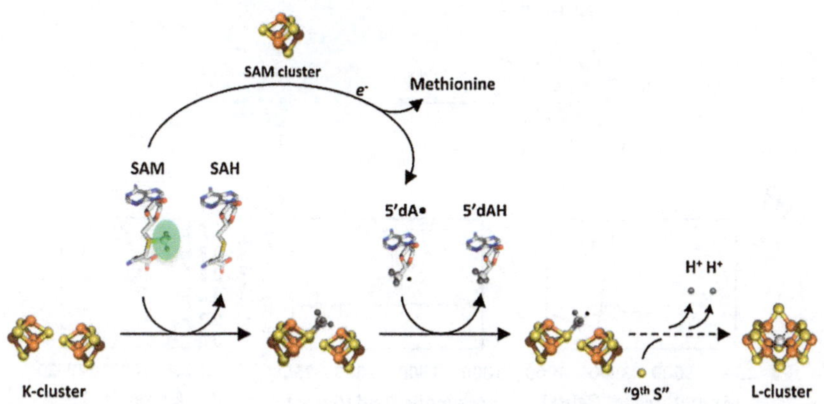

Figure 9.5 Proposed mechanism for carbide insertion concomitant with the K- to L-cluster conversion. The pathway begins with the transfer of a methyl group from one equivalent of SAM to the K-cluster, followed by hydrogen abstraction by a 5′-dA• radical generated from the homolytic cleavage of a second equivalent of SAM, which yields a K-cluster-bound methylene radical. This process continues with further dehydrogenation/deprotonation of the methylene radical into a μ_6-carbide, incorporation of a 9th sulfur and rearrangement of the two 4Fe modules, which eventually leads to the formation of an $[Fe_8S_9C]$ L-cluster. Atoms of the clusters are colored as follows: Fe, orange; S, yellow; C, grey; O, red; N, blue.

9.2.3 Maturation of the Precursor into an M-Cluster

NifEN plays a key role in maturing the L-cluster into an M-cluster. It receives the L-cluster from NifB and serves as a scaffold protein for NifH-mediated insertion of Mo and homocitrate into the L-cluster, a process involving the substitution of one terminal Fe of the cluster by Mo/homocitrate that results in a mature M-cluster (Figure 9.6).[37–39] The function of NifEN as an M-cluster maturase was first elucidated in 2005, when a unique metal cluster displaying

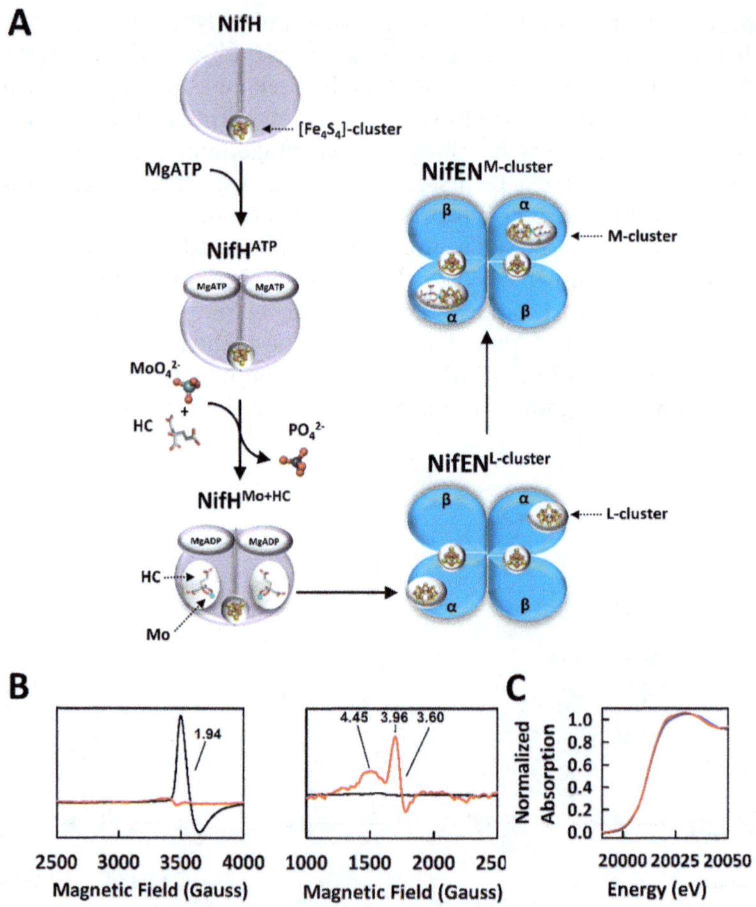

Figure 9.6 Conversion of L- to M-cluster on NifEN. (A) Schematic representation of the role of NifH as an ATP-dependent, Mo/homocitrate (HC) insertase that supplies the two missing components to the NifEN-bound L-cluster, converting it to a fully complemented M-cluster. (B) EPR spectra of $NifEN^{L\text{-cluster}}$ (black) and $NifEN^{M\text{-cluster}}$ (red) in dithionite-reduced (left) and IDS-oxidized (right) states. Upon maturation, the L-cluster-specific signal at $g = 1.94$ (left) disappears; whereas an M-cluster-like signal at $g = 4.45$, 3.96 and 3.60 (middle) appears. The g values are indicated. (C) Mo K-edge XAS spectra of the M-cluster on NifEN (blue) and NifDK (red), which are highly similar to each other.

a $g = 1.93$ EPR signal in the IDS-oxidized state was captured on this protein.[52] Subsequent biochemical studies established the identity of this novel cluster species as a precursor to the M-cluster (*i.e.* the L-cluster), as this cluster could be readily converted to an M-cluster in an *in vitro* maturation assay (*vide infra*).[52] X-ray absorption spectroscopy/extended X-ray absorption fine structure (XAS/EXAFS) studies further confirmed the absence of Mo and homocitrate from the NifEN-bound L-cluster, but revealed the structure of this cluster as an FeS core ([Fe_8S_9]) with nearly identical topology to that of the M-cluster.[38,53] Consistent with sequence-based predictions, X-ray crystallographic analyses demonstrated an overall structural homology between NifEN and NifDK; more importantly it established the presence of a pair of clusters in NifEN: a permanent O-cluster ([Fe_4S_4]) at the αβ-subunit interface, which is analogous to the P-cluster in NifDK; and a transient L-cluster ([Fe_8S_9]) within the α-subunit, which is homologous to the M-cluster in NifDK.[38,53,54] Finally, X-ray emission spectroscopy (XES) analysis of NifEN, along with biochemical studies of carbide insertion by NifEN-B (*vide supra*), defined the composition of the L-cluster as that of [Fe_8S_9C], completing its stoichiometry while further establishing this cluster as a nearly completed precursor to the M-cluster except for the absence of Mo and homocitrate.[16,55]

The NifEN-bound L-cluster can be converted to an M-cluster *in vitro* upon incubation of NifEN with NifH, dithionite, MgATP, molybdate (MoO_4^{2-}) and homocitrate (Figure 9.6(A)).[37–39] The role of NifH as a Mo/homocitrate-insertase was initially identified based on activity assay analyses.[37–39,52] When NifH is preloaded *in vitro* with Mo and homocitrate in the presence of dithionite and MgATP, it can be utilized as the source of Mo and homocitrate to mature the L-cluster to an M-cluster on NifEN (Figure 9.6(A) left).[39] Mo K-edge XAS analysis further demonstrates a modification of the Mo species in the preloaded NifH from the originally supplied MoO_4^{2-}, showing a reduction of the number of Mo=O bonds and/or the effective oxidation state of Mo;[39] whereas EPR analysis reveals an intermediary lineshape of the preloaded NifH between those of the MgATP- and MgADP-bound forms of NifH.[39] This observation, along with the original structural model of ADP-bound NifH in which Mo occupies a position corresponding to the γ-phosphate of ATP,[32] has led to the proposal of a binding pattern of NifH/ADP-Mo-homocitrate as the initial step of Mo/homocitrate mobilization by NifH.

Following insertion of Mo and homocitrate, the NifEN-bound L-cluster is converted to an M-cluster, which can then be used to reconstitute and activate apo-NifDK.[32] Accompanying the conversion from the L- to the M-cluster, an M-cluster-like signal appears at $g = 4.45$, 3.96 and 3.60 in the spectrum of dithionite-reduced NifEN (Figure 9.6(B) right); whereas the L-cluster-specific, $g = 1.94$ signal disappears in the spectrum of IDS-oxidized NifEN (Figure 9.6(B) left).[37] Fe and Mo K-edge XAS/EXAFS analyses suggest that the M-cluster formed on NifEN (Figure 9.6(C)) is nearly indistinguishable in structure from the M-cluster in NifDK, except for a somewhat asymmetric coordination of Mo due to a difference in cluster ligands, as well as protein environment, in NifEN.[37,40] Conversion of L-cluster cannot occur if ATP is replaced by ADP or non-hydrolyzable ATP analogs, or if NifH is replaced by a variant form that is specifically defective in MgATP hydrolysis.[38,39] Moreover, the efficiency of L-cluster conversion increases by 3–4-fold upon a 10-fold increase in dithionite concentration.[37] Such an ATP- and redox-dependence is consistent with the crucial role NifH plays in this process, interacting with NifEN in a manner analogous to that with the homologous NifDK and subsequently delivering Mo and homocitrate in an ATP- and redox-dependent manner to the L-cluster that results in the formation of a fully complemented M-cluster on NifEN.[37–39]

Upon formation of an M-cluster, NifEN can serve as an M-cluster donor to apo-NifDK.[32,56] Successful transfer of the matured M-cluster between the two proteins can be readily achieved by incubating the M-cluster-bound NifEN and apo-NifDK in an *in vitro* assay, implying that this process is facilitated by direct protein–protein interactions between NifEN and NifDK; moreover, it implies a ligand-determined difference in the affinities of the two binding sites for the M-cluster, allowing the cluster to diffuse from the low-affinity binding site in NifEN to the high-affinity binding site in NifDK.[57,58] Biochemical experiments further demonstrate that NifEN is only capable of forming a complex with apo-NifDK upon maturation of the M-cluster, suggesting that NifEN undergoes a conformational change in this process, which allows it to dock on apo-NifDK for the subsequent transfer of the M-cluster.[59] Interestingly, while partial maturation of the NifEN-bound L-cluster with Mo (but no homocitrate) also permits the formation of a complex between NifEN and apo-NifDK, cluster transfer fails to occur between the two proteins.[58] Together with the identification of a positively charged insertion path in apo-NifDK,[54] this observation supports the hypothesis that the anionic homocitrate supplies a crucial negative charge to the M-cluster to enable insertion of the M-cluster into its binding-site in NifDK *via* electrostatic interactions between the negatively charged cluster and the positively charged residues along the insertion path.[32,56,60]

Comparison of the X-ray crystallographic structures of the L-cluster-bound NifEN, apo-NifDK, and holo-NifDK provides important insights into the events that occur during the process of M-cluster maturation and transfer.[32,56,60,61] The L-cluster resides at a position in NifEN that is analogous to that of the M-cluster in NifDK (Figure 9.7(A)). However, contrary to the M-cluster, the L-cluster is nearly surface-exposed,[54] permitting easy access of the L-cluster for the NifH-mediated insertion of Mo and homocitrate that matures the L-cluster to an M-cluster. Upon maturation, the M-cluster is likely transferred to a location within NifEN along a positively charged cluster insertion pathway that is analogous to that found in NifDK. Such a transfer is accompanied by a conformational change of the NifEN protein to favor complexation with and cluster transfer to apo-NifDK. Once deposited on the surface of NifDK, the M-cluster interacts with three distinct regions along a cluster insertion path *en route* to its binding site within NifDK. The first region is termed the "lid loop" and consists of residues α353 to α364. This "lid loop" contains a highly conserved His$^{\alpha362}$ residue that may serve as a transient ligand for the

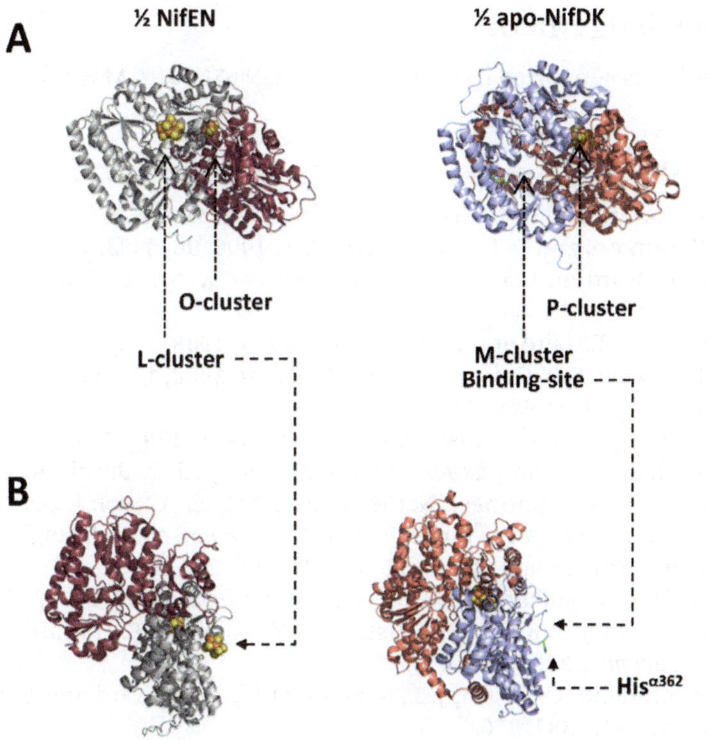

Figure 9.7 Ribbon diagrams of NifEN and apo-NifDK. (A) Structures of the αβ-dimers of NifEN (left) and apo-NifDK (right), showing the location of the L- and M-cluster sites, as well as the permanent O- and P-cluster sites, in the respective proteins. All clusters are shown in space-filling models, with the atoms colored as follows: Fe, orange; S, yellow. (B) The "lid loop" residue, His$^{\alpha 362}$, of apo NifDK, which may provide the first docking point for the M-cluster at the entrance of the insertion funnel, is illustrated.

M-cluster at the entrance of the funnel (Figure 9.7(B)). The second region is referred to as the "His triad" and consists of His$^{\alpha 274}$, His$^{\alpha 442}$ and His$^{\alpha 451}$. This "His triad" may function together as an intermediate docking point for the M-cluster while in transit to its binding site within NifDK. The third region is designated the "switch/lock" and consists of two residues, His$^{\alpha 442}$ and Trp$^{\alpha 444}$. This "switch/lock" may effectively lock the M-cluster into its final binding site *via* a swap in positions between the imidazole group of His$^{\alpha 442}$ and the bulky indole side chain of Trp$^{\alpha 444}$, with the latter preventing the M-cluster from escaping from its binding site.[44] The function of these residues in incorporating M-cluster into NifDK was initially proposed based on a detailed comparison between the X-ray crystal structures of apo- and holo-NifDK[12–14,44] and further supported by data from mutational and biochemical analyses of NifDK, shedding light on the final step of M-cluster assembly that gives a finishing touch to the structure of this unique cluster *via* the correct ligation at its target location.[62–64]

Acknowledgements

This work was supported by NIH Grant R01 GM67626 (to M.W.R.).

References

1. J. B. Howard and D. C. Rees, *Chem. Rev.*, 1996, **96**, 2965.
2. B. K. Burgess and D. J. Lowe, *Chem. Rev.*, 1996, **96**, 2983.
3. J. B. Howard and D. C. Rees, *Proc. Natl. Acad. Sci. U. S. A.*, 2006, **103**, 17088.
4. S. Hunt, S. T. Gaito and D. B. Layzell, *Planta*, 1988, **173**, 128.
5. D. B. Layzell, S. T. Gaito and S. Hunt, *Planta*, 1988, **173**, 117.
6. R. R. Eady, *Chem. Rev.*, 1996, **96**, 3013.
7. R. D. Joerger and P. E. Bishop, *J. Bacteriol.*, 1988, **170**, 1475.
8. D. Gadkari, in *Catalysts for Nitrogen Fixation*, ed. B. Smith, R. Richards and W. Newton, Springer, Netherlands, 2004, ch. 12, vol. 1, pp. 309–332.
9. D. Gadkari, in *Nitrogen Fixation at the Millennium*, ed. G. J. Leigh, Elsevier Science, Amsterdam, 2002, ch. 10, vol. 1, pp. 263–297.
10. M. Ribbe, D. Gadkari and O. Meyer, *J. Biol. Chem.*, 1997, **272**, 26627.
11. F. A. Tezcan, J. T. Kaiser, D. Mustafi, M. Y. Walton, J. B. Howard and D. C. Rees, *Science*, 2005, **309**, 1377.
12. H. Schindelin, C. Kisker, J. L. Schlessman, J. B. Howard and D. C. Rees, *Nature*, 1997, **387**, 370.
13. O. Einsle, F. A. Tezcan, S. L. Andrade, B. Schmid, M. Yoshida, J. B. Howard and D. C. Rees, *Science*, 2002, **297**, 1696.
14. T. Spatzal, M. Aksoyoglu, L. Zhang, S. L. Andrade, E. Schleicher, S. Weber, D. C. Rees and O. Einsle, *Science*, 2011, **334**, 940.
15. K. M. Lancaster, M. Roemelt, P. Ettenhuber, Y. Hu, M. W. Ribbe, F. Neese, U. Bergmann and S. DeBeer, *Science*, 2011, **334**, 974.
16. J. A. Wiig, Y. Hu, C. C. Lee and M. W. Ribbe, *Science*, 2012, **337**, 1672.
17. A. W. Fay, C. C. Lee, J. A. Wiig, Y. Hu and M. W. Ribbe, *Methods Mol. Biol.*, 2011, **766**, 239.
18. B. K. Burgess, *Chem. Rev.*, 1990, **90**, 1377.
19. V. K. Shah and W. J. Brill, *Proc. Natl. Acad. Sci. U. S. A.*, 1977, **74**, 3249.
20. C. C. Lee, Y. Hu and M. W. Ribbe, *Angew. Chem., Int. Ed. Engl.*, 2015, **54**, 1219.
21. C. C. Lee, Y. Hu and M. W. Ribbe, *Angew. Chem., Int. Ed. Engl.*, 2012, **51**, 1947.
22. Y. Ohki, Y. Ikagawa and K. Tatsumi, *J. Am. Chem. Soc.*, 2007, **129**, 10457.
23. Y. Zhang and R. H. Holm, *Inorg. Chem.*, 2004, **43**, 674.
24. Y. Zhang and R. H. Holm, *J. Am. Chem. Soc.*, 2003, **125**, 3910.
25. Y. Zhang, J.-L. L. Zuo, H.-C. C. Zhou and R. H. Holm, *J. Am. Chem. Soc.*, 2002, **124**, 14292.
26. J. A. Wiig, Y. Hu and M. W. Ribbe, *Nat. Commun.*, 2015, 6.
27. J. A. Wiig, C. C. Lee, Y. Hu and M. W. Ribbe, *J. Am. Chem. Soc.*, 2013, **135**, 4982.

28. L. M. Rubio and P. W. Ludden, *Annu. Rev. Microbiol.*, 2008, **62**, 93.
29. K. E. Brigle, W. E. Newton and D. R. Dean, *Gene*, 1985, **37**, 37.
30. M. R. Jacobson, K. E. Brigle, L. T. Bennett, R. A. Setterquist, M. S. Wilson, V. L. Cash, J. Beynon, W. E. Newton and D. R. Dean, *J. Bacteriol.*, 1989, **171**, 1017.
31. P. E. Bishop, R. Premakumar, D. R. Dean, M. R. Jacobson, J. R. Chisnell, T. M. Rizzo and J. Kopczynski, *Science*, 1986, **232**, 92.
32. M. W. Ribbe, Y. Hu, K. O. Hodgson and B. Hedman, *Chem. Rev.*, 2014, **114**, 4063.
33. C. C. Lee, M. A. Blank, A. W. Fay, J. M. Yoshizawa, Y. Hu, K. O. Hodgson, B. Hedman and M. W. Ribbe, *Proc. Natl. Acad. Sci. U. S. A.*, 2009, **106**, 18474.
34. M. S. Cotton, K. Rupnik, R. B. Broach, Y. Hu, A. W. Fay, M. W. Ribbe and B. J. Hales, *J. Am. Chem. Soc.*, 2009, **131**, 4558.
35. A. D. Smith, G. N. Jameson, P. C. Dos Santos, J. N. Agar, S. Naik, C. Krebs, J. Frazzon, D. R. Dean, B. H. Huynh and M. K. Johnson, *Biochemistry*, 2005, **44**, 12955.
36. J. A. Wiig, Y. Hu and M. W. Ribbe, *Proc. Natl. Acad. Sci. U. S. A.*, 2011, **108**, 8623.
37. J. M. Yoshizawa, M. A. Blank, A. W. Fay, C. C. Lee, J. A. Wiig, Y. Hu, K. O. Hodgson, B. Hedman and M. W. Ribbe, *J. Am. Chem. Soc.*, 2009, **131**, 9321.
38. Y. Hu, M. C. Corbett, A. W. Fay, J. A. Webber, K. O. Hodgson, B. Hedman and M. W. Ribbe, *Proc. Natl. Acad. Sci. U. S. A.*, 2006, **103**, 17119.
39. Y. Hu, M. C. Corbett, A. W. Fay, J. A. Webber, K. O. Hodgson, B. Hedman and M. W. Ribbe, *Proc. Natl. Acad. Sci. U. S. A.*, 2006, **103**, 17125.
40. Y. Hu and M. W. Ribbe, *Microbiol. Mol. Biol. Rev.*, 2011, **75**, 664.
41. L. Zheng and D. R. Dean, *J. Biol. Chem.*, 1994, **269**, 18723.
42. L. Zheng, R. H. White, V. L. Cash, R. F. Jack and D. R. Dean, *Proc. Natl. Acad. Sci. U. S. A.*, 1993, **90**, 2754.
43. J. Christiansen, P. J. Goodwin, W. N. Lanzilotta, L. C. Seefeldt and D. R. Dean, *Biochemistry*, 1998, **37**, 12611.
44. B. Schmid, M. W. Ribbe, O. Einsle, M. Yoshida, L. M. Thomas, D. R. Dean, D. C. Rees and B. K. Burgess, *Science*, 2002, **296**, 352.
45. H. J. Sofia, G. Chen, B. G. Hetzler, J. F. Reyes-Spindola and N. E. Miller, *Nucleic Acids Res.*, 2001, **29**, 1097.
46. N. D. Lanz and S. J. Booker, *Biochim. Biophys. Acta*, 2015, **1853**, 1316.
47. S. J. Booker and T. L. Grove, *F1000Prime Rep.*, 2010, **2**, 52.
48. J. Wang, R. P. Woldring, G. D. Roman-Melendez, A. M. McClain, B. R. Alzua and E. N. Marsh, *ACS Chem. Biol.*, 2014, **9**, 1929.
49. D. G. Fujimori, *Curr. Opin. Chem. Biol.*, 2013, **17**, 597.
50. A. K. Boal, T. L. Grove, M. I. McLaughlin and N. H. Yennawar, *Science*, 2011, **332**, 1089.
51. K.-H. H. Lee, L. Saleh, B. P. Anton, C. L. Madinger, J. S. Benner, D. F. Iwig, R. J. Roberts, C. Krebs and S. J. Booker, *Biochemistry*, 2009, **48**, 10162.
52. Y. Hu, A. W. Fay and M. W. Ribbe, *Proc. Natl. Acad. Sci. U. S. A.*, 2005, **102**, 3236.

53. M. C. Corbett, Y. Hu, A. W. Fay, M. W. Ribbe, B. Hedman and K. O. Hodgson, *Proc. Natl. Acad. Sci. U. S. A.*, 2006, **103**, 1238.
54. J. T. Kaiser, Y. Hu, J. A. Wiig, D. C. Rees and M. W. Ribbe, *Science*, 2011, **331**, 91.
55. A. W. Fay, M. A. Blank, C. C. Lee, Y. Hu, K. O. Hodgson, B. Hedman and M. W. Ribbe, *Angew. Chem., Int. Ed. Engl.*, 2011, **50**, 7787.
56. Y. Hu and M. W. Ribbe, *J. Biol. Chem.*, 2013, **288**, 13173.
57. Y. Hu, A. W. Fay, C. C. Lee, J. A. Wiig and M. W. Ribbe, *Dalton Trans.*, 2010, **39**, 2964.
58. A. W. Fay, M. A. Blank, J. M. Yoshizawa, C. C. Lee, J. A. Wiig, Y. Hu, K. O. Hodgson, B. Hedman and M. W. Ribbe, *Dalton Trans.*, 2010, **39**, 3124.
59. J. M. Yoshizawa, A. W. Fay, C. C. Lee, Y. Hu and M. W. Ribbe, *JBIC, J. Biol. Inorg. Chem.*, 2010, **15**, 421.
60. Y. Hu and M. W. Ribbe, *Biochim. Biophys. Acta*, 2013, **1827**, 1112.
61. Y. Hu and M. W. Ribbe, *Coord. Chem. Rev.*, 2011, **255**, 1218.
62. Y. Hu, A. W. Fay and M. W. Ribbe, *JBIC, J. Biol. Inorg. Chem.*, 2007, **12**, 449.
63. A. W. Fay, Y. Hu, B. Schmid and M. W. Ribbe, *J. Inorg. Biochem.*, 2007, **101**, 1630.
64. Y. Hu, A. W. Fay, B. Schmid, B. Makar and M. W. Ribbe, *J. Biol. Chem.*, 2006, **281**, 30534.

CHAPTER 10

Tungsten-Containing Enzymes

WILFRED R. HAGEN[a]

[a]Department of Biotechnology, Delft University of Technology, van der Maasweg 9, 2629 HZ Delft, The Netherlands
*E-mail: w.r.hagen@tudelft.nl

10.1 Introduction

In the framework of biological evolution one can define the concept of natural selection of chemical elements,[1,2] in short: why is an element used by life. For element 74 the question why (some) cells decided to adopt tungsten to become relevant, or even essential, for their functioning is a particularly appropriate one in view of W being the only bio-member of the 5d transition row, and of its apparently extensive similarity to group-6 congener, molybdenum, from which it is separated by the thin line of equalizing lanthanide contraction. It has been more than four decades now since tungstate was found as an essential growth factor for expression of enzyme activity;[3,4] more than three decades ago the first enzyme was purified to apparent homogeneity with a stoichiometric content of tungsten;[5] and two decades ago the first crystal structure of a tungstoenzyme was determined (to 2.3 Å resolution), which established the metal coordination as bis(dithiolene) from two tungstopterins.[6] In the intervening years biochemical, spectroscopic and omics-based studies have provided a data set of considerable volume and detail not only to establish tungsten-containing enzymes, or tungstoenzymes, or W-enzymes, as a bona fide entry in biocatalysis textbooks, but also to act as a reference source to ask the question all over again: but why tungsten?

RSC Metallobiology Series No. 5
Molybdenum and Tungsten Enzymes: Biochemistry
Edited by Russ Hille, Carola Schulzke, and Martin L. Kirk

Published by the Royal Society of Chemistry, www.rsc.org

Over the years this question has been typically put forth from the potentially biasing perspective of molybdenum biochemistry: OK, so we have Mo-enzymes; and what can W add to this? The use of tungsten by cells received the verdict of "a most unlikely choice for a metal with a biological function",[7,8] "tungsten is used in some early organisms which have remained in deep sea anaerobic trenches to this day",[2] "more ancient than Mo-containing enzymes",[9] "to catalyze certain conversions that are near the limits of biological systems",[10] "W-containing enzymes predated the Mo-containing enzymes",[11] "in transition from a W-world to a Mo-world",[12] "tungsten, which shared a similar role with molybdenum in certain primitive organisms, is not found to be of value in advanced living systems",[13] "its use was extremely primitive",[14] "the tungstoenzymes that are holdovers to this day".[15]

The overall picture emerging of W enzymology is one of a biological niche in space and time characterized by some or all of the following boundary conditions: strict anaerobicity, high temperature, extremely low redox potentials, high tungstate concentration, high sulfide concentration, with its finest hour very early in the history of life, and now slowly exiting from existence in favor of Mo enzymology. The disadvantage of this view is that it suggests an answer to the "why tungsten" question whose support is to be found in the non-reproducible distant past. This review takes the different preamble that we have only the present-day tungstoenzyme and cellular species and that in the present biosphere there is no such thing as a primitive cell since all living species are the result of long-term evolution and, therefore, that the answer to "why tungsten" can and should be found in the biochemistry of today.

10.2 General Properties of Tungstoenzymes

10.2.1 Tungsten-Based Enzymology: State of the Art

The subject of W and Mo biochemistry has been rather extensively covered in several reviews specific for tungsten[8–10,16–19] and in many other ones on molybdenum in which tungsten is brought in as a contrast. Of the latter we cite only the more recent ones[20–25] with additional reference to several chapters of the present book. The chemistry of Mo and W model compounds has also been reviewed recently.[26,27] The aqueous chemistry of molybdate MoO_4^{2-} and tungstate WO_4^{2-} and their polyanions is quite similar but not identical, with slight differences in coordination bond length (longer for W), free energy of solvation, tendency of polymerization (more pronounced for W). These differences do not go a long way in providing even the beginning of an explanation for the differences in Mo *versus* W biochemistry. On the other hand, most reviewers claim that reported differences in reduction potential for W(VI) and Mo(VI) complexes and metalloproteins may well be key to understanding the W/Mo differential enzymology. Another issue considered to be of importance in particular from an evolutionary point of view is the difference in water solubility of the metal sulfide (greater for W). Both issues warrant a critical evaluation, below. Yet another matter proposed to be of relevance is a difference in concentration in sea water of Mo *versus* W (W is typically two orders of magnitude less), and reported local reversal of this ratio in, *e.g.*, marine hyperthermophilic habitats. Here, data paucity stands in the way of pronouncing more than a caveat; the matter may be more meaningfully discussable when linked to remarkable biological selectivity for W *versus* Mo at the level both of active transport into the cell and of active post-translational modification in the synthesis of W and Mo holoenzymes.[19] The last W-specific review was submitted by us some eight years ago;[18] the intervening years have produced new data on tungstoenzymes, on quantum-chemical modeling and on related model compound chemistry at a rate of *circa* 10 primary papers per year. With this the reference database has been expanded and extended, but the "why tungsten" question is still on the board in majuscules. An increasingly pressing problem in addressing this issue is the lack of pre-steady-state kinetic studies on tungstoenzyme reaction intermediates.

10.2.2 Tungsto-Pterin Prosthetic-Group: Nomenclature Issues

The structures of the W-cofactor in tungstoenzymes is schematically given in Figure 10.1. Before the first crystal structures of a tungstopterin-enzyme[6] and a molybdopterin-enzyme[28] became available the prosthetic group was extracted from several molybdoenzymes, *e.g.* sulfite oxidase, xanthine oxidase and nitrate reductase, in the form of a stable, demetallated, oxidized and derivatized product whose structure was determined to be a mono-phosphorylated pterin derivative.[29] The inferred structure for the original cofactor (also without the metal) was given the name molybdopterin or MPT. Later, a more complex product, isolated from *Rhodobacter sphaeroides* dimethyl sulfoxide reductase, was identified as the guanidine monophosphate ester of the phosphorylated pterin, and the native structure was therefore inferred to be a guanidine di-phospho-pterin.[30] The name given to this cofactor (without metal) was molybdopterin guanine dinucleotide, or MGD (*cf.*, *e.g.*, ref. 31). This is somewhat of a misnomer, because a nucleotide, by definition, is composed of a heterocyclic base plus a ribose (C_5) sugar plus a phosphate, and a dinucleotide should therefore have two of each. MGD has only one ribose, which makes it a mononucleotide derivative, *i.e.* molybdopterin guanosine (mono)phosphate, or possibly pterin guanine diphosphate.

When the first X-ray structure of a tungstoenzyme, *Pyrococcus furiosus* aldehyde oxidoreductase, revealed that W was bis-dithioleno coordinated by two pterins (Figure 10.1) that were the ring-closed forms of the structure that was previously published as molybdopterin, it was proposed that molybdopterin is in fact always a three-ringed system, and thus that the coordination of W in AOR is bis-MPT. To avoid the awkwardness of having to call the ligand to a tungsten a molybdopterin, it was proposed to generalize the abbreviation MPT to read *metallo*pterin,[16] which would make the prosthetic group of *P. furiosus* AOR a tungsten bis-metallopterin, or W-bis-MPT. Others suggested

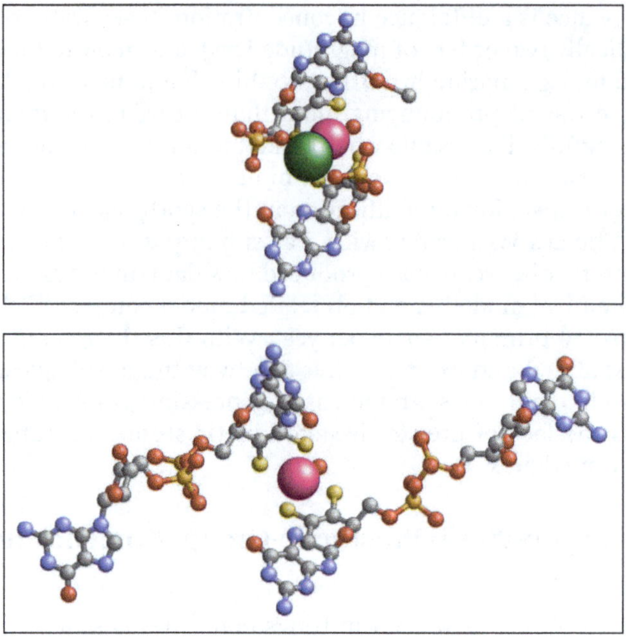

Figure 10.1 Comparison of the dense (top) and loose (bottom) W-cofactors in AOR-family *versus* DMSOR-family tungstoenzymes. The W atoms (magenta) and the Mg atom (green) are inflated for emphasis. The top structure is in *Pyrococcus furiosus* formaldehyde oxidoreductase (1B25.pdb); the gray atom is a calcium. The bottom structure is in *Desulfovibrio gigas* formate dehydrogenase (1HOH.pdb). Note that in the dense bis-tungstopterin-magnesium cofactor the edge-to-edge distance between phosphates is *circa* 4 Å while in the loose bis-tungstopterin guanosine phosphate the distance between phosphates on the pterins is *circa* 12 Å.

to use the name tungstopterin, or WTP, for the dithiolene ligand of tungsten,[6] which would then provide *P. furiosus* AOR with a W-bis-WTP center.

An additional complication is the finding that in the crystal structure of *P. furiosus* AOR the two pterins are also coordinatively connected *via* an Mg(II) ion through their phosphate moieties, and so the complete prosthetic group (excluding interactions with the protein) would rather be a W-(MPT)$_2$-Mg or a W-(WTP)$_2$-Mg unit.

In a quest for simplicity, since the W-(WTP) combination may well be a potential source of confusion, we suggest to return to the proposal of formally reading MTP as metallopterin.[16] Furthermore, we suggest to retain the abbreviation MGD (and generally MXD), but to formally read it as metallopterin guanine diphosphate. Then the prosthetic groups for the representative of Mo/W families, in Figure 10.2, and treated in the next paragraph, are simply: W-(MGD)$_2$ for W-formate dehydrogenase, Mo/W-(MGD)$_2$ for DMSO reductase, W-(MTP)$_2$-Mg for W-aldehyde oxidoreductase, Mo-MPT for xanthine oxidase and for sulfite oxidase.

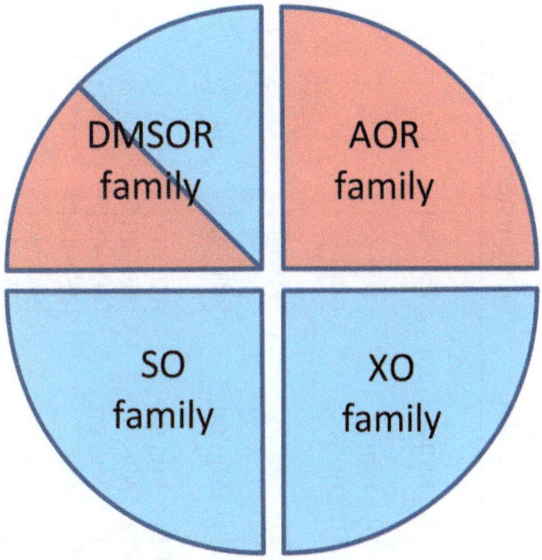

Figure 10.2 The world of tungsten (red) and molybdenum (blue) metallopterin-containing enzymes. Typically, tungstoenzymes from the AOR family contain one bis-tungstopterin-magnesium and one [4Fe–4S] cluster per catalytic subunit; they are oxygen-sensitive, and W-specific (W cannot be replaced with Mo); ferredoxin is the natural redox partner. Typically, tungstoenzymes from the DMSOR family contain one bis-tungstopterin guanosine phosphate and one [4Fe–4S] or [2Fe–2S] cluster per catalytic subunit; they are not W-specific (W can be replaced with Mo); a variety of redox partners is found.

10.2.3 Tungstoenzymes Come in Two Families

The enzymes that contain W or Mo bound to pterin are classified in families based predominantly on sequence homology and additionally on structural details of the active site, on quaternary structure, on O_2 sensitivity, on redox partners and on metal preference. The prevalent grouping is in four families labeled with the (abbreviated) name of an example enzyme as in Figure 10.2 Its relevance in the present context is that W-enzymes occur in two groups only, one in combination with Mo-variants (the DMSOR family) and one to the exclusion of Mo-variants (the AOR family). All tungstoenzymes carry bis-pterin tungsten prosthetic groups; however, members of the FDH family have guanidine monophosphate nucleotide phospho-estered to the pterin phosphate group, W-$(MGD)_2$, while the pterins in the members of the AOR family lack nucleotides, *i.e.* they have W-$(MPT)_2$-Mg.

Note in particular the magnesium ion in the prosthetic group of AOR members, since its importance appears to have been underestimated thus far. Although the number of X-ray structures determined for tungstoenzymes is still limited (Table 10.1), all structures from the AOR family have the Mg^{2+} ion bridging the two W-coordinating MPTs, while in all structures from the FDH family the ion is conspicuously absent. In 3D-structural terms this is

Table 10.1 Crystal structures of tungstoenzymes.

Enzyme	Species	PDB-file	Resolution/Å	W-pterin	Extra metal	Fe/S cluster	Reference
FDH/DMSOR family							
Formate dehydrogenase (FDH)	*Desulfovibrio gigas*	1H0H	1.80	W-bis(MGD)	none	[4Fe–4S]	12
Acetylene hydratase (ACH)	*Pelobacter acetylenicus*	2E7Z	1.26	W-bis(MGD)	none	[4Fe–4S]	32
DMSO reductase (DMSOR)	*Rhodobacter capsulatus*	1E18	2.00	W-bis(MGD)	none	none	33
AOR family							
Aldehyde oxidoreductase (AOR)	*Pyrococcus furiosus*	1AOR	2.30	W-bis(MPT)	Mg; Fe	[4Fe–4S]	6
Formaldehyde oxidoreductase (FOR)	*Pyrococcus furiosus*	1B25	1.85	W-bis(MPT)	Mg; Ca	[4Fe–4S]	34
with glutarate	*Pyrococcus furiosus*	1B4N	2.40	W-bis(MPT)	Mg; Ca	[4Fe–4S]	34
BenzoylCoA reductase (BCR)	*Geobacter metallireducens*	4Z40	2.35	W-bis(MPT)	Mg	[4Fe–4S][a]	35
with Zn	*Geobacter metallireducens*	4Z3Z	2.67	W-bis(MPT)	Mg; Zn	[4Fe–4S][a]	35
with benzoylCoA	*Geobacter metallireducens*	4Z3Y	2.36	W-bis(MPT)	Mg	[4Fe–4S][a]	35
with 1-monoenoylCoA	*Geobacter metallireducens*	4Z3X	1.85	W-bis(MPT)	Mg	[4Fe–4S][a]	35
with 1,5-dienoylCoA	*Geobacter metallireducens*	4Z3W	2.21	W-bis(MPT)	Mg	[4Fe–4S][a]	

[a]complex with second subunit containing 3[4Fe–4S].

– possibly causatively – related to a different overall "fold" of the two pterins in the members of the two families as schematically indicated in Figure 10.1: in the AOR family members the shortest distance between phosphates is two Mg coordination bonds, *circa* 4 Å, while in the FDH family members the shortest through-space distance between the phosphates of the pterin is typically some 12 Å. For future reference one should perhaps note that this difference maps with a key functional difference between the two families: substitution of W by Mo leads to inactivity in the AOR family members, while it frequently leads to a functional enzyme (be it with modified activity) in the DMSOR family members. We note in passing that the insertion of the Mg ion in the AOR family members has not yet been considered as a possibly relevant step in the biosynthesis of the holo-enzymes.

The Mg ion in the AOR family members is approximately octahedrally coordinated by two Os from the two pterin phosphate groups, two Os from water molecules and two Os from the protein backbone. The latter are from Asn93 and Aal183 in *P. furiosus* AOR,[6] from Asn92 and Ala181 in *P. furiosus* FOR[34] and from Met94 and Ser183 in *G. metallireducens* BCR.[15] In the archaeal enzymes AOR and FOR they form the only covalent bonds of the cofactor to the protein; the W is not coordinated by the protein.[6,34] In the bacterial BCR W is also cysteinato coordinated by Cys322.[35] Contrarily, in the DMSOR family members there is no Mg ion; however, the W is always coordinated by one amino acid side group, *i.e.* Se from Sec158 in *D. gigas* FDH,[12] S from Cys141 in *P. acetylenicus* ACH[32] and O_γ from Ser147 in *R. capsulatus* DMSOR.[33]

Next to the bis-dithiolene and the protein side group there is additional coordination of the tungsten by small ligands and, just as with molybdoenzymes, this is the most controversial part of the picture. As an introduction to a more detailed discussion, below, on structure and function of BCR, we note that a recent statement[35] that "AORs form a distorted square-pyramidal ligation geometry with four dithiolene sulfurs at the bottom and an oxo and a water or substrate ligand at the top" is an incorrect citation, or at least an overinterpretation, of the literature as far as the small ligand coordination is concerned. W-coordination data from described crystal structures are summarized in Table 10.2.

The crystal structures (*cf.* Table 10.1) also show that none of these enzymes has the capacity to accept or donate a *pair* of electrons from/to the active W site because their second prosthetic group is a single-electron transferring iron–sulfur cluster, which implies that W(V) is a mandatory intermediate in all redox catalysis or, in other words, that the W center acts as a switchbox between bioorganic electron pair redox catalysis and bioinorganic single-electron transfer.

10.2.4 Reactions Catalyzed by Tungstoenzymes

Catalysis by tungstoenzymes is restricted to essentially seven different types of reactions, although some of the enzymes exhibit a very wide substrate spectrum. Scheme 10.1 is an attempt to succinctly summarize the prototype

Table 10.2 Coordination of W in tungstoenzymes deduced from crystallography.

Enzyme	Pterin ligand	5th (6th; 7th) ligand		Approximate geometry
Formate dehydrogenase (FDH)	bis-dithioleno 4S	Se (SEC158)	OH or S	trigonal prismatic (or octahedral?)
Acetylene hydratase (ACH)	bis-dithioleno 4S	S (CYS141)	O (water)	octahedral
DMSO reductase (DMSOR)	bis-dithioleno 4S	O_γ (SER147)	O_A, (O_B?)	square pyramidal or trigonal prismatic
Aldehyde oxidoreductase (AOR)	bis-dithioleno 4S	2O (from glycerol and/or oxo)		trigonal prismatic
Formaldehyde oxidoreductase (FOR)	bis-dithioleno 4S	1O		square pyramidal
BenzoylCoA reductase (BCR)	bis-dithioleno 4S	S (CYS322)	X (CN or CO?)	octahedral

RCHO	+	H_2O	→	RCOOH	+	2[H]	(1. AOR)
C_6H_7(CO)-*CoA*	+	—	←	C_6H_5(CO)-*CoA*	+	2[H]	(2. BCR)
HCOOH	+	—	↔	CO_2	+	2[H]	(3. FDH)
MFR-NH-CHO	+	H_2O	←	*MFR*-NH-COOH	+	2[H]	(4. FMD)
C_2H_2	+	H_2O	→	CH_3CHO	+	—	(5. ACH)
$(CH_3)_2S$	+	H_2O	←	$(CH_3)_2SO$	+	2[H]	(6. DMSOR)
NO_2^-	+	H_2O	←	NO_3^-	+	2[H]	(7. NAR)

Scheme 10.1 Unified reaction scheme of tungstoenzyme catalysis. Reactions of known types of W-enzymes are written in a format to emphasize the role of reducing-equivalent pairs and of water as cosubstrates.

reactions from the two families in a unifying manner by putting emphasis on the central role of electron pair transfer and the importance of water as a frequent second substrate. The reaction arrows in Scheme 10.1 indicate the presumed (predominant) directions in physiology. Four of the five reactions involving water as a substrate/product are O-atom transfer reactions (AOR, MFR, DMSOR, NAR); the exception is the (apparently non-redox) hydration of acetylene by ACH. The remaining two reactions (BCR and FDH) are straightforward two-electron reductions/oxidations, but solution protons are presumably involved.

10.3 Specific Properties of AOR-Family Tungstoenzymes

10.3.1 Aldehyde Oxidoreductases

AORs catalyze the oxidative conversion of aldehydes to carboxylic acids. The hyperthermophilic marine archaeon *Pyrococcus furiosus* has become the class representative species for these enzymes as it was the source of the first AOR to be purified to homogeneity,[36] and this protein also became the first W-enzyme for which a 3D structure was determined by crystallography.[6] Four more AOR tungstoenzymes have been characterized from *P. furiosus*: formaldehyde oxidoreductase (FOR),[37] glyceraldehyde-3-phosphate dehydrogenase (GAPOR),[38] W-containing oxidoreductase number four (WOR4)[39] and W-containing oxidoreductase number five (WOR5).[40] The enzymes are either monomeric or homo-dimers or homo-tetramers; no cooperativity, or inter-subunit catalytic interaction, has been reported for the multimeric species. All catalytic subunits contain a single tungstopterin cofactor and a single [4Fe–4S] cluster.

The natural redox partner of all five enzymes is presumably ferredoxin. Only one ferredoxin has been purified from *P. furiosus*;[41,42] it is quite abundant, and the genome carries no indication for any other small e-transfer protein.[43] The ferredoxin is central to the physiology of *P. furiosus* as it interacts with a broad range of key enzymes.[44,45] Two properties of the ferredoxin, its reduction potential and its polymeric state, may be relevant to an understanding of the function(s) of AORs. The reduction potential of *P. furiosus* ferritin is not particularly low: $E_{0,7.2} = -362$ mV at 25 °C and −1.23 mV °C^{-1},[46] which would seem to imply that the AOR reaction is generally in the direction of oxidation. The ferritin was also found to be a hyperthermostable dimer at physiological ionic strength of 0.35 M.[43] Functionality as a dimer implies that the ferredoxin could accept/donate two (subsequent) reducing equivalents without having to dissociate from its enzyme partner. Note, however, that the AORs have only one [4Fe–4S], which implies that the paramagnetic W(v) state must be a functional intermediate.

P. furiosus belongs to the Euryarchaeota, one of the two main divisions of the archaea. An AOR has also been purified from *Pyrobaculum aerophylum*, which belongs to the other main division: the Crenarchaeota. This tungstoenzyme presumably has a 7Fe ferredoxin as physiological redox partner, but it is not clear whether this acts as an electron-pair or as a one-electron acceptor.[47] Of some interest is also the finding that *P. aerophylum* AOR lacks the EXXH sequence that was found to bind a single iron in the crystal structure of dimeric *P. furiosus* AOR.

A sixth type of tungstopterin containing AOR-family enzyme was recently reported to make up 2% of the cytoplasmic protein of the thermophilic bacterium *Caldicellulosiruptor bescii*. For lack of a known function and activity the protein was named XOR.[48]

10.3.1.1 Substrate Specificity

Taking *P. furiosus* as an example, a clear dichotomy is observed in the substrate specificity of the tungstoenzymes of the AOR family: some have a (very) broad substrate spectrum, others are specific for a single substrate. This means of course that for some the physiological role is not straightforwardly pinpointed, while for others it is obvious. *P. furiosus* AOR has its highest activity with the unsaturated croton aldehyde with a $K_M = 40$ μM and a $V_{max} = 67$ U mg^{-1} with methyl viologen as the artificial electron acceptor when measured at 80 °C, but butyraldehyde, glyceraldehyde, acetaldehyde and formaldehyde are also oxidized at significant rates (≥20% of crotonaldehyde oxidation activity). Using the natural electron acceptor ferredoxin at 80 °C gives a rather high $K_M = 200$ μM and a $V_{max} = 41$ U mg^{-1}.[36] The substrate spectrum of AOR may well be much broader than the small list tested.

A W-AOR with broad substrate specificity has also been isolated from the aromatic-degrading betaproteobacterium *Aromatoleum aromaticum* with comparable benzyl viologen reducing activity for acetaldehyde, benzaldehyde, crotonaldehyde, formaldehyde, glutardialdehyde, phenylacetaldehyde and propionaldehyde.[49]

P. furiosus FOR exhibits its highest activity of $k_{cat} = 1.3 \times 10^5$ s^{-1} with formaldehyde when measured with benzyl viologen as electron acceptor at 80 °C but the apparent $K_M = 15$ mM.[37] This value may require correction in view of the unusual properties of formaldehyde in water (see below). It has, however, been proposed that glutaric dialdehyde is a better substrate, because it has a $K_M = 1$ mM and with a $k_{cat} = 5.7 \times 10^4$ s^{-1} its specificity constant k_{cat}/K_M is seven times that of formaldehyde. The enzyme is also active (≥10%) with acetaldehyde, propionaldehyde, phenylpropionaldehyde, indole-3-acetaldehyde and succinic semialdehyde, and it has a minor activity with formamide, pyruvaldehyde, glyceraldehyde, crotonaldehyde, butyraldehyde, isovaleraldehyde and indole-3-acetaldehyde.[37]

P. furiosus WOR5 exhibits its highest $V_{max} = 16$ U mg^{-1} with methyl viologen as acceptor at 60 °C for the substrate hexanal with a $K_M = 180$ μM. Similar activities (and K_M values) are obtained with 2-phenylpropionaldehyde, 3-phenylbutyraldehyde, 2-ethylhexanal, isobutyraldehyde, 2-methylbutyraldehyde and 2-methylvaleraldehyde. Activity with lower specificity constants was also detected for 2-naphthaldehyde, cinnamaldehyde, 2-methoxybenzaldehyde, glutaraldehyde, crotonaldehyde, acetaldehyde and formaldehyde.[40]

The lists of putative substrates tested for the three enzymes AOR, FOR and WOR5 only partially overlap, but the general impression is that *P. furiosus* with this set is producing an aldehyde oxidizing machine with a very wide substrate spectrum in which AOR has some preference for short aliphatic aldehydes, WOR5 has some preference for aromatic aldehydes and FOR is to convert formaldehyde and/or aliphatic C_5 semi- or dialdehydes.

Glyceraldehyde-3-phosphate (GAP) is not converted at all by *P. furiosus* AOR, FOR[38] or WOR5.[40] For *P. furiosus* GAPOR, glyceraldehyde-3-phosphate dehydrogenase, GAP is the only substrate, and it is converted into

3-phosphoglycerate as an essential step in glycolysis with reducing equivalents transferred to ferredoxin.[38] Reported kinetic parameters at 70 °C are V_{max} = 350 U mg^{-1} with benzyl viologen (BV) as electron acceptor and V_{max} = 90 U mg^{-1} with ferredoxin as acceptor; also, K_M (GAP) = 30 μM, K_M (BV) = 430 μM and K_M (ferredoxin) = 6 μM.[38] These values are determined assuming no substrate inhibition; however, the kinetics are probably more complex. When the activity assay with BV as acceptor was repeated at 60 °C and analyzed assuming substrate inhibition the values changed to V_{max} = 445 U mg^{-1} and K_M (GAP) = 125 μM with K_I (GAP) = 6 μM. Furthermore, when GAP was replaced with the pure enantiomer D-GAP, the K_M and K_I values were halved, indicating that the enzyme GAPOR is fully enantioselective.[50]

Finally, *P. furiosus* WOR4 has been purified to homogeneity, but no activity has been established yet;[39] remarkably, this has not stopped chemists constructing a "functional" model for this enzyme without activity, to let it carry out a model reaction, to propose a reaction mechanism, to support this with a DFT analysis and finally to claim these results to present an efficient catalytic model system of the WOR4 enzyme.[51]

10.3.1.2 Reaction Mechanism

In terms of reaction mechanism *P. furiosus* FOR is the enzyme that has been scrutinized the most thoroughly amongst all W-AORs and perhaps even amongst all tungstoenzymes. This does not necessarily mean that we can claim to have a good understanding of how this enzyme (and, for that matter, any tungstoenzyme) operates because kinetics studies have revealed many experimental obstacles on the way to establishing clear structure-function relationships. A first attempt at casting a working mechanism was based on inspection of the crystal structures of the free enzyme (1B25.pdb), the enzyme co-crystallized with the product glutarate (from the low-K_M substrate glutaric dialdehyde) (1B4N.pdb) and the enzyme in docking complex with the ferredoxin (no pdb file deposited). Glutamate residue 308 is close to the active site in a spatial position that is very similar to Glu869 in the sequentially unrelated Mo-AOR from *D. gigas*. One carboxylate of the glutarate product is found near the tungsten and has a hydrogen bond to the Glu308 side chain carboxylate. It was proposed that Glu308 activates water to attack the carbonyl group of the bound substrate, which would be followed by hydride transfer from the CHO group to (a W-ligand of) the active site.[34] Note that 1B4N is the only available structure of a W-AOR with substrate/product and that no direct bond is observed between the glutarate and the metal.

The ferredoxin co-crystal structure supported the rather obvious suggestion of an electron transfer path from the substrate to the tungstopterin to the [4Fe–4S] in FOR to the [4Fe–4S] of ferritin. Ligands of the two iron–sulfur clusters, Cys287 in FOR and ASP14 in Fd were found to be in van der Waals contact, and the shortest distance between the clusters was *circa* 13 Å. Note, however, that the Fd structure was only well defined for the residues participating in the binding interaction with FOR, while structure was increasingly less well defined away from the interface. Since the crystallization was done with two- to four-fold excess of Fd, it is not certain whether the Fd was monomeric or dimeric.

In a subsequent redox spectroscopic study of *P. furiosus* FOR enzyme activity was found to vary over an order of magnitude for protein isolated from different batches. Two different W(v) EPR signals were found, but only one of these developed during reduction by formaldehyde or by sodium dithionite. Reduction measured as appearance of the W(v) EPR signal was very slow at ambient temperature, and was still quite slow (0.2 min^{-1}) at 60 °C.[52] A major factor that hampered data analysis was the solvation of the substrate formaldehyde, which reacts with water to form methylene glycol and subsequently polymerized to poly(oxymethylene) glycols.[53] At ambient temperature only 0.1% is indeed dissolved formaldehyde, but this increases by a factor of *circa* 90 at 80 °C (and the optimal growth temperature of *P. furiosus* is 100 °C).[54] Although this would define a K_M in the 10^{-5} M range, convincing evidence that formaldehyde is indeed the (preferred) substrate of FOR was not provided. To add to the confusion, in a study of FOR from the related (87% sequence identity) species *Thermococcus litoralis* three different W(v) EPR signals were found and the spectroscopic response to incubation with formaldehyde was different from that of *P. furiosus* FOR.[55]

Key experimental information for the development of a detailed enzymatic working mechanism usually required pre-steady-state kinetic experiments. A stopped-flow series of experiments was carried out with FOR and once more revealed several experimental problems that may well be typical for W/Mo enzymes. For one, the optical absorption of the tungstopterin overlaps with that of the [4Fe–4S] cluster with the intensity of the latter being an order of magnitude stronger. Furthermore, the kinetic data turned out to be dominated by a complex, substrate-induced reductive activation process, and this made it essentially impossible to observe specific intermediates in the active reaction cycle.[56] By combining the pre-steady-state data with steady-state data and static spectroscopic information an activation and electron-transfer model was proposed as a working model for future studies. Unfortunately, since 2008 no more pre-steady-state studies on this, or for that matter on any other tungstoenzyme, has been reported.

In the meantime a DFT study has been carried out on *P. furiosus* FOR at the B3LYP/6-311+G(d,p) level of theory with the LANL2TZ(f) basis set for metal.[57] Four different reaction scenarios have been considered including the one originally proposed on the basis of inspection of the crystal structures. Glu308 was assigned an essential role in all four mechanisms, but hydride transfer was found to be energetically highly unlikely. The strongly preferred scenario is re-drawn in Scheme 10.2, which combines and simplifies data from figures in ref. 57. Note that the proposal of ref. 57 is not a complete reaction cycle: it is only concerned with the activation and oxidation of formaldehyde. The leftmost structure in Scheme 10.2 is only one of the ES complexes; the arrows in the lower part of the Scheme are to indicate that a complete catalytic cycle additionally requires at least

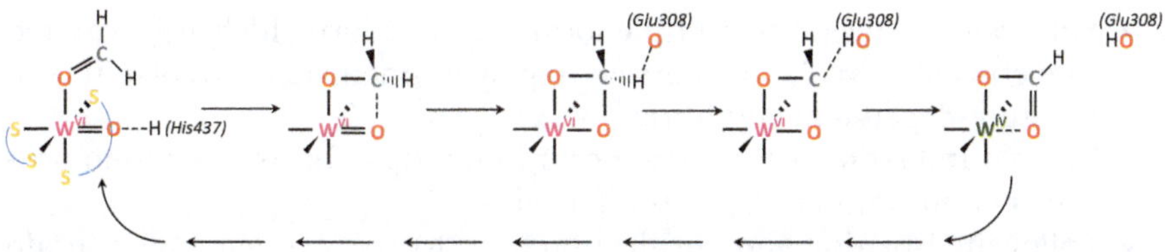

Scheme 10.2 Mechanistic proposal of *Pyrococcus furiosus* formaldehyde oxidoreductase catalysis. DFT-based calculations afford a preferred reaction path for the activation and conversion of formaldehyde.[57] Additional arrows indicate the minimum number of steps required to make a complete catalytic cycle.

(ignoring transition states and assuming dimeric ferredoxin) nine more steps, namely, starting from the lower right corner: dissociation of formate, association of ferredoxin, oxidation of W(IV) in two one-electron steps, dissociation of ferredoxin, association of water, double deprotonation in two steps and association of formaldehyde.

The substrate formaldehyde binds directly to the tungsten with a carbonyl-O to W distance of 2.40 Å. A nucleophilic attack ensues by the W=O group. One of the substrate protons points towards the Glu308 carboxylate at a distance of 2.12 Å. Transfer of this proton to Glu308 occurs concomitantly with two-electron transfer from the substrate to the tungsten to afford W(IV). In this scheme both the substrate and the product are directly bound to the tungsten, namely to W(VI) in the reactant complex and to W(IV) in the reaction product. Note that this theoretical result is the only indication that substrate and/or product actually coordinate the metal during catalysis. Both complexes, substrate-W(VI) and product-W(IV), should (but have not yet) somehow be observable by spectroscopy (but not by EPR) after rapid freeze.

To catch these and other putative intermediates in the reaction cycle would appear to be the major challenge of tungsten biochemistry development. It is perhaps useful to state a long list of obstacles to be faced by those who are willing to go down this road. The following list is mainly based on data from *P. furiosus* FOR but is likely to apply in large part to other tungstoenzymes:

- The natural substrate(s) for several AORs has not yet been clearly identified (see paragraph on substrate specificity).
- Slow reductive activation occurs by incubation with substrate and/or a non-physiological reductant.[37,52]
- Substrate inhibition is not uncommon (*e.g.* GAP with *P. furiosus* GAPOR,[38,50] formaldehyde and ferredoxin with *P. furiosus* FOR[56]).
- Species producing tungstoenzymes can have very high temperatures for optimal growth (*cf.* 100 °C for *P. furiosus*[54]), which may not be easily amenable in kinetic studies.
- Redox equilibria at high temperature may be difficult to catch by rapid freezing (*e.g. P. furiosus* ferredoxin).[46]

- Substrate can occur in multiple forms (*e.g.* free formaldehyde, hydrated to methylene glycol, polymerized to poly-oxymethylene glycols) in temperature-dependent equilibria.[52]
- The natural redox partner can occur in multiple forms (*e.g.* monomer–dimer equilibrium of *P. furiosus* ferredoxin.[43]
- Following isolation and purification tungstoenzymes generally contain substoichiometric amounts of tungsten (typically 0.3–0.9 W per protein.[40,58–61] Whether this also holds *in vivo* is not known.
- Tungstoenzymes are generally quite sensitive to inactivation by molecular oxygen.[5,58]
- The oxidation state(s) of W in crystals of tungstoenzymes used for crystallography is not known and is presumably a source of inhomogeneity (*e.g. P. furiosus* AOR was crystallized in the presence of dithionite and dithiotreitol;[6] *P. furiosus* FOR was crystallized under Ar with dithionite and dithiotreitol;[34] *G. metallireducens* BCR was crystallized under H_2/N_2 and with dithiothreitol but without the Ti(III) required in activity assays.[35]
- EPR spectroscopy of W(V) usually identifies more species than one would wish for; some of these appear to be associated with activation/inactivation processes (*e.g.* ref. 52 and 55).
- A direct bond between W and substrate and/or product has yet to be identified in crystallographic protein structures.[34,35]
- The number of W-ligands is not always clear from the crystal structure (*e.g.* does W in *P. furiosus* FOR have one or two Os?[34]).
- Optical absorption of tungstopterin is rather featureless and relatively weak, *e.g.* compared to that of iron–sulfur clusters.[33,56]

10.3.2 Benzoyl-CoA Reductase

Four different microbial strategies are known for the degradation of aromatic compounds; two are aerobic and two are anaerobic.[62] A key enzyme of one of the anaerobic routes is a tungstoenzyme: class II benzoyl-CoA reductase or BCR.[63] The reaction is given in Scheme 10.3.

The genome of the model species *Geobacter metallireducens* carries a cluster of eight genes *bamBCDEFGHI* (bam = benzoate metabolism) whose expression is induced by benzoate,[64] and whose product is presumably a complex

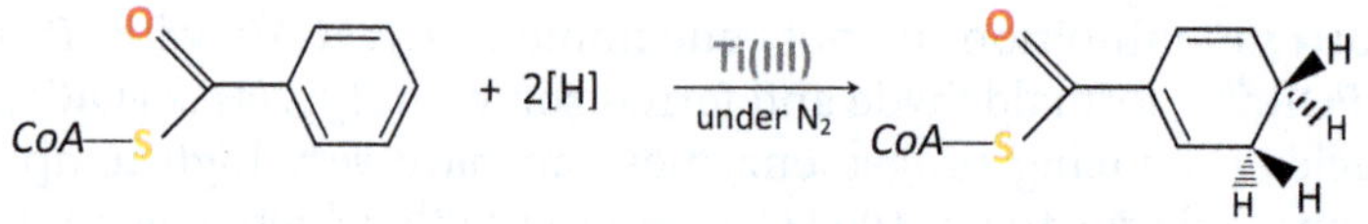

Scheme 10.3 Reaction catalyzed *in vitro* by the *Geobacter metallireducens* tungstoenzyme benzoyl- coenzyme A reductase. Mandatory presence of a strong reductant indicates W(IV) to be the active species.

of eight proteins holding a tungstopterin, numerous iron–sulfur clusters and three FADs, and carrying activity for BCR and for transfer of reducing equivalents to/from a ferredoxin and NAD^+.[65] The BCR activity is localized on bamB and has been studied in an isolated and purified bamBC heterodimer; bamC is an electron transfer protein and contains three iron–sulfur clusters. BenzoylCoA reduction activity is measured in the presence of the strong reductant Ti(III)citrate ($E^{0\prime} \approx -720$ mV) and a low-potential mediator ($E^{0\prime} = -674$ mV).[66] The physiological electron donor has not been identified yet. The reduction potential of the benzoyl-CoA/enoyl-CoA couple has been determined to be $E^{0\prime} = -622$ mV.[66]

The *G. metallireducens* bamBC dimer has been crystallized. BamB carries the tungstopterin and a [4Fe–4S] cluster, and BamC has three [4Fe–4S] clusters. The BamBC structure has been solved in five different forms (*cf.* Table 10.1): enzyme as isolated, with excess Zn(II), with substrate benzoyl-CoA, with product dienoyl-CoA and with inhibitor monoenoyl-CoA.[35] Although sequence homology and the presence of a W-MPT cofactor (including the Mg(II) ion that couples the phosphates of the two pterins) clearly places the enzyme in the AOR family, the coordination of W appears to be quite distinct from that observed in the crystal structures of *P. furiosus* AOR and FOR: there is now a direct bond from the protein to the tungsten in the form of a cysteinato ligation by Cys322. A distorted octahedral coordination is completed by a small molecule as sixth ligand. The coordinating atom was identified by EXAFS spectroscopy as C, N or O at 2.00 Å from W, and an additional atom appears to be present at *circa* 3.2 Å, which led to the proposal that W might be coordinated by CN or CO.[35] However, no clear peaks in the 2000–2200 cm^{-1} FTIR spectral region were found.[35]

From inspection of the crystal structure and with explicit reference to the documented mechanism of chemical reduction of aromatic compounds such as benzoic acid, known in organic chemistry as the Birch reaction, a preliminary reaction mechanism was discussed in which W(IV) donates a single electron to the aromatic ring in two different steps, thereby creating organic radical intermediates.[35] W(V) must then be twice re-generated to W(IV) by reducing equivalents that should eventually have to come from an external donor. In the proposed mechanism substrate and product are never directly bound to the tungsten, because no direct bond was found in the crystal structures. While this has been claimed to be a unique property of the BCR enzyme as compared to the AOR family members,[35] we should note that no direct bond between substrate/product and W has ever been found yet for any AOR family member, although the database is at this time admittedly of very limited size (*cf.* Table 10.2). In the original characterization of *G. metallireducens* BCR *circa* one Zn and two Ca were found per W, which were suggested to play a structural role[63] analogous to the Ca that was found bound to one of the pterins of W-MPT in *P. furiosus* FOR.[34] In the subsequently published structures the calcium has disappeared.[35] Zn is bound to a specific site in the BCR as isolated, but is expelled by addition of substrate leading to a conformational change of the protein, and this finding led to the proposal that Zn ion is a regulator to inactivate BCR when aromatic substrate is low to prevent futile reduction of solvent protons.[35] The question as to what happens to the Zn(II) after it is expelled from the enzyme has not been addressed. It is perhaps also relevant to note that many purified proteins are contaminated with zinc, which they pick up during purification, *e.g.* over size-exclusion filters.

Multiple key points still remain to be clarified, *e.g.*: (i) radical intermediates are yet to be identified with EPR spectroscopy; (ii) the putative diatomic ligand to W is yet to be identified; (iii) its contribution to the reaction mechanism is unclear; (iv) what (if any) is the physiological relevance of the Zn-induced conformational change?; and (v) what is the driving force for this low-potential reduction?

10.4 Specific Properties of DMSOR-Family Tungstoenzyme

10.4.1 Formate Dehydrogenase

These enzymes come as molybdoprotein or as tungstoprotein, and in some cases they can use both metals. The subject has been comprehensively reviewed very recently,[67,68] and so we can be relatively brief here. It appears to be *communis opinio* that the W and Mo systems are structurally and mechanistically quite similar. More work, in particular on the reaction mechanism, has been carried out for Mo-FDHs, and for now we can therefore assume that several proposed possible mechanisms (*cf.* ref. 67 and 68) also could be relevant for the W-FDHs, which implies that the W-FORs are co-protagonists in an open-ended discussion on how these enzymes work.

There is one crystal structure, namely of the as-isolated *Desulfovibrio gigas* FDH in the absence of substrate or inhibitor[12,69] (*cf.* Table 10.1). The oxygen-tolerant[70] enzyme is a heterodimer FdhAB of 92 and 29 kDa; FdhA carries a W-$(MGD)_2$ and a single [4Fe–4S] cluster; the small subunit FdhB has three [4Fe–4S] clusters. W is in a distorted trigonal prismatic coordination: the base of two dithiolenes is capped by a Se from Sec158 and a sulfur atom (at least in the resting state). The active site is completed by His159 (presumably for hydrogen bonding) and Arg407 (presumably for charge compensation) (*cf.* ref. 67) just like in the Mo-FDHs.

An interesting and unsolved question is the discrimination (or the lack thereof) between Mo and W by different species. When *Moorella thermoacetica* (previously: *Clostridium thermoaceticum*) is grown on equal amounts of W and Mo (as 10 μM tungstate and molybdate) it makes an FDH that is purely a tungsten enzyme.[5] Similarly, when *Syntrophobacter fumaroxidans* is grown on equal amounts of 0.1 μM W and Mo it makes two different FDHs that are both purely tungstoenzymes.[71] Note that these FDHs exhibit high "uphill" activity *in vitro* in the reversed direction of CO_2 reduction, which is what they should do *in vivo* when the acetogen *S. fumaroxidans* lives in symbiosis with methanogens *via* inter-species H_2 transfer.[71]

The situation for *D. gigas* FOR is less clear. The 1999 paper that describes the original purification of the enzyme[72] refers to a 1965 paper for the growth medium;[73] in the latter no mention is made of Mo or W, which indicated that the medium was not enriched in any way in Mo and/or W. On the other hand, the 1999 paper reads that "The growth medium was not depleted of molybdenum. [...] So it is clear that even with Mo available *D. gigas* FDH prefers to incorporate tungsten".[72] The "clear" does not apply to the present reviewer. Subsequent papers on the crystallography of *D. gigas* FDH also refer back to the previous work for growth conditions.[12,70] Whatever the available Mo and W, the crystal structure is that of a (predominantly) tungstoenzyme as evidenced by multiple-wavelength anomalous dispersion at the W-edge.[70]

A similar unclear situation exists for *Desulfovibrio alaskensis* FOR whose 2004 purification report[74] cites a 1984 paper[75] for the growth medium; unfortunately, the latter further cites an out-of-print book for a description of the medium. On the other hand, the authors of the *D. alaskensis* FOR purification paper also state that "all other metals <are> added mainly as contaminants by the various medium components, in particular yeast extract".[74] All in all, the amount of Mo and/or W offered to *D. alaskensis* remains obscure, but the purified FOR turns out to be a mixture: it has 0.4 Mo and 0.3 W per FOR molecule.[74]

Finally,[76] the genome of *Desulfovibrio vulgaris* encodes three selenocysteine-containing FDHs: FDH-1, FDH-2 and FDH-3,[77] of which the latter had been previously purified and shown to be a heterotrimeric molybdenum enzyme;[78] the second one is possibly part of a membrane-associated complex and the first one, FdhAB, is a heterodimer homologous to *D. gigas* W-FDH.[76,77] Expression of these three sets of structural genes has been studied as a function of growth medium composition, in particular addition (0.1–0.4 μM) of W or Mo, and energy source for reducing equivalents, namely lactate, formate or molecular hydrogen.[76] FdhAB was not found in Mo-grown cells, but in W-grown cells it was found to incorporate both Mo and W.[76] The nature of the metal-dependent regulation was not established. Differential activity of W-FdhAB *versus* Mo-FdhAB could not be conclusively established.

Taken together the available data on W-FDHs indicate a gliding scale of Mo-incorporation, while a paucity in metal-specific (*i.e.* W- *versus* Mo-based) activity data presently precludes a definitive assignment of these enzymes as sometimes Mo-promiscuous tungstoenzymes rather than tungstate-induced variants on the molybdoenzyme FDH theme.

10.4.2 Formylmethanofuran Dehydrogenase

The reaction catalyzed by formylmethanofuran dehydrogenase (FMD) is usually written as the reduction of CO_2 with a low-potential electron donor, such as Ti(III), and in the presence of methanofuran (MFR), to the formylated MFR. In reality the MFR reacts spontaneously with carbon dioxide to the carbamate *N*-carboxymethanofuran,[79] and therefore the catalyzed reaction is that of carboxymethanofuran to formylmethanofuran (written in protonated form as reaction 4 in Scheme 10.1 to retain unity in expressing the reducing equivalents as 2[H]). The formal oxidation state of the C atom goes from +IV to +II, and so the reaction is equivalent to the reversed formate dehydrogenation catalyzed by the *S. fumaroxidans* FHDs except that the reductive activation of carbon dioxide now affords an activated formyl group attached to the methanogens-specific carrier molecule methanofuran as a preparation for further stepwise reduction all the way down to CH_4. This is the most endergonic reaction in methane formation by H_2-consuming methanogenic archaea,[80] a process that has recently been shown to be cyclic.[81] The very low potential electron donor is a specific polyferredoxin ($E^{o\prime} \approx -500$ mV), which in turn eventually receives its reducing equivalents from molecular hydrogen *via* a hydrogenase-heterodisulfide reductase complex (electron-bifurcating) or alternatively *via* a transmembrane hydrogenase complex energetically driven by a sodium gradient (reviewed in ref. 82 and 83).

Formylmethanofuran dehydrogenases come in Mo and W varieties. The genome of the model species *Methanobacterium thermoautotrophicum* carries two FMD-related operons, namely *fwdHFGDACB* and *fmdECB* where the abbreviations fwd and fmd stand for W-associated FMD and Mo-associated FMD, respectively. From the first operon a W-FMD is expressed as the heterotetrameric complex FwdABCD with the subunit FwdB carrying the W-MGD cofactor and (presumably) a [4Fe–4S] cluster. In the heterotrimeric complex FmdABC the subunit FmdB has a Mo-MGD and (presumably) a [4Fe–4S] cluster. Note that the A-subunit of the Mo-enzyme is actually encoded by the *fwd* operon, which is perhaps interesting from a regulatory viewpoint, but also gives rise to a somewhat awkward gene-product notation.[84] The open reading frame *fwdF* encodes a polyferredoxin with eight [4Fe–4S] clusters.[85] The physiological significance of the dual FMD system is enigmatic. The Fwd protein is constitutive in the sense that it is expressed by cells grown on mineral medium either with tungstate or with molybdate, while the Fmd protein is only expressed by molybdate-grown cells. However, the constitutive Fwd system has been reported to contain Mo, not W, when growth is on molybdate.[59] Activities of Fwd and Fmd appear to be comparable, but the activity of the Mo-Fwd has not been determined.[59]

The structure of operons and Mo-dependent expression pattern of *M. thermoautotrophicum* are also found in *Methanobacterium wolfei*,[86] but the system is far from universal. *Methanosarcina barkeri* has only a Mo-FMD,[87] which is expressed from an *fmdEFACDB* operon[88] apparently as a heterohexameric complex.[89] On the other hand, the hyperthermophilic *Methanopyrus kandleri* has been reported to contain two W-FMD systems, *fwu* and *fwc*, and no Mo-FMD. The product of the *fwuB* gene apparently has selenocysteine in the active site and that of the *fwcB* has cysteine in the active site. Although only the *fwu* system is expressed when selenocysteine is supplemented in the growth medium, remarkably, both systems are expressed in the absence of selenocysteine.[90] We cautiously generalize all these findings in the conclusion that formylmethanofuran dehydrogenase is not particularly outspoken in its preference for W *versus* Mo with flexibility in adjusting to environmental availability. A crystal structure for any FMD enzyme is yet to be determined; also, no pre-steady-state kinetic study has been reported yet.

10.4.3 DMSO Reductase

DMSO respiration is induced in many species as a response to low oxygen tension. The enzyme that donates reducing equivalents to the dimethylsulfoxide/dimethylsulfide couple ($E^{0\prime}$ = +160 mV) is known as a Mo-protein with $E^{0\prime}(Mo^{VI/V})$ = +141 mV and $E^{0\prime}$ $(Mo^{V/IV})$ = +200 mV in *Rhodobacter sphaeroides*.[91] The crystal structure of the enzymes from *R. sphaeroides* and *Rhodobacter capsulatus* have been determined.[92,93] They form an attractive object of study because they are isolated from the periplasm as monomeric proteins carrying a Mo-MGD and no other metal prosthetic groups. In particular the absence of iron-sulfur clusters allows for a straightforward monitoring of the relatively weak optical spectrum of the molybdopterin.[91,92,94] When grown phototrophically on mineral medium in the presence of DMSO[95] with molybdate replaced by 0.003–10 mM tungstate *R. capsulatus* makes a W-version of its DMSOR, which is structurally virtually identical to the original Mo-version.[33] In particular the 6-coordination by the two pterin dithiolenes plus an O from Ser147 plus an extra O is fully retained according to protein crystallography and W L_{III}-EXAFS data[33] (but see also ref. 93).

Since both the W- and the Mo-enzyme are active, they together form an ideal test system for the "why tungsten?" question. An answer is not immediately obvious from differential expression experiments: grown on molybdate *R. capsulatus* makes Mo-DMSOR, and grown on tungstate it makes W-DMSOR (W : Mo > 99 : 1) but when the two metals are offered in equal concentration, the bacterium only exhibits a modest 2 : 1 preference to incorporate Mo.[33] A sharper picture emerges when redox titration data[96] are juxtaposed to activity data[33] in Table 10.3: the reduction potentials of the tungstoenzyme are 335 mV lower than those of the molybdoenzyme. DMSO reduction activity is 17 times higher for the tungstoenzyme, but the reverse activity of DMS oxidation is absent, and these results are taken to suggest a physiologically relevant functional difference between W- and Mo-variants related to a difference in reduction potential. The latter fits in a trend observed for W *versus* Mo model compounds.[96]

Table 10.3 Reduction potentials and activities of *Rhodobacter capsulatus* DMSO reductase W/Mo variants.[a]

	$E^{0\prime}$(VI/V) mV	$E^{0\prime}$(V/IV) mV	DMSO reduction[b] 1 s^{-1}	DMS oxidation[c] 1 s^{-1}
Mo-enzyme	+141	+200	27	8.5
W-enzyme	−194	−134	470	<0.05

[a]from low-temperature EPR monitored room temperature redox titrations at pH 7.[91,96]
[b]with dithionite and methyl viologen.[33]
[c]with phenazine ethosulfate and dichlorophenolindophenol.[33]

A related enzyme is trimethylamine *N*-oxide reductase, or TMAOR, or TorA, a molybdopterin containing protein produced, *e.g.* by *Escherichia coli.* When molybdate in the growth medium is replaced with 1 mM tungstate, a W-TMAOR is produced. Substitution of W for Mo halves the k_{cat} but increases the K_M for TMAO, resulting in a two-fold increase in specificity constant.[97]

10.4.4 Nitrate Reductase

Bacterial nitrate reductases exhibit a wide variety in structure and function, but they are all molybdoenzymes.[24] A few nitrate reductase enzymes have also been isolated from archaea as Mo-enzymes.[98] For example the hyperthermophile *Pyrobaculum aerophilum* expresses a heterotrimeric nitrate reductase;[99] this Mo-NarGHM is expressed as a tungstoenzyme when the medium contains 4.5 μM tungstate and molybdate is omitted.[98] Growth on 0.5 μM molybdate plus 1 μM tungstate affords roughly equal amounts of Mo- and W-protein. The two enzymes exhibit the same K_M of 57 μM for nitrate; the activity of the W-enzyme is *circa* 50% of that of the Mo-enzyme in assays with artificial e-donors.[98] Since the reduction potentials for both metals could not be determined in the purified enzymes, it is perhaps not warranted to draw a more specific conclusion beyond that of apparent indifference of the enzyme for W *versus* Mo. No crystal structure determinations and no pre-steady-state studies have been reported.

10.4.5 Acetylene Hydratase

When grown under acetylene the strictly anaerobic fermenting bacterium *Pelobacter acetylenicus* produces a tungsten-containing acetylene hydratase, ACH.[60] The product of the catalyzed, highly exergonic reaction is acetaldehyde (reaction 4 in Scheme 10.1), which is subsequently disproportionated enzymatically into ethanol and acetate. ACH is by all means the oddball in the tungstoenzyme superfamily as it apparently catalyzes a non-redox reaction. *P. acetylenicus* ACH is an oxygen-tolerant 83 kDa monomer with one W-bis-MGD and one [4Fe–4S] cluster.[32,60,100] Remarkably, the enzyme is only active *in vitro* in the presence of a strong reductant, Ti(III) citrate or dithionite, indicating W(IV) to be the active species.[100] The activity titrates with a two-electron effective reduction potential equal to −340 mV while the [4Fe–4S] cluster has $E^{0\prime} = -410$ mV, which would seem to imply that the redox state of the iron–sulfur cluster is irrelevant for the activity.[100]

Inspection of the crystal structure of the reduced (*i.e.* W(IV)) enzyme has led to the following mechanistic proposal. The tungsten is in a distorted octahedral coordination formed by the two dithiolene ligands with an additional thiolate from Cys141 and completed by a water molecule. The water is activated by a nearby aspartic acid, Asp13, whose pK_a is strongly shifted under the influence of the nearby [4Fe–4S] cluster. Although the ACH has homology to other members of the DMSOR family, its fold forms a substrate access channel from a very different direction, which ends in a hydrophobic pocket next to the coordinated water molecule involving in particular Ile142.[32,101] The protonated Asp13 is proposed to turn the coordinating water into an electrophile, which can directly attack the triple bond of acetylene affording a vinyl cation intermediate.[32] DFT calculations supported the small ligand to W to be identified as water and not hydroxo, and electrostatic calculations afforded feasibility to the concept of an Asp13 with a very high pK_a.[32] However, a subsequent theoretical study by a different group supports an alternative mechanism in which the ligand is hydroxo and is replaced by acetylene thus forming a direct organometallic η^2 intermediate, which is then subjected to nucleophilic attack by a water molecule.[102] A series of theoretical studies by a third group provided further support for this latter view.[103–106] Thus the nature of the ligand (water *versus* hydroxo) is a key point in this controversy, which for the time being remains unsettled.

A few additional comments are in order in particular with respect to the unique role of Asp13 and its alleged unusual pK_a value. A site-directed mutagenesis study on heterologously expressed ACH concludes that "Replacing residue Asp13 with either alanine or glutamate gives strong support for the crucial catalytic role of the residue, as predicted from the crystal structure and theoretical calculations".[32,101] The conclusion strikes one as odd, where the Asp13Glu mutant is reported to have unchanged activity.[101] In a similar vein this reviewer is surprised by the claim that the unusual pK_a value of Asp13 is "caused by a nearby low potential [4Fe–4S] cluster",[32] where the ACH activity had previously been reported to be unaffected by a change in the redox state of the iron–sulfur cluster.[100] Finally, in addressing the "why tungsten?" question, information on Mo substitution could be helpful. In ref. 101 it is claimed that "a molybdenum dependent active form of ACH could be obtained from *P. acetylenicus* cultivated on molybdate (2 μM) in the presence of nanomolar concentrations of tungstate" with citation of ref. 100. Unfortunately, in ref. 100 this reviewer was unable to find any experimental data and/or discussion on a Mo-substituted ACH.

10.5 Tungsten Metallomics

In classical proteomics all proteins of a cellular fraction are separated in 2D gel electrophoresis under denaturing conditions. Individual proteins sampled from color-developed spots are subsequently identified by tandem mass spectrometry. The experimental conditions are such that most metals in metalloproteins will be lost from their host – or even worse: redistributed over other proteins – in the course of the electrophoretic procedure. To avoid this serious loss of bioinorganic chemical information the method of MIRAGE was developed: Metal Isotope native Radio Autography in Gel Electrophoresis.[107–111] The method makes use of a radioactive isotope of the metal of interest with a half-life of the order of one day. This may require on-site production of the radionuclide. Typically, an extract from cells grown on a medium containing the radioactive isotope is separated into proteins by gel electrophoresis under *non*-denaturing conditions. Separation in the first dimension is by iso-electric focusing and in the second dimension on mass after charging the proteins with Coomassie Brilliant Blue. Radioactivity is then located by radioautography, and after, say, a week (some ten half-lives) when radioactivity has been reduced to below detection limit, the marked spots are excised for analysis by mass spectrometry. The complete experiment can be repeated for different environmental conditions, *e.g.* variable concentration of the metal in the growth medium.

One of the first systems that the method was applied to was the tungsten proteome of *P. furiosus* cell extract[110] using the isotope ^{187}W ($t_{1/2} \approx 24$ h) as illustrated in Figure 10.3. Salient findings were: (i) all five until then characterized AOR-family enzymes were identified in the MIRAGE, and no other tungstoprotein was found; (ii) differential expression using a cold shock resulted in changes in expression levels of the tungstoproteins that corresponded well with mRNA levels previously determined in DNA microarray experiments;[112] (iii) the five W-enzymes accounted for 70% of all tungsten in the cell; the 30% remainder was not protein-associated and consisted presumably of free tungstate. In a follow-up study the radionuclide ^{99}Mo ($t_{1/2} \approx$ 66 h) was used to show, in contrast to an earlier biochemical study,[113] that molybdenum *can* in fact be incorporated under proper conditions in all five AOR-family tungstoenzymes.[114]

An alternative approach to metalloproteomics was proposed based on conventional column chromatography but with the efflux of each column monitored off-line for particular elements by ICP-MS with protein identification by tandem MS.[115] A direct comparison with the MIRAGE method is possible because the column chromatography method was also applied to the

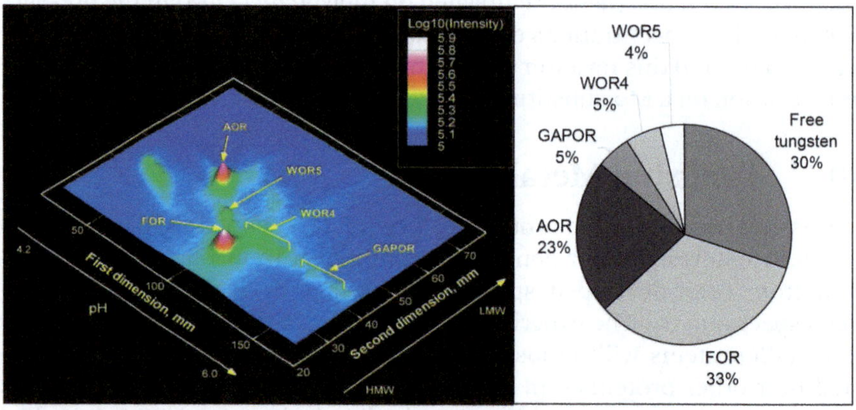

Figure 10.3 Metallomics of tungsten proteins in *Pyrococcus furiosus*. Cells were grown on the radionuclide ^{187}W and the cell-free extract was analyzed with the MIRAGE method. The left-hand panel is a 3D image of native-native 2D polyacrylamide gel electrophoresis with radioactivity as the third dimension. Reproduced from ref. 110. The right-hand panel is a percentage distribution pie of detected tungstoenzymes and free tungstate. Reproduced from ref. 110.

cytoplasm of *P. furiosus*. The latter method resulted in identification of the astronomical number of 52 W-proteins and 37 Mo-proteins. Two of the prospective Mo-proteins were subjected to multi-column chromatography: one was purified to homogeneity to an amount of 2.5 µg sufficient to determine a Mo stoichiometry of 0.8 per protein; for both proteins an unprecedented direct coordination by Cys residues was proposed.[115] The present reviewer is perplexed by these results and is eagerly awaiting more specific data from any follow-up research. Note that in these experiments *P. furiosus* was grown on 9 µM W and 0.2 µM Mo, which allegedly resulted in 6.7 µM W and 0.34 µM Mo in the cytoplasmic fraction,[115] although the cells are known to have a several-orders-of-magnitude preference for W uptake over Mo uptake.[114]

10.6 Why Tungsten?

The observation that addition of tungstate to the growth medium of *Clostridium thermoaceticum* stimulates formate dehydrogenase activity dates back to 1973[3] and so does the question "why tungsten?" In the intervening more than four decades our knowledge of the biochemistry of tungsten has increased perhaps following a classical sigmoidal pattern in time with an inflection point in the early 1990s and a subsequent steady state till today (*cf.* the reference list for this chapter). This by now constant build-up of knowledge does not appear to have produced an equal increase in our insight into the "why tungsten?" problem. Reports on tungstoenzymes typically culminate in attempts to interpret experimental findings in the frame of evolution theory, circulating around a small number of recurring assumptions: (i) in a prehistoric anoxic atmosphere and in the presence of high sulfide molybdenum was insoluble while tungsten was not; (ii) entering oxygen the world switched to oxoanions leading to similar (and high) solubility of the two metals; (iii) pedospheric occurrences of molybdenum and tungsten are roughly equal, but in marine environments Mo outdoes W by two orders of magnitude: (iv) in submarine volcanic niches the Mo/W ratio is inversed; (v) Mo complexes would be less stable at elevated temperatures than W complexes. Based on these premises the scenario is developed in which cells had no choice in early times but to go after tungsten. Then in the course of evolution they switched to molybdenum, and tungsten is leaving the bioinorganic chemical stage with only a few strongholds left in exotic, hot places.

Aside from the fundamental difficulty that evolution is not a reproducible experiment on a human time scale, the picture of Mo conquering W is not altogether convincing to this reviewer in view of the following experimental observation: (a) in a sulfide-rich environment Mo(VI) is soluble as thiomolybdates, $MoO_{4-n}S_n^{2-}$, and reduction to Mo(IV) is not straightforward, *e.g.* requiring activated hydrogen at high pressure, while the products are not simply MoS_2 but rather Mo(III)hydrosulfide and Mo(IV)hydroxide;[116] (b) many modern-time anaerobe organisms, for which molybdenum is an essential element, *e.g.* sulfate reducing bacteria, thrive happily in a self-produced sulfide-rich environment; (c) at least certain organisms can intracellularly hold considerable amounts of molybdate;[114] (d) some hyperthermophiles employ molybdenum at elevated temperatures (*e.g.* ref. 99); (e) some tungstoenzyme-producing organisms are not thermophilic at all (*e.g.* ref. 61); (f) the idea that W-dependent organisms live in W-rich habitats is unfounded, or at least poorly documented. An early review cites that some deep-sea hydrothermal vent fluids can be low in Mo,[8] but none of the thus far studied tungstoenzymes come from deep-sea species. Unfortunately, the habitats where they do come from appear not to have been chemically characterized in detail. In a not particularly systematic check this reviewer went to the site of origin of *P. furiosus* to find an elevated W level, which, however, was still outdone by Mo by a factor of 6.[114] In this environment *P. furiosus* finds it necessary to maintain an import system that has a thousand times higher affinity for tungstate over molybdate, and to maintain another order-of-magnitude selectivity (whose molecular basis is yet to be determined) of W over Mo in its intracellular synthesis of tungsten holoenzymes.[19,114] In *Campilobacter jejuni* the selectivity of the uptake system is even up to 50 000 times in favor of tungstate.[117,118] All in all the verdict on the evolutionary story of Mo and W would be that "even if it is not true, it is well conceived". Or in other words, it is perhaps not likely to be particularly helpful in addressing the "why tungsten?" question.

Of an altogether different nature is the idea[8,113] that W is selected over Mo for reactions that occur at very low reduction potential, a proposal that does not (necessarily) refer to evolutionary arguments. A case in point would be the observation that *R. capsulatus* Mo-DMSO reductase can almost equally well catalyze the reverse reaction of DMS oxidation, but when Mo is replaced with W the reduction activity increases while the oxidation activity essentially vanishes,[33] and this can be combined with an observed 335 mV difference in reduction potentials for Mo- *versus* W-enzyme.[96] A caveat, however, may be in order: The physiological function of DMSO reductase in *R. capsulatus* is reduction of DMSO, not the reverse reaction, and still, when offered equal amounts of Mo and W, the bacterium has a slight preference for Mo although the W-enzyme *in vitro* exhibits much higher activity. Furthermore, some tungstoenzymes do not appear to employ low reduction potentials at all.[97,98] An aspect that appears to have been ignored in the discussion is the fundamental difference between tungstoenzymes from the AOR family and those from the DMSOR family. While in the latter a Mo-promiscuity may well be the rule rather than the exception, in the former the requirement for W is possibly absolute. By implication, the DMSO-family tungstoproteins may generally not be proper systems to test Mo/W differential reactivity, while, unfortunately, testing AOR-member tungstoenzymes is clearly much more difficult with only one case of partial Mo incorporation reported.[114] Other aspects that hamper development of a balanced discussion are our presently limited knowledge on regulatory mechanisms in the expression and post-translational modification of these enzymes (*cf.* ref. 119) as well as on the structure of enzyme intermediates in the catalytic mechanisms.

The bottom line is that even after 40 plus years of research into tungstoenzymes it is still too early to expect a reasonable answer to the "why tungsten?" question. What *can* be concluded, however, is that in order to put us in the future in a position in which we will be able to meaningfully address the question, a number of paradigmatic re-orientations in the research are called for with some urgency. These are summarized in the following recommendations:

- Tungstoenzymes from the AOR family with their absolute requirement for W and with their compact, Mg(II) ion-tied, W-bis-MPT prosthetic group should be viewed as a set that is very different from tungstoenzymes from the DMSOR family with their gliding scale of promiscuity for Mo and with their extended, W-(Mo-)bis-MGD prosthetic group. A serious effort should be made to construct AOR-family enzymes with Mo in place of W and to study these artificial systems structurally and, especially, functionally;
- A step-up is called for in research into mechanisms of regulation of expression and mechanisms of regulation of post-translational modification of tungstoenzymes from both families;
- Tungsten biochemists should perhaps shift their attention from identifying new tungstoenzymes in more species towards scrutinizing reasonably well-characterized tungstoenzymes by means of advanced methods of pre-steady-state kinetics in order to identify structural and electronic characteristics of early intermediates in relevant reactions;
- Theoretical chemists should exhibit some reluctance in modeling detailed reaction mechanisms as long as the experimental data on intermediates to test their models on have not been collected yet;
- Microbiologists who go out to isolate novel species should always take habitat samples, *e.g.* to be put in an ICP-MS apparatus; in fact, they should perhaps go back to the original sites from where now prominent model organisms were isolated to take samples for elemental and speciation analysis.

If we are all willing to accept these paradigmatic changes in research orientation, then there is a chance that sometime in the next 40 years we will be able to say, "tungsten, why not?"

References

1. R. J. P. Williams and J. J. R. Fraústo da Silva, *The natural selection of the chemical elements*, Oxford University Press, Oxford, 1996.
2. R. J. P. Williams, *Cell. Mol. Life Sci.*, 1997, **53**, 816.
3. J. R. Andreesen and L. G. Ljungdahl, *J. Bacteriol.*, 1973, **116**, 867.
4. J. R. Andreesen, E. El Ghazzawi and G. Gottschalk, *Arch. Microbiol.*, 1974, **96**, 103.

5. I. Yamamoto, T. Saiki, S.-M. Liu and L. G. Ljungdahl, *J. Biol. Chem.*, 1983, **258**, 1826.
6. M. K. Chan, S. Mukund, A. Kletzin, M. W. W. Adams and D. C. Rees, *Science*, 1995, **267**, 1463.
7. J. J. R. Fraústo da Silva and R. J. P. Williams, *The biological chemistry of the elements; the inorganic chemistry of life*, Oxford University Press, Oxford, 1991.
8. A. Kletzin and M. W. W. Adams, *FEMS Microbiol. Rev.*, 1996, **18**, 5.
9. J. R. Andreesen and K. Makdessi, *Ann. N. Y. Acad. Sci.*, 2008, **1125**, 215.
10. M. K. Johnson, D. C. Rees and M. W. W. Adams, *Chem. Rev.*, 1996, **96**, 2817.
11. J. J. R. Fraústo da Silva and R. J. P. Williams, *The biological chemistry of the elements; the inorganic chemistry of life*, Oxford University Press, Oxford, 2nd edn, 2001.
12. H. Raaijmakers, S. Macieira, J. M. Dias, S. Teixeira, S. Bursakov, R. Huber, J. J. G. Moura, I. Moura and M. J. Romao, *Structure*, 2002, **10**, 1261.
13. R. J. P. Williams and J. J. R. Fraústo da Silva, *Coord. Chem. Rev.*, 2000, **200–202**, 247.
14. R. J. P. Williams and J. J. R. Fraústo da Silva, *Biochem. Biophys. Res. Commun.*, 2002, **292**, 293.
15. G. Wächtershäuser, *Bioinspired catalysis: metal-sulfur complexes*, ed. W. Weigand and P. Schollhammer, Wiley-VCH, Weinheim, Germany, 2015, ch. 1, pp. 3–19.
16. W. R. Hagen and A. F. Arendsen, *Struct. Bonding*, 1998, **90**, 161.
17. R. Roy and M. W. W. Adams, *Molybdenum and tungsten: their roles in biological processes*, ed. A. Siegel and H. Siegel, CRC Press, Boca Raton, FL, 2002, vol. 39, ch. 19, pp. 673–697.
18. L. E. Bevers, P.-L. Hagedoorn and W. R. Hagen, *Coord. Chem. Rev.*, 2009, **253**, 269.
19. W. R. Hagen, *Coord. Chem. Rev.*, 2011, **255**, 1117.
20. G. Schwarz, R. R. Mendel and M. W. Ribbe, *Nature*, 2009, **460**, 839.
21. M. J. Romão, *Dalton Trans.*, 2009, **2009**, 4053.
22. A. Majumdar and S. Sarkar, *Coord. Chem. Rev.*, 2011, **255**, 1039.
23. R. Hille, *Dalton Trans.*, 2013, **42**, 3029.
24. R. Hille, J. Hall and P. Basu, *Chem. Rev.*, 2014, **114**, 3963.
25. M. J. Pushie, J. J. Cotelesage and G. N. George, *Metallomics*, 2014, **6**, 15.
26. S. Groysman and R. H. Holm, *Biochemistry*, 2009, **48**, 2310.
27. R. H. Holm, E. I. Solomon, A. Majumdar and A. Tenderholt, *Coord. Chem. Rev.*, 2011, **255**, 993.
28. M. J. Romão, M. Archer, I. Moura, J. J. G. Moura, J. LeGall, R. Engh, M. Schneider, P. Hof and R. Huber, *Science*, 1995, **270**, 1170.
29. J. L. Johnson, B. E. Hainline and K. V. Rajagopalan, *J. Biol. Chem.*, 1980, **255**, 1783.
30. J. L. Johnson, N. R. Bastian and K. V. Rajagopalan, *Proc. Natl. Acad. Sci. U. S. A.*, 1990, **87**, 3194.

31. P. Basu and S. J. N. Burgmayer, *Coord. Chem. Rev.*, 2011, **255**, 1016.
32. G. B. Seiffert, G. M. Ullmann, A. Messerschmidt, B. Schink, P. M. H. Kroneck and O. Einsle, *Proc. Natl. Acad. Sci. U. S. A.*, 2007, **104**, 3073.
33. L. J. Stewart, S. Bailey, B. Bennett, J. M. Charnock, C. D. Garner and A. S. McAlpine, *J. Mol. Biol.*, 2000, **299**, 593.
34. Y. Hu, S. Faham, R. Roy, M. W. W. Adams and D. C. Rees, *J. Mol. Biol.*, 1999, **286**, 899.
35. T. Weinert, S. G. Huwiler, J. W. Kung, S. Weidenweber, P. Hellwig, H.-J. Stärk, T. Biskup, S. Weber, J. J. H. Cotelesage, G. N. George, U. Ermler and M. Boll, *Nat. Chem. Biol.*, 2015, **11**, 586.
36. S. Mukund and M. W. W. Adams, *J. Biol. Chem.*, 1991, **266**, 14208.
37. R. Roy, S. Mukund, G. J. Schut, D. M. Dunn, R. Weiss and M. W. W. Adams, *J. Bacteriol.*, 1999, **181**, 1171.
38. S. Mukund and M. W. W. Adams, *J. Biol. Chem.*, 1995, **270**, 8389.
39. R. Roy and M. W. W. Adams, *J. Bacteriol.*, 2002, **184**, 6952.
40. L. E. Bevers, E. Bol, P.-L. Hagedoorn and W. R. Hagen, *J. Bacteriol.*, 2005, **187**, 7056.
41. S. Aono, F. O. Bryant and M. W. W. Adams, *J. Bacteriol.*, 1989, **171**, 3433.
42. R. C. Conover, A. T. Kowal, W. Fu, J. B. Park, S. Aono, M. W. W. Adams and M. K. Johnson, *J. Biol. Chem.*, 1990, **265**, 8533.
43. M. N. Hasan, P.-L. Hagedoorn and W. R. Hagen, *FEBS Lett.*, 2002, **531**, 335.
44. P. J. Silva, E. C. D. van den Ban, H. Wassink, H. Haaker, B. de Castro, F. T. Robb and W. R. Hagen, *Eur. J. Biochem.*, 2000, **267**, 6541.
45. K. Ma, Z. H. Zou and M. W. W. Adams, *FEMS Microbiol. Lett.*, 1994, **122**, 245.
46. P.-L. Hagedoorn, M. C. P. F. Driessen, M. van den Bosch, I. Landa and W. R. Hagen, *FEBS Lett.*, 1998, **440**, 311.
47. P.-L. Hagedoorn, T. Chen, I. Schröder, S. R. Piersma, S. de Vries and W. R. Hagen, *JBIC, J. Biol. Inorg. Chem.*, 2005, **10**, 259.
48. I. M. Schott, G. M. Rubinstein, G. L. Lipscomb, M. Basen, G. J. Schut, A. M. Rhaesa, W. A. Lancaster, F. L. Poole II, R. M. Kelly and M. W. W. Adams, *Appl. Environ. Microbiol.*, 2015, **81**, 7339–7347.
49. C. Debnar-Daumler, A. Seubert, G. Schmitt and J. Heider, *J. Bacteriol.*, 2014, **196**, 483.
50. P.-L. Hagedoorn, *Metalloproteins containing iron and tungsten: biocatalytic links between organic and inorganic redox chemistry*, PhD thesis, Delft University of Technology, 2002 (http://repository.tudelft.nl.
51. H. Sugimoto, H. Tano, K. Toyota, R. Tajima, H. Miyake, I. Takahashi, S. Hirota and S. Itoh, *J. Amer. Chem. Soc.*, 2010, **132**, 8.
52. E. Bol, L. E. Bevers, P.-L. Hagedoorn and W. R. Hagen, *JBIC, J. Biol. Inorg. Chem.*, 2006, **11**, 999.
53. I. Hahnenstein, H. Hasse, C. G. Kreiter and G. Maurer, *Ind. Eng. Chem. Res.*, 1994, **34**, 440.
54. G. Fiala and K. O. Stetter, *Arch. Microbiol.*, 1986, **145**, 56.
55. I. K. Dhawan, R. Roy, B. P. Keohler, S. Mukund, M. W. W. Adams and M. K. Johnson, *JBIC, J. Biol. Inorg. Chem.*, 2000, **5**, 313.

56. E. Bol, N. J. Broers and W. R. Hagen, *JBIC, J. Biol. Inorg. Chem.*, 2008, **13**, 75.
57. R.-Z. Liao, J.-G. Yu and F. Himo, *J. Inorg. Biochem.*, 2011, **105**, 927.
58. R. A. Schmitz, M. Richter, D. Linder and R. K. Thauer, *Eur. J. Biochem.*, 1992, **207**, 559.
59. P. A. Bertram, R. A. Schmitz, D. Linder and R. K. Thauer, *Arch. Microbiol.*, 1994, **161**, 220.
60. B. M. Rosner and B. Schink, *J. Bacteriol.*, 1995, **177**, 5767.
61. C. M. H. Hensgens, W. R. Hagen and T. A. Hansen, *J. Bacteriol.*, 1995, **177**, 6195.
62. G. Fuchs, M. Boll and J. Heider, *Nat. Rev. Microbiol.*, 2011, **9**, 803.
63. J. W. Kung, C. Löffler, K. Dörner, D. Heintz, S. Gallien, A. van Dorsselaer, T. Friedrich and M. Boll, *Proc. Natl. Acad. Sci. U. S. A.*, 2009, **106**, 17687.
64. S. Wischgoll, D. Heintz, F. Peters, A. Erxleben, E. Sarnighausen, R. Reski, A. van Dorsselaer and M. Boll, *Mol. Microbiol.*, 2005, **58**, 1238.
65. M. Boll, C. Löffler, B. E. L. Morris and J. W. Kung, *Environ. Microbiol.*, 2014, **16**, 612.
66. J. W. Kung, S. Baumann, M. von Bergen, M. Müller, P.-L. Hagedoorn, W. R. Hagen and M. Boll, *J. Am. Chem. Soc.*, 2010, **132**, 9850.
67. L. B. Maia, J. J. G. Moura and I. Moura, *JBIC, J. Biol. Inorg. Chem.*, 2015, **20**, 287.
68. T. Hartmann, N. Schwanhold and S. Leimkühler, *Biochim. Biophys. Acta*, 2015, **1854**, 1090.
69. H. C. Raaijmakers and M. J. Romão, *JBIC, J. Biol. Inorg. Chem.*, 2006, **11**, 849.
70. H. Raaijmakers, S. Teixeira, J. M. Dias, M. J. Almendra, C. D. Brondino, I. Moura, J. J. G. Moura and M. J. Romão, *JBIC, J. Biol. Inorg. Chem.*, 2001, **6**, 398.
71. F. A. M. de Bok, P.-L. Hagedoorn, P. J. Silva, W. R. Hagen, E. Schiltz, K. Fritsche and A. J. M. Stams, *Eur. J. Biochem.*, 2003, **270**, 2476.
72. M. J. Almendra, C. D. Brondino, O. Gavel, A. S. Pereira, P. Tavares, S. Bursakov, R. Duarte, J. Caldeira, J. J. G. Moura and I. Moura, *Biochemistry*, 1999, **38**, 16366.
73. J. Le Gall, G. Mazza and N. Dragoni, *Biochim. Biophys. Acta*, 1965, **99**, 385.
74. C. D. Brondino, M. C. G. Passeggi, J. Caldeira, M. J. Almendra, M. J. Feio, J. J. G. Moura and I. Moura, *JBIC, J. Biol. Inorg. Chem.*, 2004, **9**, 145.
75. J. R. Postgate, H. M. Kent, R. L. Robson and J. A. Chesshyre, *J. Gen. Microbiol.*, 1984, **130**, 1597.
76. S. M. da Silva, C. Pimentel, F. M. A. Valente, C. Rodrigues-Pousada and I. A. C. Pereira, *J. Bacteriol.*, 2011, **193**, 2909.
77. J. F. Heidelberg, *et al.*, *Nat. Biotechnol.*, 2004, **22**, 554.
78. C. Sebban, L. Blanchard, M. Bruschi and F. Guerlesquin, *FEMS Microbiol. Lett.*, 1995, **133**, 143.
79. S. Bartoschek, J. A. Vorholt, R. K. Thauer, B. H. Geierstanger and C. Griesinger, *Eur. J. Biochem.*, 2000, **267**, 3130.

80. P. A. Bertram and R. K. Thauer, *Eur. J. Biochem.*, 1994, **226**, 811.
81. T. J. Lie, K. C. Costa, B. Lupa, S. Korpole, W. B. Whitman and J. A. Leigh, *Proc. Natl. Acad. Sci. U. S. A.*, 2012, **109**, 15473.
82. R. K. Thauer, *Nat. Rev. Microbiol.*, 2008, **6**, 579.
83. R. K. Thauer, *Proc. Natl. Acad. Sci. U. S. A.*, 2012, **109**, 15084.
84. A. Hochheimer, D. Linder, R. K. Thauer and R. Hedderich, *Eur. J. Biochem.*, 1996, **242**, 156.
85. A. Hochheimer, R. A. Schmitz, R. K. Thauer and R. Hedderich, *Eur. J. Biochem.*, 1995, **234**, 910.
86. A. Hochheimer, R. Hedderich and R. K. Thauer, *Arch. Microbiol.*, 1998, **170**, 389.
87. R. A. Schmitz, P. A. Bertram and R. K. Thauer, *Arch. Microbiol.*, 1994, **161**, 528.
88. J. A. Vorholt, M. Vaupel and R. K. Thauer, *Eur. J. Biochem.*, 1996, **236**, 309.
89. M. Karrasch, G. Börner, M. Enssle and R. K. Thauer, *Eur. J. Biochem.*, 1990, **194**, 367.
90. J. A. Vorholt, M. Vaupel and R. K. Thauer, *Mol. Microbiol.*, 1997, **23**, 1033.
91. N. R. Bastian, C. J. Kay, M. J. Barber and K. V. Rajagopalan, *J. Biol. Chem.*, 1991, **266**, 45.
92. A. S. McAlpine, A. G. McEwan and S. Bailey, *J. Mol. Biol.*, 1998, **275**, 613.
93. H.-K. Li, C. Temple, K. V. Rajagopalan and H. Schindelin, *J. Am. Chem. Soc.*, 2000, **122**, 7673.
94. N. Benson, J. A. Farrar, A. G. McEwan and A. J. Thomson, *FEBS Lett.*, 1992, **2**, 169.
95. A. G. McEwan, S. J. Ferguson and J. B. Jackson, *Biochem. J.*, 1991, **274**, 305.
96. P.-L. Hagedoorn, W. R. Hagen, L. J. Stewart, A. Docrat, S. Bailey and C. D. Garner, *FEBS Lett.*, 2003, **555**, 606.
97. J. Buc, C.-L. Santini, R. Giordani, M. Czjezek, L.-F. Wu and G. Giordano, *Mol. Microbiol.*, 1999, **32**, 159.
98. S. de Vries, M. Momcilovic, M. J. F. Strampraad, J. P. Whitelegge, A. Baghai and I. Schröder, *Biochemistry*, 2010, **49**, 9911.
99. S. Afshar, E. Johnson, S. de Vries and I. Schröder, *J. Bacteriol.*, 2001, **183**, 5491.
100. R. U. Meckenstock, R. Krieger, S. Ensign, P. M. H. Kroneck and B. Schink, *Eur. J. Biochem.*, 1999, **264**, 176.
101. F. tenBrink, B. Schink and P. M. H. Kroneck, *J. Bacteriol.*, 2011, **193**, 1229.
102. S. Antony and C. A. Bayse, *Organometallics*, 2009, **28**, 4938.
103. R.-Z. Liao, J.-G. Yu and F. Himo, *Proc. Natl. Acad. Sci. U. S. A.*, 2010, **107**, 22523.
104. R.-Z. Liao and F. Himo, *ACS Catal.*, 2011, **1**, 937.
105. R.-Z. Liao and W. Thiel, *J. Chem. Theory Comput.*, 2012, **10**, 3793.
106. R.-Z. Liao and W. Thiel, *J. Phys. Chem. B*, 2013, **117**, 3954.
107. A.-M. Sevcenco, G. C. Krijger, M. W. H. Pinkse, P. D. E. M. Verhaert, W. R. Hagen and P.-L. Hagedoorn, *JBIC, J. Biol. Inorg. Chem.*, 2009, **14**, 631.

108. A.-M. Sevcenco, M. W. H. Pinkse, H. Th. Wolterbeek, P. D. E. M. Verhaert, W. R. Hagen and P.-L. Hagedoorn, *Metallomics*, 2011, **3**, 1324.
109. A.-M. Sevcenco, W. R. Hagen and P.-L. Hagedoorn, *Chem. Biodiversity*, 2012, **9**, 1967.
110. A.-M. Sevcenco, M. W. H. Pinkse, E. Bol, G. C. Krijger, H. Th. Wolterbeek, P. D. E. M. Verhaert, P.-L. Hagedoorn and W. R. Hagen, *Metallomics*, 2009, **1**, 395.
111. A.-M. Sevcenco, *The MIRAGE Method of Metalloproteomics*, PhD Thesis, Delft University of Technology, 2010, http://repository.tudelft.nl.
112. M. V. Weinberg, G. J. Schut, S. Brehm, S. Datta and M. W. W. Adams, *J. Bacteriol.*, 2005, **187**, 336.
113. S. Mukund and M. W. W. Adams, *J. Bacteriol.*, 1996, **178**, 163.
114. A.-M. Sevcenco, L. E. Bevers, M. W. H. Pinkse, G. C. Krijger, H. Th. Wolterbeek, P. D. E. M. Verhaert, W. R. Hagen and P.-L. Hagedoorn, *J. Bacteriol.*, 2010, **192**, 4143.
115. A. Cvetkovic, A. Lal Menon, M. P. Thorgersen, J. W. Scott, F. L. Poole II, F. E. Jenney Jr, W. A. Lancaster, J. L. Praissman, S. Shanmukh, B. J. Vaccaro, S. A. Trauger, E. Kalisiak, J. V. Apon, G. Siuzdak, S. M. Yannone, J. A. Tainer and M. W. W. Adams, *Nature*, 2010, **466**, 779.
116. I. H. Warren and Y. Okita, *Ind. Eng. Chem. Fundam.*, 1973, **12**, 342.
117. J. P. Smart, M. J. Cliff and D. J. Kelly, *Mol. Microbiol.*, 2009, **74**, 742.
118. M. E. Taveirne, M. L. Sikes and J. W. Olsen, *Mol. Microbiol.*, 2009, **74**, 758.
119. A. E. Kazakov, L. Rajeev, E. G. Luning, G. M. Zane, K. Siddartha, D. A. Rodionov, I. Dubchak, A. P. Arkin, J. D. Wall, A. Mukhopadhyay and P. S. Novichkov, *J. Bacteriol.*, 2013, **195**, 4466.

Subject Index